Biogeochemistry of Trace Elements in the Rhizosphere

Members of Editorial Committee

Biogeochemistry of Trace Elements in the Rhizosphere

Edited by

P.M. Huang
University of Saskatchewan
Saskatoon, Canada

and

G.R. Gobran
Swedish University of Agricultural Sciences
Uppsala, Sweden

2005

ELSEVIER

Amsterdam • Boston • Heidelberg • London • New York • Oxford • Paris
San Diego • San Francisco • Singapore • Sydney • Tokyo

ELSEVIER B.V.
Sara Burgerhartstraat 25
P.O. Box 211, 1000 AE
Amsterdam, The Netherlands

ELSEVIER Inc.
525 B Street
Suite 1900, San Diego
CA 92101-4495, USA

ELSEVIER Ltd
The Boulevard, Langford Lane
Kidlington, Oxford OX5 1GB
UK

ELSEVIER Ltd
84 Theobalds Road
London WC1X 8RR
UK

First edition 2005
Library of Congress Cataloging in Publication Data
A catalog record is available from the Library of Congress.

British Library Cataloguing in Publication Data
A catalogue record is available from the British Library.

ISBN-13: 978-0-444-51997-9

ISBN-10: 0-444-51997-1

∞ The paper used in this publication meets the requirements of ANSI/NISO Z39.48-1992 (Permanence of Paper).

Printed in The Netherlands.

Preface

The term "rhizosphere" was first used by L. Hiltner in 1904 but has since been modified and redefined. It is the narrow zone of soil influenced by the root and exudates. The extent of the rhizosphere may vary with soil type, plant species, age, and many other factors, but is usually considered to extend from the root surface out into the soil for a few millimeters. More intense microbial activity and larger microbial populations occur in this zone than in the bulk soil, in response to the release by roots of large amounts of organic compounds. The release of exudates from roots is in turn influenced by the nature and properties of soils, e.g., bulk density, mechanical impedance, and nutrient status. Up to 18% of the C assimilated through photosynthesis can be released from roots. Microbial populations in the rhizosphere can be 10–100 times larger than the populations in the bulk soil. Therefore, the rhizosphere is bathed in root exudates and microbial metabolites and the chemistry and biology at the soil–root interface is governed by biotic (plant roots, microbes) and abiotic (physical and chemical reactions) interactions, and thus differ significantly from those in bulk soil. Consequently, to study the rhizosphere, one must deal with not only biological and biochemical aspects but also physicochemical reactions, especially the interactions of these biotic and abiotic reactions and processes. The rhizosphere is the bottleneck of trace element contamination of the terrestrial food chain. The dynamics, transformations, bioavailability, and toxicity of trace elements are influenced enormously by chemistry and biology of the rhizosphere. The research on biotic and abiotic interactions in the rhizosphere should, thus, be an issue of intense interest for years to come.

The 15 chapters in this book are largely selected from papers presented at Symposium 02 Biogeochemistry of Trace Elements in the Rhizosphere, the 7th International Conference on the Biogeochemistry of Trace Elements, Uppsala, Sweden, June 15–19, 2003. This book addresses a variety of issues on fundamentals of biogeochemistry of trace elements in the rhizosphere at the molecular and microscopic levels and the impact on food chain contamination and the terrestrial ecosystem. Section I (Chapters 1–7) addresses the fundamentals of mineral weathering reactions, characteristics of rhizosphere soils from natural and agricultural environments, role of biotic interactions in the forest rhizosphere, the influence of organic and inorganic ligands on adsorption–desorption and complexation of heavy metals and the impact on the terrestrial food chain contamination. Section II (Chapters 8–15) deals with speciation, dynamics, and bioavailability of trace metals as influenced by rhizosphere chemistry and biology, binding and electrostatic attraction of heavy metals to plasma membranes of wheat root, use of a chemical non-equilibrium approach to model metal bioavailability to a hyperaccumutor, and the influence of arbuscular mycorrhizal fungi species on the dynamics of heavy metals and radionuclides and their transfer to plants. It is hoped that this book would provide a timely publication to stimulate research and education in this extremely important and exciting area of science for years to come.

All the chapters in this book have been critically reviewed by external referees and members of the Editorial Committee. We are grateful to the authors for their contributions and to members of the Editorial Committee and the reviewers who have provided invaluable inputs to maintain the quality of this publication. Gratitude is extended to the University of Saskatchewan for providing the funding to facilitate the publication of this book.

The book will be an essential reference for chemists and biologists studying environmental systems, as well as earth, soil, and environmental scientists. It will serve as a useful reference for professors, researchers, students, and consultants, etc. in environmental science, soil sciences, ecology, and ecotoxicology.

P.M. Huang and
G.R. Gobran

About the Editors

P.M. Huang received his Ph.D. degree in Soil Science at the University of Wisconsin, Madison, in 1966. He is Professor Emeritus of Soil Science at the University of Saskatchewan, Saskatoon, Canada. His research work has significantly advanced the frontiers of knowledge on the nature and surface reactivity of mineral colloids and organomineral complexes of soils and sediments and their role in the dynamics, transformations, and fate of nutrients, toxic metals, and xenobiotics in terrestrial and aquatic environments. His research findings, embodied in over 300 refered scientific publications, including research papers, book chapters, and books, are fundamental to the development of sound strategies for managing land and water resources.

He has developed and taught courses in soil physical chemistry and mineralogy, soil analytical chemistry, and ecological toxicology. He has successfully trained and inspired M.Sc. and Ph.D. students and postdoctoral fellows, and received visiting scientists from all over the world. He has served on numerous national and international scientific and academic committees. He also has served as a member of many editorial boards such as the *Soil Science Society of America Journal, Geoderma, Chemosphere, Water, Air and Soil Pollution, and Soil Science and Plant Nutrition.* He has served as a titular member of the Commission of Fundamental Environmental Chemistry of the International Union of Pure and Applied Chemistry and is the founding and current Chairman of the Working Group MO "Interactions of Soil Minerals with Organic Components and Microorganisms" of the International Union of Soil Sciences. He received the Distinguished Researcher Award from the University of Saskatchewan and the Soil Science Research Award from the Soil Science Society of America. He is a Fellow of the Canadian Society of Soil Science, the Soil Science Society of America, the American Society of Agronomy, the American Association for the Advancement of Science, and the World Innovating Foundation.

George Gobran is Professor of Ecology with specialization in nutrient dynamics in the rhizosphere. Since 1985, Dr. Gobran has been working in the Department of Ecology and Environmental Research, Swedish University of Agricultural Sciences, Uppsala, Sweden. Dr. Gobran received his Ph.D. in soil chemistry from the Catholic University of Louvain-La-Neuve, Belgium in 1980, and his M.S. in Soil Chemistry in 1975 and B.S. in Soil and Water Sciences in 1969 from Alexandria University, Egypt. During 1981–82, he spent 6 months at the European Directorate-General for Science, Research and Development (DG XII) in Brussels, Belgium. During 1982–1984, Dr. Gobran obtained a postdoctoral fellowship from Texas A&M University, USA.

Professor Gobran has a wide experience in research dealing with biogeochemical processes, with special interest in soil–plant interactions and rhizospheric processes. Currently, Dr. Gobran focuses his research and teaching efforts on the reciprocal effects of soil–plant interactions, especially in ecosystems under environmental stress. Dr. Gobran has written many papers and book chapters, and participated in several international conferences, workshops, and symposia. In 2001, Dr. Gobran and his colleagues Drs. Walter Wenzel and Enzo Lombi edited the book "*Trace Elements in the Rhizosphere*," published by the CRC Press, p. 321. His strong interest in this field has stimulated many graduate, postgraduate students, and the initiation of a couple of national and EU research projects, such as EU COST 631 Entitled "Understanding and Modeling Plant–Soil Interactions in the Rhizosphere Environment (UMPIRE)." Dr. Gobran has frequently been invited by international universities and organizations to give lectures. He also hosted many international colleagues for short and long sabbatical leaves. Professor Gobran was included in the 1999 edition of Who's Who in the World. He was the chairman of the "International Conferences of Biogeochemistry of Trace Elements, 7th ICOBTE 2003", Uppsala, Sweden, June 15–19, 2003 [http://www-conference.slu.se/7thICOBTE/index.htm]. Dr. Gobran is a reviewer for several international journals and programs, e.g., member of the review panel of the EuroDiversity program 2004.

Contributors

A. Agnelli
Dipartimento di Scienze Ambientali e delle Produzioni Vegetali, Università Politecnica delle Marche, Ancona, Italy

M.F. Benedetti
CNRS-UPRESA, 7047 – UMPC Lab., Géochimie & Métallogénie, Paris, Cedex 05, France

J. Cao
College of Environmental Sciences, Laboratory for Earth Surface Processes, Peking University, Beijing, China

R. Capasso
Dipartimento di Scienze del Suolo, della Pianta e dell'Ambiente, Via Università, Portici (Napoli), Italy

M. Castrec-Rouelle
CNRS-UPRESA 7047 – UMPC Lab., Géochimie & Métallogénie, Paris, Cedex 05, France

Y.J. Chen
College of Environmental Sciences, Laboratory for Earth Surface Processes, Peking University, Beijing, China

M. Clairotte
INRA-ENSA.M, UMR 1222 Rhizosphère & Symbiose, Montpellier, Cedex 1, France

S. Cocco
Dipartimento de Scienze Ambientali e delle Produzioni Vegetali, Università Politecnica delle Marche, Ancona, Italy

G. Corti
Dipartimento di Scienze Ambientali e delle Produzioni Vegetali, Università Politecnica delle Marche, Ancona, Italy

F. Courchesne
Départment de géographie, Université de Montréal, Montréal, Québec, Canada

R. Cuniglio
Dipartimento di Scienze Ambientali e delle Produzioni Vegetali, Università Politecnica delle Marche, Ancona, Italy

S. Declerck
Université catholique de Louvain, Unité de microbiologie, Louvain-la-Neuve, Belgium

W.J. Fitz
Department of Forest and Soil Sciences, University of Natural Resources and Applied Life Sciences, Vienna, Austria

C. Gagnon
St. Lawrence Centre, Environment Canada, Montréal, Québec, Canada

G.R. Gobran
Department of Ecology & Environmental Research, Swedish University of Agricultural Sciences, Uppsala, Sweden

M. Greger
Department of Botany, Stockholm University, Stockholm, Sweden

M.L. Himmelbauer
Department of Water, Atmosphere and Environment, University of Natural Resources and Applied Life Sciences, Vienna, Austria

P. Hinsinger
INRA-ENSA.M, UMR 1222
Rhizosphère & Symbiose,
Montpellier, Cedex 1, France

P.M. Huang
Department of Soil Science,
University of Saskatchewan,
Saskatoon, SK Canada

T.B. Kinraide
Appalachian Farming Systems
Research Center,
United States
Department of Agriculture,
Beaver, WV USA

L.M. Kozak
Department of Soil Science,
University of Saskatchewan,
Saskatoon, SK Canada

G.S.R. Krishnamurti
Department of Soil Science,
University of Saskatchewan,
Saskatoon, SK Canada

P. Legrand
Département de géographie,
Université de Montréal,
Montréal, Québec, Canada

C. Leyval
CNRS, LIMOS-Laboratoire des
Interactions Microoganismes-
Minéraux-Matière Organique
dans les sols, Nancy, France

B.G. Li
College of Environmental Sciences,
Peking University,
Beijing, China

C. Liu
Kuo Testing Labs, Inc., Othello,
WA, USA

W.X. Liu
College of Environmental Sciences,
Peking University,
Beijing, China

W. Loiskandl
Department of Water, Atmosphere and
Environment, University of Natural
Resources and Applied Life
Sciences, Vienna, Austria

E. Lombi
CSIRO Land and Water, Glen Osmond
SA 5064, Australia

D. Mahammedi
INRA-ENSA.M, UMR 1222
Rhizosphère & Symbiose, Montpellier,
Cedex 1, France

R.R. Martin
Department of Chemistry,
University of Western Ontario,
London, Ontario, Canada

J. Martinez
CEMAGREF-UR Gestion des Effluents
d'Elevage et des Déchets Municipaux 17,
Rennes Cedes, France

D.F.E. McArthur
Department of Soil Science,
University of Saskatchewan,
Saskatoon, SK Canada

S.J. Naftel
Department of Chemistry,
University of Western Ontario,
London, Ontario, Canada

D.R. Parker
Department of Environmental Sciences,
University of California, Riverside,
CA, USA

F. Persin
Université Montpellier II,
Lab. GPSA – Equipe Génie des procédés CC024,
Montpellier, Cedex 5, France

P. Peu
CEMAGREF-UR Gestion des Effluents d'Elevage et des Déchets Municipaux
17, Rennes Cedes, France

M. Pigna
Dipartimento di Scienze del Suolo, della Pianta e dell'Ambiente,
Via Università, Portici (Napoli), Italy

M. Puschenreiter
Department of Forest and Soil Sciences, University of Natural Resources and Applied Life Sciences, Vienna, Austria

S.M. Reichman
School of Botany,
University of Melbourne,
Melbourne, Vic., Australia

M. Ricciardella
Dipartimento di Scienze del Suolo, della Pianta e dell'Ambiente,
Via Università, Portici (Napoli), Italy

G. Rufyikiri
SCK· CEN, Radiation Protection Research Department, Mol,
Belgium

M.F. Sanjurjo
Departamento de Edafología y Química Agricola, Escola Politécnica Superior,
Lugo, Spain

S. Sauvé
Département de chimie,
Université de Montréal,
Montréal, Québec, Canada

A. Schnepf
Department of Water, Atmosphere and Environment, University of Natural Resources and Applied Life Sciences,
Vienna, Austria

T. Schrefl
Institute of Solid State Physics,
University of Technology, Vienna,
Austria

V. Séguin
Départment de géographie,
Université de Montréal, Montréal,
Québec, Canada

W. Skinner
Ian Wark Research Institute,
UNISA, Mawson Lakes,
SA Australia

S.Tao
College of Environmental Sciences,
Laboratory for Earth Surface Processes,
Peking University,
Beijing, China

Y. Thiry
SCK·CEN, Radiation Protection Research Department, Mol,
Belgium

S. Thomas
INRA-ENSA.M, UMR 1222
Rhizosphère & Symbiose, Montpellier,
Cedex 1, France

M.-C. Turmel
Département de géographie,
Université de Montréal,
Montréal, Québec, Canada

M.-P. Turpault
Biogéochimie des Ecosystèmes Forestiers, INRA, Champenoux,
France

A. Violante
Dipartimento di Scienze del Suolo,
della Pianta e dell'Ambiente,
Via Università, Portici (Napoli),
Italy

M.K. Wang
Department of Agricultural Chemistry,
National Taiwan University, Taipei,
Taiwan

W.W. Wenzel
Department of Forest and Soil Sciences,
University of Natural Resources and
Applied Life Sciences, Vienna, Austria

F.L. Xu
College of Environmental Sciences,
Laboratory for Earth Surface Processes,
Peking University,
Beijing, China

U. Yermiyahu
Agricultural Research Organization,
Gilat Research Center, D.N. Negev 2
85280, Israel

TABLE OF CONTENTS

PART II. SPECIATION, BIOAVAILABILITY, AND PHYTOTOXICITY OF TRACE ELEMENTS

Part 1: FUNDAMENTALS OF TRANSFORMATIONS AND DYNAMICS OF TRACE ELEMENTS

Biogeochemistry of Trace Elements in the Rhizosphere
P.M. Huang and G.R. Gobran (Editors)

Chapter 1

Contribution of rhizospheric processes to mineral weathering in forest soils

G.R. Gobran[a], M.-P. Turpault[b], and F. Courchesne[c]

[a]Department of Ecology & Environmental Research, Swedish University of Agricultural Sciences, P.O. Box 7072, S-750 07 Uppsala, Sweden
E-mail: george.gobran@eom.slu.se

[b]Biogéochimie des Ecosystèmes Forestiers, INRA, 54280 Champenoux, France

[c]Department de géographie, Université de Montréal, C.P. 6128, Succursale Centre-Ville, Montréal H3C 3J7 Canada

ABSTRACT

A review of literature on methods to quantify weathering in forest soils revealed that there is very little *in situ* information on the impact of root-induced changes on mineral weathering dynamics in forest soils owing to the coexistence of soil, roots and associated microorganisms in the rhizosphere. The review also emphasizes the need for the quantification of mineral dissolution rates resulting from weathering in the soil, especially in the rhizosphere. In this chapter, we present a novel scientific approach for estimating weathering in the rhizosphere of forest soils. Our research results to date of two ongoing field case studies in northern and southwestern Sweden are presented to indicate how mineral weathering can be monitored using well-defined techniques. These techniques include homogeneous soil bags (HSB) and test-mineral bags (TMB) of two different meshes, 51 and 541 μm, which either allow the penetration of roots and hyphae (541 μm) or exclude root growth in the bags (51 μm). In these studies, mineral weathering is assessed by scanning electron microscopy and mineral mass losses. Our research results to date from the TMB method suggest that weathering of apatite was much faster in the coarse- than in the fine-mesh bags. These results indicate a higher level biological weathering in the presence of roots and hyphae (as was the case in the 541 μm bags) than in the absence of roots (as was the case in the 51 μm bags). Through our examination of the sites and field treatment effects, and the consequent results on tree growth, we conclude that a consideration of the combined effects of climate and nutrient inputs to forest ecosystems is necessary for a better assessment of the weathering rate of soil minerals in the rhizosphere. Indeed, a proper quantification of mineral dissolution rates resulting from weathering in the rhizosphere would significantly improve the predictive capability of existing growth and biogeochemical models.

1. INTRODUCTION

1.1. Processes of mineral weathering

A review published by Gobran *et al.* (1998) provided evidence of the existence of steep chemical, microbiological and physical gradients between the rhizosphere and bulk soil. These gradients differed substantially in soil pH (Nye, 1986; Marschner and Römheld, 1996), cation-exchange equilibria (Chung *et al.*, 1994), metal availability (Sarkar and Wyn Jones, 1982), organic acid concentration (Gardner *et al.*, 1982), microflora (Robert and Berthelin, 1986), monosaccharide content (Dormaar, 1988) and grain-size distribution (Sarkar *et al.*, 1979). These gradients were integrated into a conceptual model relating nutrient availability in the soil and plant growth to transfers of matter and energy among three soil fractions: the soil–root interface, the rhizosphere and the bulk soil (Gobran and Clegg, 1996). The field study of Courchesne and Gobran (1997) showed that the rhizosphere soil was more intensively weathered and had accumulated more acid and base cations than the bulk soil. These studies have led us to stress the fact that conventional tests using bulk soil give no information about root-induced changes in the rhizosphere, owing to the coexistence of soil and roots and the associated microorganisms. However, there is very little *in situ* information on the impact of these two important factors on mineral weathering dynamics in forest soils. The quantification of the rates of mineral dissolution as a result of weathering in the soil, especially in the rhizosphere, is lacking, and is the main objective of our ongoing fieldwork presented here as a case study.

1.2. The rhizospheric environment, the mycorrhizosphere

The functioning of roots, mycorrhiza and organic matter result in weathering of the rhizosphere soil more intensively than the bulk soil. Therefore, fieldwork that does not take into account rhizospheric processes underestimates the extent of mineral weathering. The role of ectomycorrhizal fungi in the weathering of minerals has been investigated in a number of other projects in Sweden, but quantitative information about the roles played by the different biological components of the rhizosphere and mycorrhizosphere is still unavailable. Pot studies suggest that the exudation of organic acids by mycorrhizal fungi may influence the dissolution of minerals, releasing P and K (Wallander *et al.*, 1997; Wallander, 2000). Detailed laboratory and field studies of the elemental composition of hyphal fragments suggest that the mycorrhizal hyphae (Wallander *et al.*, 2002) may be capable of mobilizing significant amounts of P and K and of transporting them to trees. However, there is now a need to complement these small-scale measurements with additional methods to estimate weathering rates under field conditions using larger sample volumes.

1.3. Potential impact of rhizosphere on mineral dissolution

There is no doubt that soil mineral weathering is an important source of nutrients to the forest ecosystem. It should therefore be properly quantified. The lack of such quantification might explain why existing models such as PROFILE have not been successful in predicting forest growth under changing environments. For example, the model developed by Sverdrup *et al.* (1994) estimated that over 80% of Sweden's forests are receiving more N and S than the "critical load," and predicted widespread mortality and growth reductions in the next two decades. The model predictions contradict survey measurements of the basal-area growth rates of Scots pine and Norway spruce across all of Sweden, which show significant increases of about 30% between 1953 and 1992 (Elfving and Tegnhammer, 1996). April and Keller (1990) attributed the resulting accelerated mineral degradation to a series of root-induced acidifying processes such as the exudation of H^+ ions, CO_2 and complexing organic acids (OA). Under field conditions, much of the OA will either be taken up and degraded by the soil's microbial biomass (Lundström, 1994; Jones *et al.*, 1996), become complexed with metals (Cline *et al.*, 1982) or become sorbed to the soil's anion-exchange sites (Gobran *et al.*, 1997; Jones and Brassington, 1998). Therefore, the role of OA in weathering could have been overestimated in laboratory experiments (Drever, 1994). Indeed, the most significant source of uncertainty in quantifying the role of OA in weathering is the quantification of rates of mineral dissolution due to weathering in the rhizosphere (Hodson *et al.*, 1997).

2. OBJECTIVES

Although it is to be expected that plant growth and the consequent elemental uptake could increase the dissolution of soil minerals, as nutrient uptake is a sink for the dissolved minerals in the rhizosphere, the quantification of weathering under field conditions is lacking. Indeed, a proper quantification of mineral dissolution rates as a result of weathering in the rhizosphere would significantly improve the predictive capability of existing growth and biogeochemical models. Thus, the objective of this chapter is to review methods for quantifying the weathering of soil minerals, in general, and to identify the contribution of rhizosphere processes to weathering reactions in forest soils, in particular. To reach our goal, we propose both a synthesis of existing literature and a summary of results from ongoing fieldwork in northern and southern Sweden, Canada and France. We also present a new scientific approach for estimating weathering in the rhizosphere of forest soils.

3. LITERATURE REVIEW ON WEATHERING

3.1. Methods to estimate the biochemical weathering of minerals in soils

Weathering is a phenomenon that results in the fragmentation, decomposition and dissolution of rocks and minerals at or near the surface of the earth due to the combined action of physical, chemical and biological processes. Frost wedging, shattering by growing salt crystals, rapid thermal shocks (induced by wildfires or daily temperature cycles in hot deserts) and abrasion by violent winds constitute the main physical processes causing the fragmentation of coherent rocks (Birkeland, 1999). The dominant chemical mechanisms involved in the congruent or incongruent dissolution of minerals are hydrolysis, oxidation–reduction cycles, carbonation, hydration and chelation. In addition to their impact on weathering through the production of H^+ ions and of chelating agents, the biota, notably tree roots, also exert considerable localized pressures in soils, causing the breakdown of rocks and minerals.

Weathering plays a key role in the biogeochemistry of terrestrial ecosystems because it has a strong and sustained effect on, among others, the production of reactive secondary solids (clays, oxides) in soils, the supply of nutrients to biota, the release of potentially toxic elements, the buffering of acidic inputs, the support capacity of mineral substrates, soil genesis and the evolution of landscapes. On a scale of a million years, the weathering of silicates is associated with the global C cycle because weathering consumes CO_2, thus exerting feedback control on the CO_2 level of the atmosphere and, consequently, on the long-term changes in climatic conditions (Drever, 1994).

A wide spectrum of field and laboratory methods is used to estimate the amount and rate of biochemical weathering and to determine the elemental fluxes due to dissolution. These approaches cover a range of spatial and temporal scales of reference, a fact that may complicate the interpretation and comparison of results (Kelly *et al.*, 1998). For example, some techniques are based on detailed study of individual rock fragments or of soil profiles, while others rely on the monitoring of whole watersheds. As for temporal considerations, some methods provide information that are applicable for short periods of time (years to decades), whereas others deliver results that integrate millennia to millions of years of weathering action. In the following section, we present five of the most commonly used methods to estimate the biochemical weathering of minerals in soils.

3.1.1. Watershed budget

The weathering rate of minerals can be calculated from elemental budgets, typically for Ca, Mg, K, Na, Al, Fe and Si, performed on an annual basis at the watershed scale. This method involves the measurement of elemental inputs to the system, such as incoming bulk precipitation (dry plus wet deposition), and of outputs in the stream, such as dissolved and suspended materials. The annual storage of elements in the vegetation and the net changes in soil-exchangeable pools must also be evaluated for a complete assessment of outputs. On the basis of the conservation of mass principle, the weathering flux is assumed to equal the difference between inputs and outputs. The main assumptions underlying this method are that (1) the catchment is impervious; (2) the ecosystem is at steady-state, neither aggrading nor degrading, with respect to biomass and soil elemental pools; and (3) precipitation and stream flow are the only input and output, respectively (Likens and Bormann, 1995). Assumptions 1 and 3 can often be accepted in granitic and vegetated catchments situated far from direct anthropic influences. The second assumption is generally false, because steady-state is seldom achieved, and changes in elemental storage in plants and soils must be measured to obtain a sound estimate of weathering (Velbel, 1985). Finally, because of the strong interannual variability in the hydroclimatic and biogeochemical behaviour of terrestrial ecosystems, the scientific reliability and usefulness of watershed budgets are largely tributary to the length and continuity of the data set. This is the case for several sites, such as the Hubbard Brook Ecosystem Forest (USA), the Turkey Lake Area (Canada), Solling (Germany), Birkenes (Norway), Gårdsjön (Sweden) and Glen Feshie (Scotland).

3.1.2. Profile balance

Elemental distribution patterns in regoliths partly result from long-term geochemical changes occurring in surficial deposits and soil profiles. Accordingly, these patterns were used to calculate weathering rates (Olsson and Melkerud, 2000). The idea is to compare the total concentration of an element in any soil horizon with that in the parent material. The difference is summed for the whole profile and this cumulative elemental quantity is divided by soil age to obtain a mean annual weathering rate. The assumptions are that (1) the properties of the parent material were initially uniform; (2) a conservative component (Ti, Zr) is present in the soil to be used as a tracer whose absolute mass does not change in time; and (3) the properties of the actual C horizon are identical to those of the initial parent material. If sites are well chosen, assumptions 1 and 2 can be met, notably in young glaciated areas. Because soil chronosequences are often used in weathering studies, the uniformity requirement may pose a problem. As for assumption 3, the answer is generally unknown. This approach provides data on historical weathering rates because it integrates changes that occurred since the deposition of the parent material. As a corollary, the method may yield weathering rates that do not correspond well with present-day rates determined using other approaches, such as the watershed budget method. Brimhall *et al.* (1991) expanded the method by integrating volume changes in the soil (dilation or collapse) and strain with differences in concentrations. The approach is also used with soil minerals (White *et al.*, 1996). In this case, the conservative component is generally quartz because quartz weathering is negligible compared to that of primary silicate minerals.

3.1.3. Isotope ratios

Since direct measurement of weathering in the field is a difficult task, the Sr isotope (Sr has four isotopes, all constant except for ^{87}Sr) method was developed to estimate the contribution of weathering to cation fluxes in streams (Åberg *et al.*, 1989). The approach is based on two facts: (1) different geological sources have distinct isotopic compositions of Sr and hence, distinct $^{87}Sr/^{86}Sr$ ratios; and (2) processes at the surface of the earth do not fractionate Sr isotopes and therefore, variations in Sr values are caused by mixing. It follows that a two-component mixing model can be used to estimate the weathering flux if the bedrock of, and the atmospheric deposition on, a given catchment have distinct $^{87}Sr/^{86}Sr$ ratios and if these ratios are constant in time. Although Sr^{2+} is neither a plant nutrient nor massively released by weathering, its chemical behaviour is very similar to that of Ca^{2+} since both are alkaline earths and have the same valence and a similar ionic radius. In fact, Sr isotope ratios can then be used as proxies for Ca, and to a lesser extent for other base cations like Mg, for studying their release by weathering. It remains that variability in the isotopic composition of bedrock and soils complicates the assessment of weathering fluxes, although promising results were obtained with this method (Bailey *et al.*, 1996).

3.1.4. Laboratory experiments

A wide array of laboratory-based methods are used to estimate weathering rates in soil materials. Most of these techniques are in fact variations of three types of approaches: batch extractions, flow experiments and column leaching. However, not only do the techniques used differ among laboratory experiments, the materials used also vary considerably and can include pure reference minerals, ground rocks, soil minerals with coatings removed and naturally coated soil minerals. Experimental conditions such as temperature, flow velocity, presence or absence of organic matter and pH also differ among experiments. In some experiments, the dissolved weathering products are evacuated from the reaction vessel, while in others they are not. Nonetheless, the advantages of the laboratory approach to measuring weathering rates are many, and include: the relative simplicity of operation of the methods; the possibility of estimating the kinetics of weathering reactions; and control over environmental conditions. The main disadvantage is the difficulty of extrapolating the results to field conditions. Although laboratory-derived rates are often adjusted for differences in temperature and water content, significant discrepancies in rates indeed persist between the field and the laboratory.

3.1.5. Numerical modelling

Mechanistic geochemical models have been formulated to describe the weathering of minerals. The model PROFILE (Sverdrup and Warfvinge, 1992) ranks among the most commonly used in the literature. The key soil variables controlling PROFILE outputs are the surface area of particles, water content, temperature, bulk density, H^+ and organic acid contents and mineralogy. Long-term (10^6-year scale) silicate weathering scenarios are also tested with global C cycle models to examine the impact of the advent of lichens, primitive algae and vascular plants on land compared with situations where only barren landscapes existed (Berner, 1992). These models are extremely useful for assessing weathering at sites where field data are unavailable, to estimate historic and future weathering fluxes and to generate new research hypotheses. The most common limitations of weathering models, as for most models, are the large input requirements, over-parametrization, the fact that some processes are poorly documented, and hence, marginally integrated in the model and the unavailability of independent field data to validate the model.

3.1.6. Other selected methods

Ugolini and Dahlgren (1991) conducted work on Spodosols and Andisols to explain the occurrence of imogolite in the B horizons of these soils. They focused on the study of the charge balance of solutes in the soil solution to determine the weathering pathways; whether imogolite formed *in situ* or migrated in the solution as a sol. Their results revealed the presence of two contrasting weathering environments: an upper compartment (O, E, Bhs) controlled by organic acids, and a lower compartment (B, BC, C) where weathering is driven primarily by H_2CO_3. Based on the fact that organic acids prevent the format ion of imogolite and that proto-imogolite sols were absent in the solution of the upper compartment, they concluded that imogolite formed *in situ* and that migration was unlikely. This method is an interesting tool to study weathering, although it is confined to the study of present-day weathering processes.

The study of the thickness and chemistry of weathering rinds on basaltic clasts from Costa Rica was conducted by Sak *et al.* (2004) to establish their potential to determine the degree of surficial-deposit weathering and landscape age, and to constrain models of basalt weathering. The microscale variability in chemical composition was determined by electron microprobe analyses along transects running from the core to the rind boundary of clasts. Elemental mass balance calculations on rinds revealed the hierarchy of cation mobility ($Ca > K > Si > Al > Fe$) and the mineral sequence (plagioclase/augite, kaolinite, gibbsite, Fe oxides) during the weathering of basalt. The long-term rate of rind advance could also be quantified.

The extent of weathering enhancement associated with the large-scale appearance of biota on earth is still a matter of debate. To resolve the respective role of temperature and rainfall on weathering in the absence and presence of lichen, Brady *et al.* (1999) decided to look directly at mineral grains, in this case plagioclase and olivine, on basalt flows in Hawaii. Their microscale approach consisted of quantifying weathering by the digital processing of images obtained by back-scattered electron (BSE) microscopy. They compared images of mineral grains from underneath lichens with images from nearby abiotic sites. The method showed that for a given rainfall regime, lichens weathered a much larger amount of rock compared with the paired abiotic system because the presence of lichens extended the stay of corrosive moisture in pores and assured the secretion of organic acids.

The *in situ* soil bag method involves the insertion of known amounts of exchange resins or of pure test-minerals in permeable bags. These are placed back in the soil profile, left to react under field conditions for years and then removed for the purpose of analyzing mineralogical changes. The method is presented in detail in the section on results from Skogaby and Flakaledin in Sweden.

3.2. Case studies on weathering in the rhizosphere of forest soils

3.2.1. In southern Sweden under Norway spruce

While pursuing work on weathering in soils (Hendershot *et al.*, 1992; Courchesne *et al.*, 1996) it became clear that very little was known about the contribution of rhizosphere processes to mineral weathering in forest soils, even though this is fundamental to understanding the biogeochemistry of elements in terrestrial ecosystems. To investigate this issue, a project on the mineralogy of the rhizosphere of forest soils was initiated. The aim of the project was to establish the impact of processes occurring at the soil–root interface on the mineralogical composition of the solid phase and, consequently, on the intensity of mineral weathering. The scientific approach was based on a comparison of the mineralogy of the bulk and rhizosphere components of two Podzol profiles collected under Norway spruce (*Picea abies* (L.) Karst) in the untreated plots of the Skogaby site, in southwestern Sweden.

In each profile, samples were collected from the E (0–5 cm), the Bh (10–25 cm) and the Bs (25–50 cm) horizons. The separation of the bulk and rhizosphere components was conducted in the field. Living roots were removed by hand, lightly shaken and stored in plastic bags. The soil (intimately associated with the root surfaces) and adhering to the roots after they had been shaken was considered as rhizosphere soil. The proximal material not colonized by the roots was regarded as the bulk component. The rhizosphere was brushed away from the roots and freed from rootlet fragments. All samples were air-dried and passed through a 2-mm sieve. The mineralogy of the clay-sized particles of the rhizo-sphere and the bulk components was determined in triplicates by X-ray diffraction (XRD) of oriented specimens after removal of surface coatings with dithionite-citrate (DC) and H_2O_2, saturation with Mg, magnesium-ethylene glycol or K and heating of the K-saturated specimens to 300 and 550°C (Whittig and Allardice, 1986). The integrated intensity of each mineral (I) was normalized relative to the intensity of the (100) peak (d = 0.426 nm) of quartz (I_{QZ}) to calculate a mineral intensity ratio (I/I_{QZ}) for comparing mineral assemblages from different horizons. Iron and Al were extracted with acid-ammonium oxalate (Al_o, Fe_o) and analyzed by atomic absorption spectrophotometry (AAS). Oxalate is considered to dissolve amorphous organic and inorganic solids, most of which are of pedogenic origin and accumulated as *in situ* weathering products.

The mineral intensity ratio (I/I_{QZ}) in the rhizosphere differed consistently from that in the bulk soil, as presented in Table 1 (Courchesne and Gobran, 1997). The amount of change was associated with the relative stability of primary minerals in a weathering environment stimulated by root activity and followed the order amphiboles > plagioclases > K-feldspars. Indeed, compared with the bulk component, the rhizosphere contained significantly lower amounts of amphiboles ($\alpha < 0.10$), the most weatherable among the primary minerals found in the Skogaby soils. Moreover, the abundance of plagioclase was lower in the rhizosphere for five of the six horizons, but the overall difference was not significant. The XRD patterns showed no rhizosphere effect for K-feldspars. The absence of a measurable rhizosphere effect for K-feldspars is not surprising and is in agreement with Kodama *et al.* (1994). Expandable phyllosilicates, probably a vermiculite–smectite intergrade, were also less abundant in the rhizosphere ($\alpha < 0.10$). Finally, oxalate-extractable Al and Fe were systematically higher in the rhizosphere than in the bulk soil, (Table 1) with differences being significant in the E and Bh horizons. Similar observations on Al and Fe were made by Chung *et al.* (1994).

The XRD results reveal that mineralogical assemblages significantly differ among soil components, that the difference is almost systematic and that the effect is most pronounced for the least stable primary minerals. The oxalate data complement the mineralogical results by suggesting the existence of a concomitant preferential accumulation of weathering products close to root surfaces. Indeed, secondary Al and Fe solid phases appear to form more abundantly in the rhizosphere, probably as coatings on grain surfaces or on organic materials. These two sets of observations jointly point towards the accelerated weathering

Tablc 1
Mean mineral composition of the bulk and rhizosphere components of the six horizons studied and oxalate-extractable Al and Fe in the Bh horizon of profile 1

Soil component		*n*	I/I_{QZ}[a]					
			Amph	Plagio	K-Feld	Expand	Al_o	Fe_o
Rhizo	mean[b]	6	0.03a	1.73a	1.28a	0.54a	22.9a	54.3a
	sd		0.04	0.37	0.28	0.25	5.1	8.8
Bulk	mean	6	0.12b	2.24a	1.29a	1.14b	14.8b	40.8b
	sd		0.07	0.78	0.54	0.70	4.6	8.6

Amph = amphibole; Plagio = plagioclase; K-Feld = K-feldspar; Expand = expandable phyllosilicate; Al_o = oxalate-extractable Al; Fe_o = oxalate-extractable Fe. Al_o and Fe_o are expressed in g kg^{-1}.
[a] Intensity of a mineral divided by the intensity of the 100 quartz peak ($d = 0.426$ nm). [b] In a given column, mean values (sd = standard deviation) for the six horizons followed by the same letter are not significantly different at the $\alpha \leq 0.10$ probability level (ANOVA). Adapted from Courchesne and Gobran (1997).

of mineral structures in the rhizosphere zone of these soils where the weathering regime seems to be stimulated by the activity of roots and of associated microorganisms.

The accelerated weathering of minerals such as biotite, phlogopite or illite, and the release of K, Mg, Ca and Fe were documented for the rhizospheres of a range of cultural plants together with the impact of the microflora, symbiotic or not (Robert and Berthelin, 1986). For example, the rapid weathering of minerals in the vicinity of roots was demonstrated by Hinsinger *et al.* (1992), who showed that vermiculitization was initiated after only 3 days of continuous contact between phlogopite and a dense root mat under ryegrass or rape. April and Keller (1990) also showed that the mineralogical changes observed in the rhizosphere could be accompanied by the physical disruption of crystals and mineral particles (such as the bending and tangential alignment of phyllosilicate minerals and increased bulk density). However, apart from the work of Arocena *et al.* (1999) in the mycorrhizosphere of subalpine fir, and of April and Keller (1990) on the preferential dissolution of biotite compared with muscovite close to root surfaces, we are aware of no other field study on the impact of roots on mineral weathering in forest soils. But our field data from forest soils and the results of controlled growth experiments with cultural plants tend to converge and indicate that the rhizosphere is a more corrosive environment for weatherable minerals than the adjacent bulk soil.

The accelerated degradation of mineral structures in the rhizosphere zone can be related to a series of root-induced acidifying processes like the release of H^+ ions and CO_2, and the exudation of a range of metal-complexing organic acids (Hinsinger *et al.*, 2003). The release, accumulation and transformation of organic compounds (organic matter of plant or microbial origin, mucilage, exudates) in the rhizosphere is known to contribute to mineral weathering. These substances are either released by roots or accumulate as the decomposition products of dead organic tissues. They represent a series of organic acids that can efficiently attack mineral structures and complex weathering products such as metals. But the net contribution of these compounds to mineral weathering has been challenged, in part because most compounds are short-lived in soils and are readily decomposed by microbes. The uptake of nutrients by roots to support plant growth is thus viewed as the main mechanism producing the acid compounds needed to accelerate weathering in the rhizosphere. The H^+ ions are released as the roots take up dissolved nutrients present in the cationic form. The imbalance between cation and anion uptake, and thus between H^+ and OH^- release to solution, determines the amount of free acidity flowing through the rhizosphere and available for reaction with mineral surfaces. Finally, it is generally accepted that soil microorganisms, in particular symbiotic fungi, play a key role on the weathering of minerals in the rhizosphere (Wallander, 2000). In this case, the effect on minerals appears to be more largely mediated by the release of organic substances, in particular, by low-molecular-mass organic acids that accelerate the decomposition of mineral structures in their microscale surroundings, as shown by Jongmans *et al.* (1997).

3.2.2. In eastern Canada under Trembling aspen

Despite their critical role in biogeochemical cycles in terrestrial ecosystems, the relationships between root activity, mineral weathering and the bioavailability of metals are largely unknown. Hence, a study was designed to establish the impact of root activity on mineral weathering, and to determine its consequences for the spatial distribution of trace metal forms in the vicinity of roots.

This field study was conducted along a soil contamination gradient extending to the southeast of a Cu smelter in the Rouyn-Noranda area (Québec, Canada), where rhizosphere and bulk soil materials were sampled under trembling aspen (*Populus tremuloïdes* Michx) trees less than 30 years of age and growing on clayey soils. Cutting-edge analytical techniques, including time-of-flight secondary-ion mass spectroscopy (TOF-SIMS), micron-scale X-ray diffraction (μXRD) and X-ray absorption near-edge structure (XANES), were used in combination with XRD analysis and acid ammonium oxalate (AAO) extractions of Al, Co, Cr, Cu, Fe, Mg, Mn, Ni, Si and Zn to identify the pathways and mechanisms of weathering in the rhizosphere.

The results from these complementary approaches showed that the rhizospheric environment accelerated the weathering of primary minerals. *In situ* weathering products contributed to the retention of trace metals at the soil–root interface. The reader is referred to Chapter 2 written by Séguin *et al.* (2005) for a detailed presentation of the methods used to measure weathering and for a comprehensive discussion of the results gathered as part of this project.

3.2.3. In France using test-minerals

In France, the test-mineral technique based on mass loss determination was first developed by Sadio (1982) and used to quantify mineral dissolution (Augusto *et al.*, 2000) and the weathering of the interfoliar layers in clay minerals (Ranger *et al.*, 1990, 1991; Augusto *et al.*, 2001). The method was also applied at field sites outside France in temperate forest soils (Nugent *et al.*, 1998) and in tropical soils (Righi *et al.*, 1990). One of the advantages of this technique is its potential for quantifying present-day soil processes, prominent seasonal changes and for comparing different terrestrial ecosystems, forest canopies or soil compartments (Augusto *et al.*, 1998). When this method is used to estimate the rate of mineral dissolution, great care must be taken for the measurement of mineral masses before and after the field experiment. Because the method relies heavily on mass determination, these measurements should ideally be performed by the same operator. Moreover, the total surface of minerals introduced in the bags must be calibrated as a function of the dissolution constant of the mineral, the duration of the experiment and of site properties such as pH, moisture content and temperature.

The test-mineral method was used to establish the impact of different forest tree species (Norway spruce, Scots pine, sessile oak, pedunculated oak and European beech) on plagioclase weathering (Augusto *et al.*, 2000). Two bags containing plagioclase were inserted into five different soils and at various depths (under litter, 5, 15 and 40 cm). The bags were then left in soils for 3 and 9 years and subsequently removed. The results showed that dissolution rates decrease with depth in the soil profile. The dissolution rate of test-minerals was also strongly dependent on environmental conditions, in particular, pH and soil type. For a given soil, plagioclase weathering proceeded faster under coniferous species than under deciduous tree species.

4. NEW APPROACHES FOR ESTIMATING WEATHERING IN THE RHIZOSPHERE

4.1. Scientific approach

Field soil variability often masks differences between soils and treatments. For better precision in detecting the changes under field conditions, we propose the use of the homogeneous soil bag (HSB) method. The HSB is useful for assessing nutrient changes over time. This method has been used and recommended for ascertaining changes in soil chemistry and for analyzing the effects of experimental manipulations and long-term environmental changes (Mitchell *et al.*, 1994). The approach in Sweden includes the use of an HSB to allow the *in situ* monitoring of processes in the rhizosphere. One important aim of the ongoing experiments is to quantify the dissolution rate of soil minerals and index pure minerals such as plagioclase and apatite using test-mineral bags (TMBs).

Nylon bags made of two different meshes 51 and 541 μm in size, and with dimensions of 21 × 7 cm (for the HSB) and 10 × 5 cm (for the TMB) were manufactured by Sintab (Malmö, Sweden). The nylon bags are: very hydrophilic, biologically inert, non-hygroscopic and low extractable; show negligible absorption and absorption of filtrate; and have exceptionally low tare weights, low trace-element levels, excellent chemical resistance and thermal stability.

The HSB and TMB use different meshes to either include (541 μm) or exclude (51 μm) the penetration of roots and hyphae. Mesh bags (51 μm) excluding roots enable an assessment of the relative contributions of roots and mycorrhizal hyphae to weathering at the upper two mineral soil horizons. The HSBs were filled with 150 g each of the field moist composite homogeneous bulk soil (HSB soil) from each horizon. Fig. 1 depicts of 541 and 51 μm HSBs both filled with bulk soil from the upper mineral soil horizon (E) in Skogaby.

The soil, free of visible roots, was collected from the bulk material and homogenized. The HSBs were placed horizontally into the soil from the front side of the soil profile and evenly spaced in the top 2–3 cm of the horizon from

Fig. 1. HSB of 541 and 51 μm used in North and South of Sweden.

Fig. 2. HSB (51 μm) filled with bulk soil from the northern site (Flakaledin) and inserted back to B horizon at time zero.

which the soil was sampled. Fig. 2 illustrates the placement of an HSB between A and B horizons in Flakaledin.

TMBs of 10 × 5 cm of the two 51- and 541-μm-sized meshes, and containing 12 ± 0.0005 g of pure 0.5–1 mm grain-size plagioclase were used. As TMBs, they are useful for monitoring and estimating the mineral weathering of pure minerals under field conditions. Two test-minerals were chosen: (1) a labradorite plagioclase from Norway and (2) a fluoro-apatite from Mexico (Durango). The test-minerals were supplied by Compagnie Générale de Madagascar (Paris). The initial plagioclase material is composed of 99.9% labradorite and 0.1% ilmenite. The structural formula of labradorite calculated from microprobe analysis is $Si_{2.49}Al_{1.49}K_{0.02}Ca_{0.52}Na_{0.45}O_8$ (Augusto *et al.*, 2000). The Durango (Mexico) apatite is crystalline and forms pyramids 0.9–2 cm high and 1.2–4.5 cm in diameter. On the basis of total chemical analyses, the structural formula of apatite can be simplified to

$$P_{5.49}\,(Na_{0.09}\,Ca_{11.27}\,Sr_{0.01})\,F_{1.75}\,OH_{0.19}\,Cl_{0.14}$$

The test minerals were chosen based on the following criteria: (1) they are present in a range of soils; (2) they constitute key Ca and P sources; (3) mineral weathering is not related to transformation reaction and to microdivision – dissolution is the main reaction; (4) the rate of mineral weathering is fast enough to allow a rapid response to treatments; (5) chemically and mineralogically pure mineral phases are avilable; and (6) they are used in other experimental forested ecosystems for comparison purposes.

Initial mineral grains were ground in a jaw crusher. The plagioclase and the apatite are sieved at 0.5–1 and 1–2 mm, respectively. The 0.5–1 mm fraction was magnetically sorted in order to remove grains containing ilmenite. Apatite and labradorite grains were also treated ultrasonically and washed with distilled water in order to remove the fine particles.

To prepare TMBs, 12 $\pm$ 0.0005 g of labradorite plagioclase (0.5–1 mm) was placed in 10 $\times$ 5 cm bags of both meshes (300 μm and 51 μm), whereas 3 $\pm$ 0.0005 g of apatite (1–2 mm) was introduced in 3.5 $\times$ 7 cm bags (541 μm and 51 μm). The TMBs were placed between the forest floor and the upper mineral soil horizon in Skogaby and Flakaledin.

4.2. Ongoing field studies in Sweden

Our fieldwork has been conducted since 1999 at two sites with different treatments. At Skogaby (southwestern Sweden), treatments included a control (C), irrigation (I) and irrigation with liquid fertilization (IF) containing a complete set of nutrients according to nutrient flux concept. At Flakaledin (north Sweden), treatments also included I and IF (corresponding to IF at Skogaby). All treatments stimulated tree growth at Flakaledin (Bergh *et al.*, 1999) and at Skogaby (Nilsson and Wiklund, 1992), except for the I treatment at the Flakaledin site, which had no significant effect on tree growth compared with the control (Bergh *et al.*, 1999). Therefore, the I treatment in Flakaledin was used as a control for IF. The selected plots in Flakaledin were I-12B and IF-7A, while in Skogaby, the selected plots were C-24, I-25 and IF-24.

4.2.1. Site description of Skogaby and field manipulation

The Skogaby site is located in southwest Sweden (13°13′E, 56°33′N). The altitude ranges from 95 to 110 m above sea level. The annual precipitation is 1100 mm. The mean annual air temperature is 7.6°C. The Norway spruce (*Picea abies* (L.) Karst.) stand was planted in 1966. The site was surveyed and the field plots (45 plots of 45 m^2) were selected during 1987. The soil was classified as a

Haplic podzol (FAO-UNESCO, 1988) with a silty loam texture throughout the profile.

The irrigation source was water from Lake Råsjön, applied by a sprinkler system that prevented water storage deficits greater than 20 mm during May to September. The nutrient content in the I and IF treatments are presented in Table 2. Irrigation treatments were initiated during the growing season of 1988. The site and treatments were described in detail by Nilsson and Wiklund (1992).

4.2.2. Site description of Flakaledin and field manipulation

Flakaledin (64°07′N, 19°27′E) is situated in northwestern Sweden. The experimental area, 310–320 m above sea level, is above the highest postglacial coastline with a sandy-silty glacial till. The soil is classified as a typic Orthic podzol (FAO-UNESCO, 1988). The annual precipitation averages 580 mm, and more than one-third of the precipitation falls as snow. The site has a harsh boreal climate, and monthly mean air temperature ranges from −8.7°C in February to 14.4°C in July (Dambrine *et al.*, 1995). The Norway spruce (*Picea abies* (L.) Karst.) stand was planted in 1963. The site was surveyed and the field plots (45 plots of 100 m^2) were selected, and the manipulation treatments started in 1987. The selected treatments are I and IF. Further details regarding treatments are described by Linder (1995).

Table 2
Nutrient contents in irrigation treatments and tree volume production in Skogaby and Flakaledin

Added Nutrients	N	P	K	Ca	Mg	Total N Deposition	Volume growth[a] (m^3 ha^{-1} yr^{-1})
	kg ha^{-1} yr^{-1}						
Irrigation						16	
Skogaby							
S-C							14
S-I	2	0	1	5	3		21
S-IF	102	17	49	11	9		25
Flakaledin						2	
F-I or F-C							4
F-IF	80	13	36	5	8		16

Note: Data recalculated from Bergholm (2001) and Persson and Nilsson (2001) for Skogaby and from Strömgren and Linder (2002) for Flakaledin.
[a] Average value of volume production over the period from 1988 to 2000.

At both sites of the irrigated-fertilized plots (IF), the spruce stand is supplied with a balanced nutrient mix dissolved in the irrigation water. This means that the annual dose of nitrogen was initially 100 kg N ha^{-1}, the other nutrients (P, K, Ca, S, Mg) being supplied in fixed proportion to N and adjusted annually against needle samples, in an attempt to attain the optimal nutrient dose. This technique allows trees to grow under steady-state conditions i.e. conditions in which the physiological state of the plants remained constant and where, consequently, the relative uptake rate of carbon and nutrients, the relative growth rate, and the allocation patterns also remained constant. The basic principle of this technique has been described by Ingestad and Ågren (1995) and Ågren and Bosatta (1996). All plant nutrients are dissolved in water and distributed by means of a sprinkler system every second day during the period from June to August. The nutrient contents in irrigation treatments and tree volume production in Skogaby and Flakaledin are listed in Table 2. Some chemical characteristics in the upper mineral horizons for Skogaby and Flakaledin are listed in Table 3.

Table 3
Mean chemical properties in the bulk and rhizosphere materials of the upper mineral horizons at Skogaby and Flakaledin

Treatment[a]	LOI (%)	pH		Na	K	Mg	Ca	Al	H	CEC	BS (%)
		H_2O	KCl	mmol kg^{-1}							
Skogaby											
S24E C	5.9	4.55	3.68	0.10	0.45	0.61	0.74	30.4	12.0	44.3	4.3
S24ER C	8.4	4.49	3.59	0.12	0.93	1.17	0.89	34.4	15.8	53.3	5.8
S22E IF	6.1	4.77	3.65	0.17	0.63	1.26	1.26	31.1	12.6	47.0	7.1
S22ER IF	7.2	4.30	3.53	0.06	0.58	1.53	1.73	35.5	18.5	57.9	6.7
S25E I	6.2	4.31	3.59	0.05	0.32	0.93	0.59	25.6	14.9	42.4	4.5
S25ER I	7.9	4.37	3.56	0.05	0.72	1.34	0.80	31.7	16.5	51.0	5.7
Flakaledin											
F7B IF	6.8	5.18	4.72	0.01	0.74	0.95	8.11	34.4	—	44.2	22.2
F7BR IF	8.6	5.18	4.64	0.01	2.02	1.29	9.07	52.4	—	64.8	19.1
F12B I	12.7	5.22	4.66	0.04	0.66	0.89	7.60	46.9	—	56.1	16.4
F12BR I	14.5	5.08	4.62	0.09	0.92	1.09	9.41	63.5	—	75.0	15.3

Note: Nutrient contents are expressed in mmol kg^{-1}.
[a] E = E horizon; B = B horizon; R = rhizosphere; C = control; I = irrigation; F = fertilization.
LOI = loss on ignition; CEC = cation exchange capacity; BS = base saturation.

4.2.3. Field manipulation and tree growth

In Skogaby, irrigation increased biomass production by 150% when compared with control in the period 1987–1993, indicating that water was still a growth–stimulating factor at this site despite high annual precipitation. An additional increased biomass production of 33% was due to IF (Table 2).

In Flakaledin, addition of irrigation (I) to the annual precipitation did not increase the annual biomass production, which was significantly stimulated by IF and increased four- or five-fold during 1988–2000 (Bergh *et al.*, 1999; Strömgren and Linder, 2002) (see also Table 2).

4.2.4. The installation, sampling and analyses of HSBs and TMBs

In September 1999, fieldwork started with the installation of the HSB and the TMB with plagioclase. In September 2000, we installed additional TMBs, this time using apatite as the test-mineral. In Skogaby, we buried a total of 24 TMBs with apatite for each of the two meshes in the E horizon of the I, IF and C treatment plots. In Flakaledin, we buried 18 TMBs containing apatite for each of the two meshes in the E horizon of the I and IF treatment plots. To reduce potential contrasts in hydrological flow associated with differences in mesh sizes, the bags were tightly placed in the soil in such a way as to assure a very close contact between the surface of the bags and the soil matrix. Moreover, the moisture content of the soil material in the HSBs that were buried close to the TMBs of both mesh sizes at the two sites was compared upon sampling and was not found to be significantly different.

5. PRELIMINARY RESULTS AND DISCUSSION

5.1. Field observation

The results of the HSB are not presented here, as it is still too early to observe changes in soils after 4 years. Therefore, only the results from the TMB are reported here. It is also worth noting that whatever the field site and treatment in the TMB experiment, apatite buried for 4 years weathered much more rapidly than plagioclases that were buried for 3years, and the total mass dissolved from apatite and plagioclases was 0.8–2% and 0.15–0.2%, respectively. Because of the slow weathering rate of plagioclases, we present only the results for apatite.

On-site observation of retrieved TMBs on September 2003 (Fig. 3) clearly revealed that the control bags with coarse mesh (541 μm, Fig. 3 left) allowed the penetration of fine roots. Hyphal bags with fine mesh (51 μm, Fig. 3 right) did not allow any visible roots to penetrate the bags, but very fine roots were growing on the surface of the bags. Similar observations were also noted when some of the HSB were retrieved at the same time. Furthermore, microscopic analyses of the HSB from Flakaledin indicated high mycorrhizal hyphae growth in both sizes of mesh bags. These observations confirm that bags with the two different meshes will enable us to examine the relative contributions of roots and mycorrhizal

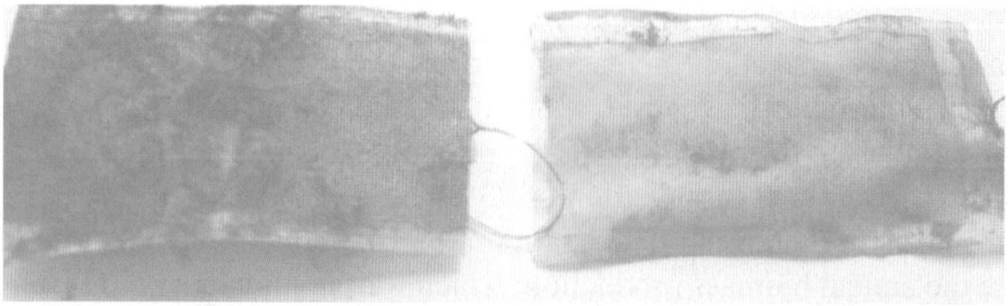

Fig. 3. TMB filled with apatite mineral in both coarse (left) and fine mesh bags.

hyphae to weathering. We therefore expect that the HSB and TMB techniques will allow us to investigate the role of the mycorrhizal hyphae themselves as well as of their associated bacteria by comparison between bags containing roots and hyphae together with those excluding roots.

5.2. SEM

The SEM images at both magnifications show apatite grains before (Fig. 4a and b) and after a 3-year period in the field at Flakaledin (Fig. 4c and d) and Skogaby (Fig. 4e and f). For each mesh size and treatments, six mineral grains were randomly selected and submitted to SEM, and images were taken. The images presented in Fig. 4 are representive of the main observations made for a given treatment and mesh size.

These observations confirm the losses of mineral mass. Grains that were exposed to field conditions at Skogaby present clear indications of more intense weathering as revealed by the abundance of dissolution features (Fig. 4e and f). The number of these features is low on the apatite surface after incubation in Flakaledin site (Fig. 4c and d) and absent in grain before the incubation in field (Fig. 4a and b).

5.3. MML

5.3.1. Mesh effects

The MML measured following mineral exposition to field conditions show that apatite in large-mesh bags (541 μm) weathered significantly faster than apatite in fine-mesh bags (51 μm). In four of five plots, apatite dissolved almost twice as fast as in the large-mesh bags compared with fine-mesh bags. The presence of roots and associated microorganisms in the large-mesh bags appeared to have accelerated the weathering of apatite. This observation is valid for both sites, except for the IF treatment at Skogaby (Fig. 5). The absence of this effect in the IF treatment at Skogaby will be discussed and compared with the IF treatment in Flakaledin (see below).

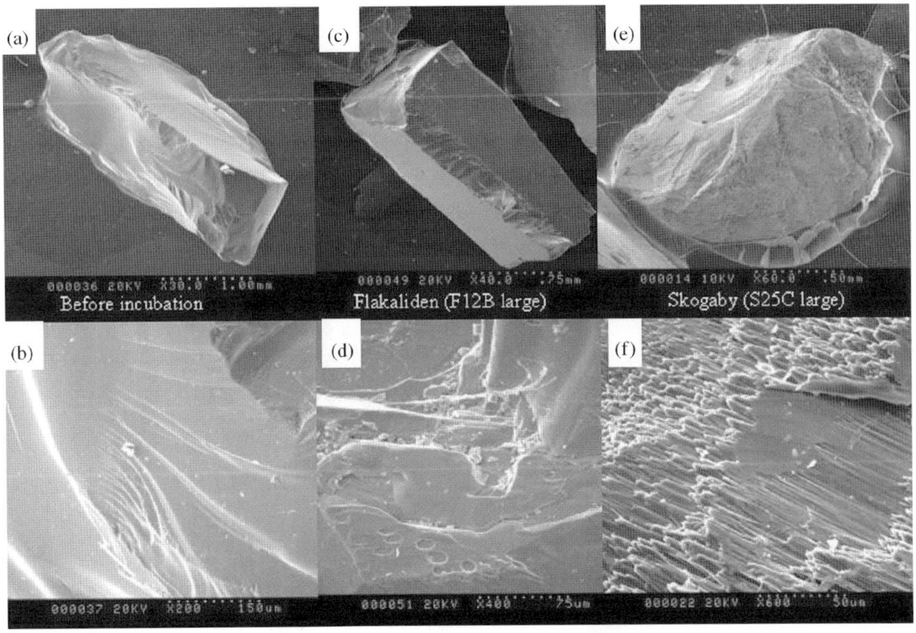

Fig. 4. SEM images showing apatite grains before (a, b) and after a 3-year period in the field at Flakaledin (c, d) and Skogaby (e, f).

We recognize that the addition of apatite grains to the soil, even in bags, could have an effect on the proximal biological activity (Wallander, 2000). Moreover, despite our SEM observations confirming losses of mineral masses recorded for apatite, it is still difficult to conclusively establish the contribution of rhizosphere processes on apatite weathering as long as this effect is not quantified.

5.3.2. Site effects

Fig. 5 shows that apatite dissolved at a faster rate at both sites, Flakaledin and Skogaby. The rate of apatite dissolution was twice as fast in southern Sweden (Skogaby) as it was in the northern site, Flakaledin. These results are in agreement with the SEM observations showing grains exposed to field conditions at Skogaby (Fig. 4e and f) and present clear evidence of more intense weathering as revealed by the greater relative abundance of dissolution features compared with the Flakaledin site. The number of dissolution features on the surface of apatite grains is lower after the field incubation at the Flakaledin site (Fig. 4c and d). These features are obviously absent from mineral surfaces before exposition to field conditions (Fig. 4a and b). The contrasts in climatic conditions such as

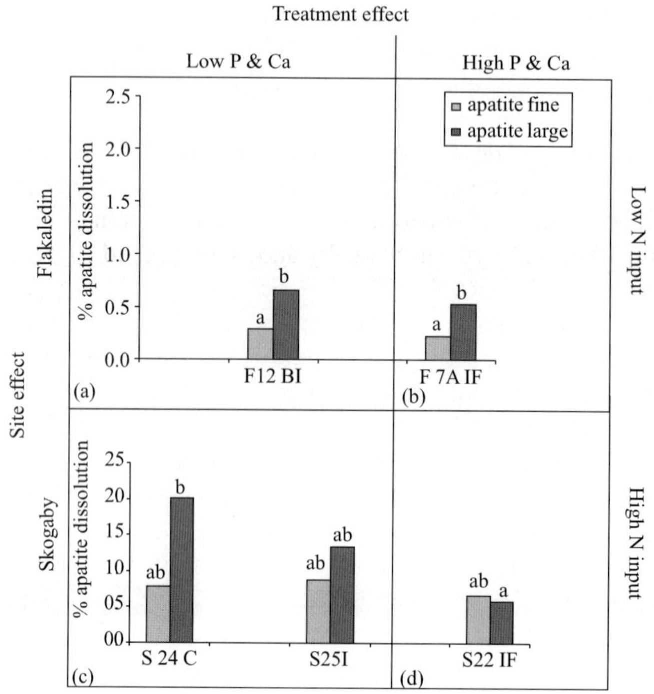

Fig. 5. Treatment effects on percent apatite dissolution in large and fine-mesh bags at the Flakaledin (plots F 12BI & F7A IF) and the Skogaby (plots S 24C, S 25I 6, S 22 IF) sites. Mean Values followed by the same letter (a, b or c) are not significantly different at the 5% level.

temperature and precipitation between the Flakaledin and the Skogaby sites could explain the differences in weathering.

5.3.3. Field treatment effects

5.3.3.1. Control and irrigation effects Soil chemical characteristics (Table 3) show that soil samples in Skogaby are more acidic and poorer in nutrients than in Flakaledin. This could be considered, together with climatic conditions, as a significant factor explaining the faster weathering in Skogaby (Fig. 5c) compared with in Flakaledin (Fig. 5a), the control treatments (I in Flakaledin and C and I in Skogaby). Clegg and Gobran (1997) found that irrigation treatment (I) in Skogaby lowered the P content in the rhizosphere of the upper soil mineral horizon relative to the control treatment. They suggested a close coupling between rapid increase in aboveground growth and relative depletion of soil P. The effect of the rapid growth observed here in Table 2 could therefore be an additional factor explaining the faster weathering rate of apatite in Skogaby compared with

Flakaledin. This effect is mainly due to the stimulated tree demands for nutrients, which is considered the most evident factor affecting acid–base conditions of the rhizosphere (Nye, 1986; Marschner and Römheld, 1996).

5.3.3.2. Irrigation–fertilization effects. Nutrients were added to the irrigation water to optimize tree growth at both sites. However, the volume production at Skogaby was much higher than at Flakaledin. Table 2, shows volume production of 25 compared with 14 $m^3 ha^{-1} yr^{-1}$ in Skogaby and 16 compared with 4 $m^3 ha^{-1} yr^{-1}$ in Flakaledin within 12 years. Fertilization did not have similar and straightforward effects on weathering at both sites. The essential difference in the amounts of apatite weathering between large and fine-mesh bags was at its maximum only at the sites of control treatment (Fig. 5a and c) and tended to decrease as the bioavailability of Ca, P and N increased (Fig. 5b and d). The absence of mesh effect in the IF treatment at Skogaby compared with IF in Flakaledin yielded lower Ca and P in Flakaledin than in Skogaby. Furthermore, the IF treatment in both sites differs greatly in terms of Ca balance with respect to N (Table 2), e.g. Ca/N are 11% in Skogaby compared with 6% in Flakaledin. Moreover, N deposition in Skogaby is far higher than at Flakaledin: 16 versus 2 kg N $ha^{-1} yr^{-1}$ (Table 2). Such high N input in the south could have a negative effect on ectomycorrhizal community. In Sweden, Taylor *et al.* (2000) compared the ectomycorrhizal community of two forest sites of Norway spruce and found highest species richness, highest diversity and highest abundance of mycorrhizal root tips at the northern site (close to Flakaledin) compared with the southern site Skogaby. It has also been demonstrated that the bacteria to fungi ratio increases with increasing N input in these both sites (Schröter *et al.*, 2003). The higher fungal community in Flakaledin compared with Skogaby suggests the biological activity, namely, the capacity of fungi to utilize and translocate carbon and nutrients (Leake and Read, 1997), is higher at this site. All these advantages of fungal activity in the northern part of Sweden has led Näsholm *et al.* (1998) and Lindahl *et al.* (2002) to conclude that the role of fungi is pivotal for boreal forest ecosystems. Schröter *et al.* (2003) revealed that as the high N inputs and bioavailable N increased at Skogaby, bacteria pathways became more important and the cycling of C and N became faster than in Flakaledin. This high bacteria to fungi ratio and consequent high turnover of organic matter could explain why weathering was low and the mesh effect was absent at Skogaby (Fig. 5d) compared with Flakaledin (Fig. 5b).

We thus suggest that the high bioavailability of Ca and P in the IF treatment at Skogaby and the elevated N inputs favoured a lower biological weathering of apatite. In contrast, the low bioavailability of Ca, P and N in IF treatment at Flakaledin and low N inputs enhanced the biological weathering of apatite and resulted in a significant difference between the large-mesh bags (541 μm) and fine-mesh bags (51 μm).

6. CONCLUSION AND FUTURE RESEARCH

There is very limited *in situ* information on the impact of root-induced changes on mineral weathering of soils and on mineral weathering dynamics in forest soils, particularly in the rhizosphere. There is also a need for the quantification of mineral dissolution rates as a result of weathering in the soil. We presented a novel scientific approach for estimating weathering in the rhizosphere of forest soils. Two ongoing field case studies in northern and southwestern Sweden were presented. Preliminary results demonstrate how mineral weathering could be monitored using well-defined techniques. These include HSB and TMB of two different meshes, 51 and 541 μm, which either allow the penetration of roots and hyphae (541 μm) or exclude root growth in the bags (51 μm). In-bag mineral weathering is being assessed by SEM and MML.

Our preliminary results from the TMB suggest that weathering of apatite was much faster in coarse-mesh than in the fine-mesh bags. A biological component that allowed roots and hyphae in the 541 μm bags rather than one that excluded roots in the 51 μm bags increased weathering. The combined effects of climate and nutrient inputs to forest ecosystems is necessary for a better assessment and interpretation of weathering of minerals in the rhizosphere. The quantification of weathering on mineral dissolution rates could be improved by using HSB and TMB techniques. More data similar to those obtained from our ongoing field studies are needed to improve the predictive capability of existing growth and biogeochemical models.

ACKNOWLEDGMENTS

The authors thank Dr. Jeff Wilson (Macaulay Institute, Scotland, UK) and Dr. Naomi Assadian (Texas A&M University, USA) for the thorough reading this chapter.

REFERENCES

Åberg, G., Jacks, G., Hamilton, P.J., 1989. Weathering rates and $^{87}Sr/^{86}Sr$ ratios: An isotopic approach. Journal Hydrol. 109, 65–78.

Ågren, G.I., Bosatta, E., 1996. Theoretical Ecosystem Ecology–Understanding Element Cycles, Cambridge University Press.

April, R., Keller, D., 1990. Mineralogy of the rhizosphere in forest soils of the eastern United States. Biogeochemistry 9, 1–18.

Arocena, J.M., Glowa, K.R., Massicotte, H.B., Lavkulich, L., 1999. Chemical and mineral composition of ectomycorrhizosphere soils of subalpine fir (Abies lasiocarpa (Hook.) Nutt.) in Ae horizon of Luvisol. Can. J. Soil Sci. 79, 25–35.

Augusto, L., Ranger, J., Turpault, M.P., Bonnaud, P., 2001. Experimental *in situ* transformation of vermiculites to study the weathering impact of tree species on the soil. Eur. J. Soil Sci. 52, 81–92.

Augusto, L., Bonnaud, P., Ranger, J., 1998. Impact of tree species on forest soil acidification. For. Ecol. Manage. 105, 67–78.

Augusto, L., Turpault, M.P., Ranger, J., 2000. Impact of forest tree species on feldspar weathering rates. Geoderma 96, 215–237.

Bailey, S.W., Hornbeck, J.W., Driscoll, C.T., Gaudette, H.E., 1996. Calcium inputs and transport in a base-poor forest ecosystem as interpreted by Sr isotopes. Water Resourc. Rese. 32, 707–719.

Bergh, J., Linder, S., Lundmark, T., Elfving, B., 1999. The effect of water and nutrient availability on the on the productivity of Norway spruce in northern and southern Sweden. Forest Ecology and Management 119, 51–62.

Bergholm, J., 2001. Long-term effects of enhanced nitrogen and sulphate additions on soil acidification and nutrient cycling in a Norway spruce stand. Doctoral thesis, Acta Universitatis Agriculturae Sueciae: Silvestria 215.

Berner, R.A., 1992. Weathering, plants, and the long-term carbon cycle. Geochim. Cosmochim. Acta 56, 3225–3231.

Birkeland, P.W., 1999. Soils and Geomorphology. Oxford University Press, Oxford, p. 430.

Brady, P.V., Dorn, R.I., Brazel, A.J., Clark, J., Moore, R.B., Glidewell, T., 1999. Direct measurement of the combined effects of lichen, rainfall, and temperature on silicate weathering. Geochim. Cosmochim. Acta 63, 3293–3300.

Brimhall, G.H., Chadwick, O.A., Lewis, C.J., Compston, W., Williams, I.S., Danti, K.J., Dietrich, W.E., Power, M.E., Hendricks, D., Bratt, J., 1991. Deformational mass transport and invasive processes in soil evolution. Science 255, 695–702.

Chung, J.-B., Zasoski, R.J., Burau, R.G., 1994. Aluminum-potassium and aluminum-calcium exchange equilibria in bulk and rhizosphere soil. Soil Sci. Soc. Am. J. 58, 1376–1382.

Clegg, S., Gobran, G.R., 1997. Rhizospheric P and K in forest soil manipulated with ammonium sulfate and water. can. J. Soil Sci. 77, 525–533.

Cline, G.R., Powell, P.E., Szaniszlo, P.J., Reid, C.P.P., 1982. Comparison of the abilities of hydroxamic, synthetic, and other natural organic acids to chelate iron and other ions in nutrient solution. Soil Sci. Soc. Am. J. 46, 1158–1164.

Courchesne, F., Gobran, G.R., 1997. Mineralogical variations of bulk and rhizosphere soils from a Norway spruce stand, southwestern Sweden. Soil Sci. Soc. Am. J. 61, 1245–1249.

Courchesne, F., Turmel, M.-C., Beauchemin, P., 1996. Magnesium and potassium release by weathering in Spodosols: Grain surface coating effects. Soil Sci. Soc. Am. J. 60, 1188–1196.

Dambrine, E., Martin, F., Carisey, N., Granier, A., Hällgren, J.-E., Bishop, K., 1995. Xylem sap composition: A tool for investigating mineral uptake and cycling in adult spruce. Plant Soil 168–169, 233–241.

Dormaar, J.F., 1988. Effect of plant roots on chemical and biochemical properties of surrounding discrete soil zones. Can. J. Soil Sci. 68, 233–242.

Drever, J.I., 1994. The effects of land plants on weathering rates of silicate minerals. Geochim. Cosmochim. Acta 56, 2325–2332.

Elfving, B., Tegnhammer, L., 1996. Trends of tree growth in Swedish forests 1953–1992: An analysis based on sample trees from the National Forest Inventory. Scand. For. Res. 11, 38–49.

FAO-UNESCO, 1988. Soil map of the world. Revised legend. FAO, Rome.

Gardner, W.K., Parbery, D.G., Barber, D.A., 1982. The acquisition of phosphorus by Lupinus albus L: I. Some characteristics of the soil/root interface. Plant Soil 68, 19–32.

Gobran, G.R., Clegg, S., 1996. A conceptual model for nutrient availability in the mineral soil-root system. Can. J. Soil Sci. 76, 125–131.

Gobran, G.R., Clegg, S., Courchesne, F., 1998. Rhizospheric processes influencing the of forest ecosystems. In: Breemen, N. van (Ed.), Plant Induced Soil Changes: Processes and Feedbacks. Biogeochemistry 42, 107–120.

Gobran, G.R., Courchesne, F., Dufresne, A., 1997. Relationships between sulfate retention and release, solution pH and DOC in the Gårdsjön soils. In: Hultberg, H., Skeffington, R.A., (Eds.), Experimental Reversal of Acid Rain Effects. John Wiley and Sons, Ltd. p. 466.

Hendershot, W.H., Courchesne, F., Schemenauer, R.S., 1992. Soil acidification along a topographic gradient on Roundtop Mountain, Québec. Canada. Water Air Soil Poll. 61, 235–242.

Hinsinger, P., Jaillard, B., Dufey, J., 1992. Rapid weathering of a trioctahedral mica by the roots of ryegrass. Soil Sci. Soc. Am. J. 56, 977–982.

Hinsinger, P., Plassard, C., Tang, C., Jaillard, B., 2003. Origins of root-mediated pH changes in the rhizosphere and their response to environmental constraints: A review. Plant Soil 248, 43–59.

Hodson, M.E., Langaan, S.J., Wilson, M.J., 1997. A critical evaluation of the use of the profile model in calculating mineral weathering rates. Water, Air Soil Poll. 98, 79–104.

Ingestad, T., Ågren, G.I., 1995. Plant nutrition and growth: Basic principles. Plant Soil 168/169, 15–20.

Jones, D.L., Brassington, D.S., 1998. Sorption of organic acids in acid soils and its implications in the rhizosphere. Eur. J. Soil Sci. 49, 447–455.

Jones, D.L., Prabowo, A.M., Kochian, L.V., 1996. Kinetics of malate transport and decomposition in acid soils and isolated bacterial populations: The effect of microorganisms on root exudation of malate under Al stress. Plant Soil 182, 239–247.

Jongmans, A.G., van Breemen, N., Lundstrom, U., van Hees, P.A.W., Finlay, R.D., Srinivasan, M., Unestam, T., Giesler, R., Melkerud P.A., Olsson, M., 1997. Rock-eating fungi. Nature 389, 682–683.

Kelly, E.F., Chadwick, O.A., Hilinski, T.E., 1998. The effects of plants on mineral weathering. Biogeochemistry 42, 21–53.

Kodama, H., Nelson, S., Yang, A.F., Kohyama, N., 1994. Mineralogy of rhizospheric and non-rhizospheric soils in corn fields. Clay Clay Miner. 42, 755–763.

Leake, J.R., Read, D.J., 1997. Mycorrhizal fungi in terrestrial habitats. In: Wicklow, D.T., Söderström, B.E., (Eds.), Environ. Microb. Relat. Springer-Verlag, New York, pp. 281–301.

Likens, G.E., Bormann, F.H., 1995. Biogeochemistry of a Forested Ecosystem, Second ed. Springer-Verlag, New York, p. 159.

Lindahl, B., Taylor, A.F.S., Finlay, R.D., 2002. Defining nutritional constraints on carbon cycling–towards a less "phytocentric" perspective. Plant Soil 242, 123–135.

Linder, S., 1995. Foliar analysis for detecting and correcting nutrient imbalance in Norway spruce. Ecological Bulletins (Copenhagen) 44, 178–190.

Lundström, U.S., 1994. Significance of organic acids for weathering and the podsolization process. Environ. Int. 20, 21–30.

Marschner, H., Römheld, V., 1996. Root induced changes in the availability of micronutrients in the rhizosphere. In: Waisel, Y., Amram, E., Kafkafi, U., (Eds.), Plant Roots: The Hidden Half. Marcel Decker Inc.

Mitchell, M.J., David, M.B., Fernandez, I.J., Fuller, R.D., Nadelhoffer, K., Rustad, L.E., Stam, A.C., 1994. Response of buried mineral bags to experimental acidification of forest ecosystem. Soil Sci. Soc. Am. J. 58, 556–563.

Näsholm, T., Ekblad, A., Nordin, A., Giesler, R., Högberg, M., Högberg, P., 1998. Boreal forest plants take up organic nitrogen. Nature 392, 914–916.

Nilsson, L.O., Wiklund, K., 1992. Influence of nutrient and water stress on Norway spruce production in south Sweden–the role of air pollutants. Plant Soil 147, 251–265.

Nugent, M.A., Brantley, S.L., Pantano, C.G., Maurice, P.A., 1998. The influence of natural mineral coating on feldspar weathering. Nature 395, 588–591.

Nye, P.H., 1986. Acid-base changes in the rhizosphere. Adv. Plant Nut. 2, 129–153.

Olsson, M., Melkerud, P.-A., 2000. Weathering in three podzolized pedons on glacial deposits in northern Sweden and central Finland. Geoderma 94, 149–161.

Persson, T., Nilsson L.-O. (Eds.), 2001. The Skogaby experiment. Effects of long-term nitrogen and sulphur application to a forest ecosystem. Swedish Environmental Protection Board, Report 5173, p. 220. (In Swedish, Skogabyförsöket – Effekter av långvarig kväve– och svaveltillförsel till ett skogsekosystem. Naturvårdsverket, Rapport 5173, sid 220).

Ranger, J., Dambrine, E., Robert, M., Righi, D., Félix, C., 1991. Study of current soil-forming processes using bags of vermiculite and resins placed within soil horizons. Geoderma 48 (3–4), 335–350.

Ranger, J., Robert, M., Bonnaud, P., Nys, C., 1990. Les minéraux-test, une approche expérimentale in situ de l'altération biologique et du fonctionnement des écosystèmes forestiers: essais des types de sols et des essences forestières feuillues et résineuses. Annes des Sciences forestières 47, 529–550.

Righi, D., Bravard, S., Chauvel, A., Ranger, J., Robert, M., 1990. *In situ* study of soil processes in an Oxisol-Spodosol sequence of Amazonia (Brazil). Soil Sci. 150 (1), 438–445.

Robert, M., Berthelin, J., 1986. Role of biological and biochemical factors in soil mineral weathering. In: Huang, P.M., Schnitzer, M., (Eds.), Interactions of Soil Minerals with Natural Organics and Microbes. SSSA, Madison, WI, pp. 453–495.

Sadio, S. 1982. Altération expérimentale de phyllosilicates-tests sous végétations forestières acidifiantes. Ph.D. Thesis, Univ. Nancy I.96p

Sak, P.B., Fisher, D.M., Gardner, T.W., Murphy, C., Brantley, S., 2004. Rates of weathering rind formation on Costa Rican basalt. Geochim. Cosmochim. Acta 68, 1453–1472.

Sarkar, A.N., Jenkins, D.A., Wyn Jones, R.G., 1979. Modifications to mechanical and mineralogical composition of soil within the rhizosphere. In: Harvey, J.L., Scott-Russell, R. (Eds.), The Soil–Root Interface. Academic Press.

Sarkar, A.N., Wyn Jones, R.G., 1982. Effect of rhizosphere pH on the availability and uptake of Fe, Mn, and Zn. Plant Soil 66, 361–372.

Schröter, D., Wolters, V., De Ruiter, P.C., 2003. C and N mineralisation in the decomposer food webs of a European forest transect. Oikos 102, 294–308.

Séguin, V., Courchesne, F., Gagnon, C., Martin, R.R., Naftel, S., Skinner, W., 2005. Mineral weathering in the rhizosphere of forested soils. In: Huang, P.M., Gobran, G.R. (Eds.), Biogeochemistry of Trace Elements in the Rhizosphere. Elsevier, New York, pp. 29–55 (in this book).

Strömgren, M., Linder, S., 2002. Effects of nutrition and soil warming on stem wood production in a boreal Norway spruce stand. Global Change Biology 8, 1195–1204.

Sverdrup, H., Warfvinge, P., 1992. Calculating field weathering rates using a mechanistic geochemical model – PROFILE. Applied geochemistry 27, 283.

Sverdrup, H., Warfvinge, P., Nihlgård, B., 1994. Assessment of soil acidification effects on forest growth in Sweden. Water, Air, Soil Poll. 78, 1–36.

Taylor, A.F.S., Martin, F., Read, D.J., 2000. Fungal diversity in ectomycorrhizal communities of Norway spruce (Picea abies [L.] Karst.) and beech (Fagus syl_atica L.) along north-south transects in Europe. In: Schulze, E.-D. (Ed.), Carbon and Nitrogen Cycling in Forest Ecosystems. Springer-Verlag, New York, pp. 343–365.

Ugolini, F.C., Dahlgren, R.A., 1991. Weathering environments and occurrence of imogolite/allophane in selected Andisols and Spodosols. Soil Sci. Soc. Am. J. 55, 1166–1171.

Velbel, M.A., 1985. Geochemical mass balances and weathering rates in forested watersheds of the southern Blue Ridge. Am. J. Sci. 285, 904–930.

Wallander, H. 2000. Uptake of P from apatite by Pinus sylvestris seedlings colonised by different ectomycorrhizal fungi. Plant Soil 218, 249–256.

Wallander, H., 2002. Use of strontium isotopes and foliar K content to estimate weathering of biotite induced by pine seedlings colonized by ectomycorrhizal fungi from two different soils. Plant Soil 222, 215–229.

Wallander, H., Johansson, L., Pallon, J., 2002. PIXE analysis to estimate the elemental composition of ectomycorrhizal rhizomorphs grown in contact with different minerals in forest soil. FEMS Microbiology Ecology 39, 147–156.

Wallander, H., Wickman, T., Jacks., G., 1997. Apatite as a P source in mycorrhizal and non-mycorrhizal Pinus sylvestris seedlings. Plant Soil 196, 123–131.

White, A.F., Blum, A.E., Schulz, M.S., Bullen, T.D., Harden, J.W., Peterson, M.L., 1996. Chemical weathering rates of a soil chronosequence on granitic alluvium: I. Quantification of mineralogical and surface area changes and calculation of primary silicate reaction rates. Geochim. Cosmochim. Acta 60, 2533–2550.

Whittig, L.D., Allardice, W.R., 1986. X-ray diffraction techniques. In: A. Klute (Ed.), Methods of Soil Analysis. Part 1. Physical and Mineralogical Methods, Second ed., Agron. Monogr. 9. ASA and SSSA, Madison, WI., pp. 331–362

Biogeochemistry of Trace Elements in the Rhizosphere
P.M. Huang and G.R. Gobran (Editors)

Chapter 2

Mineral weathering in the rhizosphere of forested soils

V. Séguin[a], F. Courchesne[a], C. Gagnon[b], R.R. Martin[c], S.J. Naftel[c], and W. Skinner[d]

[a]Départment de géographie, Université de Montréal, C.P. 6128, succursale Centre-ville, Montréal, Québec, Canada, H3C 3J7
E-mail: francois.courchesne@umontreal.ca

[b]St. Lawrence Centre, Environment Canada, 105 McGill St., 7th Floor, Montréal, Québec, Canada, H2Y 2E7

[c]Department of Chemistry, University of Western Ontario, London, Ontario, Canada, N6A 5B7

[d]Ian Wark Research Institute, UNISA, Mawson Lakes, 5095, SA Australia

ABSTRACT

The rhizosphere is a microenvironment enriched in organic matter and generally more acidic than the bulk soil. In this chapter, we submit that mineral weathering and metal fractionation differ in the rhizosphere compared to the bulk soil, a change that could impact on plant nutrition and element toxicity. The objective of the study is to establish the nature of the effect of roots on mineral weathering in the rhizosphere of forested soils based on differences in (1) mineralogical composition and (2) the chemical forms of metals between the rhizosphere and the bulk soil. The study area was located in Rouyn–Noranda (Canada), where samples were collected under *Populus tremuloïdes* growing on Luvisolic soils. X-ray diffraction (XRD), time of flight secondary-ion mass spectroscopy (TOF-SIMS) and X-ray absorption near-edge structure (XANES) analyses were performed. The concentrations of Al, Ca, Cd, Co, Cr, Cu, Fe, Li, Mg, Mn, Ni, Pb, Si and Zn were obtained from an acid ammonium oxalate (AAO) extraction. The XRD results show differences in mineralogical abundance, particularly of chlorite and amphibole, which is interpreted as an increase in mineral weathering in the rhizosphere. It was suggested in the literature that the higher alteration in the rhizosphere could be related to K uptake by roots. However, our results show greater $BaCl_2$-extractable K in the rhizosphere, an observation in opposition to this nutrient-depletion hypothesis. The AAO extraction reveals higher contentrations of Fe and Mn in the rhizosphere. These data support XRD results and suggest the formation of secondary oxides in the rhizosphere through weathering. In turn, the greater abundance of oxides creates absorption sites for trace elements such as Cu and Zn as supported by the AAO extractions. The TOF-SIMS mapping also shows an accumulation of total metals at the soil–root interface. The XANES analysis of Mn further indicates that metals tend to be oxidized in the rhizosphere, whereas they are found in organic forms in the root or as a mixture of both at the soil–root interface. The presence of oxidized forms of Mn in the rhizosphere is in agreement with the results of the AAO extraction. In summary, weathering is shown to be higher in the rhizosphere, favors the formation of oxides, notably Mn oxides and, hence, the retention of trace metals.

1. INTRODUCTION

It is well established in the literature that pH, organic matter and microorganisms affect mineral weathering. The pH of the soil solution also impacts on the chemical activity of metals, with most metals being more soluble under acidic conditions (Lindsay, 1979). As such, acidic soils will favour alteration through the solubilization of the metal atoms forming the mineral structures. Organic matter, either in the solid or the dissolved state, has the ability to dissolve and bind metals (McBride, 1994). Metal binding can also accelerate the alteration of minerals with higher organic matter contents producing greater weathering rates. Microorganisms produce organic compounds and lower the pH of the surrounding soil, and can, by these means, increase the uptake of nutrients by plants (Marschner, 1995). The nutrient uptake by plants subsequently affects the acid–base equilibrium between the liquid and the solid phases. Microorganisms can also impact directly on mineral weathering. For example, Jongmans *et al.* (1997) showed that mycorrhizae were able to create the equivalent of about 150 m of pores per dm^3 of soil in the E horizon of a podzolic soil from Sweden. This, in turn, will impact on the forms of metals and on the weathering rate.

The rhizosphere is considered to be a microenvironment that is more acidic than the bulk soil, which contains more organic matter and in which microorganisms are more abundant. Indeed, the soil under root influence generally has a lower pH, although the effect on pH depends on the nutritional status of the plant (Haynes, 1990; Nye, 1981). A plant absorbs both anions and cations, but rarely in an equivalent amount. In order to maintain its electroneutrality and that of the surrounding soil solution, the plant must either extrude OH^- or H^+ to equilibrate the anion–cation balance. In many cases, the plant will absorb more cations than anions, thus inducing the release of H^+ and the acidification of the soil zone under root influence. The rhizosphere also contains a greater amount and diversity of organic matter (Grayston *et al.*, 1996). The increased presence of organic matter is induced by the root itself, through rhizodeposition. This abundance of organic matter in turn increases the quantity and diversity of microorganisms in the rhizosphere because it is a source of nutrient and energy for these microorganisms.

It is estimated that the abundance of microorganisms in the rhizosphere is 5–50 times that of the bulk soil, depending on soil and plant characteristics (Marschner, 1995). Based on the literature relating soil acidity, organic matter content and microorganisms to the intensity of mineral weathering, it can then be expected that mineral weathering will be increased in the rhizosphere.

However, few studies have considered the dynamics of mineral weathering in the rhizosphere. The interest in mineral weathering in the rhizosphere was initiated a few decades ago with, for example, the work of Mortland *et al.* (1956) and Spyridakis *et al.* (1967). Since then, little research has been done to relate the mineralogy to the chemical characteristics of the rhizosphere. Nevertheless, April and Keller (1990) showed an association between the morphology of selected minerals in the rhizosphere and the presence of roots that could suggest a direct effect of roots on mineral weathering. Usually, there are few differences between the mineral assemblages of the rhizosphere and of the associated bulk soil. The differences between the two soil components are rather revealed by the relative abundance of some minerals. For example, the abundance of vermiculite was shown to increase in the rhizosphere as a result of biotite (Mortland *et al.*, 1956; Adamo *et al.*, 1998) or phlogopite (Hinsinger *et al.*, 1993) weathering. Other weathering products of biotite include kaolinite when the cation exchange capacity (CEC) and buffering capacity are low enough (Spyridakis *et al.*, 1967). For phlogopite, weathering can lead to the formation of interlayered hydroxides if acidic conditions are present (Hinsinger *et al.*, 1993). Moreover, the rhizosphere effect on mineral weathering can be very rapid, as indicated by Hinsinger *et al.* (1992), who documented an impact within three days at a distance of 0.5 mm from a dense root mat. In a similar experiment, Hinsinger and Jaillard (1993) observed the release of 19% of the potassium contained in the phlogopite and the complete vermiculitization of the mineral at a distance of 0.5 mm of the root mat within 32 days.

Vermiculization occurs when the interlayer potassium is released from micas to maintain the equilibrium between the solid and liquid phases of the soil. As plants take up potassium, a macro nutrient, the decrease in potassium concentration in the liquid phase induces an acceleration of potassium release by minerals such as biotite and phlogopite (Hinsinger *et al.*, 1993). In the field, April and Keller (1990) indicated that the rhizoplan (soil directly in contact with the roots) was enriched in well-crystallized muscovite and vermiculite, while the adjacent rhizosphere soil contained more degraded biotite and muscovite. These results agree with those obtained in the laboratory pertaining to the vermiculization of biotite closer to the root. The most easily weathered minerals (e.g. expandable phyllosilicates, plagioclases and amphiboles) are more impacted by root activity than resistant minerals (e.g. k-feldspars) (Courchesne and Gobran, 1997). Moreover, the fine-textured materials are more affected by the mineralogical changes taking place in the rhizosphere as they have a higher specific surface area, making them more sensitive to weathering reactions (Sarkar *et al.*, 1979).

Incidentally, the fine-textured particles are sometimes more abundant in the rhizosphere (Leyval and Berthelin, 1991).

Other impacts of roots on minerals have also been documented. For instance, Adamo *et al.* (1998) demonstrated that fine-textured particles often adopt a position tangential to the root. This is related to the mechanical repositioning induced by the growth of roots. Also, the mineral faces oriented towards the root usually present greater signs of alteration (April and Keller, 1990). In addition, the presence of organic acids related to roots increases the adherence of mineral particles to roots (Leyval and Berthelin, 1991). With respect to thermal stability, April and Keller (1990) indicated that kaolinite is more stable in the rhizosphere than in the bulk soil. This increased thermal stability could reflect the greater crystallinity of the kaolinite in the rhizosphere, a result of recrystallization or of the presence of organic acids (the solubilization of Al is necessary to the formation of kaolinite).

The increased weathering of minerals in the rhizosphere can induce changes in the abundance and the forms of metals in this soil component as compared to the bulk soil. For instance, Courchesne and Gobran (1997) measured greater contents of acid ammonium oxalate (AAO) extractable Al and Fe in the rhizosphere than in the bulk soil. AAO solutions are operationally considered to extract the inorganic amorphous solid phases (e.g. sesquioxides) together with organically complexed metals, most of which occur as weathering products in soils. As such, the results of Courchesne and Gobran (1997) suggest that the greater abundance of oxides in the rhizosphere could result from an increase in weathering rates.

Most of the work concerned with the mineralogical differences between the rhizosphere and the bulk soil has been conducted in the laboratory under conditions that maximize the impact of root activity on minerals. For example, Hinsinger *et al.* (1993) used a dense root mat to study the impact of roots on phlogopite. Laboratory experiments make rhizosphere sampling easier by using devices such as rhizoboxes (Youssef and Chino, 1987), which artificially create a dense rhizosphere. Studies on the rhizosphere were also mostly conducted with cultivated plant species, generally shrubs or herbs. Moreover, most of these studies use phyllosilicates, particularly trioctahedral micas such as phlogopite and biotite, as test minerals (Hinsinger *et al.*, 1993). However, trioctahedral micas are ten times less resistant to alteration than dioctahedral minerals (Robert and Berthelin, 1986). This selection of the test minerals apparently reflects the interest for research on major nutrients, in particular potassium, as it is released from interlayer positions by weathering (Harley and Gilkes, 1997). Minor nutrients and trace elements are comparatively much less studied. Also, an increased mineral weathering in the rhizosphere will affect the forms under which metals are present in soil through the equilibrium between the solid and liquid phase (Courchesne, and Gobran, 1997). However, the information available on metal fractionation in the rhizosphere, as opposed to that on the bulk soil, is rather fragmentary.

More data on mineral weathering in the rhizosphere are needed to get new insights on how roots influence metal fractionation. Greater weathering can change the forms of metals and render some of them morc bioavailable. This change in metal forms can affect plant nutrition or the release of toxic elements (Hinsinger and Gilkes, 1997). Other than for nutritional purposes in agriculture or sylviculture, the bioavailability of metals in soils is of great interest for soil decontamination using phytoremediation techniques. A better knowledge of the influence of roots on mineral weathering and metal forms would also help to develop new indices of soil quality (Lombi *et al.*, 2001).

Consequently, the objective of this study is to establish the nature of the effect of roots on mineral weathering in the rhizosphere of forested soils based on differences in (1) the mineralogical composition and (2) the chemical forms of metals.

2. MATERIALS AND METHODS

2.1. Study sites

The study area is situated near Rouyn-Noranda, about 600 km to the north–west of Montréal, Canada (48°14′N, 79°01′W). Three sites were sampled at a downwind distance of 0.5, 2 and 8 km from the Horne copper smelter. At each site, soil samples were collected under three trembling aspen (*Populus tremuloïdes* Michx) of similar age ($<$ 30 years old). The soils developed in post-glacial lake sediments of silt texture to form Luvisols, according to the Canadian System of Soil Classification (Soil Classification Working Group, 1998). The extent of soil contamination received through the atmospheric deposition of metals represents the main difference between the three sites as other characteristics were kept constant (e.g. climate, parent material, slope, aspect, etc.). For a more detailed description of the sites, see Séguin *et al.* (2004).

2.2. Sample collection and soil component separation

The sampling was performed at the end of September 1998. At each of the three sampling sites, three *Populus tremuloïdes* were uprooted carefully (for a total of nine trees). The three trees are the field replicates used to establish site variability. All soil samples are taken from the upper B horizon (15–20 cm under the organic-mineral interface). This horizon enables the collection of enough roots to provide sufficient rhizosphere mass for chemical analyses, while being deep enough to avoid the lack of contrast between the rhizosphere and the bulk soil that is found in organic horizons. The root diameter was between 0.5 mm and 1 cm.

Two hand-shaking operations are performed to separate soil components. The first one is done in the field. The soil falling from the roots and the remaining soil is labelled bulk soil. The soil adhering to roots is considered rhizosphere soil (Rollwagen and Zasoski, 1988). The second hand-shaking operation is performed in the laboratory and allows the separation of an outer rhizosphere (soil falling from the roots) and an inner rhizosphere (soil still adhering to roots). According to this procedure, three soil components are produced: the bulk soil, the outer rhizosphere and the inner rhizosphere. All samples were air-dried and sieved at 0.5 mm to reduce texture differences between soil components (Grinsted *et al.*, 1982). Root fragments were retrieved from samples using plastic tweezers and static electricity. Samples were stored in sterile plastic bags and contacts with metallic tools were avoided.

2.3. X-ray diffraction (XRD)

The mineral composition of the clay fraction of the three soil components was established by X-ray diffraction (Courchesne and Gobran, 1997). Sample pretreatments were performed using sodium hypochlorite at pH 9.5 to destroy organic matter and with citrate–bicarbonate to eliminate Fe and Al sesquioxydes. The clay-size particles were isolated by preserving the supernatant after centrifugation of a soil–water mixture. This operation was repeated until the supernatant became clear. Clay particles were sedimented by the addition of magnesium chloride and then washed with pure water. The clay materials were saturated with magnesium, magnesium ethylene glycol or potassium. Each of these subsamples was mounted on glass slides with a preferential orientation. After a first XRD analysis at 25°C, the potassium–saturated slides were heated at 300° and 550°C and reanalysed.

The analyses were performed on a Rigaku Miniflex diffractometer (Cu-Kα radiations) at 30 kV and 10 mA, and at a scanning speed of $2°\theta\,\mathrm{min}^{-1}$ from 2 to 30°. An intensity ratio (I) was calculated for each mineral using the height of its peak divided by the height of the quartz peak (I_q) at 0.426 nm. This ratio allows the comparison of the different slides prepared for the three components of a given soil. The calculation is based on the assumption that quartz, a mineral resistant to weathering, is present in comparable absolute amounts in the bulk soil and in the two rhizospheres of a given sample.

2.4. Barium chloride extraction

After adding 15 ml of 0.1 M barium chloride to 1.5 g of soil the mixture was reacted on an end-over-end shaker for 2 h. The mixture was subsequently centrifuged at 1400 ***g*** for 15 min before the supernatant was filtered with cellulose filters (Osmonic micronSep mixed esters 0.45 μm) in a vacuum system and acidified with 2% HNO_3. Samples were stored at 4°C prior to analysis. This extraction is assumed to dissolve the exchangeable cations (Hendershot *et al.*, 1993). The analysis of the K concentration was performed on an ICP-AES. The detection limit (0.1 mg L^{-1}) was based on the guidelines given by Centre Saint-Laurent (2001).

2.5. AAO extraction

To 250 mg of soil, 10 ml of 0.2 M AAO solution was added. The mixture was shaken on an end-over-end mixer for 4 h in the dark before centrifugation at

1400 ***g*** for 20 min. The supernatant was filtered with cellulose filters (Osmonic micronSep mixed esters 0.45 μm) on a vacuum system. The AAO extracts were acidified with 2% HNO_3 and stored at 4°C prior to analysis. This extraction is considered to dissolve amorphous oxides together with organically complexed metals (Ross and Wang, 1993).

The analysis of the samples was performed on an ICP-AES for Al, Ca, Cd, Co, Cr, Cu, Fe, Li, Mg, Mn, Ni, Pb and Zn. The interest in these elements lies in their implication in plant nutrition and ecotoxicology (McBride, 1994; Fergusson, 1990). The detection limit for each element was based on the guidelines given by Centre Saint-Laurent (2001).

2.6. Time-of-flight secondary-ion mass spectroscopy (TOF-SIMS)

Subsamples of the soil aggregates adhering to root were mounted on double-sided, conducting carbon tape on silicon wafer. A Physical Electronics PHI TRIFT II spectrometer was employed with a pulsed Ga liquid metal ion gun at 25 kV (pulse width 20 ns, beam current 60 pA). Maps obtained were 200 × 200 μm with a resolution of 1.5 μm for Al, Ca, Fe, Mg and Mn. All elemental maps were normalized for the ion yield by dividing by the total ion image. This enables the reduction of artefacts related to regions of inherently high secondary ion yield (e.g. the oxide minerals as compared to organic materials) (Martin *et al.*, 2004).

2.7. X-ray fluorescence (XRF)

Soil clusters adhering to roots after the second hand-shaking were preserved in order to perform the XRF analyses. These samples were dehydrated in a graded acetone series. They were infiltrated overnight with a 1:1 mixture of Spur's (hard) and acetone and afterwards with pure Spur's for 72 h. Samples were polymerized at 55°C for 24 h before being sectioned with a razor blade in slices of 20–50 μm. Root sections were about 1 mm in diameter with the rhizosphere mostly intact. The sections were mounted on a silicon wafer with a carbon tape for insertion into the beamline at 45° of the incident X-ray beam (Naftel *et al.*, 2002).

The XRF intensity maps were obtained at beamline 20-ID-B (PNC-CAT) of the Advanced Photon Source (Argonne National Laboratory) at room temperature. The beamline had a channel cut Si(111) crystal monochromator (3–27 keV) and a pair of Kirkpatrick–Baez mirrors, which focused the beam to 4 μm square at the sample. A 13-element Ge detector at 90° of the incident beam was employed to capture the emitted fluorescence. A step size of 4 μm was used for an image of 324 × 504 μm at an energy of 10 or 7 keV.

2.8. X-ray absorption near-edge structure (XANES)

Micro-XANES were obtained for Cu and Mn on the same beamline with the same experimental set-up as for the XRF. The Fe spectra were collected as a single scan in fluorescence yield mode at selected points within the cross-section of the root. All XANES had a linear pre-edge background removed and were normalized to an edge jump of unity.

Mn-acetate and Mn-malate as well as birnessite and MnO_2 were used as reference materials for the organic and oxidized forms of Mn, respectively. The spectra obtained for the soil samples were compared with those of the reference materials to assess the predominant form of Mn present in the samples.

2.9. Statistical analyses

Considering the n value of three (field replicates) or nine (all trees), normality cannot be assumed and a non-parametric test is required. The Friedman test was conducted to validate the significance of the differences between the three soil components (bulk soil, outer rhizosphere and inner rhizosphere). This test was chosen because the sampling design involved more than two groups of related samples (three soil components), and a variation between sites related to atmospheric deposition of metals (Legendre and Legendre, 1998). As a consequence of the low degree of freedom (df = 2), two levels of significance were employed: $p < 0.10$ and $p < 0.05$. Statistics were calculated using the software SPSS 10.0 for Windows.

3. RESULTS

3.1. X-Ray Diffraction

The XRD patterns of the <2 μm fraction (K saturation) for the site at 8 km are presented in Fig. 1. A comparison of normalized peak intensity with respect to the 0.426 nm quartz peak is also shown in Table 1. The results for the three sites indicate that the relative abundance of minerals follows on average the sequence quartz > chlorite > plagioclase > vermiculite > amphiboles > K-feldspar, although this order varies slightly from site to site. The presence of chlorite is confirmed by a peak at 1.4 nm with both Mg-25 and K-25°C treatments and by the 1.4 nm peak for the K-550°C treatment (Barnhisel and Bertsch, 1989). Vermiculite includes hydroxy-interlayered vermiculite as suggested by the absence of total collapse of the 1.2 nm peak after the K-300°C and K-550°C treatments (Allen and Hajek, 1989). There is no smectite as indicated by the absence of a 1.8 nm peak for the Mg-ethylene glycol treatment.

A comparison of the three soil components shows that the rhizosphere tends to be depleted compared to the bulk soil for many minerals and at most sites. This trend is particularly well expressed for chlorite and amphiboles (Table 1). Vermiculite and plagioclases follow the same general trend when the rhizosphere is considered as a whole, although the trend is inverted at one site for each mineral (Table 1). Micas show no specific trend. On the other hand, K-feldspars rather tend to be less weathered in the inner rhizosphere.

For all minerals, the site located at 0.5 km from the smelter shows a higher relative mineral content in the bulk soil component (Table 1). The inner and outer

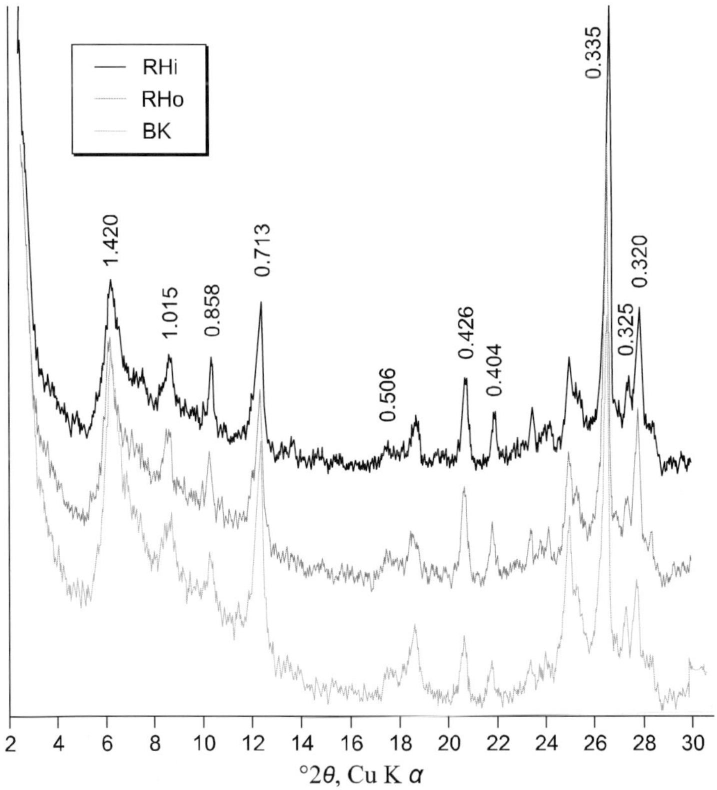

Fig. 1. X-ray diffraction patterns of the < 2 μm fraction of the inner rhizosphere (RHi), the outer rhizosphere (RHo) and the bulk soil (BK) of the site located at 8 km from the copper smelter (field triplicate B). The soil materials are saturated with K at 25°C. Spacing in nm.

rhizospheres have lower relative mineral amounts, with the outer rhizosphere being the most depleted soil component. The site at 2 km of the smelter also has a depleted zone around roots, with the exception of vermiculite. At this site, the inner rhizosphere is usually the most depleted component while the outer rhizosphere occupies an intermediate position. The last site at 8 km shows the same tendency of a relative impoverishment of the rhizosphere. The outer rhizosphere is on average the most depleted soil component. All sites present comparable results, and the rhizosphere as a whole is depleted in minerals compared with the associated bulk soil.

3.2. Barium chloride extraction

The K concentrations in the barium chloride extracts are presented in Fig. 2. The results show a highly significant ($p < 0.01$) enrichment in K following the order inner rhizosphere > outer rhizosphere > bulk soil. The K concentrations in

Table 1
Intensity ratio (I/I_q) for minerals in the inner rhizosphere (RHi), the outer rhizosphere (RHo) and the bulk soil (BK) of the three study sites

Mineral	Peak (nm)	Saturation	Site 0.5 km			Site 2 km			Site 8 km		
			RHi	RHo	BK	RHi	RHo	BK	RHi	RHo	BK
Quartz	0.426	All	1.00	1.00	1.00	1.00	1.00	1.00	1.00	1.00	1.00
Chlorite	1.42	K-25°C	1.70	1.66	2.22	0.35	1.53	1.62	1.94	1.73	3.40
Vermiculite	1.20	K-25°C	0.28	0.24	0.39	0.57	0.47	0.54	0.77	0.68	1.00
Amphiboles	0.853	Mg-25°C	0.44	0.53	0.59	0.46	0.46	0.51	0.62	0.67	0.72
Plagioclase	0.404	Mg-25°C	0.54	0.55	0.68	0.35	0.46	0.51	0.69	0.59	0.69
K-feldspars	0.325	Mg-25°C	0.54	0.39	0.39	0.54	0.46	0.54	0.64	0.48	0.62

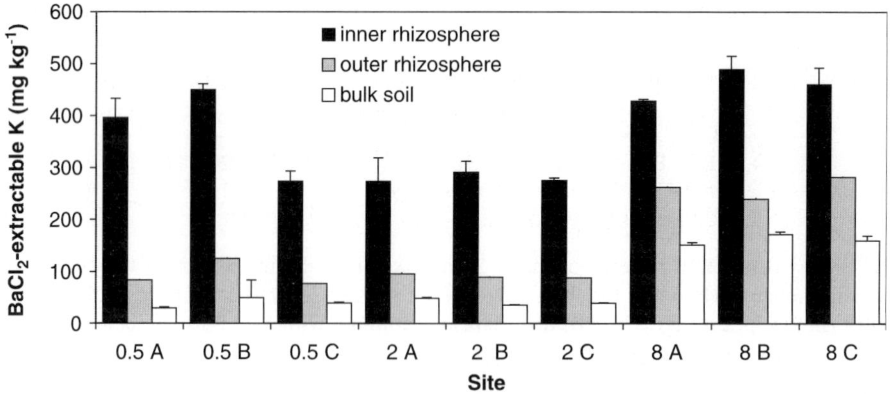

Fig. 2. Mean K concentration in the barium chloride extractions in the inner rhizosphere (RHi), the outer rhizosphere (RHo) and the bulk soil (BK) at all sites (0.5, 2 and 8 km from a smelter) and for the three field replicates (A, B and C). Values are means of two or three replicates depending on the soil mass available. Error bars represent the standard deviation when analyses were performed in triplicates.

the inner rhizosphere are clearly greater, while the outer rhizosphere and the bulk soil present smaller quantitative differences. The content in barium chloride-extractable K can be up to 13 times higher in the inner rhizosphere than in the bulk soil component, with an average of seven times the concentration. This relationship is constant throughout the three sites and for all the trees sampled. Similar trends in elemental concentrations in barium chloride extracts among soil components were observed for ten other metals (Al, Ca, Cd, Co, Cu, Fe, Mg, Mn, Ni and Zn) at

Rouyn-Noranda and for nine other tree individuals from three different species (*Abies balsamea* (L.) Mill, *Acer saccharum* Marsh and *Betula papyrifera* Marsh) at Saint-Hippolyte (about 60 km north of Montréal, Canada) (unpublished data).

3.3. AAO extraction

Of the 14 metals of interest in the acid AAO extraction, nine were detectable, and one was partly detectable. The metal concentrations for these ten metals are reported in Tables 2 and 3. When all the trees are taken into account ($n = 9$), significant trends emerge for six out of ten metals. Highly significant ($p \leq 0.01$) trends exist for Cu and Zn, while Fe, Mg, Mn and Ni differ at the $p \leq 0.05$ level. Only Al, Co, Cr and Si ($p = 0.121$ for Si) concentrations show no significant differences between the three soil components.

Although many significant relationships exist between the soil components and the metal concentrations, their direction is not always the same. Mainly, two trends can be observed. First, for the vast majority of metals (Al, Co, Cu, Fe, Mg, Mn, Ni and Zn), the tendency is towards an enrichment of the rhizosphere (inner and outer taken together) compared to the bulk soil. However, within this group of metals, the behaviour of the outer rhizosphere is not constant. For instance, in the case of Cu and Zn, the inner rhizosphere is the most enriched component. On the other hand, for Mn and Co, the outer rhizosphere shows the highest concentration of AAO-extractable metal. In all cases, the rhizosphere as a whole presents a distinctive behaviour compared to the bulk soil. The second trend is applicable only to Si and Cr, two elements that are somewhat depleted in the rhizosphere, although the trends are not significant (with the exception of the site at 0.5 km for Si concentrations).

On the other hand, because of the small n value ($n = 3$), significant relationships between soil components and metal concentrations are not abundant on a specific site basis as can be seen from Tables 2 and 3. Zn is the only metal presenting a very clear relationship between the soil components and the metal concentration with the AAO extracts increasing towards the root surface at all sites. Copper and Mg have significant tendencies for two out of three sites, while Cr presents no significant trend.

Results for Ca and Pb could not be obtained as interferences induced by the AAO extractant were noted on the ICP-AES. Concentrations for Cd and Li are not presented because they are below the detection limit in all cases. Results for Ni only partly exceed the detection limit. For all the elements measured in the AAO extraction, similar analytical results were obtained for another site located in Saint-Hippolyte (unpublished data).

3.4. TOF-SIMS and X-ray fluorescence

The maps obtained by TOF-SIMS for selected metals are presented in Fig. 3. Lighter colours depict higher relative metal concentrations. These maps show a net accumulation of metal at the soil–root interface.

Table 2
Major element concentrations in the acid ammonium oxalate extraction

Site (km)	Al ($g\ kg^{-1}$)					Fe ($g\ kg^{-1}$)					Mg ($mg\ kg^{-1}$)				
	RHi[b]	RHo[b]	BK[b]	Stats.[c]		RHi[b]	RHo[b]	BK[b]	Stats.		RHi[b]	RHo[b]	BK[b]	Stats.	
0.5 A[a]	6.87	7.52	9.01			10.4	10.6	7.54			46.7	27.0	15.1		
	(0.13)	(0.35)	(0.34)			(0.11)	(0.48)	(0.18)			(1.43)	(1.67)	(0.77)		
0.5 B	5.30	6.00	9.44	NS		11.5	13.1	9.31	**		62.8	41.9	12.0	**	
	(0.19)	(0.08)	(0.19)			(0.36)	(0.08)	(0.12)			(2.78)	(1.07)	(0.55)		
0.5 C	3.63	3.91	3.37			8.12	9.05	7.54			45.6	29.1	16.6		
	(0.17)	(0.36)	(0.07)			(0.34)	(0.78)	(0.25)			(3.87)	(3.57)	(0.63)		
2 A	3.90	3.40	3.33			6.32	7.85	6.55			153	98.0	101		
		(0.06)	(0.13)				(0.30)	(0.50)				(3.74)	(3.42)		
2 B	3.53	2.97	2.62	*	NS	6.64	6.46	5.32	NS	**	99.6	58.2	54.4	*	**
	(0.27)	(0.15)	(0.19)			(0.50)	(0.45)	(0.31)			(10.9)	(1.52)	(4.58)		
2 C	3.61	3.25	3.38			7.44	7.17	7.03			120	83.3	62.0		
		(0.08)	(0.23)				(0.17)	(0.46)				(3.66)	(2.90)		
8 A	2.87	2.67	2.24			4.97	4.76	4.08			277	247	292		
	(0.35)	(0.16)	(0.15)			(0.65)	(0.31)	(0.32)			(36.7)	(11.5)	(20.0)		
8 B	3.14	3.18	3.35	NS		5.73	6.18	7.01	NS		215	172	109	NS	
	(0.07)	(0.20)	(0.16)			(0.16)	(0.33)	(0.47)			(8.45)	(11.2)	(5.62)		
8 C	2.92	2.61	2.07			4.79	4.50	3.55			257	194	387		
		(0.01)	(0.09)				(0.04)	(0.14)				(5.96)	(15.2)		

Site (km)	Mn (mg kg^{-1})					Si (g kg^{-1})				
	RHi	RHo	BK	Stats.		RHi	RHo	BK	Stats.	
0.5 A	107 (3.72)	113 (6.34)	41.2 (2.79)			780 (6.58)	906 (57.3)	1492 (51.2)		
0.5 B	101 (4.32)	144 (7.85)	40.6 (2.21)	**		446 (25.2)	575 (17.4)	1536 (31.9)	*	
0.5 C	144 (8.22)	175 (20.8)	125 (7.62)			259 (19.5)	329 (49.4)	285 (4.88)		
2 A	283	529 (17.6)	333 (34.7)			246	292 (10.8)	296 (6.17)		
2 B	153 (16.8)	141 (15.4)	68.8 (20.0)	NS	**	240 (20.4)	237 (30.9)	229 (29.0)	NS	NS
2 C	411	415 (10.0)	233 (2.89)			247	236 (5.14)	265 (24.5)		
8 A	355 (39.6)	432 (20.3)	314 (32.0)			321 (53.3)	338 (23.1)	319 (22.5)		
8 B	330 (35.6)	450 (33.8)	260 (24.1)	NS		293 (16.8)	321 (21.5)	299 (26.4)	NS	
8 C	298	292 (10.6)	338 (38.6)			287 (7.44)	297 (7.93)	342 (23.5)		

Note: Values in parentheses represent standard deviation of laboratory triplicates. Standard deviations are not given if soil mass was insufficient to replicate analyses.
[a] The sites are coded as follows: the first part of the code refers to the distance from the copper smelter in Rouyn-Noranda (0.5 = 0.5 km; 2 = 2 km; 8 = 8 km); the letter that follows refers to the field replicates.
[b] Soil component: RHi = inner rhizosphere; RHo = outer rhizosphere; BK = bulk soil.
[c] Stats. = for each metal, the results of the non-parametric Freidman test are presented per site (first column) and for all sites (second column), NS = not significant; $^{*}p \leq 0.10$; $^{**}p \leq 0.05$; $^{***}p \leq 0.01$.

Table 3
Trace element concentrations in the acid ammonium oxalate extraction

Site (km)	Co (mg kg^{-1})					Cr (mg kg^{-1})					Cu (mg kg^{-1})				
	RHi	RHo	BK	Stats.[a]		RHi	RHo	BK	Stats.		RHi	RHo	BK	Stats.	
0.5 A	< 2.00	2.43 (0.13)	< 2.00			7.84 (0.20)	7.54 (0.59)	8.05 (0.26)			57.3 (1.12)	47.8 (3.68)	14.5 (1.29)		
0.5 B	< 2.00	3.07 (0.18)	2.42 (0.48)	**		7.97 (0.39)	8.28 (0.31)	9.07 (0.06)	NS		91.1 (3.18)	77.9 (2.03)	8.12 (0.42)	**	
0.5 C	< 2.00	2.33 (0.41)	< 2.00			5.44 (0.15)	5.64 (0.54)	5.30 (0.33)			71.5 (5.61)	49.7 (6.15)	27.3 (0.29)		
2 A	6.06	11.39 (0.46)	7.43 (0.54)			3.61	3.20 (0.32)	3.90 (0.40)			218	132 (3.92)	44.4 (2.33)		
2 B	5.04 (0.80)	4.48 (0.84)	2.99 (0.35)	NS	NS	4.68 (0.53)	4.40 (0.37)	4.34 (0.41)	NS	NS	95.2 (8.48)	71.9 (3.21)	49.9 (2.19)	**	***
2 C	6.33	6.64 (0.27)	4.97 (0.39)			4.55	4.12 (0.11)	5.17 (0.18)			163	123 (2.01)	72.8 (4.85)		
8 A	9.36 (1.38)	13.0 (0.33)	9.08 (1.60)			3.93 (0.84)	3.23 (0.41)	3.64 (0.24)			22.36 (2.80)	13.4 (0.88)	2.47 (0.29)		
8 B	8.48 (1.36)	11.2 (1.73)	6.39 (0.76)	NS		4.47 (0.19)	3.90 (0.28)	5.35 (0.58)	NS		24.0 (0.76)	17.6 (2.05)	25.6 (1.13)	NS	
8 C	7.73	7.95 (0.28)	9.05 (2.02)			3.31	3.39 (0.35)	2.85 (0.33)			27.8	13.2 (1.08)	< 2.00		

Site (km)	Ni (mg kg^{-1})				Zn (mg kg^{-1})				
	RHi	RHo	Bk	Stats.	RHi	RHo	BK	Stats.	
0.5 A	2.34 (0.27)	< 2.00	2.44 (1.18)		55.9 (2.35)	21.5 (1.35)	14.4 (0.86)		
0.5 B	2.40 (0.11)	< 2.00	< 2.00		87.5 (3.20)	38.7 (0.88)	15.8 (0.60)	**	
0.5 C	< 2.00	< 2.00	< 2.00		46.1 (2.18)	21.8 (3.60)	12.6 (0.94)		
2 A	2.96	< 2.00	2.26 (0.16)		41.7	23.5 (0.10)	20.2 (0.95)		
2 B	3.64 (0.40)	2.32 (0.12)	< 2.00	*	48.3 (7.15)	18.6 (0.62)	14.8 (0.10)	**	***
2 C	2.67	2.12 (0.05)	< 2.00		39.9	21.7	15.8 (0.91)		
8 A	< 2.00	< 2.00	< 2.00		12.3 (2.07)	8.64 (0.49)	4.54 (0.30)		
8 B	< 2.00	< 2.00	< 2.00		13.9 (0.73)	10.1 (0.55)	7.23 (0.57)	**	
8 C	2.13	< 2.00	< 2.00		12.9	7.59 (0.18)	3.34 (0.26)		

Note: Values in parenthesis represent standard deviation of laboratory triplicates. Standard deviations are not given if soil mass was insufficient to replicate analyses. The codes for sites and soil component are as in Table 2.

[a] Stats. = for each metal, the results of the non parametric Freidman test are presented per site (first column) and for all sites (second column): NS = not significant; * $p \leq 0.10$; ** $p \leq 0.05$; *** $p \leq 0.01$.

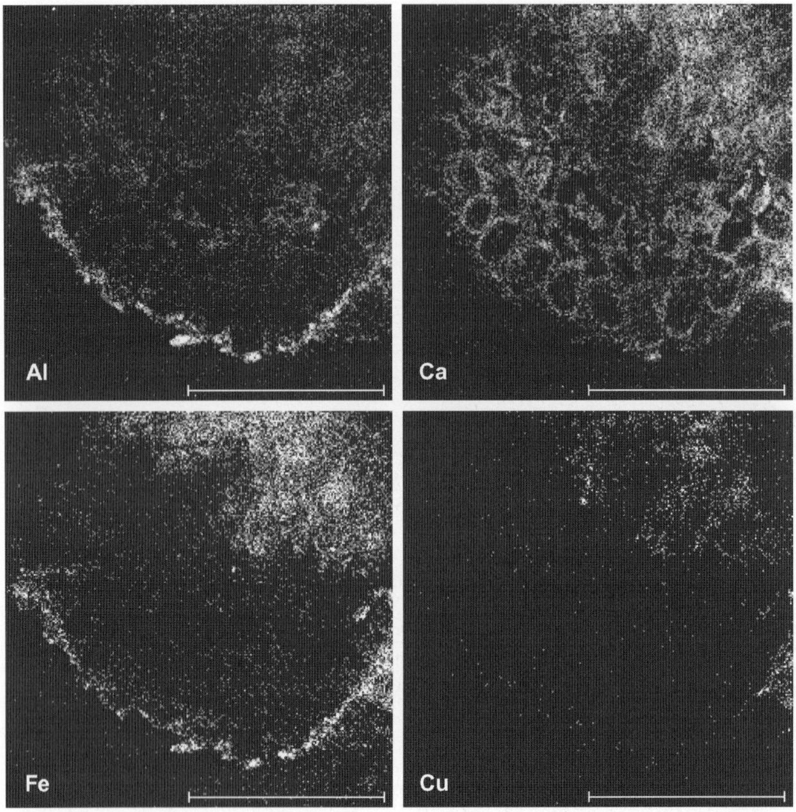

Fig. 3. Maps of the relative concentration of (a) Al, (b) Ca, (c) Fe and (d) Cu determined by tof-SIMS of the perpendicular cut of a root and its surrounding rhizosphere soil. The lighter colour represent the higher relative concentrations and are located at the soil–root interface. The bar is 100 μm and each pixel is 1.5 μm^2. Reprinted from Martin *et al.*, 2004, with permission from Elsevier.

The map of total Mn concentrations obtained by Synchrotron XRF is presented in Fig. 4(a). Relative concentrations are presented with lighter colours indicating higher metal concentrations. As for TOF-SIMS, there is clearly a higher relative content of Mn at the soil–root interface. Similar results were also obtained for other metals (e.g. Cu and Fe) (Naftel *et al.*, 2002).

Moreover, the maps obtained by both methods show that some metals display distinct accumulation patterns. For instance, Al tends to concentrate mainly at the periphery of the root. Calcium can be present in association with Al, but it is also present in tissues delimitating structures inside the root, a pattern very similar to that of Mn (Naftel *et al.*, 2002). Copper and Fe also present similar sites of high concentrations inside the root.

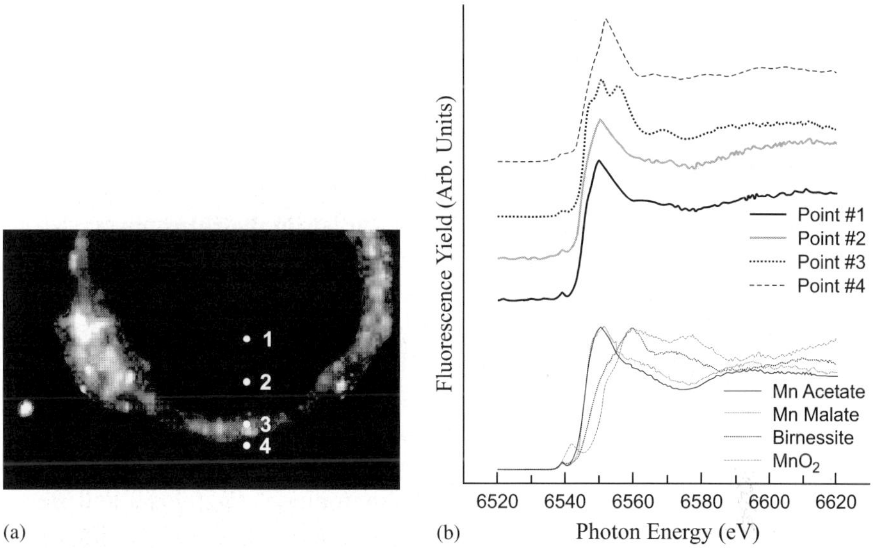

(a) (b)

Fig. 4. (a) Relative Mn concentration determined by Synchrotron XRF of the perpendicular cut of a root and its surrounding rhizosphere soil (see text for more details) and (b) XANES spectra (arbitrary units) for points 1 to 4 on Fig. 4a) as well as XANES of pure Mn-acetate, Mn-malate, birnessite ($[Na_{0.7}\ Ca_{0.3}]Mn_7O_{14}{\cdot}2{\cdot}8H_2O$) and MnO_2.

3.5. X-ray absorption near-edge structure

Eight XANES spectra for Mn are illustrated in Fig. 4(b). The interest for Mn lies on the possibility that this metal could be present in either reduced or oxidized forms in soils as well as being complexed with organic matter. These forms can be differentiated by XANES as they generate different spectra. The uppermost four spectra were obtained following the analysis of rhizosphere materials at the sampling points shown in Fig. 4(a). The other four spectra show the measurement performed on pure substances of Mn-acetate, Mn-malate, birnessite and MnO_2. These four substances serve as standards for comparison purposes with the XANES spectra obtained from the samples. The similar peaks for Mn-acetate and Mn-malate indicate that pure organic forms of Mn should peak at 6,550 eV. Pure oxidized forms of Mn peak at a value of 6,560 eV as revealed by birnessite and MnO_2.

Results obtained with the rhizosphere materials present a shift in the peak position from sampling points 1 to 4. Compared to the XANES spectra of pure substances, it appears that sampling points 1 and 2 contained some organic forms of Mn. These two points are located inside the root. On the XANES spectra of sampling point 4, the peak has moved to the right and is comparable to that of the oxidized form of Mn. Point 4 is in the rhizosphere soil. In between, the XANES spectra of sampling point 3 shows a double peak corresponding to both organic and oxidized forms of Mn. Point 3 is located at the soil–root interface where the frontier between phases is unclear and where metals tend to accumulate the most.

4. DISCUSSION

4.1. Types and relative abundance of silicate minerals in the rhizosphere

The XRD results indicate that the relative abundance of the minerals is different in the rhizosphere as compared to the bulk soil although the mineral assemblage of the rhizosphere and the bulk soil are comparable. Similar results were reported by April and Keller (1990) and Kodama *et al.* (1994). In the current study, the clay fraction of the rhizosphere is depleted in several minerals, particularly in easily weathered minerals such as chlorite and amphiboles. Amphiboles are known to weather more easily in soil environments (Huang, 1989). Chlorite is also considered to be less stable in acidic environment (Barnhisel and Bertsch, 1989); all our samples had pH values close to 5 (Séguin *et al.*, 2004). Plagioclases and vermiculite also presented a tendency to be depleted in the rhizosphere, although the trend was not as systematic as for chlorite and amphiboles. Results for phyllosilicates pertaining to the impact of roots are rather consistent with those of Sarkar *et al.* (1979) and Hinsinger *et al.* (1993). The K-feldspars do not present a clear trend, and the inner rhizosphere even seems to be less affected by weathering for this mineral.

Comparable results for plagioclases and K-feldspars were obtained by Courchesne and Gobran (1997) and April and Keller (1990). Both studies have been conducted in the field. Their results indicate that easily weathered minerals such as amphiboles and expandable phyllosilicates were depleted in the rhizosphere as compared to the bulk soil (Courchesne and Gobran, 1997). April and Keller (1990) also showed that the rhizoplan and the rhizosphere were depleted in biotite, a K-bearing mineral. Similarly, Kodama *et al.* (1994) and Hinsinger *et al.* (1993) conducted pot experiments that indicated a relative enrichment in vermiculite in the rhizosphere.

An explanation submitted by Hinsinger *et al.* (1993) to interpret the increased weathering of easily weathered minerals in the rhizosphere is the depletion of exchangeable cations in the soil under root influence. For plants, K (and also Mg) is a macro nutrient, but its abundance in soil is often low compared to plant needs. As the soil solution is progressively depleted in K by uptake, the interaction between the liquid and the solid phases requires a transfer of K from the soil minerals. A source of relatively available K is the K cations present in the interlayer space of phyllosilicates. When weathering proceeds, exchange reactions with other cations from the solution can take place and the interlayer K is removed from the mineral structure to replenish the soil solution. By doing so, the phlogopite used by Hinsinger *et al.* (1993) in the laboratory was transformed into vermiculite or hydroxy-interlayer vermiculite (HIV) in a matter of days in the rhizosphere of *Brassica napus*. The formation of HIV indicates that the partial dissolution of the micas was accompanied by the release of structural (octahedral) Al under acid conditions. The same observation was made by these authors for interlayer Mg.

Our study with acidic soils indicates the presence of HIV in the rhizosphere as evidenced by a broad peak at about 1.2 nm under K saturation. Yet, our results for barium chloride-extractable K show a clear enrichment of K in the rhizosphere (Fig. 2) rather than the depletion expected based on the Hinsinger *et al.* (1993) hypothesis. The cationic exchange capacity is also higher in the zone under root influence mostly as a consequence of organic matter accumulation. The enrichment of the rhizosphere in exchangeable K and its higher cationic exchange capacity were also reported by Chung and Zasoski (1994). April and Keller (1990) also hypothesized that a K enrichment of the rhizosphere solution could explain their results. Yet, solid–liquid phase equilibrium cannot constitute the basis for an explanation of the increased weathering of minerals in the rhizosphere of our soils.

Mechanisms other than those proposed by Hinsinger *et al.* (1993) must thus take place to explain both the weathering of minerals and the enrichment in barium chloride-extractable K at the soil–root interface. Ochs (1996) indicates that solution acidity and chelating agents are important factors controlling mineral weathering. These properties create an aggressive environment for solids where mineral weathering could be accelerated. Numerous organic components are known to increase the weathering rate because of their ability to complex metals (Banfield *et al.*, 1999). For instance, Wang *et al.* (2000) show that a greater abundance of organic acids produced higher K concentration close to roots. In the current study, all soil samples studied are acidic and the rhizosphere is generally more acid than the bulk soil, with the exception of the first field replicate at site 0.5 km (Séguin *et al.*, 2004). Despite a rhizosphere pH slightly higher than that of the bulk soil, the mineralogical data of the 0.5 km sample nonetheless follow the same trend as the other sites presented (Table 1). As such, pH cannot be regarded as the sole variable explaining the difference in mineralogy between the three soil components. In particular, the dissolved and solid-phase organic matter contents were systematically higher in the rhizosphere of all samples (Séguin *et al.*, 2004).

Some authors question the impact of pH and organic matter on mineral-weathering reactions and indicate that organic substances may have distinct effects on weathering rates. Robert and Berthelin (1986) showed that complexing organic acids are only able to dissolve micas, but that acid compounds are able to induce the formation of vermiculite. Ochs (1996) demonstrated that humic substances were unable to weather significantly aluminium oxides while low molecular mass organic acids could. Moreover, *Lolium multiflorum* Lam. has a better capacity to free interfoliar K^+ than clover (Tributh *et al.*, 1987), even if the latter lowers the pH of the rhizosphere at a greater extent (Mengel *et al.*, 1990).

Microorganisms also affect weathering. For example, mycorrhizae increased the weathering of chlorite, muscovite, montmorillonite and kaolinite in luvisolic soils of British Columbia (Canada) (Arocena and Glowa, 2000). Greater vermiculization rates of biotite and phlogopite were obtained by Mojallali and Weed (1978) when mycorrhizae were present. Other microorganisms can impact on mineral weathering. Leyval and Berthelin (1991) showed that higher alteration rates of phlogopite were attained in the presence of bacteria (*Agrobacterium* sp.) compared to mycorrhizae. Microorganisms can also affect mineral weathering mechanically by micro-dividing mineral grains after digging small pores (Robert and Berthelin, 1986).

Reversing the explanation of Hinsinger *et al.* (1993) could generate an interesting hypothesis. Indeed, instead of using K depletion in the rhizosphere as the motor for an accelerated mineral weathering in the zone under root influence, increased weathering in the rhizosphere could be viewed as the main source of the enrichment in barium chloride extractable elements. This could explain the greater content of barium chloride and water-extractable metals observed in the rhizosphere, and relates well to the presence of lower pH and higher organic matter content (Séguin *et al.*, 2004).

The current results agree with an increase in weathering rates in the rhizosphere. Nevertheless, it must be taken into account that the XRD results were normalized with respect to the quartz peak at 0.426 nm. Yet, Kodama *et al.* (1994) showed that quartz can be weathered in the rhizosphere. However, in the absence of indications of quartz enrichment in the rhizosphere and because quartz is resistant to weathering, we use quartz as the reference mineral to normalize the peaks of other minerals from XRD patterns. Differences in particle size may also exist between components. However, the soils of Rouyn are fine-textured and, as such, textural differences between the bulk soil and the rhizosphere are expected to be negligible and very difficult to measure. Also, the methodology used to acquire the inner rhizosphere materials excluded some of the finest particles that were intimately bound to the roots and that could not be detached without root fragments entering the sample. Consequently, it might be possible that the XRD results presented in the current study underestimate the actual effect of the rhizospheric environment on weathering.

4.2. Oxides and aluminosilicates

Considering the more intense alteration observed in the rhizosphere, it can be expected that more secondary minerals might form as weathering products in this soil component. Indeed, one of the common weathering products of clay minerals such as chlorite is metallic oxides. Moreover, the alteration of amphiboles can lead to the formation of Fe oxides (Allen and Hajek, 1989). In short, if weathering is increased in the rhizosphere, its impact on the difference in metal concentrations between the inner rhizosphere, outer rhizosphere and bulk soil should be detectable using the AAO extractions.

4.2.1. Iron and manganese

The results for the AAO extraction presented in Table 2 show a similar behaviour for Fe and Mn. These two metals accumulate preferentially in the rhizosphere (inner or outer). In the case of Mn, the AAO-extractable concentration in the zone under root influence is nearly twice that obtained from the bulk soil of most samples. For Fe, the quantitative difference between soil components is not as pronounced.

AAO can extract the inorganic amorphous solid phases, including sesquioxides (Ross and Wang, 1993). The higher concentrations of Fe and Mn extracted by AAO from the rhizosphere could thus be interpreted as an accumulation of metallic oxides representing secondary products of the mineral weathering taking place in soils. This means that the increased weathering suggested by the XRD results is supported by the AAO data for Fe and Mn.

The XANES results further support the presence of oxidized metal forms in the rhizosphere. Indeed, the Mn XANES spectra for sampling point 4 (Fig. 4) located in the rhizosphere compares well with the birnessite and MnO_2 spectra. The XANES spectra clearly show a shift in Mn forms along a gradient that spans from the rhizosphere to the inside of the root. While Mn is oxidized in the soil surrounding the root, the organic forms dominate close to the root centre.

4.2.2. Silicon

The results obtained for Si are collected in Table 2. The Si concentrations in the AAO extraction are higher in the bulk soil than in the rhizosphere. This result is contrary to the trend observed for Fe and Mn.

Two hypotheses can thus be drawn from these results. First, there are more non-crystalline Si-bearing minerals in the bulk soil than in the rhizosphere. Indeed, the presence of Fe and Mn oxides as well as the higher organic matter content in the rhizosphere can have a diluting effect on the Si concentrations measured in AAO extracts. Also, the main forms of Si-bearing inorganic amorphous minerals are imogolite and allophane. These two mineral groups do not form when organic matter content and acidity are high (Courchesne *et al.*, 1998) like it is the case in the rhizosphere. As such, the rhizosphere could potentially inhibit the formation of amorphous Si minerals.

The second hypothesis is that the weathering of finely divided amorphous minerals is more intense in the zone under root influence than in the bulk soil. These minerals would be subjected to an aggressive environment characterized by a greater organic acids content and lower pH values (Séguin *et al.*, 2004). The XRD results are in phase with this hypothesis as they indicate a depletion of most minerals in the rhizosphere. The XRD analyses were performed after the destruction of organic matter and oxides, thus accounting for the dissolution effect.

Interestingly, the Si concentrations at 0.5 km from the copper smelter are significantly higher than at the two other sites. This could be explained by the dust emissions produced by the smelter. These dust particles are rich in Si and are deposited near the smelter (Knight and Henderson, in press). Some of this dust has apparently migrated and percolated to the B horizon.

4.2.3. Aluminum

Aluminum is one of the three elements not presenting a significant trend as shown in Table 2. However, six cases out of the nine show an increase in AAO-extractable Al in the rhizosphere when compared to the bulk soil. In this sense, the explanation put forward for Fe and Mn could also apply to Al, which could form secondary oxides in the rhizosphere in response to a higher weathering rate in the zone under root influence. For example, April and Keller (1990) found indication of a precipitation of Al at the root surface in a non-crystallized form likely to be oxides.

On the other hand, Al is a major component of allophane, imogolite and alumino-silicate minerals. Some of these minerals are easily weathered in the rhizosphere. As such, it could be anticipated that Al concentrations extracted with AAO should be lower in the zone under root influence. Thus, the explanation used for Si could also be considered for Al.

In sum, Al apparently presents a mixed behaviour between Fe/Mn and Si. The lack of a strong trend for Al might be the consequence of antagonistic processes resulting from the intense weathering occurring in the rhizosphere.

4.2.4. Trace elements

Of all elements, zinc and copper emerge as those presenting the most significant relationships between metal concentrations and soil components. The inner rhizosphere is clearly the zone of accumulation of Zn and Cu associated with inorganic amorphous solid phase (Cu presents one case where the inner rhizosphere is not the most enriched component). The strength of this trend is highlighted by the average Zn concentration in the rhizosphere being three times higher than that of the bulk soil, even reaching a five-fold difference at 0.5 km. In the case of Cu, the mean difference between the zone under root influence and the bulk soil reaches a factor of six with the largest difference reaching a factor of 14 at the 8 km site.

The amphoteric behaviour of Fe and Mn oxides could help explain the increased concentration of Cu and Zn in the zone under root influence. Indeed, the surface of these Fe and Mn oxides bears variable charges. At the acidic pH measured in these soils (Séguin *et al.*, 2004), surface sites with negative charges are likely to abound. Hence, trace elements like Cu and Zn could be adsorbed to these surfaces. Both trace elements present a high affinity for surface adsorption on Fe and Mn oxides or can even be incorporated in their structure through co-precipitation (McKenzie, 1989; Schwertmann and Taylor, 1989). Similar exchange sites are also present on the surface of clay minerals and organic matter, which were previously shown to be present in greater abundance in the rhizosphere.

Thus, the increased weathering rate, and the associated formation of Fe and Mn oxides as weathering products, apparently stimulates the adsorption of Cu and Zn onto oxides surfaces in the rhizosphere.

4.3. Total concentration distribution of metals around roots

Metals preferentially accumulate at the soil–root interface (Fig. 3). Identifying the precise location of the interface between the root and the soil is, however, not easy, and the transition from the soil to the root is gradual (Naftel *et al.*, 2002). Moreover, mineral particles can be incorporated in the external tissues of roots (Adamo *et al.*, 1998).

The metals studied accumulate in root cells and form different patterns. For example, Al tends to build up mostly on the exterior part of the root whereas Si is found further inward. Ca is mainly located inside the root underlining many internal structures. Similar results were obtained by April and Keller, (1990) who observed a gradient in elements from the exterior to the interior of the root following the sequence: Al (precipitate, oxidized) to Si (precipitate) to Ca (oxalate). Adamo *et al.* (1998) also presented similar results on the Al-Si-Ca gradient, while Kodama *et al.* (1994) indicated that Si tends to accumulate in roots. The accumulation of Al at the root surface is generally greater when soils are acidic (Adamo *et al.*, 1998) as it is the case in this study (Séguin *et al.*, 2004).

Similar to Al, Fe also accumulates at the soil–root interface, probably in the form of oxides. Inside the root, there is a close association between the Cu and the Fe distributions. These metals could be associated with organic matter to form complexes. It is known that Cu and Fe are relatively immobile elements in plants and that they are not readily transferred between tissues during growth. (Alloway, 1990).

5. CONCLUSION

The results taken as a whole converge and indicate a more intense mineral weathering in the rhizosphere compared to the bulk soil. The XRD results show a change in mineral assemblage that reflects increased mineral weathering in the rhizosphere, a reaction that mostly affects easily weathered minerals such as chlorite and amphiboles. A hypothesis submitted in the literature to explain the greater alteration in the rhizosphere links nutrient uptake by plants to a disequilibrium between the liquid and the solid phases that induces the release of interlayer elements such as K and Mg. However, the current study showed that K and Mg concentrations extracted with barium chloride (associated to exchangeable form) are higher in the rhizosphere. Other mechanisms, such as changes in pH or organic matter content, must then take place in addition to the disequilibrium between the liquid and the solid phase to explain the higher weathering in the rhizosphere.

Moreover, the AAO extractions provide additional support to the XRD data. The Fe and Mn concentrations extractable with AAO are higher in the zone under root influence and could thus indicate the neoformation of secondary oxides resulting from mineral weathering. The tendency, although not significant, to have lower concentrations of AAO-extractable Si in the rhizosphere could be attributed to the increased weathering of amorphous Si-bearing minerals or to the existence of chemical conditions that inhibit the formation of amorphous Si solid phases. The neoformation of secondary oxides and the weathering of aluminosilicate have opposite effects on the accumulation of some metals, an effect that could explain the weak trend for AAO extractable Al. Consistent with the results from AAO extraction, XANES spectra show the presence of oxidized metal form in the rhizosphere, and organic form in the root, while a mixture of both forms is present in between. The Synchrotron XRF mapping further confirms the accumulation of metals at the soil–root interface. The convergence of results from different techniques supports the existence of more intense mineral weathering reactions in the rhizosphere compared to the bulk soil, particularly of easily weathered minerals, and the associated neoformation of metallic oxides.

ACKNOWLEDGMENTS

We thank Patrice Turcotte, Julie Turgeon and Marie–Claude Turmel for technical assistance as well as Marc Girard for help with the figures. Financial support for this research was provided by the Natural Science and Engineering Research Council of Canada (NSERC), by the Fonds Québécois de la Recherche sur la Nature et les Technologies (FQRNT) and by Metals in the Environment – Research Network (MITE-RN).

ABBREVIATIONS

AAO	acid ammonium oxalate
CEC	cation exchange capacity
HIV	hydroxy-interlayered vermiculite
ICP-AES	inductively coupled plasma – atomic emission spectroscopy
I/Iq	peak intensity ratio of mineral intensity (I) over quartz peak intensity (I_q)
TOF-SIMS	time-of-flight secondary-ion mass spectroscopy
XANES	X-ray absorption near-edge structure
XRD	X-ray diffraction
XRF	X-ray fluorescence

REFERENCES

Adamo, P., Mchardy, W.J., Edwards, A.C., 1998. SEM observations in the back-scattered mode of the soil-root zone of *Brassica Napus* (Cv. Rafal) plants grown at a range of soil pH values. Geoderma 85, 357–370.

Allen, B.L., Hajek, B.F., 1989. Mineral occurence in soil environment. In: Dixon, J.B. Weed, S.B. (Eds.), Minerals in Soil Environments. Soil Science Society of America, Madison, pp. 199–278.

Alloway, B.J., 1990. Metals in Soils. Blackie, Glasgow.

April, R., Keller, D., 1990. Mineralogy of the rhizosphere in forest soils of the Eastern United-States—Mineralogic studies of the rhizosphere. Biogeochemistry 9, 1–18.

Arocena, J.M., Glowa, K.R., 2000. Mineral weathering in ectomycorrhizosphere of subalpine fir (*Abies Lasiocarpa* (Hook.) Nutt.) as revealed by soil solution composition. Forest Ecol. Manag. 133, 61–70.

Banfield, J.F., Barker, W.W., Welch, S.A., Taunton, A., 1999. Biological impact on mineral dissolution: Application of the lichen model to understanding mineral weathering in the rhizosphere. Proceed. Nat. Acad. Sci. USA, 96, 3404–3411.

Barnhisel, R.I., Bertsch, P.M., 1989. Chlorites and hydroxy-interlayered vermiculite and smectite. In: Dixon, J.B., Weed, S.B. (Eds.), Minerals in Soil Environments. Soil Science Society of America, Madison, pp. 729–788.

Centre Saint-Laurent, 2001. Détermination de la Limite de Détection d'une Méthode d'Analyse. Centre Saint-Laurent, Environnement Canada, Montréal.

Chung, J.B., Zasoski, R.J., 1994. Ammonium-potassium and ammonium-calcium exchange equilibria in bulk and rhizosphere soil. Soil Sci. Soc. Am. J. 58, 1368–1375.

Courchesne, F., Gobran, G.R., 1997. Mineralogical Variations of Bulk and Rhizosphere Soils From a Norway Spruce Stand. Soil Sci. Soc. Am. J. 61, 1245–1249.

Courchesne, F., Laberge, J.F., Dufresne, A., 1998. Influence of soil organic matter on sulfate retention in two podzols in Quebec, Canada, Can. J. Soil Sci. 79, 103–109.

Fergusson, J.E., 1990. The Heavy Metals: Chemistry, Environmental Impacts, and Health Effects. Oxford, University Press, New York.

Grayston, S.J., Vaughan, D., Jones, D., 1996. Rhizosphere carbon flow in trees, in comparison with annual plants: the importance of root exudation and its impact on microbial activity and nutrient availability. Appl. Soil Ecol. 5, 29–56.

Grinsted, M.J., Hedley, M.J., White, R.E., Nye, P.H., 1982. Plant-induced changes in the rhizosphere of rape (Brassica-Napus Var - Emerald) Seedlings .1. pH change and the increase in *p* Concentration in the Soil Solution. New Phyt. 91, 19–29.

Harley, A.D., Gilkes, R.J., 2000. Factors influencing the release of plant nutrient elements from silicate rock powders: a geochemical overview. Nutr. Cycl Agroecosystems 56, 11–36.

Haynes, R.J., 1990. Active ion uptake and maintenance of cation–Anion balance—a critical-examination of their role in regulating rhizosphere pH. Plant Soil 126, 247–264.

Hendershot, W.H., Lalande, H., Duquette, M. 1993. Ion exchange and exchangeable cations. In: Carter, M.R., (Ed.), Soil Sampling and Methods of Analysis. Lewis Publishers, Boca Raton, FL pp. 167–176.

Hinsinger, P., Gilkes, R.J., 1997. Dissolution of phosphate rock in the rhizosphere of five plant species grown in an acid, p-fixing mineral substrate. Geoderma 75, 231–249.

Hinsinger, P., Jaillard, B., 1993. Root-induced release of interlayer potassium and vermiculitization of phlogopite as related to potassium-depletion in the rhizosphere of ryegrass. J. Soil Sci. 44, 525–534.

Hinsinger, P., Jaillard, B., Dufey, J.E., 1992. Rapid weathering of a trioctahedral mica by the roots of ryegrass. Soil Sci. Soc. Am. J. 56, 977–982.

Hinsinger, P., Elsass, F., Jaillard, B., Robert, M., 1993. Root-induced irreversible transformation of a trioctahedral mica in the rhizosphere of rape. J. Soil Sci. 44, 535–545.

Huang, P.M., 1989. Feldspars, olivines, pyroxenes and amphiboles. In: Dixon, J.B., Weed, S.B. (Eds.), Minerals in Soil Environments. Soil Science Society of America, Madison, pp. 975–1050.

Jongmans, A.G., van Breemen, N., Lundstrom, U., van Hees, P.A.W., Finlay, R.D., Srinivasan, M., Unestam, T., Giesler, R., Melkerud, P.-A., Olsson, M., 1997. Rock-eating fungi. Nature 389, 682–683.

Knight, R.D., Henderson, P.J., In press. Characterization of smelter dust from the mineral fraction of humus collected around Rouyn-Noranda, Québec. In: Bonham-Carter, G. (Ed.), Metals in the Environment around Smelters at Rouyn-Noranda, Quebec, and Belledune, New Brunswick, and conclusions of the GSC-MITE Point Source Project. Geological Survey Canada Bulletin, Ottawa (in press).

Kodama, H., Nelson, S., Yang, A.F., Kohyama, N., 1994. Mineralogy of rhizospheric and non-rhizospheric soils in corn fields. Clays Clay Min. 42: 755–763.

Legendre, P., Legendre, L., 1998. Numerical Ecology. Elsevier, Amsterdam.

Leyval, C., Berthelin, J., 1991. Weathering of a mica by roots and rhizospheric microorganisms of pine. Soil Sci. Soc. Am. J. 55, 1009–1016.

Lindsay, W.L., 1979. Chemical Equilibria in Soils. Wiley, New York.

Lombi, E., Wenzel, W.W., Gobran, G.R., Adriano, D.C., 2001. Dependency of phytovailability of metal on indigenous and induced rhizosphere processes: a review. In: Gobran, G.R., Wenzel, W.W., Lombi, E. (Eds.), Trace Elements in the Rhizosphere. CRC Press, Boca Raton, FL, pp. 165–185.

Marschner, H., 1995. Mineral Nutrition of Higher Plants. Academic Press, London.

Martin, R.R., Naftel, S.J., Macfie, S., Skinner, W., Courchesne, F., Seguin, V., 2004. Time of flight secondary ion mass spectrometry studies of the distribution of metals between the soil, rhizosphere and roots of *Populus Tremuloides Minchx* growing in forest soil. Chemosphere 54, 1121–1125.

McBride, M.B., 1994. Environmental Chemistry of Soils. Oxford University Press, New York.

McKenzie, R.M., 1989. Manganese oxides and hydroxides. In: Dixon, J.B., Weed, S.B. (Eds.), Minerals in Soil Environments, Soil Science Society of America, Madison, pp. 439–465.

Mengel, K., Horn, D., Tributh, H., 1990. Availability of interlayer ammonium as related to root vicinity and mineral type. Soil Sci. 149, 131–137.

Mojallali, H., Weed, S.B., 1978. Weathering of micas by mycorrhizal soybean plants. Soil Sci. Soc. Am. J. 42, 367–372.

Mortland, M.M., Lawton, K., Uehara, G., 1956. Alteration of biotite to vermiculite by plant growth. Soil Sci. 82, 477–481.

Naftel, S.J., Martin, R.R., Courchesne, F., Seguin, V., Protz, R., 2002. Studies of the effect of soil biota on metal bioavailability. Can. J. Anal. Sci. Spec. 47, 36–40.

Nye, P.H., 1981. Changes of pH across the rhizosphere induced by roots. Plant Soil 61, 7–26.

Ochs, M., 1996. Influence of humified and non-humified natural organic compounds on mineral dissolution. Chem. Geol. 132, 119–124.

Robert, M. Berthelin, J., 1986. Role of biological and biochemical factors in soil mineral weathering. In: Huang, P.M., Schnitzer, M., (Eds.), Interactions of Soil Minerals with Natural Organics and Microbes. Soil Science Society of America, Madison, pp. 453–495.

Rollwagen, B.A. Zasoski, R.J., 1988. Nitrogen-source effects on rhizosphere pH and nutrient accumulation by pacific northwest Conifers. Plant Soil 105, 79–86.

Ross, G.J., Wang, C., 1993. Extractable Al, Fe, Mn and Si. In: Carter, M.R. (Ed.), Soil Sampling and Methods of Analyses. Lewis Publishers, Boca Raton, FL, pp. 239–246.

Sarkar, A.N., Jenkins, D.A., Wyn Jones, R.G., 1979. Modifications to mechanical and mineral composition of soil within the rhizosphere. In: Harley, J.L., Russel, R.S., (Eds.), The Soil–Root Interface. Academic Press, Oxford, pp. 125–136.

Schwertmann, U. Taylor, R.M., 1989. Iron oxides. In: Dixon, J.B., Weed, S.B. (Eds.), Minerals in Soil Environments. Soil Science Society of America, Madison, pp. 379–438.

Séguin, V., Gagnon, C., Courchesne, F., 2004. Changes in water extractable metals, pH and organic carbon concentrations at the soil-root interface of forested soils. Plant Soil 260, 1–17.

Soil Classification Working Group, 1998. The Canadian System of Soil Classification. Agriculture and Agri-Food Canada, Ottawa.

Spyridakis, D.E., G. Chesters, S.A. Wilde. 1967. Kaolinization of biotite as a result of coniferous and deciduous seedling growth. Soil Sci. Soc. Am. Proc. 31, 203–210.

Tributh, H., Vonboguslawski, E., Vonlieres, A. Steffens, D., Mengel. K., 1987. Effect of potassium removal by crops on transformation of illitic clay-minerals. Soil Sci. 143, 404–409.

Wang, J.G., Zhang, F.S., Zhang, X.L., Cao, Y.P., 2000. Release of potassium from K-bearing minerals: effect of plant roots under *p* deficiency. Nutr. Cycl. Agroecosystems 56, 45–52.

Youssef, R.A. Chino, M., 1987. Studies on the behavior of nutrients in the rhizosphere: 1. establishment of a new rhizobox system to study nutrient status in the rhizosphere. J. Plant Nutr. 10, 1185–1195.

Biogeochemistry of Trace Elements in the Rhizosphere
P.M. Huang and G.R. Gobran (Editors)

Chapter 3

Characteristics of rhizosphere soil from natural and agricultural environments

G. Corti[a], A. Agnelli[a], R. Cuniglio[a], M.F. Sanjurjo[b], and S. Cocco[a]

[a]Dipartimento di Scienze Ambientali e delle Produzioni Vegetali, Università Politecnica delle Marche, Via Brecce Bianche, 60131 Ancona, Italy
E-mail: g.corti@univpm.it

[b]Departamento de Edafología y Química Agricola, Escola Politécnica Superior, Universidade de Santiago de Compostela, 27002 Lugo, Spain

ABSTRACT

In the first part of this chapter, we present an overview of the methodologies adopted to study the rhizosphere soil, focusing on the protocols devised for its separation. For these methodologies and protocols, we also discuss the advantages and disadvantages inherent in their use. The sections of the chapter are dedicated to reports on three case studies, where the bulk and rhizosphere soil of three arboreal species are compared and contrasted. The first case study deals with the strategy used by *Genista aetnensis* Biv. to colonize the inhospitable volcanic soils on the flanks of Mount Etna (Sicily, Italy). In this environment, *Genista* are able to overcome the low availability of nutrients through forcing the roots to excrete oxalic acid, and to preserve P by the hosting of a microbial population in the rhizosphere soil that is responsible for the biological cycling of P. As a by-product of the weathering promoted by the roots, the yellowish-coloured collar around them, which is due to the presence of amorphous Fe-oxyhydroxides, reveals the thickness of the rhizosphere soil. The second case presented deals with the ability of *Erica arborea* L. to colonize a soil derived from alkaline marine deposits in Central Italy. The *Erica* plants, which are established in this environment due to the formation of superficial acid horizons, have been able to modify the upper 60 cm of soil through root excretion of organic acids until the differences between bulk and rhizosphere are removed. The roots of *Erica* are now colonizing the horizon underneath, where the rhizosphere soil is more acidic than the bulk. At deeper levels, carbonates persist and roots of *Erica* are rare. The final case study reports on the chemical fractionation of lanthanides in bulk and rhizosphere soil of adult vines (*Vitis vinifera* L.) from two vineyards, one in Tuscany (Italy) and the other in Galicia (Spain). In these soils, the presence of lanthanides has been ascribed mostly to the long-lasting practices of cultivation and, in particular, to the use of fertilizers and the deep mechanical working of the soil, which have greatly affected the soil characteristics over the centuries. The chemical fractions more involved in the binding of lanthanides have resulted in the organic matter and the Fe-oxyhydroxides. In both soils, root activity since the planting of the vineyard (some decades ago) has been able to modify the chemical fractionation of lanthanides within the horizons, with a small effect on the redistribution throughout the profile.

1. METHODOLOGIES FOR SAMPLING RHIZOSPHERE SOIL

1.1. Introduction

Soil is a natural body constituted by solids (minerals and organic matter), liquid and gases, where the soil-forming forces have acted so as to organize it into horizons and to make it able to support the life of rooted plants (Soil Survey Staff, 1999). The soil-forming forces responsible for soil genesis were defined by Jenny (1941), who reported them in the form of the mathematical equation:

$$S = \int (\text{pm}, \text{cl}, o, r, t, ...)$$

where S is the type of soil or any other soil property, pm the parent material, cl the climate, o the organisms, r the relief, t the time, and … represents other factors, not recognized or generally negligible.

Among these forces, biota may accelerate soil genesis. In natural soils, the trophic chain made up of plants, microfauna and microbes is the major source of protons, which are mainly responsible for mineral weathering. In cultivated soils, human activity has a relevant impact on their evolution.

Relationships among plants, microorganisms and soil mostly occur in the vicinity of the roots, where chemical and biochemical reactions are concentrated. This soil portion is referred to as the *rhizosphere*, which is usually subdivided into three ecological niches (Lynch, 1990):

- *Endorhizosphere*: The thin layer that spans from the root surface to the near-surface cells, which is colonized or potentially colonizable by microbes
- *Rhizoplane*: First introduced by Clark (1949), this denotes the external plant-root surface, namely the two-dimensional interface between root and soil
- *Ectorhizosphere*: The soil layer surrounding roots and affected by the activity of roots themselves, and the microorganisms. The thickness of this soil portion usually ranges from one to a few millimetres. The ectorhizosphere was initially defined by Hiltner (1904), who referred to it simply as the rhizosphere. The same definition and term were reported by Curl and Truelove (1986).

The soil matrix that envelops the rhizosphere and that is less affected by the activity of roots and microorganisms, is termed the *bulk soil*.

The ectorhizosphere is of prime importance for the whole soil–plant–microorganism environment as depending on soil type, plant species and environmental stresses, plants can release from 10% to 40% of their total net C assimilation per year from their roots (Lynch and Whipps, 1990; Van Veen *et al.*, 1991; Uren, 2000; Bertin *et al.*, 2003) through respiration and rhizodeposition; the latter is responsible for generating the "rhizosphere effect" (Lynch and Whipps, 1990).

One of the main problems encountered by soil scientists who have investigated the ectorhizosphere is the absence of a precise limitation between it and the bulk soil. Indeed, the spatial extent of the ectorhizosphere is hard to identify in the field, as each soil property has its own radial gradient (Nye, 1984; Jungk, 1991; Darrah, 1993; Uren, 2000); in addition, morphological features between ectorhizosphere and bulk soil are often similar and, hence, useless for separating the two soil portions. This difficulty in collecting the ectorhizosphere has resulted in the devising of a number of procedures to solve the problems they have encountered. All the procedures are able to concentrate the ectorhizosphere over the bulk soil, but they can also introduce experimental artefacts that may lead to erroneous considerations (Norvel and Cary, 1992; Wenzel *et al.*, 2001). These problems evidently make it difficult to compare the results obtained by different researchers. Nonetheless, the studies of the ectorhizosphere at field, greenhouse and bench levels have greatly increased the knowledge of this particular soil portion, which may show great differences with respect to the bulk soil in terms of chemical, mineralogical, biochemical and biological properties. Even though present in very small amounts in the soil, the ectorhizosphere plays a fundamental functional role in soil–plant relationships, and this allowed Lombi *et al.* (2001) to define it as the soil micro-environment characterized by feedback loops of interactions between root activity, soil properties and the dynamics of the associated microbial community. As with human body, the benefits gained from having eyes or a liver cannot be quantified by the percentage of the entire body that they represent!

We provide here a review of the protocols used to collect ectorhizosphere and three case studies dealing with the differences between bulk soil and ectorhizosphere (hereafter referred to as rhizosphere soil).

1.2. Sampling rhizosphere soil

1.2.1. Background

Non-destructive methods to study the rhizosphere soil have been conducted either on natural or on laboratory-cultivated plants. These methodologies include determinations, by autoradiographic techniques, of elements that have radioisotopes with suitable properties, such as ^{90}Sr (Barber, 1962), ^{35}S (Barber *et al.*, 1963; Wray and Tinker, 1968), ^{45}Ca (Wilkinson *et al.*, 1968), ^{32}P (Lewis and Quirk, 1967; Bhat and Nye, 1973) and ^{86}Rb (Claassen *et al.*, 1981a, b). Such an approach is rather intriguing, but it does not give information on the form in which the elements exist (soluble, exchangeable, in living tissues or microorganisms, etc.). To achieve this goal, it is mandatory to physically collect rhizosphere soil.

Because of the lack of a morphological delimitation between rhizosphere and bulk soil, collection of such samples is unavoidably subjective. Studies on the microbial communities of the rhizosphere by biochemical (biolog, gene sequences, rDNA fingerprints) and traditional (plating) methods have established that plant species, cultivars and genotypes may have distinctive microbial populations (i.e. Mozofar *et al.*, 1992; Lemanceau *et al.*, 1995; Latour *et al.*, 1996; Massol-Deya *et al.*, 1997; Munson *et al.*, 1997; Grayston *et al.*, 1998; Di Giovanni *et al.*, 1999; Elo *et al.*, 2000). However, this knowledge has been gained by growing plants in hydroponic systems or in soil pots, which provide a poor characterization of the soil used without ensuring that a rhizosphere sample has been obtained. Few studies have taken the care to separate the three ecological niches indicated above. For example, Elo *et al.* (2000) studied the bacteria of bulk and rhizosphere soil plating homogenized roots so as to explore the microbial populations living in the endorhizosphere and on the rhizoplane. In practice, the term "rhizosphere soil" is used in a very loose way in these studies, and it is difficult to compare or combine results obtained by the different authors.

Many protocols for collecting rhizosphere soil have been developed, although the lack of a precise delimitation within a continuum means that subdivision has been performed as a function of the research aims. Soil scientists of different disciplines may use different procedures, and the rhizosphere soil so obtained has to be considered as an *operationally defined fraction*. According to Lynch (1990), this may indicate different things to different researchers. The procedure followed to obtain rhizosphere and bulk soils obviously depends on the type of plants investigated and whether they are cultivated (in containers or in the field) or are growing naturally.

1.2.2. Sampling rhizosphere soil from containers

1.2.2.1. Rhizotrons. Many researchers have been aware of tremendous soil variabilities, and have overcome the problem by growing seedlings in pots or cans filled with a homogeneous soil substrate (i.e. Metzger, 1928; Starkey, 1931, 1958; Papavizas and Davey, 1961; Riley and Barber, 1969, 1970; Hoffmann and Barber, 1971; Smiley, 1974; Soon and Miller, 1977). The same approach has been followed over the last two decades (i.e. Grinsted *et al.*, 1982; Sarkar and Wyn Jones, 1982; Pena-Cabriales and Alexander, 1983; Linehan *et al.*, 1985; Thomas *et al.*, 1986; Gillespie and Pope, 1990; Olsthoorn *et al.*, 1991; Chung *et al.*, 1994; Gorissen and Cotrufo, 1999; Veneklaas *et al.*, 2003; Warembourg *et al.*, 2003). The pots are of plastic material and have a cylindrical or sub-conical form, but containers with particular shapes have also been adopted to facilitate viewing and sampling of the roots. For example, Gollany *et al.* (1997) used pots of 45 cm length × 2.5 cm width × 13 cm height to grow seedlings of *Sorghum bicolor* (L.) Monech and *Sorghum sudanese* L., while Bakker *et al.* (1999) used containers of 55 cm × 3.5 cm × 40 cm to grow seedlings of *Quercus petraea* (Liebl.) M. These pots have also been called rhizotrons, but they should more properly be termed bench-rhizotrons to distinguish them from the full-scale rhizotrons in covered underground rooms or walkways that are provided with windows on one or both sides to ensure viewing of the roots (e.g. that of the College of Agriculture, Auburn University, Alabama), or the containers of 1 m × 1 m × 2 m used by Ludovici and Morris (1997). The term mini-rhizotron has also been used to define a thin tube (2–5 cm in diameter) of glass or plastic material, fitted with a microvideo camera placed within the soil to follow growth, phenology and demography of roots of herbaceous and arboreal species. For an up-to-date overview on mini-rhizotrons, see Withington *et al.* (2003).

In the case of plants grown in pots filled with soil, they are harvested at a certain stage of their development, with the substrate being loosened so as to reveal the roots, which are then removed and gently shaken or vibrated. The soil particles detached during shaking are considered to be bulk (and eventually added to the non-root soil portion), while the soil particles that remain adhere to the roots, and in particular to the root hairs, are considered to be the rhizosphere soil.

Removal of the rhizosphere soil particles is critical, and several procedures have been developed that have depended on the particular aim of each investigation. A simple way to collect rhizosphere soil is by shaking, or vibrating vigorously, the root-soil mass over a surface (Metzger, 1928; Starkey, 1931, 1958; Smiley, 1974; Linhean *et al.*, 1985) or directly into a solution (Pena-Cabriales and Alexander, 1983). Ultimately, it is possible to remove all of the visible root debris in this separation by hand (Smiley, 1974). However, other procedures have also been followed.

Papavizas and Davey (1961) used a micro-sampler to withdraw serial 3-mm samples from the rhizoplane of *Lupinus angustifolius* L. Sarkar and Wyn Jones (1982) used a sampler to collect the rhizosphere soil from the root surface in the form of a cylindrical core 1 cm thick. Later, to collect the rhizosphere soil exactly from radial rings at 1, 2 and 3 mm from the roots of *Sorghum bicolor* L. and *Sorghum sudanese* L., Gollany *et al.* (1997) used a particular dissecting knife in order to slice such thin soil layers. A thin sectioning technique was also used by Smith and Pooley (1989) to selectively sample the rhizosphere soil of *Picea rubens* Sarg.

Riley and Barber (1969, 1970) used pots with a preincubated soil substrate to grow seedlings of *Glycine max* (L.) Merr. After carefully removing the roots, they considered the loosely adhering particles that could be detached after a gentle shaking as rhizosphere soil, and the material that remained adhered to roots as rhizoplane soil. These authors also stated that with respect to the root surface, the rhizosphere soil was present from a distance of about 1 to 4 mm, while the rhizoplane soil was present from 0 to 2 mm. However, the rhizoplane soil was not separated from the roots, and so both were treated as a whole sample. To study the effect of ion accumulation on P availability in close proximity to the roots, Hoffmann and Barber (1971) used pots with preincubated soil to grow seedlings of *Glycine max* (L.) Merr. After harvesting, the roots with the adhering soil were carefully removed from the substrate and soaked as they were with distilled water in a Buchner funnel; bulk soil was also treated in the same way. After 1 h, both soils were leached with a known amount of water.

Soon and Miller (1977) removed 11-day-old roots of *Zea mays* L. seedlings from pots and shook off the loosely adhering soil so as to leave a cylinder of soil 2–2.5 cm in diameter all around the roots. The roots with this soil collar (termed a rhizocylinder) were placed into a filtering centrifuge tube and centrifuged so as to obtain the rhizocylinder solution, which was compared with a bulk soil solution similarly obtained.

Olsthoorn *et al.* (1991) used stainless-steel pots, into which they transplanted seedlings of *Pseudotsuga meziesii* (Mirb.) Franco. After 8 months, the plants were harvested and the pots frozen to solidify the soil–root mass, which was then cut into four horizontal layers, each more than 10 cm thick. After thawing, the roots were collected from each layer, allowed to dry for 30 min at room temperature, and then shaken on a tray to obtain the rhizosphere soil.

Chung *et al.* (1994) removed the entire soil mass from pots where they had grown seedlings of *Prunus persica* L. Batsch var. persica. The soil mass was gently crushed to separate the roots from the loosely adhering soil, which was then shaken off. The tightly adhering particles, the rhizosphere soil, were collected by shaking in a plastic bag after the earth-covered roots had been oven-dried at 60°C. Chung and Zasoski (1994) purified the thus-collected rhizosphere soil from root debris by blowing on them after spreading the sample onto paper.

Bakker *et al.* (1999) destroyed the pots with seedlings of *Quercus petraea* Liebl. M. and separated the roots from the soil by wet sieving, at 4 and 2 mm, under a stream of water. Roots collected from the sieves were floated, separated into living and dead roots, and subdivided into those less than and those larger than 2 mm in diameter. The finer roots were then gently shaken until a very small amount of soil remained adhering to the roots. This soil was considered to be the rhizosphere soil, and after the roots had been dried for a few hours, it was removed by brushing. Similarly, Boyle and Shann (1998) used a paintbrush to remove the rhizosphere soil.

Gorissen and Cotrufo (1999) grew separate seedlings of grasses (*Lolium perenne* L., *Agrostis capillaris* L. and *Festuca ovina* L.) in pots filled with a loamy sand soil. At a certain stage in their development, the plants were removed from the pots with lumps of soil. The roots were separated from the bulk soil, and all of the particles adhering to them were considered to be rhizosphere soil. The roots plus the rhizosphere soil were sieved at 2 mm, and the resulting earth-covered roots were washed. The suspension obtained was centrifuged, and the deposited particles, also considered to be rhizosphere soil, were collected and mixed with the rhizosphere soil obtained by sieving.

1.2.2.2. Rhizoboxes. As also recognized by Helal and Sauerbeck (1981), in all of the procedures dealing with plants grown in pots, the main problem in obtaining a rhizosphere soil sample is the subjectivity of the separation. The use of pots filled with artificial substrates, such as calcined sand and clay (Warembourg *et al.*, 2003), facilitates separation of the rhizosphere soil from the bulk, but it is not very real to life.

To overcome the problems of a subjective separation of rhizosphere soil from bulk, another type of bench container has been devised: the rhizobox. This was initially designed by Cappy and Brown (1980), and then adopted, with a few modifications, by several investigators (Kuchenbuch and Jungk, 1982; Brown and Ul-Haq, 1984; Dormaar, 1988; Gahoonia and Nielsen, 1991, 1992; Fritz *et al.*, 1994). These containers have dimensions similar to pots, but are able to physically separate the soil from direct contact with the roots with no limitations on the circulation of solutions. In the first design, the soil–root compartment stands on top of the soil compartment, separated by means of porous membranes of steel or a plastic material. Because of this, upon contact with the membrane, the roots form a root mat. In the system conceived by Helal and Sauerbeck (1981, 1983, 1984) and Youssef and Chino (1987, 1988a, b), and adopted by Jianguo and Shuman (1991), Liao *et al.* (1993), Awad *et al.* (1994), McKenzie *et al.* (1995) and Drew *et al.* (2003), porous membranes separate a central soil-root compartment from two or more vertical soil compartments. Membranes subdivide the soil compartments into layers that are parallel to the root plane, so that the soil is at different distances from the roots. These first- and second-generation rhizoboxes have limitations and disadvantages that were reported by Wenzel *et al.* (2001), and that can be partly solved by using a system that they devised. This new system also allows the monitoring of root morphology and elongation by photographs, but it is complicated to assemble. To grow lowland rice, Kirk and Saleque (1995) and Li *et al.* (2002) used special rhizoboxes to maintain the soil in submerged conditions. In all cases, the soil of the soil-only compartment(s) affected by root activity during plant growth was considered to be rhizosphere soil. The bulk soil was represented by soil samples subjected to the same treatment, but not in the presence of the plants. In some cases, to obtain samples at different distances from the root mat, the rhizosphere soil was sliced with a microtome after having been frozen in liquid nitrogen; the same was done with the bulk soil (Jianguo and Shuman, 1991; Gahoonia and Nielsen, 1992; Fritz *et al.*, 1994; Kirk and Saleque, 1995; Wenzel *et al.*, 2001; Li *et al.*, 2002).

McGrath *et al.* (1997) used their own rhizoboxes to obtain a rhizosphere soil. These authors put soil into a bag made of nylon mesh that was able to exclude root trespassing (called a rhizobag). The rhizobag was then placed at the centre of a bigger plastic pot, which was filled with the same soil. Seeds of *Thlaspi caerulescens* J. & C. Presl. and *Thlaspi ochroleucum* Boiss. & Heldr. were sown separately in the rhizobag. After 105 days, the soil inside the rhizobag was considered to be rhizosphere soil, while that outside was considered to be non-rhizosphere (bulk) soil.

With limits on the availability of space, the approach of growing plants in bench rhizotrons and rhizoboxes has allowed the study of rhizosphere soil of seedlings, mostly of herbaceous species. In addition to those already reported above, among the many other species studied there are: *Allium cepa* L. (Farr *et al.*, 1969); *Triticum aestivum* L. (Hoffmann and Barber, 1971; Smiley, 1974); *Hordeum vulgare* L., *Avena sativa* L., *Lolium rigidum* Gaud., *Lactuca sativa* L., and *Pisum sativum* L. (Smiley, 1974; Højberg *et al.*, 1999); *Medicago sativa* L. (Blanchar and Lipton, 1986; Lipton *et al.*, 1987); *Secale cereale* L. and *Bouteloua gracilis* (H.B.K.) Lag. (Dormaar, 1988); *Lupinus albus* L. (Dinkelaker *et al.*, 1989; Veneklaas *et al.*, 2003); *Arachis hypogea* L. (Zhang *et al.*, 2002); *Brassica napus* L. (Kuchenbuch and Jungk, 1982; Gahoonia and Nielsen, 1992; Wenzel *et al.*, 2001); and *Cicer arietinum* L. (Veneklaas *et al.*, 2003). Despite the large number of studies conducted on grasses, only a few forest species have been investigated in this way: *Quercus petraea* Liebl. M. (Bakker *et al.*, 1999); and *Robinia pseudacacia* L. (Gillespie and Pope, 1990).

1.2.3. Sampling rhizosphere soil in the field

1.2.3.1. Grasses and seedlings. Studies of rhizosphere soil from plants grown in rhizotrons or rhizoboxes have been criticized because the conditions of growth are very different from those existing *in vivo* (Drever and Stillings, 1997; Jones, 1998; Parker and Pedler, 1998). Furthermore, it is well known that the activities of roots vary as a function of plant age, so that results obtained by growing seedlings for weeks or months run the risk that they may not accurately represent the conditions of the rhizosphere soil of the adult plant. This is especially true of arboreal species, the growth period of which is very short in bench containers when compared with their expected life. Some have tried to overcome such difficulties by working under natural conditions, considering the whole soil where rooted plants grow as the rhizosphere soil and that from barren areas as bulk soil (Reynolds *et al.*, 1999). Studies on soil decontamination, artificial watersheds and weed control have used a similar, rather diffuse concept, although these were conducted in containers (Dommergues *et al.*, 1973; Reddy and Sethunathan, 1983; Aprill and Sims, 1990; Anderson *et al.*, 1994; Günther *et al.*, 1996; Nichols *et al.*, 1997). In most of these studies, the bulk soil is considered to be the soil in the non-planted containers, while the rhizosphere soil is that attached to the roots after the plants have been uprooted.

Researchers have tried to overcome this problem by conducting outdoor studies in field plots of some dozens of square metres so as to study plant growth under natural conditions in reworked and/or fertilized soils. In one such case, Martin and Kemp (1986) considered the soil-root mass enclosed by a pipe of 10 cm in diameter with a single plant of *Triticum aestivum* L. in the center to be rhizosphere soil. Persmark and Jansson (1997) grew plants of *Pisum sativum* L., *Sinapis alba* L. and *Hordeum vulgare* L. for a few weeks and considered the particles still attached after gentle shaking as their rhizosphere soil and that outside the plots as bulk soil. Pal (1998) sowed *Elusine coracana* (L.) Gaertn., *Amaranthus hypochondriacus* L., *Fagopyrum esculentum* Moench., *Phaseolus vulgaris* L. and *Zea mays* L., which were carefully uprooted 30 or 60 days after sowing. The loosely adhering particles were removed, and the rhizosphere soil recovered by dipping the roots in distilled water.

This approach of considering the soil as a substrate neglects the natural organization of the soil into horizons, and might be regarded as acceptable for agricultural species. However, it has been found that each horizon plays its own role in plant growth. For example, Massee (1990) and Tanaka (1995) found that *Triticum aestivum* L. production (grain and straw) was decreased after they artificially removed the Ap horizon prior to sowing. Similarly, Thompson *et al.* (1991) found that yields of *Zea mays* L. and *Glycine max* (L.) Merr. decreased with decreasing thickness of the A horizon. These relationships that roots establish with each soil horizon are highlighted by production studies. For example, the rate of root growth, elongation and penetration are reduced in the presence of compact horizons (Blanchar *et al.*, 1978; Bar-Yosef and Lamert, 1981; Longsdon *et al.*, 1986), and also owing to the clay content and the decreased availability of oxygen and nutrients (Gerard *et al.*, 1982). Nonetheless, mechanical impedance induces roots to modify the rates and chemical compositions of their exudates (Marschner, 1995), with evident repercussions on the microbial populations.

Examples of field studies dealing with horizons and rhizosphere soil are those of Yang *et al.* (1996) and Wang and Zabowski (1998). In Yang *et al.* (1996), beans of *Glycine max* (L.) Merr. were sown in field plots where different thicknesses of the A horizon had been obtained artificially. Collection of the rhizosphere soil occurred at 1, 2 and 3 months of age using a core sampler of 10 cm in diameter. The sampler was placed on the soil so that the stem of the plant was at its center, and it was pushed 30 cm into the soil. The collected sample was wrapped and frozen, and then thawed at the moment of separation of the rhizosphere soil. The roots were exposed without disturbing the surrounding soil, and the pH measured on the rhizosphere soil at 0.5 and 5 mm from the root surface.

Wang and Zabowski (1998) planted 1-year-old seedlings of *Pseudotsuga meziesii* Mirb. Franco in two soils: either natural or fertilized. Harvesting occurred several times over 3 to 11 months, and each time, the seedlings were taken out of the soil after it had been loosened. The roots were shaken to remove all of the particles that were not tightly adherent, while those more closely associated (the rhizosphere soil) were collected by putting the roots into a bag and shaking them vigorously. The soil outside the root system was considered to be the bulk soil. In these studies, the seedlings were transplanted from another soil, and conditions of the root system at the moment of planting were not indicated.

1.2.3.2. Arboreal species. Information on the role and characteristics of the rhizosphere soil of long-living plants and on their impact on soil formation can be achieved only by working outdoors, in "natural" conditions. In this case, even though the plants were once planted, after a number of decades the characteristics of the rhizosphere soil can be considered to be in equilibrium with the environmental conditions. With this approach, the major problem is to recognize areas where the soil has low variability, and this can be obtained from previous geomorphologic and pedological surveys. A secondary problem is the collecting of the rhizosphere soil after the roots have been extracted from the soil without too much disturbance.

Even in these cases, various approaches have been followed. For example, to collect the rhizosphere soil of *Pinus Taeda* L., Sanchez and Bursey (2002) used a sampling tube of 8 mm in diameter to collect 6 cm^3 of soil every 1.5 cm from the roots of first, second and third order from the superficial horizon. To reduce sampling variability, samples were collected from two positions along the root itself. In a wood of *Quercus berberidifolia* Liebm., Klamer *et al.* (2002) used a soil corer of 5 cm in diameter to take bulk soil, rhizosphere soil and roots up to a mean depth of 10 cm. The soil fractions were obtained by passing the collected samples through a 2-mm pore-size sieve. The sieved soil was considered to be the bulk soil, while the soil attached to the roots remaining on the sieve (rhizosphere soil) was washed off into a bucket with sterile deionized water. The collected soil slurry was centrifuged at 500 ***g*** for 20 min to separate the water from the rhizosphere soil.

Roots and rhizosphere soil can be collected after loosening the soil to a certain depth (Hendriks and Jungk, 1981; Courchesne and Gobran, 1997; Marschner *et al.*, 2002; Watrud *et al.*, 2003; Yanai *et al.*, 2003), or from the face of a soil profile after the digging of a pit (Kunito *et al.*, 2001; Fernández-Sanjurjo *et al.*, 2003; Corti *et al.*, 2004, 2005; Ricci *et al.*, 2004). The first of these is less time-consuming than the second, although the second appears to be less disturbing and offers the opportunity of collecting roots according to genetic horizon rather than at depth intervals.

Once root segments have been dissected in the field, they are gently shaken; the soil particles that detach are considered to be bulk (and added to the free-roots soil portion), while those remaining adhering to the roots, and in particular to the root hairs, are considered to be rhizosphere soil (Hendriks and Jungk, 1981; Courschesne and Gobran, 1997; Fernández-Sanjurjo *et al.*, 2003; Watrud *et al.*, 2003; Corti *et al.*, 2004, 2005; Ricci *et al.*, 2004). Following this procedure, it is hard to define the thickness of the rhizosphere soil. Nonetheless, Courschesne and Gobran (1997) were able to estimate a thickness of less than 3 mm for the rhizosphere soil of *Picea abies* (L.) Karst. Fernández-Sanjurjo *et al.* (2003) took advantage of an evident colour change (yellowish) caused by the root activity of a broom on a reddish (volcanic) bulk soil, and thus could collect rhizosphere soil samples from a 2–3-cm-thick soil collar. More details about this approach are given in Chapter 3.

The successive step of removing rhizosphere soil has been solved in several ways. Hendriks and Jungk (1981) left the roots to dry for a short period of time and then separated the rhizosphere soil by gentle sieving. Others (Courschesne and Gobran, 1997; Fernández Sanjurjo *et al.*, 2003) have collected the rhizosphere soil by brushing the roots. To obtain samples enriched in rhizosphere soil from roots of *Vitis vinifera* L. grown in silty-clay loam soils, Corti *et al.* (2004, 2005) and Ricci *et al.* (2004) used a dissecting knife to remove the 2–3 mm of soil surrounding the hairs and finer roots. After a few hours of drying, the rhizosphere soil was removed from around the roots.

In other cases, soil samples have been collected to a depth of 30 cm using a soil corer (Häussling and Marschner, 1989; Clemensson-Lindell and Persson, 1992; Gobran and Clegg, 1996) or at 10 cm intervals to a depth of 30 cm (Yanai *et al.*, 2003). In these circumstances, all the root fragments were then collected from the soil retrieved from the cores by picking them up with tweezers. The roots were then gently shaken in a plastic bag to further remove the bulk soil, while the adhering soil (rhizosphere soil) was removed by brushing.

2. STRATEGY ADOPTED BY *GENISTA AETNENSIS* (BIV.) DC. TO COLONIZE PYROCLASTIC DEPOSITS ON MOUNT ETNA, ITALY

2.1. Introduction

Mount Etna (Sicily, Italy) is an active volcano that produces pyroclastic deposits because of the frequent explosive activity of the summit craters or the ephemeral lateral mouths. These pyroclastites are vesicular and form unconsolidated deposits that are colonized to about 2100 m a.s.l. by an endemic broom, *Genista aetnensis* (Biv.) DC. The aim of this study was to investigate the mechanisms that allow this broom to colonize the soils on the flanks of the Etna volcano. The study took into consideration changes in rhizosphere soil induced by the broom during the early stages of pedogenesis of the pyroclastic substrata, both in the fine earth (the less than 2 mm fraction) and in the skeleton (the greater than 2 mm fraction).

2.2. Material and methods

The study was conducted on Mount Vetore (Fig. 1), a cinder cone on the southeast flank of Mount Etna (Sicily, Italy), at about 7 km distance from the active

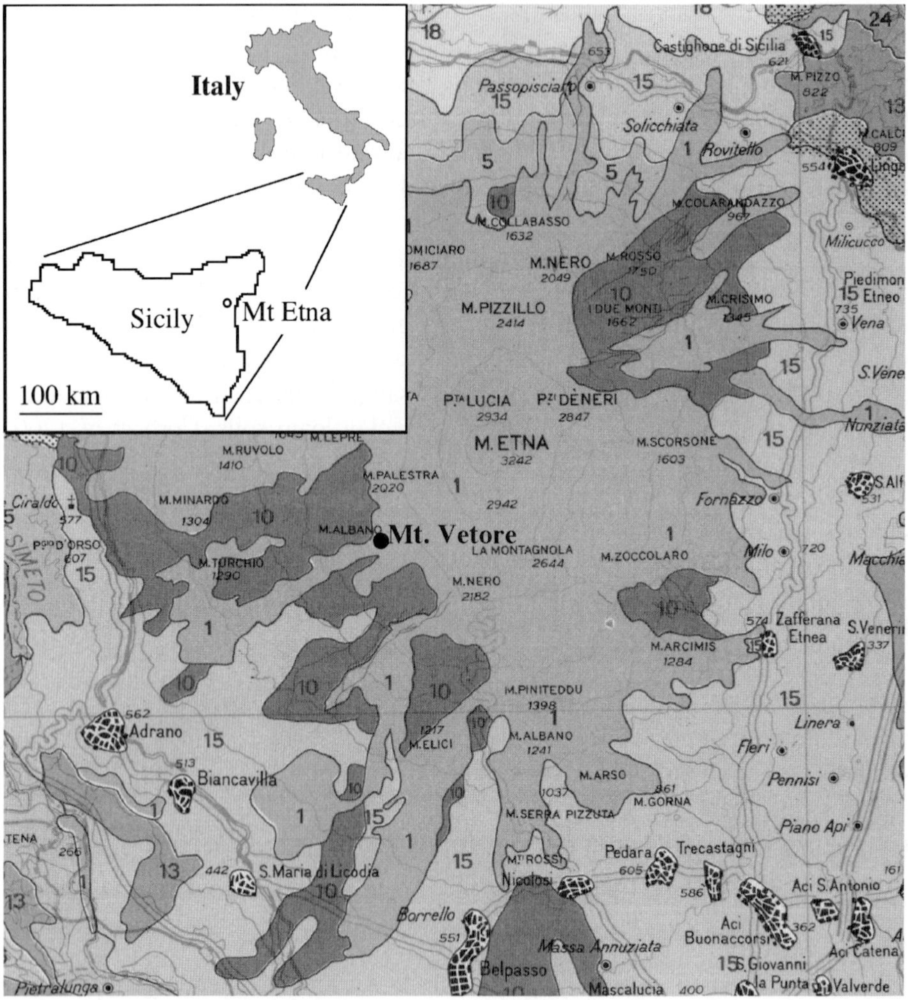

Fig. 1. Map of the area of Mount Etna with indicated the study site, Mount Vetore (Sicily, Italy).

craters of the volcano. Mount Vetore reaches 1823 m a.s.l. and the investigation site spanned from 1800 to 1820 m a.s.l. The mean annual air temperature of the area was 9.3°C, and the mean annual precipitation was 1950 mm. The surface of the cone was basically barren until the 1960s, at which time it was afforested mostly with pure plantations of Etnean broom [*Genista aetnensis* (Biv.) DC.] and Corsican pine (*Pinus laricio* Poiret).

Several profiles were opened, and the soil was described according to Soil Survey Division Staff (1993). The soil, a Vitrandic Udorthent (Soil Survey Staff,

1999), was rather loose and showed two organic horizons, Oi and Oe, and the following sequence of mineral horizons: AC, A/C, 2E (patches in contact with the broom stems), 2C1 and 2C2. The AC and A/C horizons, which were very thin and were collected together, developed from added black ash coming from the summit craters, while the 2E and 2C horizons were formed from the *in situ* pyroclastic material that constituted the cone. The roots of the broom were only in the 2C horizons and penetrated the soil to about 2 m in depth, with no nodules of symbionts. Samples were collected by horizons from two profiles. Bulk and rhizosphere soils were collected from 2C1 and 2C2 horizons, taking advantage of the different color of the soil near roots with respect to that of the bulk (Fig. 2). Indeed, the bulk had a reddish-brown colour, while the soil in the form of a collar all around the roots, with a thickness of about 3 cm, had a yellowish colour.

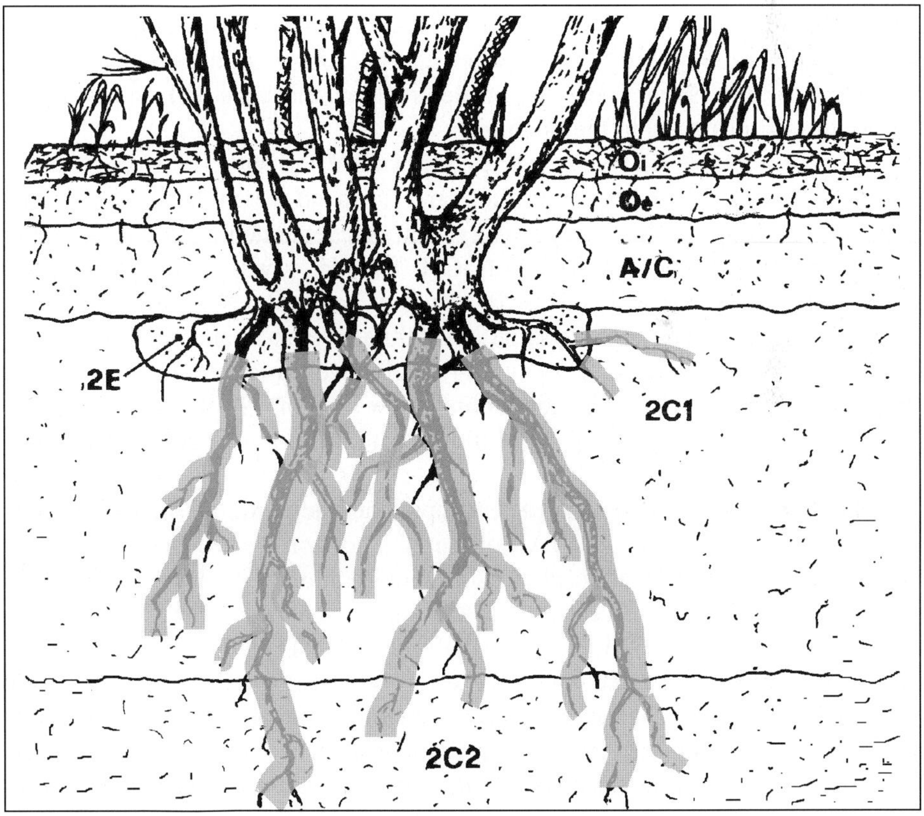

Fig. 2. Draft of the root system of the *Genista aetnensis* (Biv.) DC, which showed the soil horizons of the profile opened at Mount Vetore (Sicily, Italy). The grey shades around the roots indicated the yellowish-brown-coloured rhizosphere soil.

This colour variation was interpreted as being caused by root activity, and the yellowish material was considered to be rhizosphere soil. This was separated from the bulk in the field, collecting the two materials separately from the face of the profile.The roots were picked up from the yellowish material and gently shaken. The loose fraction plus the particles detached during shaking were put together to obtain the loosely adhering rhizosphere (LAR) soil. The particles remaining attached to the roots were removed by vigorous shaking and by brushing with a toothbrush; this fraction was called the tightly adhering rhizosphere (TAR) soil.

Bulk, LAR and TAR were separated into fine earth and rock fragments by sieving at 2 mm. Rock fragments were then sieved under a distilled water flow to obtain another < 2-mm fraction, the rock fragment washings (hereafter called "washings"), which was made up of the fine material adhering to the rock fragments. The rock fragments were then ranked into three size classes: 2–4, 4–10 and > 10 mm. To express some of the measured parameters on a volume basis, the volume and the bulk density of all of the fractions were determined according to Corti *et al.* (1998). These authors detailed that the same bulk density as the fine earth was attributed to the washings.

For particle-size analysis, the fine earth was treated with 3 M H_2O_2 and sonicated (15 min, 15 kHz); coarse, medium and fine sand were retrieved by sieving at 0.25, 0.10 and 0.05 mm, respectively; silt was separated from clay by sedimentation after dispersion in 0.01 M NaOH. The pH_{H_2O} was measured potentiometrically (solid/liquid ratio of 1:2.5). Organic C and total N were measured on acidified samples using a Carlo Erba NA1500 analyser. Available P was determined according to Olsen *et al.* (1954). Effective cation exchange capacity (ECEC) was determined by summation of the cations displaced with 0.2 M $BaCl_2$ and analyzed by atomic absorption with a Perkin-Elmer 1100B spectrophotometer.

For solution ^{31}P-NMR analyses, samples of fine earth, washings and rock fragments from bulk, LAR and TAR of the 2C1 horizon were treated with 0.1 M NaOH solution (under N_2 atmosphere, solid/liquid ratio, 1:10). After 24 h of shaking, the suspensions were centrifuged; the supernatant was filtered at 0.45 μm, taken to pH < 4.0 (with 6 M HCl), and dialyzed at 100 Da molecular mass cut-off (Spectra/Por Biotec CE). The dialyzed extracts were freeze-dried and dissolved in 2 mL of 0.5 M NaOD. The ^{31}P spectra were obtained using a 300-MHz NMR spectrometer (Varian VXR 300) operating at 121.4 MHz.

To extract different forms of Fe, the clay fractions from the fine earth of bulk, LAR and TAR were dispersed in NaOH at pH 8.5 and treated with acidic (pH 3) NH_4-oxalate (Blakemore *et al.*, 1981), citrate-bicarbonate-dithionite (CBD) (Mehra and Jackson, 1960) and 0.1 M hydroxylamine hydrochloride (HAHC) (Chao, 1972). The Fe extracted was measured by atomic absorption with a Perkin-Elmer 1100B spectrophotometer.

To investigate the presence of organo-mineral compounds, specimens of fine earth and the five separates (coarse, medium and fine sand, silt and clay) from bulk, LAR and TAR of the 2C horizons were shaken for 30 min in 1.2 M HCl solution (solid/liquid ratio, 1:10); the suspensions were centrifuged and the supernatants filtered at 0.45 μm. Aliquots of 2 mL were oven-dried (120°C) on glass slides and analyzed by X-ray diffraction using a Philips PW 1710 diffractometer (Fe-filtered Co-Kα_1 radiation). The acid extracts were also analyzed by gas chromatography, following the procedure of Fernández Sanjurjo *et al.* (2003). To assess the role of organic acids on the release of Ca, Mg, K and P, the fine earth from the bulk of the 2C horizons was treated with water and 500 μM oxalic acid solution (Jones and Darrah, 1994) for 1 h (solid/liquid ratio, 1:10). The released Ca, Mg and K were measured by atomic absorption, while P was determined according to Bray and Kurtz (1945).

The values reported are the means of at least 2 replicates from the two profiles. In the tables, the means are associated with the standard errors (Webster, 2001), and the discussion takes into consideration the intervals and overlaps of intervals.

2.3. Results

2.3.1. General characteristics of the soil

The amount of skeleton was around 100 g kg^{-1} in the O horizons, and ranged between 554 and 409 g kg^{-1} in the mineral horizons (Table 1). With the exception of the Oe horizon, the mineral part of which has a silt texture, in all the horizons the fine earth had a sandy texture, with a prevalence of coarse sand. The pH increased with increasing depth, reaching neutrality in the 2C horizons (Table 1). Organic C and total N were present in considerable amounts in the O and AC+A/C horizons, but became scarce in the 2E and 2C horizons. The ECEC followed the trend of organic C and total N, ranging from about 30 cmol(+) kg^{-1} in the O horizons to 7–8 cmol(+) kg^{-1} in the 2C ones.

2.3.2. Characteristics of bulk and rhizosphere soil

In the 2C horizons, the fine earth was the most abundant fraction on a volume basis in both bulk and LAR, whereas the rock fragments, and in particular those larger than 10 mm, were abundant in the TAR (Table 2). Still on a volume basis, the washings were present in amounts of less than 1%.

The bulk density of the different fractions of bulk, LAR and TAR (Table 2) increased from fine earth and washings to skeleton and, in this last fraction, from the 2–4 mm rock fragments to those > 10 mm, according to that previously described by Corti *et al.* (1998). The particle-size distribution of the fine earth from bulk and rhizosphere soil (Table 3) showed a relatively higher amount of clay in the TAR of both horizons. The pH of fine earth and washings tended to

Table 1
Amounts of rock fragments, particle-size distribution of the fine earth, content of glass, pH in water, organic C and total N contents, and ECEC for the soil under *Genista aetnensis* at Mount Vetore (Sicily, Italy). Standard errors in parentheses ($n = 2$)

Horizons	Rock fragments	Particle-size distribution					Glass	pH	Organic C	Total N	ECEC
		Sand			Silt	Clay					
		Coarse	Medium	Fine							
Oi	87(15)	585(15)	112(9)	124(6)	167(10)	12(2)	92(4)	4.8(0.2)	278.6(6.7)	17.1(0.8)	32.5(0.7)
Oe	110(21)	8(2)	8(3)	11(5)	955(13)	18(3)	60(2)	5.8(0.2)	157.2(3.5)	11.1(0.6)	30.4(0.9)
AC + A/C	518(44)	858(11)	155(7)	145(7)	129(9)	13(2)	65(3)	6.0(0.1)	54.5(2.9)	4.2(0.3)	25.2(0.6)
2E	570(20)	774(11)	77(3)	70(7)	70(6)	9(1)	68(8)	6.2(0.1)	13.1(0.4)	0.9(0.1)	8.8(0.3)
2C1	531(27)	878(9)	69(2)	32(4)	17(3)	4(0)	84(3)	7.0(0.1)	1.7(0.1)	0.1(0.0)	8.0(0.4)
2C2	435(27)	843(5)	96(2)	49(5)	10(1)	2(1)	85(7)	7.0(0.1)	0.4(0.0)	0.0(—)	7.3(0.4)

Note: Rock fragments, Particle-size distribution and Glass are expressed as g kg^{-1}; Organic C and Total N are expressed as g kg^{-1}; and ECEC is expressed as cmol(+) kg^{-1}.

Table 2
Percent distribution of fine earth, washings and rock fragment classes, and their bulk density, pH, organic C and total N contents, available P and effective cation exchange capacity in bulk and rhizosphere soil of the 2C1 and 2C2 horizons under *Genista aetnensis* at Mount Vetore (Sicily, Italy). Standard errors in parentheses (*n* is the number of replicates)

	Volume (%) (n = 2)	Bulk density (g dm^{-3}) (n = 3)	pH$_{water}$ (n = 3)	Organic C (g dm^{-3}) (n = 3)	Available P (mg dm^{-3}) (n = 3)	ECEC (cmol(+) dm^{-3}) (n = 3)
2C1 horizon						
Bulk						
Fine earth	61.0(3.4)	0.91(0.03)	6.96(0.15)	1.30(0.11)	2.9(0.4)	3.0(0.3)
R.f. washings	0.4(0.1)	0.91(0.03)	6.34(0.14)	8.19(0.89)	20.7(2.2)	13.4(0.9)
R.f. 2–4 mm	16.7(1.3)	1.71(0.05)	6.66(0.16)	0.94(0.07)	3.1(0.5)	2.7(0.3)
R.f. 4–10 mm	12.7(1.2)	1.77(0.02)	6.73(0.11)	0.80(0.06)	3.6(0.6)	2.0(0.3)
R.f. >10 mm	9.2(0.8)	1.87(0.03)	6.57(0.10)	0.93(0.06)	1.3(0.3)	0.9(0.1)
LAR						
Fine earth	66.8(3.9)	0.91(0.03)	6.75(0.14)	1.97(0.15)	4.0(0.7)	4.4(0.4)
R.f. washings	0.2(0.0)	0.91(0.03)	6.34(0.15)	12.92(1.08)	20.5(1.8)	18.8(0.8)
R.f. 2–4 mm	17.1(2.1)	1.71(0.02)	6.48(0.12)	0.86(0.09)	2.8(0.7)	3.1(0.4)
R.f. 4–10 mm	11.6(1.2)	1.74(0.01)	6.79(0.13)	0.44(0.07)	2.0(0.4)	2.3(0.3)
R.f. >10 mm	4.3(0.6)	1.87(0.04)	6.68(0.12)	0.84(0.08)	1.4(0.4)	1.6(0.2)
TAR						
Fine earth	39.0(2.1)	0.91(0.03)	6.46(0.12)	4.58(0.35)	2.4(0.3)	3.8(1.0)
R.f. washings	0.3(0.0)	0.91(0.03)	5.70(0.14)	24.75(1.79)	35.0(2.3)	11.6(0.6)
R.f. 2–4 mm	7.7(1.6)	1.69(0.02)	6.39(0.10)	1.44(0.12)	2.9(0.5)	3.7(0.4)
R.f. 4–10 mm	4.9(0.8)	1.75(0.03)	6.39(0.14)	0.96(0.11)	2.4(0.6)	2.5(0.2)
R.f. >10 mm	48.1(2.9)	1.80(0.02)	6.45(0.13)	0.63(0.08)	1.0(0.3)	0.9(0.1)
2C2 horizon						
Bulk						
Fine earth	60.9(3.5)	1.12(0.02)	6.97(0.18)	0.52(0.05)	1.5(0.4)	2.1(0.4)
R.f. washings	0.1(0.0)	1.12(0.02)	6.63(0.15)	4.93(0.67)	7.8(0.7)	10.1(1.1)
R.f. 2–4 mm	20.5(0.9)	1.79(0.04)	6.71(0.17)	0.54(0.08)	1.1(0.2)	2.1(0.4)
R.f. 4–10 mm	11.5(1.5)	1.89(0.03)	6.63(0.12)	0.38(0.05)	1.0(0.3)	1.3(0.2)
R.f. >10 mm	7.0(1.1)	1.96(0.04)	6.54(0.15)	0.39(0.03)	1.1(0.3)	1.4(0.2)
LAR						
Fine earth	71.9(2.8)	1.12(0.02)	6.90(0.15)	1.72(0.09)	2.3(0.6)	4.2(1.0)
R.f. washings	0.1(0.0)	1.12(0.02)	6.56(0.17)	44.69(1.67)	8.8(0.9)	29.4(2.1)
R.f. 2–4 mm	13.8(2.0)	1.79(0.04)	6.83(0.11)	0.72(0.08)	2.4(0.6)	4.9(0.7)
R.f. 4–10 mm	11.0(0.4)	1.87(0.04)	6.65(0.15)	0.75(0.08)	2.3(0.4)	2.1(0.5)
R.f. >10 mm	3.2(0.4)	1.96(0.06)	6.68(0.14)	0.39(0.07)	2.0(0.4)	1.6(0.4)
TAR						
Fine earth	26.7(2.0)	1.12(0.02)	6.73(0.10)	11.69(1.21)	5.1(0.5)	6.2(0.8)
R.f. washings	0.3(0.1)	1.12(0.02)	5.91(0.15)	26.54(1.56)	8.2(0.6)	25.8(2.6)
R.f. 2–4 mm	5.8(1.2)	1.78(0.03)	6.52(0.16)	1.78(0.19)	1.9(0.2)	4.3(0.9)
R.f. 4–10 mm	5.5(0.0)	1.86(0.05)	6.37(0.12)	0.90(0.05)	1.9(0.1)	3.8(0.9)
R.f. >10 mm	61.7(3.1)	1.93(0.04)	6.47(0.11)	0.77(0.08)	1.9(0.1)	1.7(0.7)

Table 3
Particle-size distribution of the fine earth from the bulk and the rhizosphere soil of the 2C1 and 2C2 horizons under *Genista aetnensis* at Mount Vetore (Sicily, Italy). Standard errors in parentheses (n = 3)

	Particle-size distribution (g kg^{-1})				
	Sand			Silt	Clay
	Coarse	Medium	Fine		
2C1 horizon					
Bulk	891(6)	66(3)	25(5)	15(4)	3(0)
LAR	836(6)	77(5)	55(6)	28(4)	4(1)
TAR	835(12)	70(2)	59(4)	25(4)	11(2)
2C2 horizon					
Bulk	850(11)	90(4)	48(5)	11(2)	1(0)
LAR	823(7)	107(4)	58(5)	10(2)	2(0)
TAR	830(7)	99(4)	45(6)	13(2)	13(3)

decrease from bulk to LAR to TAR, while for the rock fragment classes, such a trend was not clear (Table 2). Organic C of the fine earth increased from the bulk to the TAR in both horizons. The largest contents of organic C, however, were in the washings of the TAR from the 2C1 horizon and of the LAR from the 2C2 horizon. Among the rock fragments, those 2–4 mm in size had the major concentration of organic C (Table 2). For available P and ECEC, the washings were the fraction that showed the highest values in both horizons. In these cases, all the fractions of LAR and TAR appeared richer in available P and exchangeable cations than those of the bulk.

The ^{31}P-NMR spectra (Fig. 3) showed an increase in the complexity of the signal patterns for all of the fractions from the bulk to the TAR. For the fine earth, the spectrum of the bulk only presented a signal in the area between 6.1 and 6.7 ppm, due to inorganic orthophosphate (Newman and Tate, 1980). In the spectrum of the LAR, as well as this peak, another signal appeared between 3.0 and 6.1 ppm, which is characteristic of orthophosphate monoesters (Turner *et al.*, 2003), such as inositol phosphates, sugar phosphates and mononucleotides. For the TAR, in addition to those mentioned above, the spectrum displayed signals between −2.0 and 2.8 ppm owing to orthophosphate diesters; in this last region, it was possible to distinguish signals in the range of 0.8–2.8 ppm due to teichoic acid, a complex of compounds composed of sugar units linked by phosphate

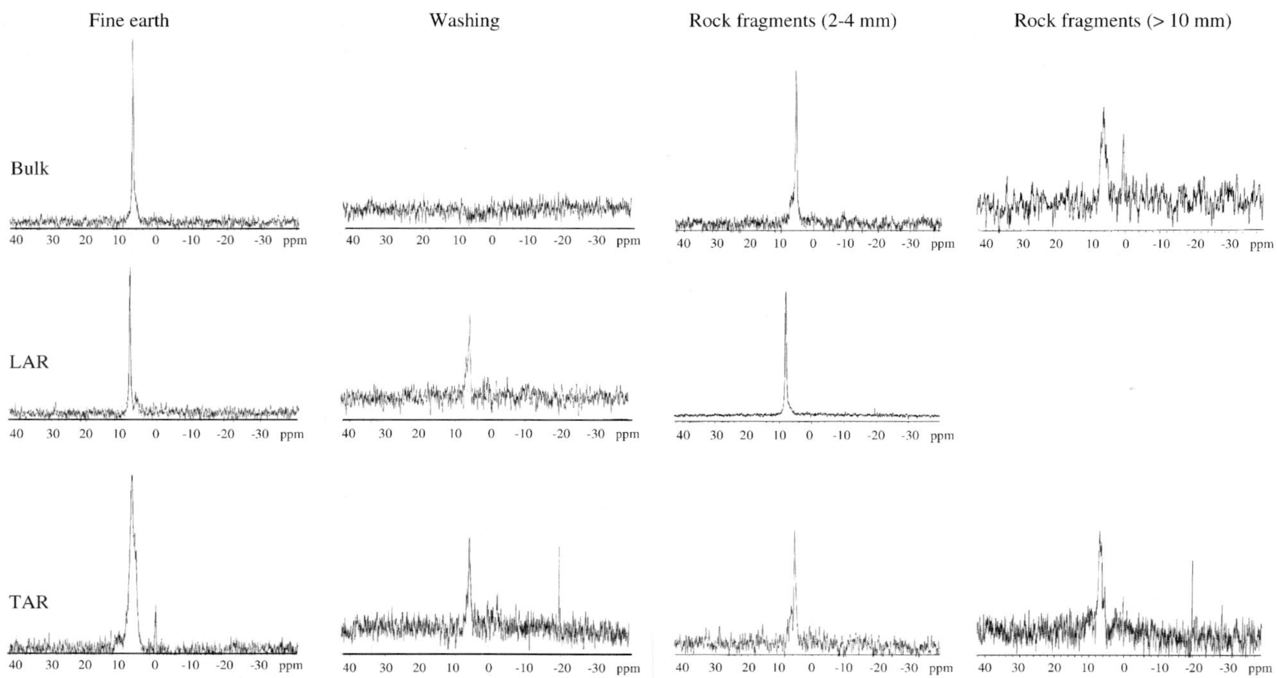

Fig. 3. ^{31}P liquid-state NMR spectra of the NaOH extracts obtained by fine earth, washing and rock fragments (2–4 mm and > 10 mm) fractions from bulk, LAR and TAR of 2C1 horizon. Mount Vetore (Sicily, Italy).

groups, and typical of prokaryotic cell walls (Guggenberger *et al.*, 1996; Rubæk *et al.*, 1999). The washings from the bulk did not show any recognizable ^{31}P-NMR signals. In the spectrum of the LAR washings, the main signal was at 6.0 ppm, owing to inorganic orthophosphate; other signals were owing to orthophosphate monoesters and orthophosphate diesters, such as teichoic acid, phospholipids and DNA (-2 and 0.8 ppm) (Condron *et al.*, 1990; Makarov *et al.*, 2002). The spectrum of the TAR washings showed the same signals seen for those of LAR and a peak at -19.5 ppm owing to polyphosphates (Newman and Tate, 1980). The ^{31}P-NMR spectra of the 2–4 mm rock fragments appeared similar to those of the fine earth, although no signals attributable to diester-P were found in the TAR. The spectrum of the $>$ 10 mm rock fragments from the bulk displayed signals owing to inorganic orthophosphate, orthophosphate monoester and, in the diester-P region, a sharp peak at 0.8 ppm attributed to teichoic acid. In the TAR, a peak at -23 ppm was attributed to polyphosphates (Dai *et al.*, 1996) also appeared.

The separates obtained from the bulk had a reddish-brown colour, while the clays of the LAR and TAR were yellowish brown, and remained so after a treatment with NaClO. The NH_4-oxalate-treated clays gave similar amounts of extracted Fe-oxyhydroxides (Table 4), even though possible differences could have been hidden by the presence of magnetite, which is slightly soluble in acid oxalate (Schwertmann and Taylor, 1989). The Fe extracted by CBD (Table 4) was lower than that extracted by acid oxalate, as CBD fails to dissolve magnetite completely (Jackson *et al.*, 1986). The HAHC solution, which is able to extract

Table 4
Fe extracted by acid NH_4-oxalate, citrate–bicarbonate–dithionite (CBD), and hydroxylamine hydrochloride (HAHC) from the clay from bulk and rhizosphere soil of the 2C1 and 2C2 horizons under *Genista aetnensis* at Mount Vetore (Sicily, Italy). Standard errors in parentheses (n = 3)

	NH_4-oxalate (g kg^{-1})	CBD (g kg^{-1})	HAHC (g kg^{-1})
2C1 horizon			
Bulk	81(8)	50(4)	32(5)
LAR	83(6)	72(6)	60(9)
TAR	85(8)	77(6)	82(11)
2C2 horizon			
Bulk	74(7)	68(6)	11(2)
LAR	56(5)	47(4)	48(7)
TAR	70(8)	65(5)	54(6)

Mn oxides and easily reducible amorphous Fe-oxyhydroxides (Chao, 1972), recovered more Fe from the rhizosphere clay than from the bulk clay (Table 5). Even if part of this Fe could be adsorbed on Mn oxides, in comparing the amounts of Fe extracted by the three solutions, it appears that most of the "free" Fe in the rhizosphere clay was in the form of easily reducible amorphous Fe-oxyhydroxides, while in the bulk clay they represented a minor aliquot. These data suggested that the yellowish-brown colour of the rhizosphere soil might be due to small amounts of secondary amorphous Fe-oxyhydroxides that coat the sand grains.

The diffraction patterns of the dried extract, obtained by treating the fine earth of bulk, LAR and TAR with 1.2 M HCl, did not display any peaks ascribable to oxalate minerals. The same happened with the separates of bulk and LAR, while fine sand and silt of the TAR showed peaks of calcium oxalate minerals: 0.616, 0.592, 0.447, 0.365, 0.296 and 0.278 nm (data not shown). Similarly, the gas-chromatography analyses of the acid extracts also indicated the presence of oxalic acid (peak at 12.50 min) only in the fine sand and silt of the TAR (data not shown). These results indicated that oxalate minerals were scarce and mainly concentrated in the fine sand and silt of the TAR, where the two separates represented about 8% of the fine earth for the 2C1 horizon, and about 6% of the fine earth for the 2C2 horizon. The experiment on the release of nutrients in water and oxalic acid solution from the fine earths (Table 5) showed scarce solubilities of Ca, Mg, K and P, while the contact with oxalic acid markedly increased the solubilities of these nutrients.

2.4. Discussion

The soil material at the base of the stems and that surrounding the roots of the 2C horizons were chemically and chromatically different from the bulk. The relatively high precipitation of the site and the fast drainage, due to the coarse texture, favoured the bleaching of the soil in contact with the stems and roots, as stem-flow and soil solutions passing through the soil used the coarse roots as

Table 5
Ca, Mg, K and P extracted by water and 0.5 mM oxalic acid solution from the fine earth of the bulk soil of the 2C1 and 2C2 horizons under *Genista aetnensis* at Mount Vetore (Sicily, Italy). Standard errors in parentheses (n = 3)

Horizons	Water (mg kg^{-1})				Oxalic acid (0.5 mM) (mg kg^{-1})			
	Ca	Mg	K	P	Ca	Mg	K	P
2C1	0(0)	2(1)	8(0)	3(1)	21(1)	34(8)	15(5)	9(0)
2C2	0(0)	1(0)	5(1)	2(1)	20(1)	19(0)	17(0)	64(7)

preferred flow paths. This could also be a cause of the lower pH recorded in the TAR with respect to the LAR and the bulk in the 2C horizons. In addition to exudates, substances transported by solutions enriched the soil portion in contact with the roots, thus inducing the development of a microenvironment favourable for microorganisms and roots. Over about 35 years, this has modified the soil near the roots so much that the contents of clay, organic carbon, available P and exchangeable cations in the LAR and TAR were higher than in the bulk soil.

Among the fractions separated from bulk and rhizosphere soil, the washings, made up of alteration products of the rock fragment surfaces and alluvial soil particles (Agnelli *et al.*, 2002), showed the highest contents for all of the considered chemical parameters, with increasing values from bulk to TAR. This could be due to the high presence of rock fragments in the TAR (from which the washings fraction was partly derived), colonized preferentially by the broom roots. The root activity evidently promoted the weathering of the rock fragment surfaces, with the consequent production of washings. In this way, this fraction would be enriched in organic C deriving from the root exudates and the microbial community harboured in the rhizosphere soil, while P and nutrient cations would derive from the alteration processes of the mineral phase.

The ^{31}P-NMR analyses showed an increase in the complexity of the phosphatic molecules from the bulk to the TAR. In the bulk, P was represented only by inorganic orthophosphate, whereas the P-monoesters (inositol phosphates, sugar phosphates, mononucleotides) and P-diesters (teichoic acid, phospholipids, DNA, RNA) were present in the rhizosphere soil. This trend, together with the presence of polyphosphate only in the washings and in the rock fragments > 10 mm of the TAR, suggested a higher presence and activity of the microflora in the TAR than in the LAR and in the bulk. The occurrence of polyphosphates could be a clue to the biological cycling of P (Adams and Byrne, 1989) and, hence, an indication of the presence of an active microbial population in the TAR. Ghonsikar and Miller (1973) found that pure cultures of bacteria, algae and fungi could accumulate intracellular polyphosphates under conditions of nutritional disequilibria. Furthermore, orthophosphate diesters are more labile than monoesters (Hinedi *et al.*, 1989; Makarow *et al.*, 2002), and can be a source of available (inorganic) P through mineralization (Tate and Newman, 1982).

Calcium oxalate minerals were found only in the rhizosphere soil. In soil, oxalates come from the acid dissociation of oxalic acid produced during the decomposition of organic matter or by the metabolism of microorganisms and fungi (Cromack *et al.*, 1979; Malajczuk and Cromack, 1982; Fox, 1995; Jones, 1998; Caviglia and Modenesi, 1999; Tait *et al.*, 1999). Plant roots can also excrete oxalic acid to increase the availability of Fe and P (Graustein *et al.*, 1977; Jurinak *et al.*, 1986; Bar-Yosef, 1991; Staunton and Leprince, 1996) or to inactivate toxic elements (Kochian, 1995). Once released, oxalic acid participates in mineral alteration (Shotyk and Nesbitt, 1992; Jones, 1998). The observation that at Mount

Vetore the oxalates were mainly in the broom rhizosphere indicated that the roots were responsible for releasing oxalic acid. With respect to the Ca, Mg, K and P extracted from the bulk fine earth by water, the extraction with oxalic acid solution is from 2- to 32-fold higher, indicating that in this soil the excretion of oxalic acid from roots may represent a mechanism for the broom to take up nutrients through mineral alteration. Even though in small amounts, exudation of oxalic acid has contributed, together with the substances carried by the soil solutions through the roots, to the alteration of the minerals around the roots themselves.

2.5. Conclusions

In the soil studied, the coarse roots represented preferential flow pathways through which most of the soil solution was discharged; as a consequence, the organic and inorganic substances transported by solution tend to accumulate in the proximity of the roots. This soil also had a low availability of nutrients, and this forced the roots to excrete oxalic acid to increase nutrient availability.

The amount of skeleton, increasing from bulk to TAR, indicated that the roots of the *Genista aetnensis* (Biv.) DC. and rock fragments are intimately connected. The rock fragment washings, partly formed by the root-induced weathering of the clasts, showed considerable concentrations of organic C, available P and exchangeable cations, with an increasing trend from bulk to LAR and TAR. Although present in extremely low amounts, the washings could represent a limited soil portion where roots, microflora and minerals interact.

The rhizosphere soil of *Genista aetnensis* (Biv.) DC. hosts a microbial population that is responsible for biological P cycling. This may be considered to be the final stage of a strategy adopted by brooms to preserve P, the most limiting nutritive element of this soil. This strategy is probably what makes the broom plants able to colonize the inhospitable soils on the flanks of Mount Etna.

As a by-product, the weathering processes occurring in the rhizosphere have produced yellowish-coloured amorphous Fe-oxyhydroxides that revealed the thickness of the soil affected by root activity, where chemical, mineralogical and biological properties were changed with respect to the bulk.

3. ROLE OF THE ROOTS OF *ERICA ARBOREA* L. IN THE GENESIS OF ACID SOIL FROM ALKALINE MARINE SEDIMENTS

3.1. Introduction

The Selva di Gallignano is a protected floristic area of about 8 ha located a few kilometres from Ancona (Central Italy). The dominant plants are *Quercus cerris* L. and *Quercus pubescens* Willd, which are almost coeval and around 70 years old. In the wood studied, there is an area of about 3 ha covered by vegetation made up of *Quercus cerris* L., *Fraxinus ornus* L., *Sorbus torminalis* (L.) Crantz, *Ostrya carpinifolia* Scop., and *Acer campestre* L., with shrub and vine-like layers of *Lonicera xylosteum* L., *Lonicera caprifolium* L. and *Smilax aspera* L., and a grass layer comprized of *Cyclamen repandum* S.S., *Ruscus aculeatus* L., *Rubia peregrina* L. and *Festuca heterophylla* Lam. (Allegrezza and Biondi, 2002). Within this area, there is a zone of about 5400 m^2, where in addition to the above-mentioned shrubs, the understory is mostly due to plants of *Erica arborea* L. The soil with diffuse *Lonicera* plants was classified as Typic Eutrudept, fine loamy, mixed and calcareous (Soil Survey Staff, 1999), while that with *Erica arborea* L. was classified as Typic Dystrudept, fine, mixed and acidic.

In a territory dominated by alkaline soils, the presence of an acid soil with related plant species is important, as they contribute to the biodiversity of an otherwise uniform environment. The presence of *Erica arborea* is rather rare in the region, and as the species is heliophile, its presence in the wood appears to be due to a past period during which the surface remained open, and consequently, subjected to colonization by this species. Although there are few in-depth studies, *Erica arborea* is a pyrophitic species (GCIAR, 1999), so it is able to colonize recently burnt-out areas. The *Erica* plants at the Selva di Gallignano are rather rachitic, possibly because the dominant species have reduced light penetration. The age of the plants, established by sections of stems, has been estimated to range from 20 to 40 years, even though there is the possibility that the stems have starved in the soil for decades before vegetating during a period of suitable conditions (C. Urbinati, 2003, personal communication). Whatever the event that favoured the entry of this species into the wood, the presence of the acid soil has enabled its continued establishment.

The aim of this study was to evaluate the contribution of the *Erica* plants to the development of the acid soil, in order to clarify the soil–plant relationships that have allowed this species to colonize the area. This study was accomplished through a geomorphologic survey and a characterization of the bulk and rhizosphere soil.

3.2. Material and methods

The site under study was located in a forest called Selva di Gallignano, Ancona, Italy (Fig. 4), at an elevation of 180 m a.s.l. and with a southeasterly exposure. The mean annual air temperature of the site was 13.6°C, and the mean annual precipitation was about 800 mm.

Following a geological survey in the area of the Selva (Nanni, 1997), a geomorphological study recognized local tilting of the sediments and the presence of mass movements and landslides. Taking advantage of the presence of escarpments, the morphology of various unweathered strata was recorded and samples were collected. Through morphology and textural analyses, some marker strata were established, enabling reconstruction of the local layering of the sediments.

The extent of the area with acid soil was assessed by auger holes and control profiles. In the area, two profiles at 12 m from each other were opened to about 40 cm from a stem of *Erica arborea*. The soil description according to the

Fig. 4. Map of Italy showing the location of the investigated site, Selva di Gallignano (Ancona, Italy).

Soil Survey Division Staff (1993) and classification are reported in Table 6. A large amount of sample (a few kilograms) was collected by horizons and carried to the laboratory. In the laboratory, the rhizosphere soil was separated from the samples at field moisture by picking up the roots together with the adhering soil. The roots were gently shaken and the detached particles added to the bulk soil. To enrich the rhizosphere fraction, the soil adhering to the roots was reduced to a collar of about 3–6 mm by a dissecting knife; in this case, the removed material was added to the bulk. The rhizosphere soil was obtained by separating the earthy material from the roots by shaking and brushing with a toothbrush, and by taking away the finer roots using tweezers.

Particle-size distribution was determined before and after dissolution of organic cements (Lavkulich and Wiens, 1970); coarse (2.00–0.50 mm), medium (0.50–0.25) and fine (0.25–0.05 mm) sands were retrieved by sieving, while silt was separated from clay by sedimentation. The pH was determined potentiometrically in water and 1 M KCl solution (solid/liquid ratio, 1:2.5). Total exchangeable

Table 6

Morphological description of the soil under *Erica arborea* from Selva di Gallignano (Ancona, Italy). For symbols see legend

Landform: steep slope (10–15%) with diffuse soil cracks due to creeping phenomena; Exposure: SE; Altitude: 180 m; Mean annual air temperature: 13.6°C; Mean annual precipitation: ≈800 mm; Parent material: Pleistocene sediments.
Vegetation: *Quercus cerris* L., *Fraxinus ornus* L., *Sorbus torminalis* (L.) Crantz, *Ostrya carpinifolia* Scop., *Acer campestre* L.; Understory: *Erica arborea* L., *Juniperus communis* L., *Lonicera xylosteum* L., *Lonicera caprifolium* L., *Cyclamen repandum* S.S., *Ruscus aculeatus* L., *Smilax aspera* L., *Rubia peregrina* L., *Festuca heterophylla* Lam.
Soil: Typic Dystrudept, fine, mixed and acidic (Soil Survey Staff, 1999).

	Depth (cm)	Colour[a]	Texture[b]	Structure[c]	Consistency[d]	Plasticity[e]	Roots[f]		Mycelium[g]	Boundary[h]	Thickness (cm)	Other observations[i]
							Total	of *Erica*				
Oi	3–1	—	—	—	—	—	0	0	0	cw	2–3	Undecomposed leaves of *Q. cerris*, *E. arborea* and *F. ornus*
Oe	1–0	—	—	—	—	—	0	0	0	cb	0–1	
A	0–2	2.5Y 3/1	sil	3f, m, c cr	mfr, wss	wps	$v_1$mi,vf	0	0	ab	0–4	
E	0–3 or 2–5	10YR 4/2 10YR 5/2 10YR 6/4	sil	2f, m cr 2f, m abk	cr=mfi, wss abk=mfr, wss	wps	2mi,vf, f, m; 3co	2mi,vf,f,m; 3co	0	cw	3–4	Roots abound into the crcr
EB	3–7 or 5–10	10YR 6/4 2.5Y 5/6	sil	3m, c abk-sbk	mfi-fr, wss	wps	2mi,vf, f, m; 3co	2mi,vf, f, m; 3co	0	cw	3–6	Crcr fulfilled of A and E materils colonized by few mycelium

Bw	7–26 or 10–26	10YR 4/4 10YR 5/6	sil	3m, c abk-sbk	mfi-fr, wss	wps	2mi,vf, f, m, co	2mi,vf, f, m, co	0	cw	15–20	Crcr fulfilled of A and E materials colonized by mycelium and 3mi,vf,f,m roots. Few Mn nodules. Few clay cutans on peds and roots
2Bw1	26–42	10YR 5/4 2.5Y 5/6	sic	2f, m abk	mfi, wss	wps	2-3mi,vf, f, m, co	2mi,vf, f, m, co	+++	cs	15–17	Few Mn nodules, mycelium
2Bw2	42–62	10YR 5/6	sic	2m, c abk	mfi, wss	wps	2-3mi,vf, f, m, co	2mi,vf, f, m, co	++	cw	18–21	Few Mn nodules, mycelium
2Bw3	62–73	10YR 4/4	sic	2m, c abk	mfi, wss	wps	2mi,vf, f, m; 3co	2mi,vf, f, m, co	+	cs	10–12	Few Mn nodules
3Bw	73–83	2.5Y 5/4	sicl	2f, m sbk	mfi, wss	wps	1mi,vf, f; 2m; 3co	1mi, vf	0	as	9–11	Few concretions of $CaCO_3$
4BC	83–91+	2.5Y 7/4	sil	3m sbk→1th pl	mfr, wss	wps	3mi,vf, f, m, co	$v_1$mi, vf	0	—	—	Plentiful concretions of $CaCO_3$

[a] Moist and crushed, according to the Munsell charts. [b] c=clay, si=silt or silty, l=loam. [c] 1=weak, 2=moderate, 3=strong; th=thin, f=fine, m=medium, c=coarse; cr=crumb, abk=angular blocky, sbk=subangular blocky, pl=platy; → breaking into. [d] m=moist, fr=friable, fi=firm; w=wet, ss=slightly sticky. [e] w=wet, s=slightly plastic. [f] 0=absent, v_1=very few, 1=few, 2=plentiful, 3=abundant; mi=micro, vf=very fine, f=fine, m=medium, co=coarse. [g] We referred to the mycelium visible at naked eyes. 0=absent, +=few, ++=plentiful, +++ = abundant. [h] a=abrupt, c=clear; s=smooth, w=wavy, b=broken.
[i] crcr=creeping cracks.

acidity and exchangeable Al were determined by titration according to Sims (1996), while the exchangeable H was obtained by subtraction. Total C and N contents were determined for specimens deprived of rootlets visible under a 2x magnifying lens, using a dry combustion analyzer, while the organic C content was estimated by the Walkley–Black method without application of heat (Allison, 1965). On the same specimens, available P was determined according to Olsen *et al.* (1954).

Particle-size distribution was conducted in duplicate using a sample from each of the two profiles ($n = 2$). For every horizon, pH determinations were accomplished on four specimens from both profiles ($n = 8$). All of the other measurements were conducted in triplicate, using two samples from one profile and one from the other. In all cases, standard errors were calculated and the discussion took into consideration intervals and overlaps of intervals.

Mineralogical investigations were conducted using a Philips PW1710 diffractometer (Cu-Kα radiation) on powders treated with Mg, K and glycerol, and heated to 550°C.

3.3. Results and Discussion

3.3.1. *Geomorphology and pedology*

The soils of the Selva derive from Pleistocene marine sediments. This parent material is made up of lithological units of arenaceous-pelitic or pelitic-arenaceous composition, which are made of different strata of clay, silt-clay, marl, silt and sands with various degrees of cementation, often intercalated with organogenic limestone such as travertine (Nanni, 1997). Geomorphological recognition and soil surveys allowed reconstruction of the principal phenomena responsible for the genesis of acid soil (Fig. 5). Originally, all of the strata were alkaline because of the presence of carbonates. The soils developed from such sediments tend to assume a horizon organization that repeats the layering of the parent material (phase 1, Fig. 5). However, at a certain depth, lenses of material with a lesser content of carbonates and a relatively coarser texture occur. From such lenses, dissolution of carbonate by organic acids or ligands carried down by soil solutions produce neutral to sub-acid horizons. At the same time, soil erosion processes make deep horizons closer to the surface (phase 2, Fig. 5). In time, when the neutral to sub-acid horizon reaches the surface after further rapid acidification, it becomes colonizable by acidophile species, even though the acid horizon rests on alkaline horizons. The acidification of the soil at depth is promoted by the activity of the vegetation (phase 3, Fig. 5).

The final step in this process is the genesis of an acid soil with morphology like that reported in Table 6. In this soil, two thin organic horizons (Oi and Oe), for a total of 3–4 cm, rest on the mineral soil. The Oi horizon is mostly made up of dead leaves that have recently fallen from the dominant trees, whereas the Oe

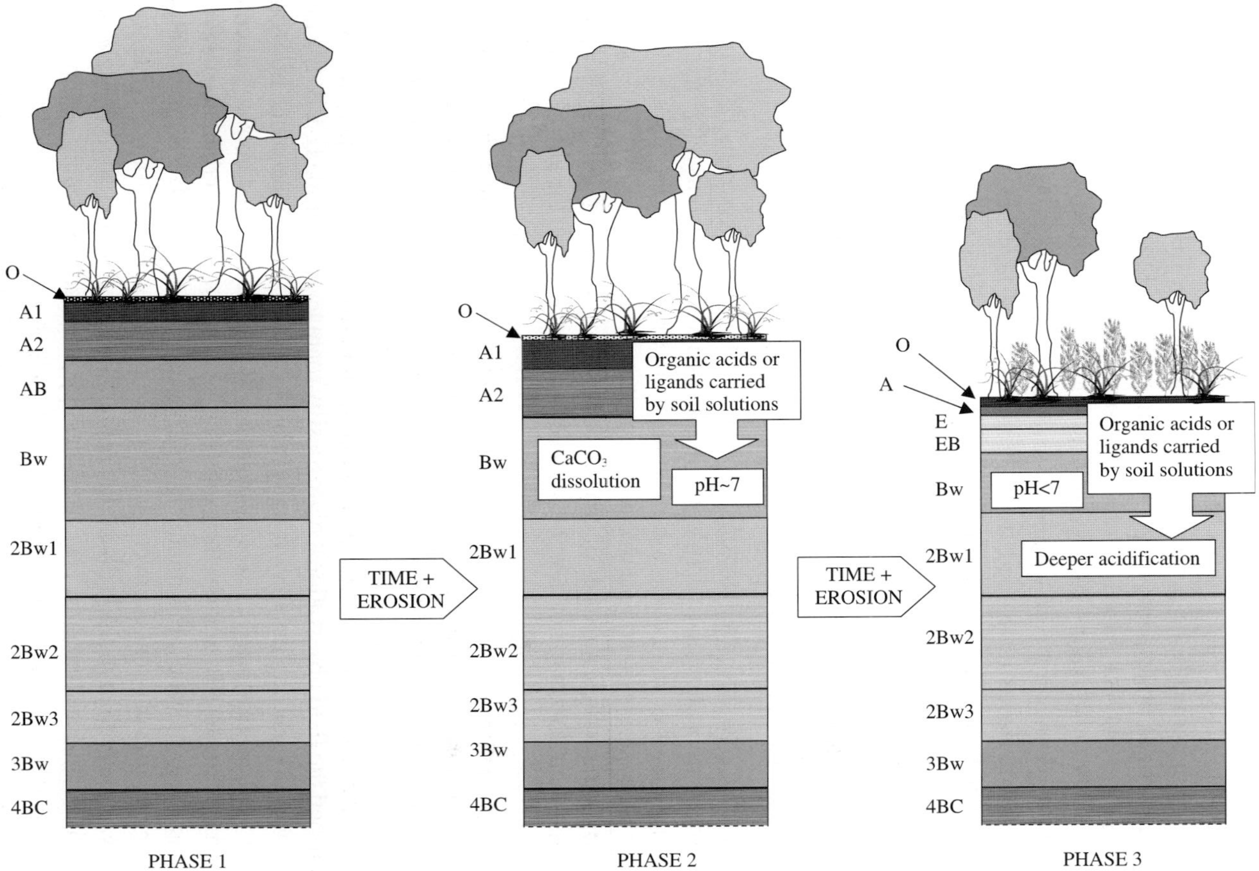

Fig. 5. Schematic representation of the genesis of the acid soil at the Selva di Gallignano (Ancona, Italy).

horizon is made up of finer organic detritus, and is not continuous owing to erosion. The same is true for the A horizon, which is present only in small flat areas (of a few square metres) where accumulation of materials coming down from the more elevated areas can also occur. A thin E horizon can be found either at the surface or underneath the A horizon, resting on a thin EB horizon. The successive Bw horizon presumably developed from the same lithological layer in which the E and EB also developed. Roots of *Erica* have colonized these three horizons, probably taking advantage of the creeping cracks, which were filled with a mixture of terrigenous material (A and E) invaded by mycelium. At a greater depth, three 2Bw horizons developed from a finer textured layer. In these three horizons, the peds were covered by mycelia, which were more abundant in the 2Bw1 horizon and tended to reduce toward the 2Bw3, which reached a depth of about 70 cm. The three 2Bw horizons were similarly colonized by *Erica* roots. At still greater depths, there was a 3Bw horizon that had been poorly colonized by roots of *Erica*; on the contrary, roots of other species were more represented. In this horizon, there were few concretions of carbonates and no trace of mycelia. The successive 4BC horizon derived from a layer with a texture similar to that from which the E, EB and Bw horizons developed. The 4BC horizon had a structure with traces of sedimentation (platy), and hosted very few roots of *Erica*, even though roots of other species were abundant; in this horizon, there were diffuse carbonate concretions and no traces of mycelia.

In both profiles, no traces of charcoal or other residues of fires were observed, indicating that fire may not have been the event that favoured the entry of *Erica*.

3.3.2. Physical, chemical and mineralogical characteristics of bulk and rhizosphere soil

Particle-size distribution without cement dissolution showed a low content of clay in the bulk, and an even lower content in the rhizosphere soil, where it was almost absent (Table 7). The silt also tended to be lower in the rhizosphere soil than in the bulk. Particle-size distribution after dissolution of organic cements confirmed the field observations, giving a silty loam texture for the E, EB, Bw and 4BC horizons and a silty clay to silty clay loam texture for the other horizons (Table 7). This supported the reconstructed soil genesis reported above, and confirmed that the E, EB, and Bw horizons derived from a single sedimentary layer. Among these horizons, the relatively high content of clay in Bw supported the findings of fine clay cutans over its peds. In the rhizosphere soil, the content of clay was lower than in the bulk in the E and EB horizons, and slightly higher in Bw, where an enrichment of clay from the upper horizons probably occurred via *lessivage,* using the roots as preferential channels. A similar situation occurred in the 2Bw horizons: in the 2Bw1 and 2Bw2, the rhizosphere soil contained less clay than the bulk, but in the 2Bw3 these differences disappeared. With respect to the particle-size distribution without cement dissolution, the clay content obtained after removal of organic cement increased by a factor of 7–22 in the bulk and by a factor of 46–270 in the rhizosphere soil; the 4BC horizon was excluded from these calculations as it was poor in *Erica* roots. Similarly, silt contents increased by 1.3–2.5-times in the bulk and 2.1–3.8 in the rhizosphere soil.

Values of pH_{H_2O} are similar in bulk and rhizosphere soil, even though it is generally expected that rhizosphere soil has a lower pH than that of the bulk (Hinsinger *et al.*, 2003). In both fractions, the pH was 5.1–5.3 in the E and EB horizon, and increased with depth, reaching values of around 7 in the 2Bw3 and 8.2–8.5 in the 3Bw and 4BC horizons. The same trend occurred for the values of pH_{KCl}, with negligible differences between bulk and rhizosphere soil, except in the 2Bw3 horizon, where the pH_{KCl} of rhizosphere soil was about 1 pH unit lower than that of the bulk. What also appeared remarkable was the difference between pH_{H_2O} and pH_{KCl}, which, considering bulk and rhizosphere soil, amounted to about 1.6–1.8 units of pH in the E and EB horizons, 1.8–2.0 in the Bw, 2Bw1 and 2Bw2, and 0.9–1.3 in the 3Bw and 4BC. In 2Bw3, the differences are 1.1 units of pH in the bulk and about 2.3 in the rhizosphere soil. All of these levels were interpreted as the result of a strong soil acidification until the 2Bw2 horizon, induced by the activity of the *Erica* roots. Indeed, the pH was similar in bulk and rhizosphere soil not only for pH_{H_2O}, but also for pH_{KCl}, and in our interpretation, this occurs because radial differences between rhizosphere and bulk soils were no longer present. From this point of view, different ΔpH values between E+EB horizons and Bw+2Bw1+2Bw2 were ascribed to the different natures of the layer from which they developed. This meant that the Bw horizon probably did not form from the same layer as E and EB, but rather from that of the 2Bw or from another with intermediate characteristics. Radial differences between rhizosphere and bulk soil still remained in the 2Bw3 horizon, where pH_{KCl} was lower near the roots. The 2Bw3 the horizon in which acidification had still not reached a maximum, indicating that the bulk soil maintained a buffer capacity that prevented the further acidification of the exchangeable complex. In the 3Bw and 4BC horizons, where there were few roots of *Erica*, the values of pH_{H_2O} showed the presence of carbonates in both bulk and rhizosphere soil, while pH_{KCl} indicated that protonation of the exchangeable complex had been initiated. The pH trend of the horizons underneath the Bw was able to explain the possible presence of mycelium, which decreased from the 2Bw1 to the 2Bw3 horizon and disappeared in the 3Bw and 4BC, where the pH was higher than 7.

The *X*-ray diffraction patterns of the fine earth from bulk and rhizosphere soil (Fig. 6) confirmed the presence of calcite (peaks at 0.303, 0.229 and 0.209 nm) in both fractions of the 3Bw and 4BC horizons. In bulk and rhizosphere soil, primary minerals were represented by quartz (peaks at 0.335, 0.425 and 0.181 nm), plagioclases (0.318, 0.320 and 0.326 nm), and micas (1.012 and 0.499 nm). Phyllosilicates comprised kaolinite (0.718 nm) and 2:1 clay minerals, with the interlayer occupied by hydroxy-Al polymers with varying degrees of polymerization (peak at 1.472 nm); among these minerals, the most represented was the hydroxy-Al interlayered vermiculite (HIV). In all horizons, diffractograms of the

Table 7
Particle-size distribution, values of pH in water and KCl, and differences (Δ) between pH values of the soil under *Erica arborea* from the Selva di Gallignano (Ancona, Italy). Numbers in parentheses are the standard errors

	Particle-size distribution (n=2) (g kg^{-1})										pH (n=8)		
	In water					After NaClO treatment					H_2O	KCl	ΔpH
	Coarse sand	Medium sand	Fine sand	Silt	Clay	Coarse sand	Medium sand	Fine sand	Silt	Clay			
Bulk													
E	415(21)	147(16)	153(22)	260(18)	25(3)	5(2)	4(2)	152(13)	659(23)	180(14)	5.20(0.03)	3.39(0.02)	1.81
EB	264(19)	165(17)	278(20)	284(20)	9(2)	2(0)	3(1)	150(12)	648(3)	197(14)	5.32(0.03)	3.62(0.03)	1.70
Bw	177(19)	174(17)	297(16)	324(23)	28(3)	3(1)	4(1)	126(10)	633(20)	234(12)	5.81(0.02)	3.77(0.02)	2.04
2Bw1	299(20)	175(13)	246(16)	258(13)	22(4)	0(-)	4(1)	126(11)	472(3)	398(15)	5.82(0.02)	3.85(0.03)	1.97
2Bw2	245(14)	197(11)	255(18)	276(12)	27(3)	4(0)	2(0)	130(11)	373(21)	491(10)	5.67(0.02)	3.87(0.02)	1.80
2Bw3	232(16)	171(11)	328(18)	237(14)	32(5)	2(0)	1(0)	113(11)	536(21)	348(10)	7.06(0.01)	5.94(0.02)	1.12
3Bw	209(10)	205(9)	373(15)	200(12)	13(2)	5(1)	6(0)	149(10)	556(22)	284(13)	8.32(0.02)	7.29(0.01)	1.03
4BC	93(13)	157(13)	287(17)	432(13)	31(4)	19(2)	14(1)	170(15)	590(9)	207(9)	8.48(0.02)	7.16(0.02)	1.32

Rhizosphere													
E	456(16)	146(13)	205(19)	192(17)	1(1)	84(5)	99(8)	170(15)	541(13)	106(15)	5.14(0.02)	3.33(0.02)	1.81
EB	356(22)	165(15)	274(20)	204(17)	1(0)	31(3)	16(3)	147(14)	665(7)	141(15)	5.20(0.01)	3.64(0.02)	1.56
Bw	168(17)	203(16)	395(17)	234(18)	0(-)	3(1)	3(1)	91(10)	635(9)	268(17)	5.83(0.02)	3.80(0.01)	2.03
2Bw1	269(16)	196(11)	283(20)	250(15)	2(0)	1(0)	4(0)	117(12)	535(6)	343(18)	5.72(0.01)	3.90(0.02)	1.82
2Bw2	255(6)	251(9)	327(14)	161(10)	6(1)	2(0)	4(0)	95(13)	619(26)	280(13)	5.59(0.02)	3.81(0.02)	1.78
2Bw3	271(12)	230(9)	335(18)	159(14)	5(1)	1(0)	2(0)	131(14)	524(26)	342(12)	7.25(0.02)	4.99(0.02)	2.26
3Bw	128(13)	162(8)	438(19)	266(15)	6(1)	19(2)	6(1)	143(10)	557(5)	275(12)	8.22(0.03)	7.29(0.02)	0.93
4BC	103(13)	126(10)	481(15)	283(11)	57(7)	14(1)	18(2)	176(15)	566(5)	226(12)	8.31(0.02)	7.16(0.01)	1.15

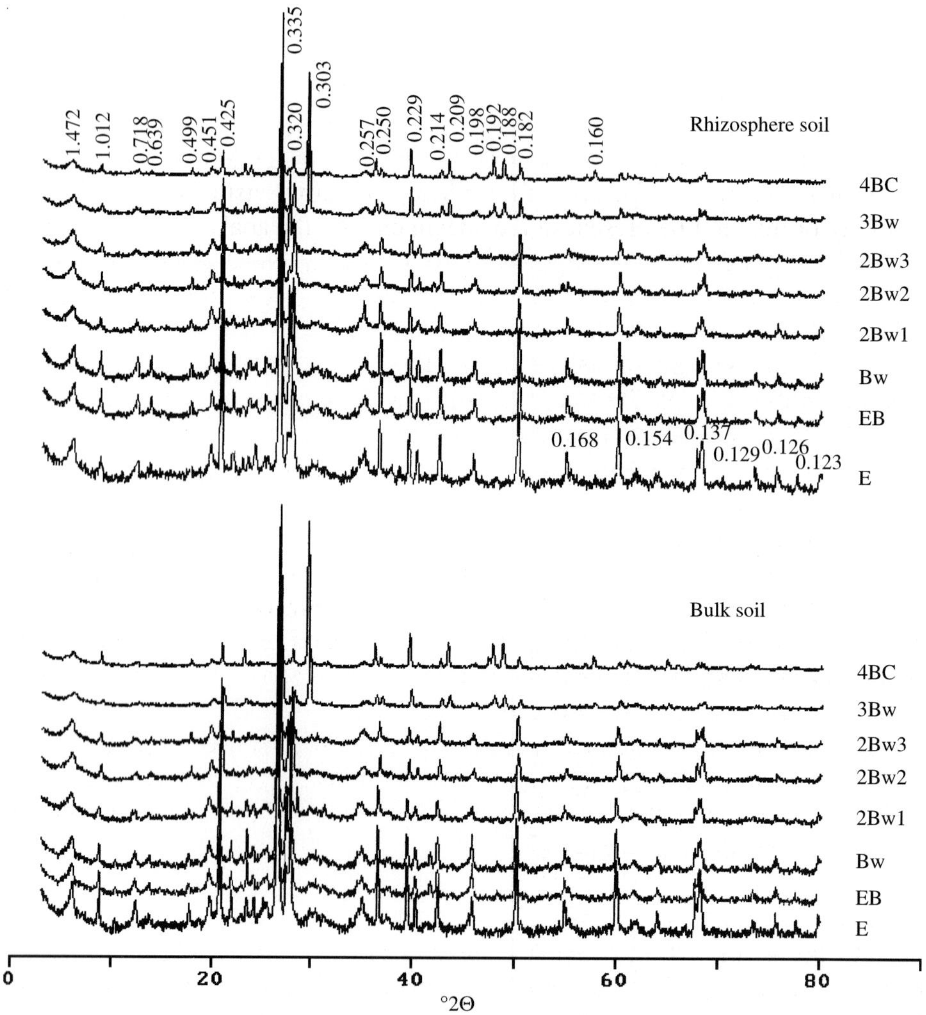

Fig. 6. X-ray diffraction patterns of the fine earth from bulk and rhizosphere soil of the Selva di Gallignano (Ancona, Italy). The d-values are in nanometer.

bulk were very similar to those of the rhizosphere soil. For both soil fractions, increasing depth was associated with a decrease in the first-order peak area of HIV, micas and kaolinite, independent of the nature of the sedimentary layer from which horizons developed. The higher presence of these minerals in the more acidified horizons was due to a high resistance to acid alteration in case of micas (Corti *et al.*, 1999) or to a greater stability in conditions of high Al^{3+} activity for HIV and kaolinite (Lindsay, 1979; Karathanasis and Hajek, 1983; Wada and Kakuto, 1983).

The content of total C tended to decrease with depth in both bulk and rhizosphere soil (Table 8). In the 3Bw and 4BC horizons, total C also included that forming carbonates. Values of the rhizosphere soil appeared to be higher than those of the bulk, although in taking into consideration the standard errors, these differences were not clear cut. The larger variability of the total C values of the rhizosphere soil was attributed to the presence of hardly visible organic detritus that may or may not be present in the specimens. As the analyzed samples were rootlet free, this detritus was represented by fungal hyphae. Values of organic C followed the same trend as total C (Table 8). In some cases, organic C assumed values slightly higher than those of total C, but this was attributed to the different analytical methods adopted and to the above-mentioned variabilities in organic detritus content. The total N (Table 8) was higher in rhizosphere soil than in the bulk of the E and EB horizons, while in the other horizons the two fractions showed similar contents. The C/N ratios, independent of total or organic C, showed a relatively high content of N throughout the soil; we do not have an explanation for this N enrichment at this stage.

The amounts of available P are very low in all of the samples analyzed, except in the rhizosphere soil of the 3Bw horizon where, on the other hand, there were few roots of *Erica*. These values are so low that one doubts that plants could survive in this soil, but evidently, *Erica* and the other plants have developed a strategy to take up P (Hinsinger, 2001).

In all of the acid horizons, exchangeable Al ions prevailed over H, with no remarkable differences between bulk and rhizosphere soil. As expected from their pH value, in the 3Bw and 4BC horizons, neither Al nor H ions were present on the exchangeable complex.

3.4. Conclusions

In the Selva di Gallignano, the genesis of superficial acid horizons from alkaline marine deposits was caused by a combined action of erosion and acidification processes. Indeed, lenses with a minor content of carbonates were progressively decarbonated and acidified during their approach to the surface because of erosion. Once at the surface, these acid horizons favoured the establishment of plants of *Erica arborea*, presumably in a period in which the wood maintained conditions as an open area. Since then, by means of their roots, *Erica* plants have started to colonize the deeper soils through prior acidification. The process of acidification possibly progressed, and continues to progress, through root excretion of organic acids (Grinsted *et al*., 1982; Youssef and Chino, 1988b; Olsthoorn *et al*., 1991; Dieffenback and Matzner, 2000; Wang *et al*., 2001a; Hinsinger *et al*., 2003). Indeed, acidification has mostly impacted on the Bw, 2Bw1 and 2Bw2 horizons, where differences between bulk and rhizosphere soil disappeared. Differences between bulk and rhizosphere soil only remained in the 2Bw3 horizon, where carbonates had already been dissolved in both fractions,

Table 8
Contents of total C, organic C, total N, available P, and exchangeable Al and H in the soil under *Erica arborea* from the Selva di Gallignano (Italy). Numbers in parentheses are the standard errors ($n = 3$)

	Total C	Organic C	Total N	Total C/N	Organic C/N	Available P	Exchangeable acidity	
							Al	H[a]
Bulk								
E	25.1(4.7)	23.2(1.0)	3.0(0.6)	8.4	7.7	1.9(0.3)	4.9(0.2)	1.2(0.3)
EB	12.8(3.3)	11.9(0.2)	2.1(0.3)	6.1	5.7	0.6(0.2)	4.6(0.2)	1.4(0.3)
Bw	6.7(1.5)	5.5(0.2)	2.5(0.6)	2.7	2.2	0.6(0.2)	3.6(0.1)	0.8(0.2)
2Bw1	5.9(1.6)	3.5(0.2)	2.5(0.5)	2.4	1.4	0.3(0.2)	3.1(0.1)	1.0(0.2)
2Bw2	2.1(0.5)	2.2(0.0)	1.3(0.4)	1.6	1.7	0.6(0.2)	1.8(0.1)	1.8(0.1)
2Bw3	2.6(0.6)	2.6(0.1)	1.2(0.5)	2.2	2.2	1.6(0.2)	0.1(0.0)	0.0(0.0)
3Bw	3.1(0.4)	2.9(0.2)	1.2(0.4)	2.6	2.4	1.1(0.2)	0.0(—)	0.0(—)
4BC	2.0(0.4)	1.8(0.3)	1.0(0.3)	2.0	1.8	1.9(0.2)	0.0(—)	0.0(—)

Rhizosphere								
E	35.1(12.6)	25.3(4.1)	5.9(0.1)	5.9	4.3	1.6(0.2)	5.2(0.2)	1.1(0.3)
EB	13.6(6.3)	13.3(1.4)	3.6(0.6)	3.8	3.7	0.3(0.1)	4.7(0.2)	1.2(0.3)
Bw	6.6(2.3)	6.1(0.4)	1.7(1.3)	3.9	3.6	0.3(0.1)	3.6(0.1)	
2Bw1	3.7(1.2)	3.9(0.1)	1.2(0.8)	3.1	3.3	0.6(0.1)	3.6(0.1)	0.5(0.2)
2Bw2	2.7(1.2)	2.5(0.3)	1.3(0.9)	2.1	1.9	0.7(0.2)	2.2(0.1)	0.6(0.1)
2Bw3	3.2(0.3)	3.1(0.2)	1.5(0.4)	2.1	2.1	0.6(0.1)	0.2(0.0)	0.2(0.0)
3Bw	4.1(0.9)	3.5(0.1)	0.3(0.3)	13.7	11.7	17.8(1.9)	0.0(—)	0.0(—)
4BC	2.3(0.4)	2.1(0.5)	0.3(0.1)	7.7	7.0	nd	0.0(—)	0.0(—)

[a]standard errors obtained by error propagation technique (Skoog and West, 1987).

nd = not determined.

Note: Total C, Organic C, Total N, Total C/N and Organic C/N are expressed as g kg^{-1}; Available P is expressed as mg kg^{-1}; and exchangeable acidity is expressed as cmol (+) kg^{-1}.

but in the rhizosphere soil, acidification of the exchangeable complex progressed more than in the bulk. In the 3Bw and 4BC horizons, carbonates still persisted in both bulk and rhizosphere soil, evidently as a result of the as yet scarce presence of *Erica* roots. In colonizing and converting soil to acid conditions, the *Erica* plants were probably helped by fungi, the mycelia of which were observed until the 2Bw2 horizons.

With respect to the survival of *Erica arborea*, acidification of a soil mass (Bw, 2Bw1 and 2Bw2 horizons) containing clay from about 25 to 50% and silt from about 35 to 60%, to a pH_{KCl} of around 3.8, appeared to have required more than the 20–40 years of age that the plants actually showed. Consequently, we hypothesized that the *Erica* plants became established in this soil many years ago (probably centuries), and that they had passed through periods of abundance and of decline, during which, even at different intensities, they had continuously acidified the soil.

4. ROLE OF THE ROOTS OF *VITIS VINIFERA* L. FOR THE MOBILIZING OF SELECTED RARE EARTH ELEMENTS IN SOIL

4.1. Introduction

Knowledge of the chemistry of the rare earth elements (REEs) in soil is limited and the behaviour of these elements are affected by their low solubilities (Weltje, 1997). However, in the soil aqueous phase, the REEs can form complexes with a number of inorganic compounds (fluorides, carbonates, phosphates and hydroxides) and organic ligands (humic and fulvic acids) or, at high pH, with hydroxyl ions. Rainfall and content of organic matter and clay can also influence the distribution of REEs in the profile (Pang *et al.*, 2002). The mineral fertilizers, and in particular the phosphoric ones (Todorovsky *et al.*, 1997), represent an additional source of REEs for cultivated soils (Wang *et al.*, 2001b), as well as soils of geological origin coming from the parent rock. Wang *et al.* (2001b) found that in the rhizosphere soil of wheat grown in a rhizobox, REEs have a higher bioavailability for plants than in the nonrhizosphere soil because of exudation of low molecular weight acids, active uptake, etc.

Nowadays, vines (*Vitis vinifera* L.) are cultivated all over the world. This reflects the high adaptability of these vines to a wide range of pedoclimatic conditions. Vines should have an efficient and "flexible" root system; however, the pedological, chemical and biochemical characteristics of rhizosphere soil of this species are still unknown. The main problems in studying the rhizosphere soil of the vine in the field are due to the difficulties of recognizing and separating it from the bulk, and to the wide and deep soil volume colonized by the vine's root. The study of the distribution of REEs in the soil under vines represents the first step towards an assessment of the hypothesis that REEs, in addition to Sr (Horn *et al.*, 1993), could provide a sort of "fingerprint" of the final product, thus linking the wine to its production area. Marisa *et al.* (2003) found that the trace element composition of soil is correlated with the elemental composition of the wine, even though they analyzed few wines and collected soil only within the first 20 cm. Taylor *et al.* (2003) found a soil–wine linkage only for alkaline earth elements, in particular for Sr, and stated that to use trace elements as a fingerprint of the origin of wine it is necessary: (1) to sample the soil to a greater depth than the top 50 cm; and (2) to subdivide elements into chemical fractions with relatively higher or lower bioavailabilities.

The aim of this work was to investigate the distribution of selected REEs in different chemical fractions from bulk and rhizosphere soil of adult vines from two vineyards (one in Tuscany, Italy, and the other in Galicia, Spain) that differed in their environmental conditions.

4.2. Materials and methods

The soils considered for this study came from profiles opened under vines at two different sites (Fig. 7):

(1) Donnini (Florence, Tuscany, Central Italy) is located at 337 m a.s.l., on a slope of 3–4%. The mean annual air temperature was 12.7°C and the mean annual precipitation was 1289 mm. The soil, developed on sandstone, was classified as Dystric Eutrudept, loamy-skeletal, mixed and mesic (Soil Survey Staff, 1999). At the time of sampling, the vines were 29 years old.
(2) Cambados (Pontevedra, Galicia, North-Western Spain) is located at 65 m a.s.l., on a slope of 3%. The mean annual air temperature was 14.6°C and the mean annual precipitation was about 1600 mm. The soil, developed on granite, was classified as Typic Dystrudept, loamy, mixed and mesic (Soil Survey Staff, 1999). At the time of sampling, the vineyard was 18 years old.

For each site, two profiles were opened close to vine plants of *sangiovese* at Donnini and *albariño* at Cambados. Profiles were described according to Soil Survey Division Staff (1993) and sampled by horizons to a depth of 1.5 m at Donnini and 1 m at Cambados. For each horizon, the roots were picked up from the face of the profiles together with the adhering soil and gently shaken. The detached particles were joined to the bulk, while the earthy material adhering to the roots was considered rhizosphere soil and separated from the roots in the laboratory by strong shaking and a toothbrush. The rhizosphere soil from the A1 horizon of Cambados was not collected because of the large presence of herbaceous roots, which prevented isolating the vine's rhizosphere soil from the soil of the grasses.

The bulk and rhizosphere soil samples were air-dried and then sieved at 2 mm. The pH was determined potentiometrically in water and 1 M KCl solution, with a solid:liquid ratio of 1:2.5. Organic carbon content was estimated using the Walkley–Black method without heating (Allison, 1965), whereas total N contents

Fig. 7. Map showing the location of the investigated sites, Donnini (Firenze, Italy) and Cambados (Pontevedra, Spain).

were determined using a Carlo Erba NA 1500 analyzer. Available P was determined according to the Mehelich 3 procedure (Mehelich, 1984).

To determine the contents of REEs (La, Ce, Pr, Nd, Sm, Eu) associated with different chemical compartments of the soil, both bulk and rhizosphere soil were subjected to "selective" sequential dissolution according to the following procedure, modified from Berna *et al.* (2000):

(1) Soluble and exchangeable elements: 1 g of soil was placed in a polyethylene centrifuge tube with 20 mL of 1M $MgCl_2$ and shaken for 30 min at room temperature. After the shaking, the mixture was centrifuged and the supernatant filtered, collected and stored at 2°C.
(2) Elements associated with easily reducible Fe-Mn-oxyhydroxides: 50 mL of 0.1 M $NH_2OH{\bullet}HCl$ (adjusted to around pH 2 with 0.5 M HCl solution) was added to the residue of the previous treatment, and this was shaken for 30 min at room temperature. The mixture was centrifuged and the supernatant filtered, collected and stored at 2°C.
(3) Elements bound to organic matter: 40 mL of 0.1 M $Na_4P_2O_7$ solution was added to the residue of the previous treatment, and this was shaken for 24 h at room temperature. The mixture was centrifuged and the supernatant filtered, collected and stored at 2°C.
(4) Elements associated with amorphous Fe-oxyhydroxides: 50 mL of 0.25 M $NH_2OH{\bullet}HCl$ + 0.25 M HCl solution was added to the residue of the previous treatment, and this was shaken for 30 min at 50°C. After centrifuging, the supernatant was filtered, collected and stored at 2°C.

(5) Elements associated with crystalline Fe-oxyhydroxides: 50 mL of 0.2 M $(NH_4)C_2O_4$ + 0.2 M $H_2C_2O_4$ + 0.1 M $C_6H_8O_6$ solution (adjusted to pH 3 with 3 M NH_4OH solution) was added to the residue of the previous treatment, and this was shaken for 30 min at 95°C in a hot bath. The mixture was centrifuged and the supernatant filtered, collected and stored at 2°C.
(6) Elements constituting the residue: 5 mL of 5 M HF:1 M HCl + 0.5 M HNO_3 solution was added to the residue of the previous treatment, and this was heated to 95°C in a hot bath to dry. This treatment was repeated till the disappearance of any trace of residue. The residue was then recovered with 40 mL of 0.5 M HNO_3 solution and left overnight. The mixture was centrifuged and the supernatant filtered, collected and stored at 2°C.

The supernatant obtained after centrifugation was filtrated with Whatman 42 filters that had been washed previously in dilute HNO_3. After each step of the sequential extraction, the samples were washed with distilled water (three times) and acetone, and dried at 40°C. The extracts were analyzed for La, Ce, Pr, Nd, Sm and Eu using an ICP-AES (Perkin Elmer Optima 3200 XL).

From each horizon, two sub-samples were submitted to the sequential dissolution, each extract was analyzed in duplicate, and the values reported are the means of the four replicates. The standard errors were calculated, and the discussion took into consideration intervals and overlaps of intervals.

4.3. Results and discussion

4.3.1. General characteristics of bulk and rhizosphere soil

At Donnini, the soil had developed on an Oligocene sandstone. The soil was broken up at the time of plantation (29 years before the study) to a depth of 80–90 cm. Now, the soil comprises two Ap horizons, about 25 cm thick, produced by the annual mechanical working of the soil. These horizons had a sandy-clay loam texture, an angular to sub-angular (Ap1) or angular (Ap2) blocky structure and few to plentiful roots, mostly of grasses; roots of vine were abundant only near the stem. Underneath, pedogenetic processes formed a thin Bw horizon with a finer texture and an angular blocky structure, and a thick Bh horizon, the primary structure of which tended to break into crumbs; the development of a secondary crumb structure was ascribed to the effect of a relatively high content of organic matter and roots that were mainly of vines. The deeper BC horizon had a texture and structure similar to those of the Bh; the tendency of the blocks to break into crumbs was here attributed to the abundance of vine roots. In this horizon, peds were covered by clay cutans, indicating that a translocation of colloids from the upper horizons had occurred. The soil management included three mechanical workings, one to about 25 cm, while the others were around 10 cm in depth. Fertilizers consisted of 2–3 distributions of sewage sludge per year (7–8 Mg ha^{-1})

and a single distribution of 100 kg ha^{-1} of a triple fertilizer (equivalent to about 20 kg ha^{-1} of N, about 4.5 kg ha^{-1} of P and 8.5 kg ha^{-1} of K).

At Cambados, the soil had developed from granite. The soil was broken up to a depth of 70 cm 2–3 years before the plantation of the vineyard, 18 years ago. During the pre-plantation period, 3 Mg ha^{-1} per year of limestone was distributed to increase soil pH. Since plantation, 1.5 Mg ha^{-1} of limestone was placed locally around the vine stems. At the soil surface, there was a thin Oi horizon that had developed because of the absence of mechanical working of the soil, grass covering and periodical additions of organic materials, such as sawdust and vine shoots. In all the horizons, crumb structure prevailed. The A horizons had a sandy loam texture and were mostly occupied by abundant roots of grasses (in the A1 horizon), or vines (in the A2 and A3 horizons). A transitional AB horizon with a loamy sand texture and few vine roots was positioned on two Bw horizons and a BC horizon. In addition to liming, the fertilization of this soil included a single distribution per year of 800 kg ha^{-1} of a triple fertilizer (equivalent to about 70 kg ha^{-1} of N, about 60 kg ha^{-1} of P and 180 kg ha^{-1} of K).

4.3.2. Chemical characteristics of bulk and rhizosphere soil (Table 9)

In the soil from Donnini, values of pH_{H_2O} for the bulk soil increased from 6.0 in the Ap1 horizon to 6.7 in the BC horizon, while for the rhizosphere soil the values were around 7 at surface and increased slightly downwards. The pH_{KCl} showed a similar trend in both bulk and rhizosphere soils, with higher values in the latter. This higher pH of the rhizosphere soil could be attributed to the excretion of hydroxyl ions to balance the uptake of anions (Grinsted *et al.*, 1982; Marschner and Romheld, 1983; Haynes, 1990). This could have occurred because of the NO_3^- produced by nitrification after the distribution of the sewage sludge. The lower content of N in the rhizosphere soil with respect to the bulk is consistent with this hypothesis. The depletion of N in the rhizosphere soil could also be ascribed to the possibility that, at some time during the year, vines take up more of this element. The organic C contents, both in bulk and rhizosphere soil, decreased from the Ap1 to the Bw horizon, increased in the Bh and decreased in the BC. The relatively high content of organic C in the Bh horizon was ascribed to soluble and colloidal organics translocated by soil solution (Cuniglio *et al.*, 2004). The amount of available P, higher in the Ap1 horizon, showed similar values in bulk and rhizosphere soil. The Fe and Mn that formed the easily reducible Fe–Mn-oxyhydroxides did not follow any specific depth trend, with no differences between bulk and rhizosphere soil. In all the samples, the Mn of the easily reducible Fe–Mn-oxyhydroxides was higher than the Fe. The Fe that constituted the amorphous Fe-oxyhydroxides showed similar values for bulk and rhizosphere soil, while the Fe of the crystalline Fe-oxyhydroxides was higher in the bulk throughout the profile, even though with no specific depth trend. In all of the horizons and in both soil compartments, the amount of crystalline oxyhydroxides was larger than the amorphous and easily reducible ones. The lower content of crystalline Fe-oxyhydroxides in the rhizosphere as compared with the bulk could be attributed to the high biological activity occurring in the soil in contact with the roots, where plant-microbe-soil interactions take place.

In the soil from Cambados, values of pH_{H_2O} of the bulk decreased from 6.8 at the surface to 4.6–4.7 at depth. For the rhizosphere soil, the values were around 5 in the entire profile. The effect of liming was evident from the pH_{KCl} of the A1 and A2 horizons, of which values of the bulk were similar or higher than those of pH_{H_2O}. In all of the other cases, the opposite was true. Total N content was larger in the bulk than in the rhizosphere soil, and decreased with increasing depth. The organic C contents were higher in the bulk than in the rhizosphere soil, and followed the same trend as N. The available P showed very high values in the A1, probably due to fertilizers; relatively high amounts of available P were in the A2 horizon, either in the bulk or in the rhizosphere soil, while in the rest of the profile the available P content was scarce. The Fe and Mn that form the easily reducible Fe–Mn-oxyhydroxides did not follow a depth trend either in the bulk or in the rhizosphere soil, although the lowest values were at depth; except in the A1 horizon, the Fe content was higher than Mn. Also, for the Fe of the amorphous Fe-oxyhydroxides, no depth trend or differences between bulk and rhizosphere soil were observed. The Fe of the crystalline Fe-oxyhydroxides was more abundant in the bulk than in the rhizosphere soil.

In all of the horizons and in the bulk and rhizosphere soil, the amount of oxyhydroxides followed the trend crystalline $>$ amorphous $>$ easily reducible.

4.3.3. Total content and distribution in the chemical fraction of REEs of bulk and rhizosphere soil

Lanthanum (Table 10)
The total contents of La were similar in the bulk and in the rhizosphere soil throughout both of the profiles. At Donnini, the total amounts of La in the bulk and rhizosphere soil tended to increase with depth, with a reverse depth trend in the Bh horizon. The values obtained for the bulk were in agreement with those found in the literature, which ranged from 1.4 to 55 mg kg^{-1} (Li *et al.*, 1998; Markert, 1987a; Land *et al.*, 1999; Kabata-Pendias and Pendias, 2001; Wang *et al.*, 2001b; Zhang *et al.*, 2001). In the rhizosphere soil, La showed values higher than those reported by Wang *et al.* (2001b). At Cambados, the total content of La in the bulk tended to decrease with depth to the Bw1 horizon and to increase in the horizons below. The La values of all of the horizons agreed with those reported in the above-mentioned literature. In the rhizosphere soil, La assumed values lower than those reported by Wang *et al.* (2001b).

In the bulk and in the rhizosphere soil of Donnini, La was almost evenly distributed in all of the fractions, except in the residue, where it was very scarce;

Table 9
Values of pH, contents of total N, organic C, available P, Fe and Mn as easily reducible oxyhydroxides, and Fe as amorphous and crystalline oxyhydroxides in bulk and rhizosphere soil from the vineyards of Donnini (Firenze, Italy) and Cambados (Pontevedra, Spain). Numbers in parentheses are the standard errors

	pH		Total N	Organic C	Available P	Easily reducible Fe–Mn-oxyhydroxides		Amorphous Fe-oxyhydroxides	Crystalline Fe-oxyhydroxides
	H_2O	KCl				Fe	Mn	Fe	Fe
Donnini									
Bulk									
Ap1	6.0(0.1)	5.6(0.1)	1.0(0.0)	8.9(0.5)	4.2(0.1)	61.6(7.1)	242.0(11.3)	333.3 (14.1)	11363(820)
Ap2	6.2(0.1)	5.6(0.1)	0.9(0.0)	7.5(1.2)	1.8(0.0)	65.4(7.0)	306.8(13.2)	385.8 (20.2)	12940(876)
Bw	6.4(0.1)	5.5(0.1)	0.5(0.0)	2.7(0.6)	1.2(0.0)	33.5(5.2)	183.1(11.0)	412.9 (13.0)	13844(867)
Bh	6.5(0.1)	5.6(0.1)	0.7(0.0)	6.2(0.1)	1.1(0.0)	52.1(6.2)	280.7(12.1)	372.4 (16.1)	12038(851)
BC	6.7(0.1)	6.3(0.1)	0.4(0.0)	0.6(0.0)	1.2(0.0)	32.8(6.3)	203.8(14.1)	451.0 (16.2)	12309(838)
Rhizosphere									
Ap1	7.0(0.1)	6.4(0.1)	0.3(0.1)	8.2(0.8)	2.8(0.3)	71.3(9.2)	246.8(13.1)	326.8(16.2)	9518(599)
Ap2	7.2(0.1)	6.3(0.2)	0.2(0.1)	6.3(1.3)	1.7(0.0)	52.8(7.1)	224.4(14.0)	343.8(20.2)	10126(666)
Bw	7.2(0.1)	6.3(0.1)	0.0(0.0)	3.6(0.4)	1.3(0.0)	77.2(8.2)	342.5(14.1)	458.2(16.0)	12036(747)
Bh	7.2(0.1)	6.5(0.1)	0.1(0.1)	5.7(1.3)	1.1(0.0)	45.8(6.0)	229.9(15.1)	378.6(12.1)	10871(629)
BC	7.3(0.2)	6.6(0.1)	0.0(0.0)	1.4(0.4)	1.1(0.0)	46.6(6.1)	167.8(12.2)	474.3(14.2)	11871(631)

Cambados									
Bulk									
A1	6.8(0.1)	6.8(0.1)	4.4(0.3)	67.3(4.1)	62.8(0.5)	23.2(4.1)	49.6(3.2)	222.4(12.1)	6694(501)
A2	5.9(0.1)	6.9(0.1)	3.0(0.3)	40.2(1.9)	13.3(2.0)	20.9(5.2)	8.9(2.1)	139.8(15.0)	6341(513)
A3	5.0(0.1)	4.3(0.0)	2.7(0.3)	33.8(5.3)	1.0(0.0)	44.0(6.0)	1.4(0.0)	106.3(11.2)	7397(521)
AB	4.7(0.1)	4.3(0.0)	1.1(0.2)	11.2(0.7)	1.1(0.0)	18.5(4.2)	0.8(1.0)	149.7(12.1)	8474(540)
Bw1	4.7(0.1)	4.2(0.0)	0.5(0.0)	4.5(0.2)	1.3(0.0)	12.5(2.1)	4.2(1.1)	174.5(10.3)	7842(435)
Bw2	4.6(0.2)	4.1(0.1)	0.4(0.1)	2.8(0.0)	1.1(0.0)	16.5(2.0)	0.9(0.0)	246.5(13.2)	9614(531)
BC	4.7(0.1)	4.0(0.0)	0.3(0.0)	1.8(0.2)	1.0(0.0)	13.6(3.1)	0.3(0.1)	188.0(9.1)	10293(465)
Rhizosphere									
A2	5.1(0.1)	4.6(0.1)	1.8(0.1)	31.9(2.6)	19.4(0.3)	31.2(4.0)	13.1(2.1)	172.9(14.0)	5381(416)
A3	5.1(0.1)	4.2(0.0)	1.5(0.1)	28.5(1.9)	1.0(0.0)	50.8(6.1)	2.1(1.0)	114.6(11.2)	5054(462)
AB	5.1(0.1)	4.2(0.0)	0.6(0.0)	10.3(1.7)	1.0(0.0)	18.3(3.2)	2.1(1.1)	132.1(11.1)	5124(373)
Bw1	5.1(0.1)	4.2(0.1)	0.4(0.1)	5.4(0.6)	1.2(0.0)	15.7(3.0)	4.0(1.2)	176.0(20.0)	5736(397)
Bw2	5.0(0.1)	4.1(0.1)	0.2(0.0)	3.9(1.0)	1.2(0.0)	17.0(2.1)	2.5(1.0)	201.2(19.2)	7008(323)
BC	5.0(0.0)	4.0(0.0)	0.2(0.0)	2.8(1.1)	1.0(0.0)	14.7(3.2)	5.4(2.1)	195.0(20.2)	8350(454)

Note: Total N and Organic C are expressed as g kg^{-1}; Easily reducible Fe–Mn–Oxyhydroxides (Fe, Mn), Amorphous Fe-oxyhydroxides-Fe and crystalline Fe-oxyhydroxides Fe are expressed as mg kg^{-1}.

Table 10
Content of lanthanum in bulk and rhizosphere soil from the vineyards of Donnini (Firenze, Italy) and Cambados (Pontevedra, Spain). Numbers in parentheses are the standard errors ($n = 2$)

	Soluble and exchangeable	Easily reducible Fe–Mn-oxyhydroxides	Organic matter	Amorphous Fe-oxyhydroxides	Crystalline Fe-oxyhydroxides	Residue	Total
Donnini							
Bulk							
Ap1	6.5(1.9)	2.4(0.2)	6.0(0.3)	5.5(0.3)	5.8(0.7)	1.1(0.1)	27.3(2.7)
Ap2	7.2(1.7)	5.1(0.5)	7.1(0.3)	6.1(0.3)	8.8(0.9)	1.2(0.1)	35.5(2.6)
Bw	7.5(1.7)	4.4(0.5)	6.7(0.3)	7.2(0.3)	11.5(1.3)	1.1(0.1)	38.4(2.6)
Bh	7.1(1.5)	4.9(0.5)	7.0(0.3)	4.7(0.2)	6.4(0.7)	1.2(0.1)	31.3(1.3)
BC	7.0(1.5)	6.5(0.5)	7.6(0.3)	10.5(0.6)	11.8(1.0)	1.2(0.0)	44.6(0.9)
Rhizosphere							
Ap1	7.0(1.9)	6.3(0.6)	8.3(0.9)	5.6(0.5)	5.0(0.6)	1.2(0.1)	33.4(3.4)
Ap2	8.2(1.4)	4.9(0.5)	8.5(0.8)	6.5(0.5)	7.0(0.7)	1.0(0.2)	36.1(2.5)
Bw	8.6(1.5)	6.7(0.6)	8.9(0.8)	6.1(0.5)	10.7(0.9)	1.3(0.1)	42.3(2.0)
Bh	8.1(1.0)	5.3(0.4)	8.9(0.6)	5.2(0.4)	6.3(0.7)	1.1(0.1)	34.9(3.2)
BC	8.0(1.5)	5.9(0.5)	8.5(0.3)	10.1(0.7)	13.3(1.0)	1.0(0.1)	46.8(1.9)

Cambados							
Bulk							
A1	7.2(1.6)	10.6(0.9)	7.9(0.8)	2.2(0.3)	1.9(0.3)	1.3(0.1)	31.1(2.8)
A2	3.8(1.0)	3.7(0.5)	6.1(0.7)	1.1(0.1)	1.4(0.2)	1.3(0.1)	17.4(1.2)
A3	0.7(0.2)	1.9(0.3)	5.5(0.5)	0.9(0.2)	1.4(0.2)	1.2(0.1)	11.6(0.5)
AB	0.5(0.3)	1.7(0.2)	3.0(0.4)	0.8(0.1)	1.6(0.2)	1.2(0.1)	8.8(0.7)
Bw1	0.5(0.1)	1.4(0.1)	2.0(0.2)	0.3(0.1)	1.5(0.1)	1.1(0.1)	6.8(0.5)
Bw2	1.0(0.2)	1.4(0.2)	1.9(0.2)	0.6(0.1)	1.1(0.1)	1.3(0.0)	7.3(0.4)
BC	3.6(0.6)	1.7(0.1)	2.2(0.2)	0.4(0.1)	1.2(0.1)	2.2(0.1)	11.3(0.6)
Rhizosphere							
A2	5.2(1.3)	4.0(0.6)	5.6(0.8)	1.3(0.2)	1.0(0.2)	1.1(0.1)	13.4(2.2)
A3	1.0(0.3)	1.8(0.3)	6.4(0.6)	1.0(0.2)	0.9(0.1)	1.1(0.1)	11.2(1.0)
AB	0.6(0.3)	1.5(0.2)	2.8(0.3)	0.7(0.1)	1.0(0.1)	1.1(0.1)	7.7(0.9)
Bw1	0.6(0.3)	1.4(0.1)	2.5(0.2)	0.6(0.1)	1.2(0.1)	1.1(0.1)	7.4(0.9)
Bw2	1.1(0.4)	1.4(0.1)	2.3(0.2)	0.6(0.1)	1.0(0.1)	1.1(0.1)	7.5(0.8)
BC	3.3(0.3)	1.8(0.1)	2.5(0.2)	0.8(0.1)	1.0(0.1)	1.1(0.1)	10.5(0.7)

Note: All values are expressed as mg kg^{-1}.

such a distribution was not in agreement with the amounts of crystalline Fe-oxyhydroxides (Table 10), which were much more abundant than the others. In the soil of Cambados, La was mainly associated with organic matter and easily reducible Fe–Mn-oxyhydroxides, even though these latter were less abundant than the Fe-oxyhydroxides (Table 9).

Results from Donnini did not agree with those obtained by Li *et al.* (1998), Land *et al.* (1999), Xinde *et al.* (2000) and Wang *et al.* (2001b), who found that both in bulk and in the rhizosphere soil, La was more associated with organic matter and, secondarily, with easily reducible Fe–Mn-oxyhydroxides. However, on soils with different degrees of acidity, Zhang and Shan (2001) found that in the more acid soil the addition of a lanthanides mixture produced a fractionation of La similar to that we obtained for the bulk of Donnini. In contrast, the La fractionation in the rhizosphere soil from Cambados was consistent with the results of Wang *et al.* (2001b).

At Donnini, for bulk and rhizosphere soil, there were two results of note: (1) La associated with the residue fraction was constant throughout the profile and represented a minor portion of the total; and (2) total La tended to increase with depth. The first can be explained by the history of the soil. The low residue/total ratio can be explained as this soil was cultivated for centuries and, in time, it has probably been homogenized in terms of mineral constituents. At the same time, La has possibly been added to the soil by the use of fertilizers; indeed, for example, cow manure has been recognized as containing considerable levels of lanthanides (Furr *et al.*, 1976). The added La has evidently fed all the chemical fractions except the residue. The increase with depth of the total amount of La was probably the result of translocations of this element, which was progressively adsorbed by the amorphous and crystalline Fe-oxyhydroxides. The reverse depth trend observed in the Bh horizon can be ascribed to its relatively high content of organic C (Table 9) and high concentration of roots, which, in combination, can produce a high complexation of Fe^{3+} (Inskeep and Comfort, 1986), so preventing the formation of Fe-oxyhydroxides and, hence, adsorption of La.

For the soil of Cambados, the use of considerable amounts of fertilizers and the absence of mechanical working of the soil were probably responsible for the elevated quantities of La in the A horizons, mostly bound to organic matter and easily reducible Fe–Mn-oxyhydroxides.

Cerium (Table 11)

In both soils, the total contents of Ce were slightly higher in the bulk than in the rhizosphere soil, in contrast to Wang *et al.* (2001b), who found the contrary in wheat. However, these values were in the ranges usually reported in the literature (Markert, 1987a; Li *et al.*, 1998; Wyttembach *et al.*, 1998; Land *et al.*, 1999; Wang *et al.*, 2001b; Zhang *et al.*, 2001). Nevertheless, Barnard and Halbig (1985) reported a mean content of Ce in Hawaiian soils of 146 mg kg^{-1}.

In the bulk of Donnini, Ce was abundant in the crystalline Fe-oxyhydroxides, residue and organic matter, while in the rhizosphere soil it was mostly associated with crystalline Fe-oxyhydroxides, organic matter and residue. At Cambados, Ce followed the same distribution order in both bulk and rhizosphere soil: crystalline Fe-oxyhydroxides > residue > organic fraction. Li *et al.* (1998) and Wang *et al.* (2001b) reported a different trend, where the fraction richest in Ce was the organic matter, followed by easily reducible Fe–Mn-oxyhydroxides and soluble + exchangeable + carbonate fraction.

The contribution of the residue to the total amount of Ce was higher than that found for La, and this was ascribed to a higher content of Ce than La in the parent materials, sandstone and granite (Kabata-Pendias and Pendias, 2001). However, it was also evident that, in time, the external addition of Ce had been lower than that of La; such an explanation takes into account the fact that Ce has a similar solubility to La, and so the possibility that Ce has been added and then discharged through drainage solutions is unlikely. Thus, the same fertilizers that contained enough La to be able to increase its content in soil were poor in Ce. This hypothesis is in contrast to the findings of Furr *et al.* (1976), who showed that manure contains more Ce than La. At Donnini, vine roots were responsible for the shift of Ce fractionation in the rhizosphere soil, the Ce of which was probably complexed by organics, bleached away and/or absorbed.

Praseodymium (Table 12)
Considering bulk and rhizosphere soil, the total contents of Pr ranged from 6 to 11 mg kg^{-1} at Donnini and from 4 to 7 mg kg^{-1} at Cambados. These values agreed with those found by Li *et al.* (1998), Markert (1987a, b), Land *et al.* (1999), Wang *et al.* (2001b) and Zhang *et al.* (2001).

At Donnini, the bulk contained Pr mostly in the crystalline Fe-oxyhydroxides and, secondarily, in organic matter; in the rhizosphere soil the contrary was true, possibly because of a lower content of crystalline Fe-oxyhydroxides (Table 11). At Cambados, the organic matter was largely responsible for the accumulation of Pr, both in the bulk and in the rhizosphere soil. The low residue-bound Pr and the constant distribution of the total Pr with depth further supported the idea that the main external source of this element and the soil homogenization due to agricultural practices occurred since cultivation. At Donnini, the different fractionation of Pr between bulk and rhizosphere soil showed the influence of the root activity. At Cambados, the root-induced redistribution of Pr was minimal.

Neodymium (Table 13)
At Donnini and Cambados, total Nd was higher in the rhizosphere soil than in the bulk. This agreed with Wang *et al.* (2001b), who, however, reported values five times lower than ours. In both soils, Nd showed a similar distribution in both bulk and rhizosphere soil, with the major amounts associated with the soluble +

Table 11
Content of cerium in bulk and rhizosphere soil from the vineyards of Donnini (Firenze, Italy) and Cambados (Pontevedra, Spain). Numbers in parentheses are the standard errors ($n = 2$)

	Soluble and exchangeable	Easily reducible Fe–Mn-oxyhydroxides	Organic matter	Amorphous Fe-oxyhydroxides	Crystalline Fe-oxyhydroxides	Residue	Total
Donnini							
Bulk							
Ap1	0.0(—)	0.6(0.1)	8.1(0.4)	1.7(0.1)	24.6(1.0)	21.6(0.8)	56.6(1.6)
Ap2	0.0(—)	0.6(0.1)	11.8(0.6)	2.8(0.2)	31.5(1.2)	23.1(0.8)	69.8(1.5)
Bw	0.1(0.0)	0.7(0.1)	11.3(0.6)	2.9(0.1)	41.2(1.7)	26.2(1.0)	82.4(1.1)
Bh	0.0(—)	0.7(0.1)	11.4(0.6)	1.0(0.1)	26.6(1.0)	21.9(0.9)	61.6(1.5)
BC	0.0(—)	0.4(0.0)	11.1(0.6)	6.8(0.2)	32.9(1.0)	25.6(0.9)	76.8(1.1)
Rhizosphere							
Ap1	0.0(—)	0.0(—)	10.6(0.6)	4.1(0.2)	20.0(1.1)	7.5(0.2)	42.2(1.7)
Ap2	0.0(—)	0.5(0.1)	12.5(0.6)	4.8(0.2)	24.3(1.1)	7.6(0.2)	49.7(1.7)
Bw	0.1(0.0)	0.1(0.0)	18.8(0.8)	6.9(0.3)	33.9(1.1)	8.3(0.2)	68.1(1.4)
Bh	0.0(—)	0.2(0.0)	13.0(0.5)	3.2(0.1)	22.0(1.0)	8.0(0.2)	46.4(1.3)
BC	0.0(—)	0.0(—)	11.0(0.4)	8.2(0.3)	31.8(1.1)	9.0(0.2)	60.0(1.0)

Cambados							
Bulk							
A1	0.0(—)	0.0(—)	10.6(0.5)	0.0(—)	21.3(1.4)	10.2(0.3)	42.1(1.2)
A2	0.0(—)	0.5(0.1)	12.6(0.6)	0.0(—)	22.7(1.4)	10.4(0.3)	46.2(1.6)
A3	0.4(0.1)	1.2(0.1)	15.1(0.7)	0.0(—)	29.3(1.5)	11.6(0.3)	57.6(1.7)
AB	0.5(0.1)	2.0(0.2)	13.7(0.5)	0.0(—)	30.9(1.5)	13.4(0.3)	60.5(1.4)
Bw1	0.7(0.1)	2.4(0.2)	10.2(0.5)	0.5(0.1)	27.1(1.3)	13.6(0.3)	54.5(1.5)
Bw2	1.9(0.2)	2.9(0.2)	10.3(0.5)	0.3(0.0)	23.4(1.3)	16.6(0.4)	55.4(1.8)
BC	3.6(0.2)	2.9(0.2)	11.6(0.5)	2.5(0.2)	20.8(1.0)	31.2(0.8)	72.6(1.3)
Rhizosphere							
A2	0.0(—)	0.2(0.0)	14.3(0.7)	0.0(—)	17.3(1.0)	11.4(0.3)	43.2(1.4)
A3	0.4(0.1)	1.5(0.1)	12.6(0.7)	0.0(—)	17.8(1.1)	11.6(0.3)	43.9(0.9)
AB	0.4(0.1)	2.0(0.1)	8.8(0.5)	0.0(—)	19.0(1.0)	12.8(0.3)	43.0(1.2)
Bw1	0.6(0.1)	2.9(0.2)	9.5(0.5)	0.0(—)	19.9(1.0)	14.1(0.4)	47.0(1.0)
Bw2	1.4(0.2)	3.5(0.2)	10.4(0.4)	0.0(—)	17.8(0.8)	18.9(0.5)	52.0(1.1)
BC	2.8(0.2)	2.3(0.2)	11.5(0.5)	0.0(—)	16.3(0.6)	20.9(0.5)	53.8(0.8)

Note: All Values are expressed as mg kg^{-1}.

Table 12

Content of praseodymium in bulk and rhizosphere soil from the vineyards of Donnini (Firenze, Italy) and Cambados (Pontevedra, Spain). Numbers in parentheses are the standard errors ($n = 2$)

	Soluble and exchangeable	Easily reducible Fe–Mn-oxyhydroxides	Organic matter	Amorphous Fe-oxyhydroxides	Crystalline Fe-oxyhydroxides	Residue	Total
Donnini							
Bulk							
Ap1	0.9(0.2)	0.0(—)	1.4(0.1)	0.7(0.2)	2.0(0.4)	0.7(0.1)	5.7(0.8)
Ap2	1.0(0.2)	0.0(—)	1.9(0.2)	0.8(0.2)	2.7(0.4)	0.7(0.1)	7.1(0.9)
Bw	1.0(0.2)	0.1(0.1)	1.8(0.2)	1.0(0.2)	3.6(0.4)	0.7(0.1)	8.2(0.8)
Bh	0.9(0.2)	0.1(0.0)	1.9(0.2)	0.6(0.1)	2.1(0.2)	0.7(0.1)	6.3(0.6)
BC	0.9(0.1)	0.0(—)	1.8(0.1)	1.6(0.2)	2.9(0.2)	0.7(0.1)	7.9(0.5)
Rhizosphere							
Ap1	1.3(0.2)	0.0(—)	3.1(0.4)	1.1(0.2)	1.8(0.3)	0.9(0.1)	8.2(0.8)
Ap2	1.4(0.2)	0.1(0.0)	3.3(0.3)	1.3(0.2)	2.3(0.3)	0.9(0.1)	9.3(0.5)
Bw	1.4(0.2)	0.0(—)	4.1(0.3)	1.5(0.2)	3.1(0.4)	0.9(0.1)	11.0(0.4)
Bh	1.4(0.2)	0.1(0.0)	3.4(0.3)	1.1(0.2)	1.8(0.2)	0.9(0.1)	8.7(0.6)
BC	1.4(0.2)	0.0(—)	3.2(0.3)	1.9(0.2)	3.1(0.3)	0.8(0.1)	10.4(0.5)

Cambados							
Bulk							
A1	0.9(0.2)	0.0(—)	1.5(0.2)	0.2(0.1)	1.1(0.2)	0.8(0.1)	4.5(0.6)
A2	0.9(0.2)	0.1(0.0)	1.5(0.3)	0.2(0.1)	0.9(0.1)	0.8(0.1)	4.4(0.4)
A3	0.8(0.1)	0.3(0.1)	1.7(0.2)	0.2(0.0)	0.9(0.1)	0.8(0.1)	4.7(0.6)
AB	0.8(0.1)	0.5(0.1)	1.6(0.2)	0.2(0.0)	1.2(0.1)	0.7(0.1)	5.0(0.4)
Bw1	0.8(0.1)	0.6(0.1)	1.3(0.2)	0.3(0.1)	1.4(0.2)	0.7(0.1)	5.1(0.6)
Bw2	1.0(0.1)	0.7(0.1)	1.4(0.2)	0.4(0.0)	1.3(0.1)	0.7(0.1)	5.5(0.4)
BC	1.2(0.1)	0.6(0.1)	1.7(0.2)	0.5(0.1)	1.5(0.1)	0.8(0.1)	6.3(0.3)
Rhizosphere							
A2	1.3(0.2)	0.0(—)	2.8(0.4)	0.2(0.0)	0.8(0.2)	0.7(0.1)	5.8(0.7)
A3	1.1(0.2)	0.4(0.1)	2.8(0.3)	0.2(0.0)	0.7(0.1)	0.7(0.1)	5.9(0.6)
AB	1.2(0.1)	0.5(0.1)	2.5(0.3)	0.2(0.0)	0.8(0.1)	0.7(0.1)	5.9(0.5)
Bw1	1.2(0.1)	0.6(0.2)	2.6(0.2)	0.2(0.0)	1.0(0.1)	0.8(0.1)	6.4(0.5)
Bw2	1.3(0.1)	0.6(0.1)	2.8(0.2)	0.3(0.0)	1.0(0.1)	0.7(0.1)	6.7(0.4)
BC	1.5(0.2)	0.6(0.1)	3.0(0.2)	0.3(0.0)	1.0(0.1)	0.7(0.1)	7.1(0.3)

Note: All values are expressed as mg kg^{-1}.

Table 13
Content of neodymium in bulk and rhizosphere soil from the vineyards of Donnini (Firenze, Italy) and Cambados (Pontevedra, Spain). Numbers in parentheses are the standard errors (n = 2)

	Soluble and exchangeable	Easily reducible Fe–Mn-oxyhydroxides	Organic matter	Amorphous Fe-oxyhydroxides	Crystalline Fe-oxyhydroxides	Residue	Total
Donnini							
Bulk							
Ap1	34.0(3.2)	3.7(0.5)	5.5(0.7)	15.7(1.6)	18.7(1.8)	0.8(0.1)	78.4(3.5)
Ap2	34.9(3.0)	2.6(0.4)	6.7(0.6)	15.9(1.6)	23.5(2.1)	0.8(0.1)	84.4(1.8)
Bw	35.9(2.8)	2.0(0.3)	5.6(0.7)	16.9(1.6)	26.5(2.3)	0.7(0.1)	87.6(4.6)
Bh	34.5(2.5)	3.4(0.4)	7.2(0.7)	15.7(1.6)	19.5(1.8)	0.8(0.1)	81.1(3.9)
BC	35.4(2.3)	4.6(0.4)	7.0(0.7)	22.3(1.8)	25.7(2.0)	0.7(0.0)	95.8(3.2)
Rhizosphere							
Ap1	59.0(4.5)	3.3(0.5)	8.1(0.8)	15.5(1.5)	13.6(1.4)	0.9(0.1)	100.4(4.2)
Ap2	58.4(4.7)	3.1(0.5)	8.3(0.9)	16.7(1.6)	16.8(1.7)	0.9(0.1)	104.2(2.9)
Bw	57.4(4.5)	3.6(0.5)	9.3(0.9)	18.3(1.7)	22.7(1.9)	0.9(0.1)	112.2(4.8)
Bh	56.8(4.5)	2.9(0.4)	8.9(0.9)	15.5(1.3)	16.4(1.4)	0.9(0.1)	101.4(3.8)
BC	55.9(4.5)	3.7(0.4)	8.3(0.7)	20.7(1.5)	25.5(1.9)	0.9(0.1)	115(5.3)

Cambados							
Bulk							
A1	34.0(3.0)	3.0(0.5)	4.1(0.6)	12.7(1.3)	11.2(1.1)	1.0(0.1)	66.0(3.8)
A2	34.8(2.9)	4.7(0.6)	5.2(0.7)	12.7(1.3)	9.9(1.0)	0.9(0.1)	68.2(3.4)
A3	34.7(2.9)	2.3(0.4)	5.9(0.7)	12.6(1.3)	10.6(1.1)	0.9(0.1)	67.0(3.3)
AB	35.5(3.0)	4.0(0.5)	3.4(0.5)	12.7(1.2)	11.9(1.0)	0.8(0.0)	68.3(4.2)
Bw1	34.9(2.8)	2.6(0.4)	1.2(0.3)	26.7(1.6)	10.8(0.9)	0.8(0.0)	77.0(4.2)
Bw2	35.7(2.9)	3.4(0.4)	1.1(0.3)	13.6(1.1)	11.9(0.9)	0.7(0.1)	66.4(3.7)
BC	37.3(2.9)	4.2(0.4)	0.7(0.2)	29.0(1.6)	11.9(0.9)	1.7(0.1)	84.8(3.5)
Rhizosphere							
A2	58.4(5.2)	3.1(0.6)	8.1(0.8)	12.7(1.1)	7.8(0.8)	0.9(0.1)	91.0(5.8)
A3	61.0(5.5)	3.6(0.5)	7.9(0.8)	13.3(1.3)	8.6(0.8)	1.0(0.1)	95.4(4.8)
AB	59.6(5.0)	1.9(0.3)	6.3(0.6)	12.3(1.1)	8.8(0.7)	0.9(0.1)	89.8(5.0)
Bw1	59.8(4.7)	2.5(0.3)	5.8(0.5)	12.5(1.2)	9.7(0.7)	0.9(0.1)	91.2(4.1)
Bw2	58.4(4.4)	3.3(0.3)	5.6(0.5)	12.7(1.2)	8.9(0.6)	0.9(0.1)	89.8(4.1)
BC	57.2(4.4)	3.2(0.3)	5.6(0.5)	12.3(1.1)	9.9(0.6)	0.9(0.1)	89.1(2.8)

Note: All values are expressed as mg kg^{-1}.

exchangeable fraction, followed by Fe-oxyhydroxides and organic matter. Land *et al.* (1999) reported an opposite Nd fractionation. Considerations about external additions and the effects of soil homogenization described above are valid for Nd too. In the rhizosphere soil, the amounts of Nd in the soluble+exchangeable fraction were higher, in absolute values, than those found in the bulk, suggesting that vines absorb very low amounts of Nd. In contrast, Xinde *et al.* (2000) found that the exchangeable Nd was the most available form for plants of *Medicago sativa* L.

Samarium (Table 14)
At Donnini and Cambados, the total contents of Sm were similar in the bulk and in the rhizosphere soil. The amounts of Sm both in the bulk and rhizosphere soil were higher than those reported in the literature, which ranged from 0.2 to 6.8 mg kg^{-1} (Markert, 1987a; Yoshida *et al.*, 1998; Li *et al.*, 1998; Wyttembach *et al.*, 1998; Land *et al.*, 1999). At Donnini, Sm was mostly concentrated in the crystalline Fe-oxyhydroxides, and then in similar amounts in the organic matter, amorphous Fe-oxyhydroxides and soluble + exchangeable forms. At Cambados, Sm was mainly bound to the organic fraction and crystalline Fe-oxyhydroxides, in agreement with the findings of Li *et al.* (1998) and Land *et al.* (1999). In the rhizosphere soil of both soils, the lower total quantity of Sm with respect to the bulk and the lower concentration in the organic form indicated that Sm was bleached away or absorbed by vines. The low residue-bound content and the almost even soil distribution of the total element supported the hypothesis of external addition and soil homogenization.

Europium (Table 15)
At Donnini, Eu showed similar amounts and distributions in bulk and rhizosphere soil. In both soil compartments, Eu was mainly bound to crystalline Fe-oxyhydroxides and organic matter. At Cambados, we also had similar amounts and distributions in bulk and rhizosphere soil, but here the chemical fractions more involved were organic matter, crystalline Fe-oxyhydroxides and soluble + exchangeable. The amounts of total Eu for both soils agreed with the data reported by Barnard and Halbig (1985) and Markert (1987a). The same amount of Eu in bulk and rhizosphere soil could indicate that the roots of the vine do not provide a redistribution of this element in the rhizosphere. In both soil compartments, the low residue-bound Eu and the rather constant total concentration throughout the profile indicated an external addition and soil homogenization.

4.4. Conclusions

In the soils from the Donnini and Cambados vineyards, the total contents of lanthanides were probably not due to an indirect enrichment caused by pedogenetic processes acting on the parent material. Indeed, the rather constant ratio between the residue and the total contents throughout both profiles suggested that the long-lasting practices of cultivation were the major source of lanthanides. In particular, the responsibility of the observed trends was attributed to the use of fertilizers and deep mechanical working of the soil that, over the centuries, have greatly affected soil composition.

Under these conditions, the chemical fractions more involved in the binding of lanthanides appeared to be organic matter and Fe-oxyhydroxides. With regard to the latter, evidence indicated that lanthanides are usually bound to crystalline Fe-oxyhydroxides; however, this fraction was also the most abundant. Only in the case of La was the chemical distribution not related to the contents of the different forms of oxyhydroxides. In both soils, since vineyard plantation (29 years for Donnini and 18 years for Cambados), a slight redistribution of lanthanides in the profile occurred because of a translocation of suspensions that use roots as the preferential way for discharge. However, in each horizon, the rhizosphere soil of the vine showed a chemical fractionation of lanthanides often different with respect to that of the bulk, as in the vicinity of the roots, the components responsible for adsorption of lanthanides (Fe-oxyhydroxides, organic matter and exchangeable complex) are altered or degraded by the root-microorganism activity. In terms of bioavailability, the elements bound to organic matter would be expected to be more available than those associated with crystalline Fe-oxyhydroxides. Because of this, we conclude that in these vineyard soils, all the processes related to the translocation and accumulation of organic matter are able to mobilize REEs, and hence, to increase their availability for plants. Distribution of fertilizers containing REEs is another way of increasing the soil content of the available forms of REEs.

5. GENERAL CONCLUSIONS

In terms of pedogenesis, plants are the soil-forming force that accelerates the development of the soil through an accumulation of organic matter at its surface and a deep modification of the soil surrounding the roots, the rhizosphere soil. Many approaches have been followed to obtain information about the characteristics of the rhizosphere soil, either in indoor or outdoor studies. The former have used rhizoboxes and mini-rhizotrones, and have the advantage that they ensure the control of the environmental conditions and soil variability; however, they are a mere representation of the natural conditions, so that the characteristics of the rhizosphere soil so obtained may be at some distance from those occurring in the field. In contrast, studies on plants living outdoors have suffered from the lack of control over soil variability, and from the difficulty of collecting the rhizosphere soil; however, this approach is able to give information on the characteristics of the natural rhizosphere soil. To reduce the problems involved in the field studies, we believe that a previous geomorphologic and pedological survey can help in assessing soil

Table 14
Content of samarium in bulk and rhizosphere soil from the vineyards of Donnini (Firenze, Italy) and Cambados (Pontevedra, Spain). Numbers in parentheses are the standard errors ($n = 2$)

	Soluble and exchangeable	Easily reducible Fe–Mn-oxyhydroxides	Organic matter	Amorphous Fe-oxyhydroxides	Crystalline Fe-oxyhydroxides	Residue	Total
Donnini							
Bulk							
Ap1	2.2(0.4)	0.0(—)	2.4(0.4)	2.2(0.3)	5.8(0.8)	1.9(0.2)	14.5(1.3)
Ap2	2.5(0.3)	0.0(—)	3.0(0.6)	2.7(0.4)	6.9(0.9)	2.0(0.2)	17.1(1.4)
Bw	2.5(0.4)	0.0(—)	2.7(0.4)	2.6(0.4)	7.4(0.9)	2.4(0.1)	17.6(1.4)
Bh	2.4(0.3)	0.0(—)	3.2(0.4)	2.1(0.2)	6.1(0.7)	1.9(0.1)	15.6(1.3)
BC	2.4(0.3)	0.0(—)	3.0(0.4)	4.3(0.4)	6.7(0.5)	2.3(0.1)	18.7(1.1)
Rhizosphere							
Ap1	1.4(0.3)	0.0(—)	4.0(0.7)	2.2(0.3)	4.7(0.8)	4.3(0.2)	16.6(1.3)
Ap2	1.3(0.3)	0.0(—)	4.2(0.7)	2.6(0.3)	5.2(0.7)	4.2(0.2)	17.5(1.0)
Bw	1.4(0.3)	0.0(—)	4.9(0.7)	3.1(0.5)	6.7(0.8)	4.9(0.2)	21.0(1.5)
Bh	1.5(0.3)	0.0(—)	4.5(0.5)	2.3(0.3)	5.6(0.7)	4.3(0.2)	18.2(0.6)
BC	1.5(0.2)	0.0(—)	4.0(0.4)	4.0(0.4)	6.8(0.7)	6.2(0.2)	22.5(0.7)

Cambados							
Bulk							
A1	2.4(0.4)	0.0(—)	4.1(0.8)	1.1(0.3)	3.1(0.5)	1.2(0.2)	11.9(1.4)
A2	1.9(0.3)	0.0(—)	5.0(0.8)	1.0(0.3)	2.9(0.4)	1.2(0.1)	12.0(0.9)
A3	1.7(0.2)	0.0(—)	6.1(0.8)	1.0(0.3)	3.2(0.4)	1.3(0.1)	13.3(1.0)
AB	1.7(0.2)	0.0(—)	4.6(0.5)	1.0(0.2)	3.7(0.6)	1.2(0.1)	12.2(1.4)
Bw1	1.8(0.2)	0.0(—)	2.6(0.3)	1.6(0.3)	3.7(0.5)	1.3(0.1)	11.0(1.2)
Bw2	1.8(0.2)	0.0(—)	2.0(0.3)	1.5(0.3)	4.3(0.5)	1.7(0.1)	11.3(0.8)
BC	2.0(0.2)	0.0(—)	1.8(0.3)	1.8(0.3)	4.8(0.6)	3.1(0.2)	13.5(0.6)
Rhizosphere							
A2	0.8(0.3)	0.0(—)	6.1(0.9)	0.7(0.3)	2.4(0.4)	1.2(0.1)	11.2(1.4)
A3	0.8(0.3)	0.0(—)	5.3(0.7)	0.8(0.3)	2.3(0.4)	1.2(0.1)	10.4(0.8)
AB	0.9(0.3)	0.0(—)	3.9(0.4)	0.8(0.3)	2.3(0.3)	1.4(0.1)	9.3(0.6)
Bw1	1.0(0.3)	0.0(—)	3.8(0.4)	0.6(0.3)	2.7(0.3)	1.5(0.1)	9.6(0.0)
Bw2	1.1(0.3)	0.0(—)	3.3(0.4)	0.9(0.4)	3.3(0.4)	1.8(0.1)	10.4(0.2)
BC	1.3(0.3)	0.0(—)	3.2(0.4)	0.8(0.2)	3.9(0.4)	2.0(0.1)	11.2(0.4)

Note: All values are expressed as mg kg^{-1}.

Table 15
Content of europium in bulk and rhizosphere soil from the vineyards of Donnini (Firenze, Italy) and Cambados (Pontevedra, Spain). Numbers in parentheses are the standard errors ($n = 2$)

	Soluble and exchangeable	Easily reducible Fe–Mn-oxyhydroxides	Organic matter	Amorphous Fe-oxyhydroxides	Crystalline Fe-oxyhydroxides	Residue	Total
Donnini							
Bulk							
Ap1	0.3(0.1)	0.0(—)	0.8(0.2)	0.3(0.1)	1.5(0.2)	0.1(0.0)	3.0(0.2)
Ap2	0.3(0.1)	0.0(—)	0.7(0.2)	0.4(0.2)	1.8(0.2)	0.1(0.0)	3.3(0.1)
Bw	0.3(0.1)	0.1(0.0)	0.6(0.2)	0.4(0.1)	2.0(0.2)	0.1(0.0)	3.5(0.2)
Bh	0.2(0.0)	0.0(—)	0.7(0.2)	0.3(0.2)	1.5(0.2)	0.1(0.0)	2.8(0.2)
BC	0.2(0.1)	0.1(—)	0.7(0.1)	0.7(0.2)	1.8(0.2)	0.1(0.0)	3.6(0.2)
Rhizosphere							
Ap1	0.3(0.1)	0.1(0.0)	0.8(0.2)	0.3(0.2)	1.0(0.2)	0.1(0.0)	2.6(0.3)
Ap2	0.3(0.1)	0.1(0.0)	0.8(0.2)	0.4(0.2)	1.3(0.1)	0.1(0.0)	3.0(0.2)
Bw1	0.3(0.0)	0.1(0.0)	0.8(0.2)	0.5(0.2)	1.7(0.1)	0.1(0.0)	3.5(0.1)
Bw2	0.3(0.0)	0.1(0.0)	0.8(0.2)	0.4(0.1)	1.3(0.1)	0.1(0.0)	3.0(0.2)
BC	0.3(0.1)	0.3(0.1)	0.8(0.2)	0.6(0.1)	1.8(0.1)	0.1(0.0)	3.9(0.2)

Cambados							
Bulk							
A1	0.2(0.1)	0.0(—)	0.5(0.1)	0.1(0.0)	0.3(0.1)	0.1(0.0)	1.2(0.1)
A2	0.2(0.1)	0.0(—)	0.5(0.1)	0.1(0.1)	0.3(0.1)	0.1(0.0)	1.3(0.0)
A3	0.2(0.1)	0.0(—)	0.6(0.1)	0.1(0.0)	0.4(0.1)	0.1(0.0)	1.5(0.1)
AB	0.2(0.0)	0.1(0.0)	0.5(0.0)	0.1(0.0)	0.5(0.2)	0.1(0.0)	1.5(0.2)
Bw1	0.2(0.0)	0.1(0.0)	0.2(0.0)	0.0(—)	0.5(0.1)	0.1(0.0)	1.2(0.1)
Bw2	0.2(0.1)	0.1(0.0)	0.2(0.0)	0.1(0.0)	0.7(0.2)	0.1(0.0)	1.4(0.1)
BC	0.2(0.0)	0.1(0.0)	0.2(0.0)	0.0(—)	0.8(0.2)	0.1(0.0)	1.5(0.2)
Rhizosphere							
A2	0.3(0.1)	0.0(—)	0.7(0.1)	0.1(0.0)	0.2(0.1)	0.1(0.0)	1.4(0.1)
A3	0.3(0.1)	0.0(—)	0.7(0.1)	0.1(0.0)	0.1(0.1)	0.1(0.0)	1.3(0.1)
AB	0.3(0.1)	0.1(0.0)	0.6(0.0)	0.1(0.0)	0.2(0.1)	0.1(0.0)	1.4(0.0)
Bw1	0.3(0.0)	0.1(0.0)	0.6(0.0)	0.1(0.0)	0.3(0.1)	0.1(0.0)	1.5(0.1)
Bw2	0.3(0.0)	0.1(0.0)	0.6(0.1)	0.1(0.0)	0.5(0.1)	0.1(0.0)	1.7(0.0)
BC	0.3(0.1)	0.1(0.0)	0.6(0.1)	0.1(0.0)	0.5(0.1)	0.1(0.0)	1.7(0.1)

Note: All values are expressed as mg kg^{-1}.

variability, thus focusing the sampling in areas where the soil can be considered to be relatively homogeneous. Such a pedological approach may be completed by opening soil profiles, so as to allow the easy collection of the rhizosphere soil and provide the opportunity of sampling by genetic horizons, even at depth.

In the three case studies presented here, we have collected and analyzed bulk and rhizosphere soil from natural and agricultural environments following the pedological approach and, in all cases, it has been fruitful.

In the study of the rhizosphere soil, an aspect that has been neglected until now is that the coarse roots may act as avenues for soil solutions, which may translocate organic and inorganic substances; as a consequence, this material tends to accumulate in the proximity of the roots at a certain depth, where it is further reworked. This has been shown to be true on both a natural (*Genista aetnensis*) and a cultivated (*Vitis vinifera*) species. This is particularly important, as in absence of a pedological investigation, one could attribute differences between bulk and rhizosphere soil solely to root activity, while they would have been due to the flushing of the soil solutions. Nonetheless, the bulk soil can be heavily modified by root activity and by the ancillary microbial populations. For example, to colonize the volcanic and gravely soils of Mount Etna, where the major limiting nutrient is P, *Genista aetnensis* sustains a microbial population that is able to recycle P. The effect of root-microorganism association is limited to the rhizosphere soil, the only soil portion enriched in organic forms of P and calcium oxalate. This finding was made possible due to the presence of a rhizosphere soil that is modified by the roots of *Genista aetnensis* to acquire a yellowish tinge, contrasting with the reddish-brown colour of the bulk.

In the case of *Erica arborea*, which is an acidophilic species, the colonization of an alkaline soil derived from marine sediments had been so efficient that it had converted to acid conditions a mass of soil to about 60 cm in depth. As indicated by the absence of differences between the two soil compartments of this soil thickness, the process of acidification was able to transform all the bulk soil into rhizosphere soil. Indeed, *Erica* roots are now colonizing the horizon underneath, where dissolution of carbonates has already been accomplished, but where the bulk is still less acidic than the rhizosphere soil. At greater depths, carbonates persist and roots of *Erica* are rare.

With regard to the content of lanthanides, in addition to the amounts inherited from the parent material, the soil of the vineyards appeared enriched in these elements because of the addition of fertilizers, independent of whether they were organic (sewage sludge) or inorganic (lime or synthetic fertilizers). In the case studies presented here, such an enrichment was detected because both soils had been cultivated for long periods of time (decades or centuries), and consequently, there has been time to accumulate elements like the lanthanides, which are not very mobile in soil. In a soil where the cultivation of vines is repeated, the breaking up of the soil before each new planting has produced a soil that, in terms of lanthanide content, is homogeneous to a depth of 1 m or more. In the relatively short period of permanence of each plantation, the vines have not been able to significantly modify the redistribution of lanthanides within the profile, with the exception of small amounts mobilized through translocation of solids by the flushing of soil solutions that use coarse roots as preferential routes of discharge. The translocated material tends to accumulate in the rhizosphere soil of the vines, where it is chemically reworked by the combined effect of the activities of the roots and microorganisms. Because of this, in a micro-environment such as that of the rhizosphere soil, the components mainly responsible for adsorption of lanthanides (Fe-oxyhydroxides, organic matter and exchangeable complex) are altered or degraded, and consequently, the chemical fractionation of lanthanides changes with respect to the bulk.

ACKNOWLEDGMENTS

We are grateful to Dr. Jeff Wilson of the Macaulay Institute, Scotland, UK and Dr. Jose Torrent of the Universidad de Cordoba, Spain for their contribution during reading of the manuscript of the chapter.

REFERENCES

Adams, M.A., Byrne, L.T., 1989. ^{31}P-NMR analysis of phosphorus compounds in extracts of surface soils from selected karri (*Eucalyptus diversicolor* F. Muell.) forests. Soil Biol. Biochem. 21, 523–528.

Agnelli, A., Celi, L., Degl'Innocenti, A., Corti, G., Ugolini, F.C., 2002. The changes with depth of humic and fulvic acids extracted from fine earth and rock fragments of a forest soil. Soil Sci. 167, 524–538.

Allegrezza, M., Biondi, E., 2002. Excursion to the "Selva di Gallignano". In: Biondi, E., Blasi, C. (Eds.), Guide to the Excursion of the Fédération Internationale de Phytosociologie to the Natural Parks of Conero, Gran Sasso, Monti della Laga, and Circeo. Fitosociologia 39(3), 33–40.

Allison, L.E., 1965. Organic carbon. In: Black, C.A., Evans, D.D., Ensminger, L.E., White, J.L., Clarck, F.E. (Eds.), Methods of Soil Analysis, Part 2. Agronomy Monograph, 9. American Society of Agronomy, Madison, WI, pp. 1367–1378.

Anderson, T.A., Kruger, E.L., Coats, J.R., 1994. Enhanced degradation of a mixture of three herbicides in the rhizosphere of a herbicide-tolerant plant. Chemosphere 28, 1551–1557.

Aprill, W., Sims, R.C., 1990. Evaluation of the use of prairie grasses for stimulating polycyclic aromatic hydrocarbon treatment in soil. Chemosphere 20, 253–256.

Awad, F., Römheld, V., Marschner, H., 1994. Effect of root exudates on mobilization in the rhizosphere and uptake of iron by wheat plants. Plant Soil 165, 213–218.

Bakker, M.R., Kerisit, R., Verbist, K., Nys, C., 1999. Effects of liming on rhizosphere chemistry and growth of fine roots and of shoots of sessile oak (*Quercus petraea*). Plant Soil 217, 243–255.

Barber, S.A., 1962. A diffusion and mass-flow concept of soil nutrient availability. Soil Sci. 93, 39–49.

Barber, S.A., Walker, J.M., Vasey, E.H., 1963. Mechanisms for the movement of plant nutrients from the soil and fertilizer to the plant root. J. Agr. Food Chem. 11, 204–207.

Barnard, W.M., Halbig, J.B., 1985. Rare earth elements in soils from selected areas on the Island of Hawaii. Pac. Sci. 39, 241–251.

Bar-Yosef, B., 1991. Root excretions and their environmental effects. In: Waisel, Y., Eshel, A. (Eds.), Plant Roots. The Hidden Half. Marcel Dekker, Inc., New York, pp. 529–556.
Bar-Yosef, B., Lamert, J.R., 1981. Corn and cotton root growth in response to soil impedance and water potential. Soil Sci. Soc. Am. J. 45, 930–935.
Berna, F., Corti, G., Ugolini, F.C., Agnelli, A., 2000. Assessment of the role of rock fragments in the retention of cadmium and lead in irrigated arid stony soils. Ann. Chim-Rome, 90, 209–217.
Bertin, C., Xiaohan, Y., Weston, L.A., 2003. The role of root exudates and allelochemicals in the rhizosphere. Plant Soil 256, 67–83.
Bhat, K.K.S., Nye, P.H., 1973. Diffusion of phosphate to plant roots in soil. I. Quantitative autoradiography of the depletion zone. Plant Soil 38, 161–175.
Blakemore, L.C., Searle, P.L., Daly, B.K., 1981. Methods for chemical analysis of soils. New Zealand Soil Bureau Scientific, Report 10a, Department Scientific and Industrial Research, New Zealand.
Blanchar, R.W., Lipton, D.S., 1986. The pE and pH in alfalfa seedling rhizosphere. Agron. J. 78, 216–218.
Blanchar, R.W., Edmonds, C.R., Bradford, J.M., 1978. Root growth in cores formed from fragipan and B2 horizons of Hobson soil. Soil Sci. Soc. Am. J. 42, 437–440.
Boyle, J.J., Shann, J.R., 1998. The influence of planting and soil characteristics on mineralization of 2,4,5-T in rhizosphere soil. J. Env. Qual. 27, 704–709.
Bray, R.H., Kurtz, L.T., 1945. Determination of total, organic, and available forms of phosphorus in soils. Soil Sci. 59, 39–45.
Brown, D.A., Ul-Haq, A., 1984. A porous membrane-root culture technique for growing plants under controlled soil conditions. Soil Sci. Soc. Am. J. 48, 692–695.
Cappy, J.J., Brown, D.A., 1980. A method for obtaining soil-free soil-solution grown plant root systems. Soil Sci. Soc. Am. J. 44, 1321–1323.
Caviglia, A.M., Modenesi, P., 1999. Oxidative stress and ascorbic acid contents in *Parmotrema reticulatum* and *Parmelia sulcata* thalli. Lichenologist 31, 105–110.
Chao, T.T., 1972. Selective dissolution of manganese oxides from soils and sediments with acidified hydroxylamine hydrochloride. Soil Sci. Soc. Am. J. 36, 764–768.
Chung, J.B., Zasoski, R.J.,1994. Ammonium-potassium and ammonium calcium exchange equilibria in bulk and rhizosphere soil. Soil Sci. Soc. Am. J. 58, 1368–1375.
Chung, J.B., Zasoski, R.J., Burau, R.G., 1994. Aluminum-potassium and aluminum-calcium exchange equilibria in bulk and rhizosphere soil. Soil Sci. Soc. Am. J. 58, 1376–1382.
Claassen, N., Hendricks, L., Jungk, A., 1981a. Erfassung der Mineralstoffverteilung im wurzelnahen Boden durch Autoradiographie. Z. Pflanzen. Boden. 144, 306–316.
Claassen, N., Hendricks, L., Jungk, A., 1981b. Rubidium-Verarmung des wurzelnahen Bodens durch Maispflanzen. Z. Pflanzen. Boden. 144, 533–545.
Clark, F.E. 1949. Soil microorganisms and plant roots. Advan. Agron. 1, 241–288.
Clemensson-Lindell, A., Persson, H., 1992. Effects of freezing on rhizosphere and root nutrient content using two soil sampling methods. Plant Soil 139, 39–45.
Condron, L.M., Frossard, E., Tiessen, H., Newman, R.H., Steward, J.W.B., 1990. Chemical nature of organic phosphorus in cultivated and uncultivated soils under different environmental conditions. J. Soil Sci. 41, 41–50.
Corti, G., Agnelli, A., Ugolini, F.C., 1999. A modified Kjeldahl procedure for determining strongly fixed NH_4^+-N. Eur. J. Soil Sci. 50, 523–534.
Corti, G., Cuniglio, R., Agnelli, A., Ricci, F., Panichi, M., 2004. Suolo e *mal dell'esca* della vite: il punto a 5 anni dall'inizio delle indagini. L'Informatore Agrario 12, 79–83.
Corti, G., Ricci, F., Cuniglio, R., Agnelli, A., 2005. Study of soils developed on Pliocene deposits under vines showing or not symptoms of *esca disease.* Tuscany, Italy. Comm. Soil Sci. Plant Anal. (submitted).

Corti, G., Ugolini, F.C., Agnelli, A., 1998. Classing the soil skeleton (greater than two millimeters): proposed approach and procedure. Soil Sci. Soc. Am. J. 62, 1620–1629.

Courchesne, F., Gobran, G., 1997. Mineralogical variation of bulk and rhizosphere soils from a Norway spruce stand. Soil Sci. Soc. Am. J. 61,1245–1249.

Cuniglio, R., Corti, G., Agnelli, A., 2004. Individuazione di orizzonti *sombrici* in alcuni vigneti della Toscana e valutazione del loro possibile ruolo nel sequestro del carbonio organico. In: O. Grasselli (Ed.), Bollettino della Società Italiana della Scienza del Suolo, Atti del Convegno Annuale: Qualità del Suolo, Impatto Antropico e Qualità dei Prodotti Agricoli. L' Agricoltura che produce Paesaggio, 9–12 June, 2003, Siena, Italy, pp. 292–296.

Cromack, K. Jr., Sollins, P., Graustein, W.C., Speidel, K., Todd, A.W., Spycher, G., Li, C.Y., Todd, R.L., 1979. Calcium oxalate accumulation and soil weathering in mats of the hypogeous fungus *Hysterangium crassum*. Soil Biol. Biochem. 11, 463–468.

Curl, E.A., Truelove, B., 1986. The Rhizosphere. Springer, Berlin.

Dai, K.H., David, M.B., Vance, G.F., Krzyszowska, A.J., 1996. Characterization of phosphorus in a spruce-fir Spodosol by phosphorus-31 nuclear magnetic resonance spectroscopy. Soil Sci. Soc. Am. J. 60, 1943–1950.

Darrah, P.R., 1993. The rhizosphere and plant nutrition: a quantitative approach. In: Barrow, N.J., (Ed.), Plant Nutrition: From Genetic Engineering to Field Practice. Kluwer, Dordrecht, The Netherlands, pp. 3–22.

Di Giovanni, G.D., Watrud, L.S., Seidler, R.J., Widmer, F., 1999. Comparison of parental and transgenic alfalfa rhizosphere bacterial communities using Biolog GN metabolic fingerprinting and enterobacterial repetitive intergenic consensus sequence-PCR (ERIC-PCR). Mic. Ecol. 37, 129–139.

Dieffenback, A., Matzner, E., 2000. *In situ* soil solution chemistry in the rhizosphere of mature Norway spruce (*Picea abies* [L.] Karst.) trees. Plant Soil 222, 149–161.

Dinkelaker, B., Römheld, V., Marschner, H.,1989. Citric acid excretion and precipitation of calcium citrate in the rhizosphere of white lupin (*Lupinus albus* L.). Plant Cell Env. 12, 285–292.

Dommergues, Y., Balandreau, J., Rinaudo, G., Weinhard, P., 1973. Nonsymbiotic nitrogen fixation in the rhizosphere of rice, maize and different tropical grasses. Soil Biol. Biochem. 5, 83–89.

Dormaar, J.F., 1988. Effect of plant roots on chemical and biochemical properties of surrounding discrete soil zones. Can. J. Soil Sci. 68, 233–242.

Drever, J.I., Stillings, L.L., 1997. The role of organic acids in mineral weathering. Colloids Surfaces 120, 167–182.

Drew, E.A., Murray, R.S., Smith, S.E., Jakobsen, I., 2003. Beyond the rhizosphere: growth and function of arbuscular mycorrhizal external hyphae in sands of varying pore sizes. Plant Soil 251, 105–114.

Elo, S., Maunuksela, L., Salkinoja-Salonen, M., Smolander, A., Haahtela, K., 2000. Humus bacteria of Norway spruce stands: plant growth promoting properties and birch, red fescue and alder colonizing capacity. FEMS Microb. Ecol. 31, 143–152.

Farr, L.E., Vaidyanathan, L.V., Nye, P.H., 1969. Measurement of ionic concentration gradients in soil near roots. Soil Sci. 107, 85–91.

Fernández-Sanjurjo, M.J., Corti, G., Certini, G., Ugolini, F.C., 2003. Pedogenesis induced by *Genista aetnensis* (Biv.) DC. on basaltic pyroclastic deposits at different altitudes, Mt Etna, Italy. Geoderma 115, 223–243.

Fox, T.R., 1995. The influence of low-molecular-weight organic acids on properties and processes in forest soils. In: McFee, W.W., Kelly, J.M. (Eds.), Carbon Forms and Functions in Forest Soils. Soil Science Society of America, Inc., Madison, WI, pp. 43–62.

Fritz, E., Knoche, D., Meyer, D., 1994. A new approach for rhizosphere research by X-ray microanalysis of microliter soil solutions. Plant Soil 161, 219–233.

Furr, A.K., Lawrence, A.W., Tong, S.S.C., Grandolfo, M.C., Hofstader, R.A., Bache, C.A., Gutenmann, W.H., Lisk, D.J., 1976. Multielement and chlorinated hydrocarbon analysis of municipal sewage sludges of American cities. Env. Sci. Tech. 10, 683–687.

Gahoonia, T.S., Nielsen, N.E., 1991. A method to study rhizosphere processes in thin soil layers of different proximity to roots. Plant Soil 135, 143–146.

Gahoonia, T.S., Nielsen, N.E., 1992. The effects of root-induced pH changes on the depletion of inorganic and organic phosphorus in the rhizosphere. Plant Soil 143, 185–191.

GCIAR (Government of the Canary Island Autonomous Region). 1999. The Canary Island Forestry Plan.

Gerard, C.J., Sexton, P., Shaw, G., 1982. Physical factors influencing soil strength and root growth. Agron. J. 74, 875–879.

Ghonsikar, C.P., Miller, R.H., 1973. Soil inorganic polyphosphates of microbial origin. Plant Soil 38, 651–655.

Gillespie, A.R., Pope, P.E., 1990. Rhizosphere acidification increases phosphorus recovery of black locust: I. Induced acidification and soil response. Soil Sci. Soc. Am. J. 54, 533–537.

Gobran, G.R., Clegg, S., 1996. A conceptual model for nutrient availability in the mineral soil–root system. Can. J. Soil Sci. 76, 125–132.

Gollany, H.T., Bloom, P.R., Schumacher, T.E., 1997. Rhizosphere soil-water collection by immiscible displacement-centrifugation technique. Plant Soil 188, 59–64.

Gorissen, A., Cotrufo, M.F., 1999. Elevated carbon dioxide effects on nitrogen dynamics in grasses, with emphasis on rhizosphere processes. Soil Sci. Soc. Am. J. 63,1695–1702.

Graustein, W.C., Cromack, K.Jr., Sollins, P., 1977. Calcium oxalate: occurrence in soils and effect on nutrient and geochemical cycles. Science 198,1252–1254.

Grayston, S.J., Wang, S., Campbell, C.D., Edwards, A.C., 1998. Selective influence of plant species on microbial diversity in the rhizosphere. Soil Biol. Biochem. 30, 369–378.

Grinsted, M.J., Hedley, M.J., White, R.E., Nye, P.H., 1982. Plant-induced changes in the rhizosphere of rape (*Brassica napus* var. emerald) seedlings. I. pH change and the increase in P concentration in soil solution. New Phytol. 91, 19–29.

Guggenberger, G., Haumaier, L., Thomas, R.J., Zech, W., 1996. Assessing the organic phosphorus status of an oxisol under tropical pastures following native savanna using ^{31}P NMR spectroscopy. Biol. Fert. Soil 23, 332–339.

Günther, T., Dornberger, U., Fritsche, W., 1996. Effects of ryegrass on biodegradation of hydrocarbons in soil. Chemosphere 33, 203–215.

Häussling, M., Marschner, H., 1989. Organic and inorganic soil phosphatases and acid phosphatase activity in the rhizosphere of 80-year-old Norway [*Picea abies* (L.) Karst.] spruce trees. Biol. Fert. Soil 8, 128–133.

Haynes, R.J., 1990. Active ion uptake and maintenance of cation-anion balance. A critical examination of their role in regulating rhizosphere pH. Plant Soil 126, 247–264.

Helal, H.M., Sauerbeck, D.R., 1981. Kurzmitteilung Ein Verfahren zur Trennung von Bodenzonen unterschiedlicher Wurzelnähe. Z. Pflanzen. Boden. 144, 524–527.

Helal, H.M., Sauerbeck, D.R., 1983. Method to study turnover processes in soil layers of different proximity to roots. Soil Biol. Biochem. 15, 223–225.

Helal, H.M., Sauerbeck, D.R., 1984. Influence of plant roots on C and P metabolism in soil. Plant Soil 76, 175–182.

Hendriks, L., Jungk, A., 1981. Erfassung der Mineralstoffverteilung in Wurzelnähe durch getrennte Analyse von Rhizo- und Restboden. Z. Pflanzen. Boden. 144, 276–282.

Hiltner, L., 1904. Über neuere Erfahrungen und Probleme auf dem Gebiet der Bodenbakteriologie und unter besonderer Berücksichtigung der Gründüngung und Brache. Arbeiten der Deutschen Landwirtschaftlichen Gesellschaft 98, 59–78.

Hinedi, Z.R., Chang, A.C., Lee, R.W.K., 1989. Characterization of phosphorus in sludge extracts using phosphorus-31 nuclear magnetic resonance spectroscopy. J. Env. Qual. 18, 323–329.

Hinsinger, P., 2001. Bioavailability of soil inorganic P in the rhizosphere as affected by root-induced chemical changes: a review. Plant Soil 237, 173–195.

Hinsinger, P., Plassard, C., Tang, C., Jaillard, B., 2003. Origins of root-mediated pH changes in the rhizosphere and their responses to environmental constraints: a review. Plant Soil 248, 43–59.

Hoffmann, W.F., Barber, S.A., 1971. Phosphorus uptake by wheat (*Triticum aestivum*) as influenced by ion accumulation in the rhizocylinder. Soil Sci. 112, 256–262.

Højberg, O., Schnider, U., Winteler, H.V., Sørensen, J., Haas, D., 1999. Oxygen-sensing reported strain of *Pseudomonas fluorescens* for monitoring the distribution of low-oxygen habitats in soil. Appl. Env. Microb. 65, 4085–4093.

Horn, P., Schaaf, P., Holbach, B., Holzl, S., Eschnauer, H., 1993. $^{87}Sr/^{86}Sr$ from rock and soil into vine and wine. Z. Lebensm. Unters Forsch. 196, 407–409.

Inskeep, W.P., Comfort, S.D., 1986. Thermodynamic predictions for the effects of root exudates on metal speciation in the rhizosphere. J. Plant Nutr. 9, 567–576.

Jackson, M.J., Lim, C.H., Zelazny, L.W., 1986. Oxides, Hydroxides and aluminosilicates. In: Klute, A. (Ed.), Methods of Soil Analysis. Part I. Physical and Mineralogical Methods. Soil Science Society of America, Inc., Madison, WI, pp. 101–150.

Jenny, H., 1941. Factors of Soil Formation: a System of Quantitative Pedology. McGraw-Hill, New York.

Jianguo, H., Shuman, L.M., 1991. Phosphorus status and utilization in the rhizosphere of rice. Soil Sci. 152, 360–364.

Jones, D.L., 1998. Organic acids in the rhizosphere: a critical review. Plant Soil 205, 25–44.

Jones, D.L., Darrah, P.R., 1994. Role of root derived organic acids in the mobilization from the rhizosphere. Plant Soil 166, 247–257.

Jungk, A.O., 1991. Dynamics of nutrient movement at the soil-root interface. In: Waisel, Y., Eshel, A., Kafkafi, U. (Eds.), Plant Roots. The Hidden Half. Marcel Dekker, Inc., New York, pp. 455–481.

Jurinak, J.J., Dudley, L.M., Allen, M.F., Knight, W.G., 1986. The role of calcium oxalate in the availability of phosphorus in soils of semiarid regions: a thermodynamic study. Soil Sci. 142, 255–261.

Kabata-Pendias, A., Pendias, H., 2001. Trace Elements in Soils and Plants. Third ed. CRC Press. Boca Raton, FL.

Karathanasis, A.D., Hajek, B.F., 1983. Transformation of smectite to kaolinite in naturally acid soil systems: structural and thermodynamic consideration. Soil Sci. Soc. Am. J. 47, 158–163.

Kirk, G.J.D., Saleque, M.A., 1995. Solubilization of phosphate by rice plants growing in reduced soil: prediction of the amounts solubilized and the resultant increase in uptake. Eur. J. Soil Sci. 46, 247–255.

Klamer, M., Roberts, M.S., Levine, L.H., Drake, B.G., Garland, J.L., 2002. Influence of elevated CO_2 on the fungal community in a coastal scrub oak forest soil investigated with terminal-restriction fragment length polymorphism analysis. Appl. Env. Microb. 68, 4370–4376.

Kochian, L.V., 1995. Cellular mechanisms of aluminium toxicity and resistance in plants. Annu. Rev. Plant Phys. Plant Mol. Biol. 46, 237–260.

Kuchenbuch, R., Jungk, A., 1982. A method for determining concentration profiles at the soil-root interface by thin slicing rhizospheric soil. Plant Soil 68, 391–394.

Kunito, T., Saeki, K., Nagaoka, K., Oyaizu, H., Matsumoto, S., 2001. Characterization of copper-resistant bacterial community in rhizosphere of highly copper-contaminated soil. Eur. J. Soil Biol. 37, 95–102.

Land, M., Öhlander, B., Ingri, J., Thunberg, J., 1999. Solid speciation and fractionation of rare earth elements in a spodosol profile from northern Sweden as revealed by sequential extraction. Chem. Geol. 160, 121–138.

Latour, X., Corberand, T., Laguerre, G., Allard, F., Lemanceau, P., 1996. The composition of fluorescent Pseudomonad populations associated with roots is influenced by plant and soil type. Appl. Env. Microb. 62, 2449–2456.

Lavkulich, L.M., Wiens, J.H., 1970. Comparison of organic matter destruction by hydrogen peroxide and sodium hypochlorite and its effect on selected mineral constituents. Soil Sci. Soc. Am. Proc. 34, 755–758.

Lemanceau, P., Corberand, T., Gardan, L., Latour, X., Laguerre, G., Boefgrasand, J.M., Alabouvette, C., 1995. Effect of two plant species, flax (*Linum usitatissimum* L.) and tomato (*Lycopersicum esculentum* Mill.), on the diversity of soilborne population of fluorescent *Pseudomonas*. Appl. Env. Microb. 61, 1004–1012.

Lewis, D.G., Quirk, J.P., 1967. Phosphate diffusion in soil and uptake by plants. III. P^{31}-movement and uptake by plants as indicated by P^{32}-autoradiography. Plant Soil 27, 445–453.

Li, F., Shan, X.-Q., Zhang, T., Zhang, S., 1998. Evaluation of plant availability of rare earth elements in soils by chemical fractionation and multiple regression analysis. Env. Poll. 102, 269–277.

Li, H., Yang, X., Kirk, G.J.D., Dobbermann, A., 2002. Root-induced changes of potassium in the rhizosphere of lowland rice (*Oriza sativa* L.). In: 17th World Congress of Soil Science, 14–21 August 2002, Bangkok, Thailand, pp. 1282–1–1282–11.

Liao, Z.W., Wang, J.L., Liu, Z.Y., 1993. Si, Fe and Mn distributions in rice (*Oryza sativa* L.) rhizosphere of red earths and paddy soils. Pedosphere 3, 1–6.

Lindsay, W.L., 1979. Chemical Equilibria in Soils. Wiley, New York.

Linhean, D.J., Sinclair, A.H., Mitchell, M.C., 1985. Mobilisation of Cu, Mn and Zn in the soil solutions of barley rhizospheres. Plant Soil 86, 147–149.

Lipton, D.S., Blanchar, R.W., Blevins, D.G., 1987. Citrate, malate, and succinate concentration in exudates from P-sufficient and P-stressed *Medicago sativa* L. seedlings. Plant Phys. 85, 315–317.

Lombi, E., Wenzel, W.W., Gobran, G.R., Adriano, D.C., 2001. Dependency of phytoavailability of metals on indigenous and induced rhizosphere processes: a review. In: Gobran, G.R., Wenzel, W.W., Lombi, E. (Eds.), Trace Elements in the Rhizosphere. CRC Press, Boca Raton, FL, pp. 3–24.

Longsdon, S.D., Reman, R.B.Jr., Parker, J.C., 1986. Corn seedling root growth as influenced by soil physical properties. Agron. J. 79, 221–224.

Ludovici, K.H., Morris, L.A., 1997. Competition-induced reductions in soil water availability reduced pine root extension rates. Soil Sci. Soc. Am. J. 61, 1196–1202.

Lynch, J.M., 1990. Introduction: some consequences of microbial rhizosphere competence for plant and soil. In: Lynch, J.M. (Ed.), The Rhizosphere. Wiley, Chichester, West Sussex, UK, pp. 1–9.

Lynch, J.M., Whipps, J.M., 1990. Substrate flow in the rhizosphere. Plant Soil 129, 1–10.

Makarow, M.I., Haumaier, L., Zech, W., 2002. The nature and origins of diester phosphate in soil: a 31P-NMR study. Biol. Fert. Soil 35, 136–146.

Malajczuk, N., Cromack, K. Jr., 1982. Accumulation of calcium oxalate in the mantle of ectomycorrhizal roots of *Pinus radiata* and *Eucalyptus marginata*. New Phytol. 92, 527–531.

Marisa, C., Almeida, R., Vasconcelos, M.T.S.D., 2003. Multielement composition of wine and their precursors including provenance soil and their potentialities as fingerprints of wine origin. J. Agric. Food Chem. 51, 4788–4798.

Markert, B., 1987a. The pattern of distribution of lanthanide elements in soils and plants. Phytochem. 26, 3167–3170.

Markert, B. 1987b. Multielementanalytic: Mögliighe Darstellungsweisen von Messdaten. Fresenius J. Anal. Chem. 327, 329– 336.

Marschner, H., 1995. Mineral nutrition of higher plants, second ed. Academic Press, San Diego, CA.

Marschner, H., Romheld, V., 1983. In vivo measurement of root-induced pH changes at the soil interface: Effect of plant species and nitrogen source. J. Plant Physiol. 111, 241–251.

Marschner, P., Marino, W., Lieberei, R., 2002. Seasonal effects on microorganisms in the rhizosphere of two tropical plants in a polyculture agroforestry system in Central Amazonia, Brazil. Biol. Fert. Soil 35, 68–71.

Martin, J.K., Kemp, J.R., 1986. The measurements of C transfers within the rhizosphere of wheat grown in field plots. Soil Biol. Biochem. 18, 103–107.

Massee, T.W., 1990. Simulated erosion and fertilizer effects on winter wheat cropping intermountain dryland area. Soil Sci. Soc. Am. J. 54, 1720–1725.

Massol-Deya, A., Weller, R., Rios-Hernandez, L., Zhou, J.Z., Hickey, R.F., Tiedje, J.M., 1997. Succession and convergence of biofilm communities in fixed-film reactors treating aromatic hydrocarbons in groundwater. Appl. Env. Microb. 63, 270–276.

McGrath, S.P., Shen, Z.G., Zhao, F.J., 1997. Heavy metal uptake and chemical changes in the rhizosphere of *Thlaspi caerulescens* and *Thlaspi ochroleucum* grown in contaminated soils. Plant Soil 188, 153–159.

McKenzie, R.H., Dormaar, J.F., Schaalje, G.B., Stewart, J.W.B., 1995. Chemical and biochemical changes in the rhizospheres of wheat and canola. Can. J. Soil Sci. 75, 439–447.

Mehelich, A., 1984. Mehelic 3 soil text extractant: a modification of Mehelic 2 extractant. Comm. Soil Sci. Plant Anal. 6, 113–128.

Mehra, O.P., Jackson, M.L., 1960. Iron oxide removal from soils and clays by dithionite-citrate system buffered with sodium bicarbonate. In: Proceedings of the 7th National Conference on Clays and Clay Minerals, Washington, DC, 1958. Pergamon Press, New York, pp. 313–327.

Metzger, W.H., 1928. The effect of growing plants on solubility of soil nutrient. Soil Sci. 25, 273–280.

Mozofar, A., Duss, F., Oertli, J.J., 1992. Effect of *Pseudomonas fluorescens* on the root exudates of two tomato mutants differently sensitive to Fe chlorosis. Plant Soil 144, 167–176.

Munson, M.A., Nedwell, D.B., Embley, T.M., 1997. Phylogenetic diversity of *Archaea* in sediment samples from a coastal salt marsh. Appl. Env. Microb. 63, 4729–4733.

Nanni, T., 1997. Caratteri geologici del bacino idrografico del fiume Musone. In: T. Nanni (Ed.), Il bacino del fiume Musone. 76th Congress of the Società Geologica Italiana, Firenze 20–26 September 1992. Communities of Osimo, Castelfidardo, Offagna. Grafiche Scarponi, Osimo, AN.

Newman, R.H., Tate, K.R., 1980. Soil phosphorus characterization by ^{31}P nuclear magnetic resonance. Comm. Soil Sci. Plant Anal. 31, 835–842.

Nichols, T.D., Wolf, D.C., Rogers, H.B., Beyrouty, C.A., Reynolds, C.M., 1997. Rhizosphere microbial population in contaminated soils. Water Air Soil Poll. 95, 165–178.

Norvel, W.A., Cary, E.E., 1992. Potential errors caused by roots in analyses of rhizosphere soil. Plant Soil 143, 223–231.

Nye, P.H., 1984. On estimating the uptake of nutrients solubilized near roots or other surfaces. J. Soil Sci. 35, 439–445.

Olsen, S.R., Cole, C.V., Watanabe, F.S., Dean, L.A., 1954. Estimation of available phosphorus in soils by extraction with sodium bicarbonate. USDA Circ. 939. USDA, Washington, DC.

Olsthoorn, A.F.M., Keltjens, W.G., Van Baren, B., Hopman, M.C.G., 1991. Influence of ammonium on fine root development and rhizosphere pH of Douglas-fir seedlings in sand. Plant Soil 133, 75–81.

Pal, S.S., 1998. Interactions of an acid tolerant strain of phosphate solubilizing bacteria with a few acid tolerant crops. Plant Soil 198, 169–177.

Pang, X., Li, D., Peng, A., 2002. Application of rare-earth elements in the agriculture of China and its environmental behavior in soil. Env. Sci. Poll. Res. 9, 143–148.
Papavizas, G.C., Davey, C.B., 1961. Extent and nature of the rhizosphere of lupinus. Plant Soil 14, 215–236.
Parker, D.R., Pedler, J.F., 1998. Probing the 'malate hypothesis' of differential aluminium tolerance in wheat by using other rhizotoxic ions as proxies for Al. Planta 205, 389–396.
Pena-Cabriales, J.J., Alexander, M., 1983. Growth of *Rhizobium* in unamended soil. Soil Sci. Soc. Am. J. 47, 81–84.
Persmark, L., Jansson, H.-B., 1997. Nematophagous fungi in the rhizosphere of agricultural crops. FEMS Microb. Ecol. 22, 303–312.
Reddy, B.R., Sethunathan, N., 1983. Mineralization of parathion in the rice rhizosphere. Appl. Env. Microb. 45, 826–829.
Reynolds, C.M., Koenen, B.A., Carnahan, J.B., Walworth, J.L., Bhunia, P., 1999. Rhizosphere and nutrient effects on remediating subarctic soils. In: R.E. Hinchee, D.B. Anderson, R.E. Hoeppel (Eds.), In situ and On-Site Bioremediation, 4th International Symposium. New Orleans, April 28-May 1, 1997. Battelle Memorial Institute, Vol. 3, pp. 297–302.
Ricci, F., Corti, G., Agnelli, A., Cuniglio, R., Pietramellara, G., Ascher, J., Ceccherini, M.T., 2004. Mal dell'esca e comunità batteriche e fungine del suolo. L'informatore Agrario 17, 59–62.
Riley, D., Barber, S.A., 1969. Bicarbonate accumulation and pH changes at the soybean (*Glycine max* (L.) Merr.) root-soil interface. Soil Sci. Soc. Am. Proc. 33, 905–908.
Riley, D., Barber, S.A., 1970. Salt accumulation at the soybean (*Glycine max* (L.) Merr.) root-soil interface. Soil Sci. Soc. Am. Proc. 34, 154–155.
Rubæk, G.H., Guggenberger, G., Zech, W., Christensen, B.T., 1999. Organic phosphorus in soil size separates characterized by phosphorus-31 nuclear magnetic resonance and resin extraction. Soil Sci. Soc. Am. J. 63, 1123–1132.
Sanchez, F.G., Bursey, M.M., 2002. Transient nature of rhizosphere carbon elucidated by supercritical freon-22 extraction and 13C NMR analysis. Forest Ecol. Manag. 169, 177–185.
Sarkar, A.N., Wyn Jones, R.G., 1982. Influence of rhizosphere on the nutrient status of dwarf French beans. Plant Soil 64, 369–380.
Schwertmann, U., Taylor, R.M., 1989. Iron oxides. In: J.B. Dixon, S.B. Weed (Eds.), Minerals in Soil Environment, second ed. Soil Sci. Soc. Am., Madison, WI, pp. 380–438.
Shotyk, W., Nesbitt, H.W., 1992. Incongruent and congruent dissolution of plagioclase feldspar: effect of feldspar composition and ligand complexation. Geoderma 55, 55–78.
Sims, J.T., 1996. Lime requirement. In: Sparks, D.L. (Ed.), Methods of Soil Analysis: Chemical Methods. Part 3. Soil Science Society of America, Inc., Madison, WI, pp. 491–515.
Skoog, D., West, D., 1987. Analytical Chemistry: An Introduction, second Italian edition. W.B. Saunders Co., Philadelphia, PA.
Smiley, R.W. 1974. Rhizosphere pH as influenced by plants, soils, and nitrogen fertilizers. Soil Sci. Soc. Am. Proc. 38, 795–799.
Smith, W.H., Pooley, A.S., 1989. Red spruce rhizosphere dynamics: spatial distribution of aluminum and zinc in the near-root soil zone. Forest Sci. 35, 1114–1124.
Soil Survey Division Staff, 1993. Soil Survey Manual, United States Department of Agriculture Handbook No. 18. Government Printing Office, Washington.
Soil Survey Staff, 1999. Soil Taxonomy. A Basic System of Soil Classification for Making and Interpreting Soil Surveys. Agriculture Handbook Number 436, U.S. Dept. Agric. Nat. Res. Conserv. Serv., second ed. U.S. Government Printing Office, Washington, DC.
Soon, Y.K., Miller, M.H., 1977. Changes in the rhizosphere due to NH_4^+ and NO_3^- fertilization and phosphate uptake by corn seedlings (*Zea mais* L.). Soil Sci. Soc. Am. J. 41, 77–80.

Starkey, R.L., 1931. Some influences of the development of higher plants upon the microrganisms in the soil: IV. Influence of proximity to roots on abundance and activity of microrganisms. Soil Sci. 32, 367–393.

Starkey, R.L., 1958. Interrelationships between microorganisms and plant roots in the rhizosphere. Bacteriol. Rev. 22, 154–172.

Staunton, S., Leprince, F., 1996. Effect of pH and some organic anions on the solubility of soil phosphate: implications for P bioavailability. Eur. J. Soil Sci. 47, 231–239.

Tait, D., Sayer, J.A., Mohammed, M.G., Gadd, G.M., 1999. Fungal production of calcium oxalate in leaf litter microcosms. Soil Biol. Biochem. 31, 1189–1192.

Tanaka, D.L., 1995. Spring wheat straw production and composition as influenced by topsoil removal. Soil Sci. Soc. Am. J. 59, 649–654.

Tate, K.R., Newman, R.H., 1982. Phosphorus fractions of a climosequence of soils in New Zeland tussok grassland. Soil Biol. Biochem. 14, 191–196.

Taylor, V.F., Longerich, H.P., Greenough, J.D., 2003. Multielement analysis of Canadian wines by inductively coupled plasma mass spectrometry (ICP-MS) and multivariate statistics. J. Agric. Food Chem. 51, 856–860.

Thomas, R.S., Dakessian, S., Ames, R.N., Brown, M.S., Bethlenfalvay, G.J., 1986. Aggregation of a silty clay loam soil by mycorrhizal onion roots. Soil Sci. Soc. Am. J. 50, 1494–1499.

Thompson, A.L., Gantzer, C.J., Anderson, S.H., 1991. Topsoil depth, fertility, water management, and weather influences on yield. Soil Sci. Soc. Am. J. 55, 1085–1091.

Todorovsky, D.S., Minkova, N.L., Bakalova, D.P., 1997. Effect of the application of superphosphate on rare earths'content in the soil. Sci. Tot. Env. 203, 13–16.

Turner, B.J., Mahieu, N., Condron, L.M., 2003. Phosphorus-31 nuclear magnetic resonance spectral assignments of phosphorus compounds in soil NaOH-ETDA extracts. Soil Sci. Soc. Am. J. 67, 497–510.

Uren, N., 2000. Types, amount, and possible functions of compounds released into the rhizosphere by soil-grown plants. In: Pinton, R., Varanini, Z., Nannipieri, P. (Eds.), The Rhizosphere: Biochemistry and Organic Substances at the Soil–Plant Interface. Marcel Dekker, Inc., New York, pp. 19–40.

Van Veen, J.A., Liljeroth, E., Lekkerkerk, J.A., 1991. Carbon fluxes in plant-soil systems at elevated atmospheric CO_2 levels. Ecol. Appl. 1, 175–181.

Veneklaas, EL., Stevens, J., Cawthray, G.R., Turner, S., Grigg, A.M., Lambers, H., 2003. Chickpea and white lupin rhizosphere carboxylates vary with soil properties and enhance phosphorus uptake. Plant Soil 248, 187–197.

Wada, K., Kakuto, Y., 1983. Intergradient vermiculite-kaolin mineral in a Korean Ultisol. Clays Clay Min. 31, 183–190.

Wang, X., Zabowski, D., 1998. Nutrient composition of Douglas-fir rhizosphere and bulk soil solution. Plant Soil 200, 13–20.

Wang, Z., Gottlein, A., Bartonek, G., 2001a. Effects of growth roots of Norway spruce (*Picea abies* [L.] Karst.) and European beech (*Fagus sylvatica* L.) on rhizosphere soil solution chemistry. Plant Nutr. Soil Sci. 164, 35–41.

Wang, Z., Shan, X.-Q., Zhang, S., 2001b. Comparison of speciation and bioavailability of rare earth elements between wet rhizosphere soil and air-dried bulk soil. Anal. Chim. Acta 44, 147–156.

Warembourg, F.R., Roumet, C., Lafont, F., 2003. Differences in rhizosphere carbon-partitioning among plant species of different families. Plant Soil 256, 347–357.

Watrud, L.S., Maggard, S., Shiroyama, T., Coleman, C.G., Johnson, M.G., Donegan, K.K., Di Giovanni, G., Porteous, L.A., Lee, E.H., 2003. Bracken (*Pteridium aquilinum* L.) frond biomass and rhizosphere microbial community characteristics are correlated to edaphic factors. Plant Soil 249, 359–371.

Webster R., 2001. Statistics to support soil research and their presentation. Eur. J. Soil Sci. 52, 331–340.

Weltje, L., 1997. Uptake and bio-concentration of lanthanides in higher plants: linking terrestrial and aquatic studies. Proceedings of the 2nd Sino-Dutch Workshop on the Environmental Behaviour and Ecotoxicology of REEs and Heavy Metals. Delft, The Netherland, pp. 1–12.

Wenzel, W.W., Wieshammer, G., Fitz, W.J., Puschenreiter, M., 2001. Novel rhizobox design to assess rhizosphere characteristics at high spatial resolution. Plant Soil 237, 37–45.

Wilkinson, H.F., Loneragan, J.F., Quirck, J.P., 1968. Calcium supply to plant roots. Science 161, 1245–1246.

Withington, J.M., Elkin, A.D., Bulay, B., Olesinski, J., Tracy, K.N., Bouma, T.J., Oleksyn, J., Anderson, L.J., Modrzynski, J., Reich, P.B., Eissenstat, D.M., 2003. The impact of material used for minirhizotron tubes for root research. New Phytol. 160, 533–544.

Wray, F.J., Tinker, P.B., 1968. A scanning apparatus for detecting concentration gradients around single plant roots. In: Whittington, W.J. (Ed.), Root Growth University of Nottingham, 15th Easter School in Agricultural Sciences.

Wyttenbach, A., Furrer, V., Schleppi, P., Tobler, L., 1998. Rare earth elements in soil and in soil-grown plants. Plant Soil 199, 267–273.

Xinde, C., Xiaorong, W., Guiwen, Z., 2000. Assessment of the bioavailability of rare earth elements in soils by chemical fractionation and multiple regression analysis. Chemosphere 40, 23–28.

Yanai, R.D., Majdi, H., Bae Park, B., 2003. Measured and modelled differences in nutrient concentrations between rhizosphere and bulk soil in a Norway spruce stand. Plant Soil 257, 133–142.

Yang, J., Blanchar, R.W., Hammer, R.D., Thompson, A.L., 1996. Soybean growth and rhizosphere pH as influenced by A horizon thickness. Soil Sci. Soc. Am. J. 60, 1901–1907.

Yoshida, S., Muramatsu, Y., Tagami, K., Uchida, S., 1998. Concentrations of lanthanide elements, Th and U in 77 Japanese surface soils. Env. Int. 24, 275–286.

Youssef, R.A., Chino, M., 1987. Studies on the behavior of nutrients in the rhizosphere. 1. Establishment of a new rhizobox system to study nutrient status in the rhizosphere. J. Plant Nutr. 10, 1185–1195.

Youssef, R.A., Chino, M., 1988a. Development of a new rhizobox system to study the nutrient status in the rhizosphere. Soil Sci. Plant Nutr. 34, 461–465.

Youssef, R.A., Chino, M., 1988b. Root-induced changes in the rhizosphere of plants. I. pH changes in relation to the bulk soil. Soil Sci. Plant Nutr. 35, 461–468.

Zhang, F., Shen, J., Li, L., Liu, X., 2002. An approach to study rhizosphere processes under major cropping systems in China. In: 17th World Congress of Soil Science, 14–21 August 2002, Bangkok, Thailand, pp. 938–1–938–10.

Zhang, F.S., Yamasaki, S., Kimura, K., 2001. Rare earth element content in various waste ashes and the potential risk to Japanese soils. Env. Int. 27, 393–398.

Zhang, S., Shan, X.-Q., 2001. Speciation of rare earth elements in soil and accumulation by wheat with rare earth fertilizer application. Env. Poll. 112, 395–405.

Biogeochemistry of Trace Elements in the Rhizosphere
P.M. Huang and G.R. Gobran (Editors)

Chapter 4

Metal complexation by phytosiderophores in the rhizosphere

S.M. Reichman[a] and D.R. Parker[b]

[a]School of Botany, University of Melbourne, Melbourne, Vic., 3010, Australia
E-mail: reichman@unimelb.edu.au

[b]Department of Environmental Sciences, University of California, Riverside, CA 92521, USA

ABSTRACT

Phytosiderophores (PS) are Fe (III)-solubilizing compounds released by the roots of Poaceae. Although still poorly quantified, the available evidence suggests that concentrations of PS in the rhizosphere are between a few and 100 μM. Daily cycles of PS in the rhizosphere occur with inputs from plants roots in the morning, the majority of PS being degraded by microorganisms, reabsorbed by plants, or adsorbed to soil solids throughout each day. Release of PS due to Fe deficiency, and direct uptake of the Fe(III)–PS complex by grass roots have been well demonstrated, but the release of PS in response to other metal deficiencies and the uptake of other metal–PS complexes is more controversial. However, irrespective of whether field-grown plants excrete PS in response to deficiencies of micronutrients other than Fe or whether they absorb significant quantities of other metal–PS complexes, metal-binding by PS in the rhizosphere may have more general implications for increased solubilization and mobilization of metals. Studies have shown that PS can solubilize other metals (e.g. Cu, Cd, Ni, Zn) in both contaminated and uncontaminated substrates. However, careful assessment of these studies reveals the use of unrealistically high concentrations of PS, poor quantification of the PS remaining in the aqueous phase, utilization of strong artificial metal sinks such as chelating resins and dialysis tubes, or microbial degradation, all of which may produce confounded results. Clearly, there is a need for more direct measurements of PS in rhizosphere soil, but in their absence, we investigated the issues by using a geochemical approach and conceptual modeling. The evidence suggests that PS could provide a soluble pool of metals for uptake with a number of potential implications:

(1) increasing the bioavailability of trace metals where deficiency is the primary concern;

(2) increasing the bioavailability of metals where phytotoxicity is the main concern;
(3) increasing the potential for food-chain transfer hazards for metals such as Cd, where toxicity to humans and animals is the chief concern; and
(4) increasing competition between Fe(III) and other metals for binding to PS complexes that may reduce solubilized Fe(III), leading to Fe deficiency.

1. INTRODUCTION

The release of phytosiderophores (PS) by the roots of Poaceae into the rhizosphere is an important process influencing the availability, uptake and food-chain transfer of Fe and, perhaps, other metals. As Fe solubility in soil solution is low, particularly in neutral to alkaline soils, plants have evolved mechanisms to maximize the potential uptake of Fe (Welch, 1995). The iron-solubilizing properties of grass-root exudates were first demonstrated by Takagi (1976) with Fe-deficient barley (*Hordeum vulgare*) and rice (*Oryza sativa*). Grasses including many crop and wild grass species have subsequently been shown to excrete PS, suggesting that they are a ubiquitous response to Fe deficiency throughout the Poaceae (Fushiya *et al.*, 1982; Sugiura and Nomoto, 1984; Nomoto *et al.*, 1987; Kawai *et al.*, 1988; Gries and Runge, 1992; Cakmak *et al.*, 1996a; Singh *et al.*, 2000). A number of different PS belonging to the mugineic acid (MA) family have been described, including MA, 2′-deoxymugineic acid (DMA), 3-hydroxymugineic acid (HMA) and 3-epi-hydroxymugineic acid (epi-HMA) (Takemoto *et al.*, 1978; Nomoto *et al.*, 1981; Ma and Nomoto, 1996). Phytosiderophores are nonprotein amino acids characterized by six functional groups that can octahedrally coordinate a metal-ion center (Fig. 1). While the putative PS precursor, nicotianamine (NA), is found throughout the plant kingdom (Noma and Noguchi, 1976; Scholz *et al.*, 1992), PS have not been found in dicots or in nongrass monocots (Römheld and Marschner, 1986). This gave rise to the nomenclature of "Strategy I" to include Fe uptake for dicots and nongrass monocots that utilize Fe reduction and proton excretion, and "Strategy II" to include the phytosiderophore producing grasses (Römheld and Marschner, 1986). Because Strategy I plants absorb Fe(II) and do not appear to absorb intact Fe-chelates or Fe-complexes (Chaney *et al.*, 1972; Welch *et al.*, 1993), it was hypothesized that, in the Poaceae, Fe(III) was cleaved from the PS molecule and reduced to Fe(II) before uptake. However, it has since been demonstrated that the intact Fe(III)–PS complex is absorbed by Strategy II plants (Römheld and Marschner, 1986; Grusak *et al.*, 1999). Fe uptake in the Poaceae is mediated by a specific root–cell plasma-membrane transporter, known as YS1 in maize (*Zea mays*) and also identified in sugar cane (*Saccharum* sp.), that recognizes the Fe(III)–PS complex (Curie *et al.*, 2001; Figueira *et al.*, 2001).

While the majority of research on PS has concentrated on their implications for Fe nutrition, PS are not highly selective for Fe and can complex other transition metals with comparable affinity (Sugiura *et al.*, 1981; Murakami *et al.*, 1989).

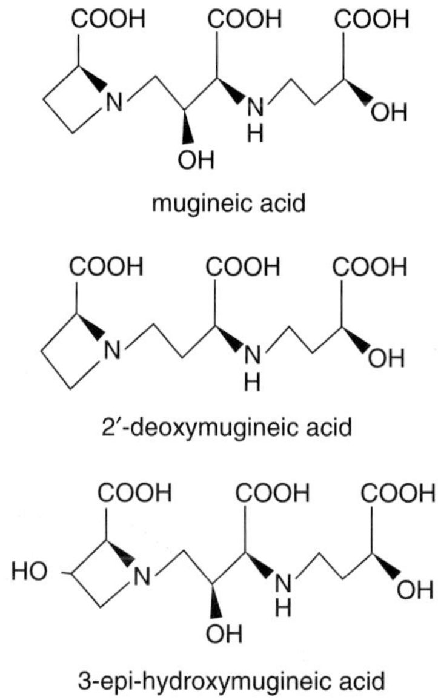

Fig. 1. Structures of some common phytosiderophores (adapted from Ma and Nomoto, 1996).

This has a number of potential implications of agronomic and environmental importance, including (i) the phytoavailability and uptake of metals in general; (ii) competition between metals and Fe interfering with the ability of PS to sequester and supply Fe; and (iii) the mobilization of metals such as Cd, where food-chain transfer is the primary concern. This chapter outlines our knowledge on PS, combining evidence from plant biology, geochemistry, and modeling to create a picture of metal–PS interactions in the rhizosphere and their consequences for the environment and human health.

2. DISTRIBUTION OF PHYTOSIDEROPHORES IN THE RHIZOSPHERE

Concentrations of PS at any point in the rhizosphere are determined by a number of different factors. Inputs of PS will vary depending on the Fe (and possibly other metal) status of the plant (Shi *et al.*, 1988), the plant species and genotype (Römheld, 1991), the time of day (Takagi *et al.*, 1984), the type of root (Bernards *et al.*, 2002) and the age of the plant (Cakmak *et al.*, 1998). Plant uptake, biodegradation, diffusion and adsorption to the soil solid phase will affect removal of PS from the rhizosphere. While we will discuss effective concentrations within the rhizosphere, the concentration of PS at any point in the rhizosphere will vary and depend on a number of factors (Fig. 2). Thus, the PS-defined rhizosphere can be thought of as being composed of four parameters:

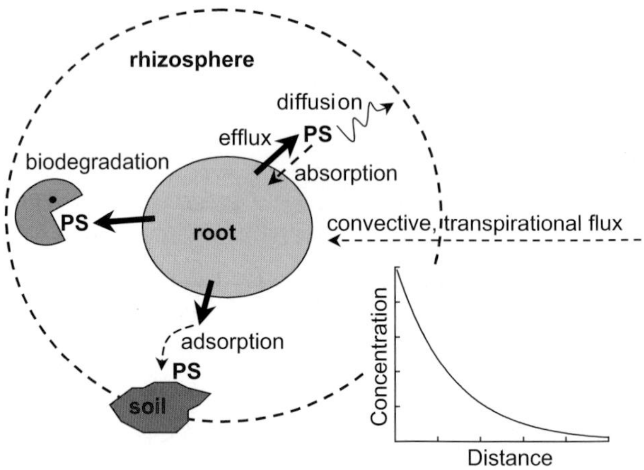

Fig. 2. Diagrammatic representation of processes affecting the concentration of phytosiderophores (PS) in the rhizosphere. The graph on the right-hand corner illustrates the reduction in PS concentration as the distance from the root surface increases.

1. The root parameter – where along the root are PS excreted and located?
2. The diffusion parameter – how far out from the root does the effective concentration extend? What is the radius of the rhizocylinder?
3. The concentration parameter – what are typical PS concentrations in the rhizosphere?
4. The time parameter – how do the above parameters change over time?

2.1. Location of phytosiderophores along the root

The initial work on Fe uptake in Poaceae was undertaken by Weavind and Hodgson (1971), who placed Fe-deficient wheat (*Triticum aestivum*) roots on ^{59}Fe impregnated agar for 4–6 days. Autoradiographs showed a pattern of Fe depletion corresponding to a clear zone from 3.5 to 13 cm behind the root tip, which roughly corresponded to the completion of suberization. Further research by Clarkson and Sanderson (1978) isolating ~3.5 mm segments of barley roots and supplying ^{59}Fe to them found that most Fe was absorbed in a zone from ~1–2 cm behind the root tip to ~5–8 cm toward the basal end of the root. Negligible amounts of Fe were absorbed in the more basal regions (i.e. ~20 cm from the root tip). More Fe was absorbed by Fe-deficient roots, and the absorption was significant further along the root axis, with some absorption occurring up to 20 cm. Thus, from the initial studies on Fe uptake in cereals, the consensus seemed to be that Fe absorption, and hence presumably, PS release, occurred in the more apical regions of the root but not at the root apex itself.

However, Marschner *et al.* (1987) conducted a study in which Fe-deficient and Fe-sufficient barley roots were placed on agar for 4 h (starting 2 h after the start of the light cycle). The agar was cut into 2–3 cm long segments and PS excretion was measured. For Fe-deficient plants, the results showed a peak in PS release in the apical 2 cm, decreasing sharply toward the basal zone, with an increase again in the zone of lateral root proliferation. In Fe-sufficient plants, PS release was at a low but constant level along the root axis. However, PS release in the basal zone of Fe-deficient plants was still approximately 10 times higher than PS release from healthy plants, and at about 20% of maximum release near the root apex in Fe-deficient plants; Fe uptake paralleled PS release. This led Marschner *et al.* (1987) to emphasize the importance of the root tip in PS production and Fe uptake. They discounted the Weavind and Hodgson (1971) results because of the length of their experiment, and stated that the 3.5 cm gap in uptake corresponded to one day's growth. They suggested that uptake in more basal zones reflected prior uptake from the root tip, as it grew through the agar. However, closer inspection reveals that the Weavind and Hodgson (1971) results are not in line with the Marschner *et al.* suggestions. The lack of uptake in the Weavind and Hodgson (1971) 0–3.5 zone means that the root tip, i.e. the Marschner *et al.* (1987) zone of highest uptake had already grown beyond this zone before uptake commenced. However, we believe that the Marschner *et al.* (1987) paper should be assessed in the light of the coarse scale of their work, with PS release only being quantified in sections of 2–3 cm. Thus, the Marschner *et al.* (1987) results could suggest maximal PS release from the root apex and from the root zone behind the root apex. But it is difficult to conclude from the Marschner *et al.* (1987) paper if the increase in the lateral root zone is confined to the primary root studied or corresponds to the lateral root apical zones. Indeed, from the results, the sum of PS release along the entire basal zone may be comparable to that released immediately behind the root tip.

Thus, on the basis of our review of the literature, we suggest that the area along the root axis where most PS release and Fe uptake occur corresponds to the apical root zone *behind* the root tip of Fe-deficient plants, although PS release at lower, but still meaningful, levels occurs along most of the root length. This is consistent with the research finding buildup of PS in the apical zone of barley roots just prior to daily secretion into the rhizosphere (Walter *et al.*, 1995). The precise placement of the zone of PS production and release depends on many factors, including Fe status and plant species.

2.2. Diffusion of phytosiderophores away from the root

Using the presence of fungi and bacteria as a guide, the rhizosphere can be generally defined as extending to an effective distance of 1–2 mm from the root surface (Rovira, 1969b). Before the discovery of PS, work was undertaken on the diffusion of C compounds away from wheat roots (Rovira, 1969a). Only 30% of C was found to diffuse farther than 1 mm from the root tip, whereas in the older root regions, up to 70% of C diffused beyond 1 mm. More recently, Shi *et al.*

(1988) detected MA at a distance of 1 mm from both healthy and Fe-deficient soil-grown barley plants. However, despite direct and indirect evidence for the rhizosphere in relation to PS, having an effective radius of at least 1 mm, more recent modeling work has utilized a rhizosphere radius of 0.5 mm (Römheld, 1991; Crowley and Gries, 1994). While this is only a small difference in an absolute sense, when used for modeling work, it could affect rhizosphere concentrations by a factor of 0.5^2 (=0.25). Hence, until more specific research is undertaken to determine the radius of the rhizocylinder with respect to PS, we suggest a 1 mm diffusion distance.

2.3. Typical rhizosphere phytosiderophore concentrations

The concentrations of PS in the rhizosphere will follow a trend similar to that shown in the graph in Fig. 2. Thus, as distance from the root increases, the concentration of PS decreases. In addition, factors such as sorption and biodegradation effects localized concentrations. Thus, when discussing "rhizosphere concentrations", we are in effect analyzing and calculating average concentrations within the rhizosphere.

To our knowledge, very few quantitative measurements have been made of PS concentrations in the rhizosphere. Shi *et al.* (1988) grew Fe-deficient barley, using a "rhizobox" technique to isolate rhizosphere soil. Phytosiderophores were extracted with dilute ammonium carbonate, and quantified by HPLC. From these measurements, they estimated an aqueous rhizosphere concentration of MA of the order of a few μM. However, their approach seemed to assume that PS production occurred uniformly along the root axis. This is in contrast to research showing peak PS production in the apical zone and decreasing production toward the basal root zone (Weavind and Hodgson, 1971; Clarkson and Sanderson, 1978; Marschner *et al.*, 1987). Thus, while still in the low μM range, concentrations of PS at the apical zone are likely to be somewhat higher than those suggested by Shi *et al.* (1988).

Rhizosphere PS concentrations can be estimated by modeling the rhizosphere as a cylinder with a hole in the center. Römheld (1991) determined PS production in soil-grown Fe-deficient barley to be 0.074 μmol $plant^{-1}$ day^{-1}. In the same study while calculating rhizosphere PS concentration (assuming that the rhizocylinder was composed of 25% water, that PS effectively diffused at 0.5 mm from the root surface and a root radius of 0.5 mm, as well as calculating a root length of 2325 mm $plant^{-1}$), he inexplicably used a much higher solution-culture concentration (0.54 μmol^{-1} $plant^{-1}$ day^{-1}) than the soil-determined value. Thus, Römheld (1991) calculated a rhizosphere concentration under non-sterile conditions of ~1 mM. However, he assumed that effective diffusion of PS into the soil only occurred to a distance of 0.5 mm (despite evidence as listed above), and that production only occurred along the apical 20% of the root surface, which is a conservative estimate when compared with other research on PS release (Weavind and Hodgson, 1971; Clarkson and Sanderson, 1978; Marschner *et al.*, 1987). Using Römheld's (1991) soil-based PS production and a more realistic effective PS diffusion of 1 mm, but keeping other factors the same, we estimate a rhizosphere root concentration of ~100 μM under nonsterile conditions. Compared with the Shi *et al.* (1988) estimate, 100 μM is still at least a factor of 10 higher. Römheld (1991) suggested that the plants used by Shi *et al.* (1988) may have recovered from much of their Fe chlorosis by the time of measurement; Shi *et al.* do mention regreening of the leaves and a drop in PS production with time. However, while Römheld's (1991) measurements correctly assume uneven PS production along the root, evidence suggests that PS production, while maximized in a small region just behind the apex, still occurs in significant quantities in other root zones (Clarkson and Sanderson, 1978; Marschner *et al.*, 1987). Thus, a truer reflection of rhizosphere PS concentrations is probably intermediate between the Shi *et al.* (1988) low μM concentration and the ~100 μM that we have determined from the Römheld (1991) soil data.

Moreover, we note that PS production on a per plant basis may be much greater in solution-cultured than in soil-grown plants (Shi *et al.*, 1988; Römheld, 1991), for reasons that remain obscure. Thus, we suggest that it is risky to use PS concentrations from solution culture to estimate concentrations in rhizosphere soil, even if reasonable allowances are made for adsorption and biodegradation. Better direct measurements of PS concentrations are clearly needed.

2.4. Changes in phytosiderophore concentrations with time

Rhizosphere concentrations of PS are in a continual state of flux. Inputs occur from the roots of grasses, which are regulated by a distinct diurnal cycle induced by changes in light or temperature (Mori, 1994; Cakmak *et al.*, 1998). A peak in PS release occurs between approximately 2 and 5 h after the commencement of the light period, with release dropping to minimal levels after a few hours (Takagi *et al.*, 1984; Gries and Runge, 1992). The only grass species yet found to not have a diurnal pattern of PS release under Fe deficiency is maize, and it has been suggested that this might be the reason for the low Fe efficiency of the species (Yehuda *et al.*, 1996). Crowley and Gries (1994) postulated that the high PS concentrations present at the peak of the diurnal cycle enables plants to maximize Fe-dissolution reactions by PS on mineral surfaces, allow temporary accumulation of PS in the presence of degrader organisms, and prevent the extracellular loss of Fe, which dissociates from PS and nonspecifically adsorbs on the root surface. In addition, peak production in the morning might reflect coordination with transpirational water flux, which would peak several hours later, and could help transport PS-solubilized Fe from the rhizosphere soil to the root surface for subsequent uptake and utilization (Parker *et al.*, 2005).

Assuming 1:1 FePS uptake, daily peak uptake rates of Fe, and hence PS, just behind the root apex in barley have been found to be ~15% of the PS production in the same zone (Marschner *et al.*, 1987). However, this does not account for potential differences in uptake patterns between solution-cultured and soil-grown plants, uptake of other metal–PS complexes, and absorption of Fe by other, as yet undiscovered, uptake paths in Strategy II species.

Biodegradation is a major cause of PS loss from the rhizosphere. In solution culture studies, yields of PS can be up to 10 times higher under sterile conditions than under nonsterile conditions (Römheld, 1991). Studies of PS in soils have regularly been confounded by degradation by microorganisms (e.g. Takagi *et al.*, 1988; Alvarez Fernandez *et al.*, 1997; Hiradate and Inoue, 1998). Römheld (1991) suggested a slower biodegradation rate in soil than solution culture because of the spatial separation between PS release and the main sites of microbial communities. That is, the root tends to remain relatively sterile for some time behind the root apex, presumably where PS release occurs, and bacteria tend to exist in discrete colonies rather than over the entire root surface. However, bacteria can be found throughout the soil and are not limited to the rhizosphere. Research tracking bacteria with flurochromes suggests rapid colonization of the root apex (Jones, 1998), and PS release, although concentrated directly behind the root apex, occurs at significant rates along most of the root surface (Weavind and Hodgson, 1971; Clarkson and Sanderson, 1978; Marschner *et al.*, 1987).

To our knowledge, few studies have dealt directly with loss of PS via adsorption. In a study designed to investigate Fe dissolution, EDTA and deferri-ferrioxamine B mesylate salt added to Fe-hydroxide was fully recovered, whereas only approximately 50% of MA was recovered (Takagi *et al.*, 1988). The authors suggested that the loss was from adsorption of MA onto the Fe-hydroxide surface. However, the study did not control for biodegradation, and this could account for much of the discrepancy. Inoue *et al.* (1993) found that MA adsorbed onto Fe-hydroxides decreased as pH increased, with negligible adsorption $>$ pH 10. The amount of MA adsorbed was related to the initial MA concentration, amount of Fe-hydroxide, sorptive capacity and specific surface area of Fe-hydroxide. Between pH 7 and 8 the addition of 100 μM MA resulted in ~50% adsorption onto ferrihydrite and less than 20% adsorption onto goethite, hematite or lepidocrocite. The soil nutrient status has also been demonstrated to affect PS adsorption, and both phosphate and sulfate in solution have been shown to strongly inhibit adsorption of MA onto Fe hydroxides, presumably through competition for binding sites (Hiradate and Inoue, 1998). Thus, while factors such as pH and soil nutrient status will affect PS adsorption onto the soil solid phase, research indicates that adsorption will be a significant loss of PS from the rhizosphere at relevant pHs.

Thus, daily cycles of PS in the rhizosphere will occur, with daily inputs from plant roots in the morning, and the majority of PS production being degraded by microorganisms, reabsorbed by plants as the Fe(III)–PS complex, or adsorbed to soil solids during the course of each day.

3. THE IMPORTANCE OF PHYTOSIDEROPHORES FOR THE ABSORPTION OF OTHER METALS BESIDES Fe

It has been proposed that PS may also act as general "metallophores" in trace-metal nutrition (Welch, 1995), although conclusive evidence is lacking (Grusak *et al.*, 1999). Increased exudation of PS by Zn-deficient wheat and barley was first reported by Zhang *et al.* (1989). Subsequently, Zn-efficient wheat genotypes and wild grasses were also shown to release more PS under Zn deficiency than Zn-inefficient genotypes (Cakmak *et al.*, 1996a; Cakmak *et al.*, 1996b). Cu deficiency, but not deficiencies of Zn or Mn, resulted in increased PS exudation in the European calcicole grass, *Hordelymus europaeus* (Gries *et al.*, 1998).

Other researchers have found evidence suggesting that the excretion of PS is limited to an Fe deficiency response. In barley cv. CM72, Zn deficiency resulted in only minor increases in PS release as compared with Fe deficiency, and deficiencies of Cu and Mn resulted in no increase (Gries *et al.*, 1995). Furthermore, increases in PS exudation in response to Zn deficiency were delayed and only commenced after severe reduction in growth, as compared to the rapid feedback mechanism in response to Fe deficiency. In response, Cakmak *et al.* (1996b) suggested that the barley cultivar used might have been Zn-inefficient, and thus, incapable of responding to Zn deficiency via increased PS release. In more recent research, no significant Zn deficiency-induced PS excretion was observed in several cultivars of both wheat and barley, including wheat cultivars that had previously been found to excrete PS under Zn deficiency (Pedler *et al.*, 2000). Walter *et al.* (1994) were able to halt PS release in Zn-deficient wheat by the application of Fe-citrate to the leaves, suggesting that PS release under Zn deficiency is due to an induced physiological Fe deficiency. However, a previous study in the same laboratory had found that Fe-foliar sprays had no detrimental effect on enhanced PS exudation in Zn deficient wheat (Zhang *et al.*, 1989). Thus, Pedler *et al.* (2000) suggested that PS release under Zn deficiency appeared to be highly method-dependent and an artifact of solution culture. The only study to attempt to probe methodology differences found greater PS release for a variety of Zn-deficient wheat cultivars grown in chelator-buffered compared with conventional nutrient solution (Rengel, 1999). However, we wonder about the validity of comparing solution culture systems using free ion activities, where metal activities are buffered in one system, whereas in the other system metal activities may change significantly between solution renewals. We suggest that a better method to compare solution culture techniques would be to compare systems with similar levels of deficiency, e.g. 10% decrease in growth.

While direct uptake of the Fe(III)–PS complex has been clearly demonstrated, the direct uptake of other metals bound to PS is more controversial. von Wirén *et al.* (1996) demonstrated that doubly radiolabeled ^{65}Zn–(^{14}C)PS complexes were directly absorbed by maize roots. Despite this, Cu–MA, and to a lesser extent Zn–MA, Co(II)–MA, and Co(III)–MA have been demonstrated to inhibit uptake of Fe(III)–MA (Ma, 1993). The authors suggested that this may have occurred owing to metal-MA complexes deforming the specific Fe(III)–PS transport system in the plasma. In addition, Fe(III) complexed by ligands of similar structure to PS, for example, NA, are taken up only slowly if at all, thus providing further evidence for the specific nature of the Fe(III)–PS transporter (Römheld and Marschner, 1986; Ma and Nomoto, 1996). Importantly, uptake rates for PS complexed to metals other than Fe have been found to be substantially lower than for Fe(III)–PS, although, the implications have been interpreted differently by various authors with respect to "specificity" (Ma *et al.*, 1993; von Wirén *et al.*, 1996). Thus, it seems as though the definitive test of the importance of PS as a general metal acquisition strategy will need validation with soil-grown plants, which has not yet been undertaken.

4. SOLUBILIZATION AND MOBILIZATION OF METALS BY PHYTOSIDEROPHORES

Irrespective of whether field-grown plants excrete PS in response to deficiencies of micronutrients other than Fe or absorb significant quantities of metal–PS complexes other than the Fe(III) complex, metal binding by PS in the rhizosphere may have implications for increased solubilization and mobilization of metals in general.

Studies in both uncontaminated and contaminated substrates have demonstrated that PS are efficient, when compared with synthetic chelators and microbial siderophores, at mobilizing soil and hydroxide-bound Fe, Cu, Zn, Ni, and Cd (Takagi *et al.*, 1988; Treeby *et al.*, 1989; Zhang *et al.*, 1991; Singh *et al.*, 1992; Inoue *et al.*, 1993; Awad and Römheld, 2000). PS consistently show a preference for solubilizing soil–Fe(III) over Zn and Mn, but Cu is sometimes bound in excess of Fe(III) (Treeby *et al.*, 1989; Singh *et al.*, 1992; Awad and Römheld, 2000). In comparison, particularly Cu, and to a lesser extent Zn, in *solution* had a negative effect on Fe dissolution from a Fe(III) hydroxide by PS (Zhang *et al.*, 1991). However, the ratios of metal: PS used suggest an extremely contaminated soil, and may not be relevant under field conditions.

In simple systems, amorphous ferrihydrite is consistently a good source of Fe for Strategy II species, while crystalline-structured goethite, hematite and lepidocrocite are much less so (Inoue *et al.*, 1993; Bertrand and Hinsinger, 2000; Gerke, 2000). Soil fertility status has also been shown to affect the ability of PS to solubilize Fe from hydroxides, with phosphate and chloride inhibiting the solubilization process (Awad *et al.*, 1988; Hiradate and Inoue, 1998). However, Gerke (2000) considered the phosphate rate used by Hiradate and Inoue (1998) to be at least 10 times higher than fertilized agricultural soils, and therefore, of little relevance. The results for chloride raise questions about the efficacy of PS release by Poaceae growing in saline soils.

Using a dynamic rhizosphere model, Gerke (1997, 2000) has emphasized the potential importance of organic matter, particularly humic substances, as a source of Fe(III) (and presumably other metals) for complexation by PS. Thus, MA combined with humic-Fe and citric acid resulted in almost complete complexation of the MA by Fe(III) (Gerke, 1997). Comparisons of different sources of Fe(III) found humic-Fe to be a consistently good source of Fe compared with goethite and ferrihydrite, irrespective of PS influx rate (Gerke, 2000). Indeed, Fe was efficiently mobilized from humic substances by HMA in a dialysis tube (Cesco *et al.*, 2000), and humic substances were a viable Fe source for barley grown in solution culture (Cesco *et al.*, 2002) and rice grown in soil (Pandeya *et al.*, 1998). In addition, other organic sources of Fe have been found to be efficient sources of Fe for Strategy II plants grown in solution-culture, e.g. microbial and fungal siderophores (Yehuda *et al.*, 1996, 2000). Gerke (2000) suggested that humic–Fe complexes provide an easily mobilizable source of Fe(III) because of the relatively low stability of Fe–humic complexes compared with PS. He also suggested that this valuable source of Fe for plants has largely been overlooked because of problems in determining the humic fraction in soils. However, humic substances tend to be relatively scarce in well-drained alkaline soils where Fe deficiency is common. Thus, the importance of humic and other organic substances as Fe sources requires testing in soil-grown plants.

When critically assessing the value of the soil studies for information on rhizosphere processes, some researchers have used unrealistically high concentrations of PS (Singh *et al.*, 1992; Inoue *et al.*, 1993; Hiradate and Inoue, 1998), and have not quantified the PS recovered in the aqueous phase (Takagi *et al.*, 1988; Singh *et al.*, 1992; Awad and Römheld, 2000). Some studies have utilized dialysis tubes and chelating resins to measure metals mobilized by PS (Awad *et al.*, 1988; Treeby *et al.*, 1989; Zhang *et al.*, 1991; Cesco *et al.*, 2000). While these studies demonstrate that PS are efficient in mobilizing metals, the inclusion of a strong metal sink (the chelating resin), and the need for solutes to diffuse across a dialysis membrane introduce artifacts not present in rhizosphere soil. Many of the studies utilized Takagi's (1976) Fe-solubilizing assay to quantify PS concentration (e.g. Awad *et al.*, 1988; Treeby *et al.*, 1989; Zhang *et al.*, 1991). However, the Fe-solubilizing assay is nonstoichiometric, and changes in temperature affect the solubilizing capacity of the assay (Takagi, 1976, 1993). Furthermore, microbial decomposition of PS has confounded most studies of PS in soil. Rapid biodegradation of PS has also been demonstrated in solution culture studies (von Wirén *et al.*, 1994). Hence, antimicrobial agents, such as Micropur™, are routinely used in solution culture research (e.g. Pedler *et al.*, 2000). In contrast, in studies of the effects of PS on the soil solid phase there has been little use of sterile conditions, even though many researchers acknowledge that their results have potentially been confounded by biodegradation of the PS complex (e.g. Takagi *et al.*, 1988; Alvarez Fernandez *et al.*, 1997; Hiradate and Inoue, 1998).

Clearly, there is a need for more rigorous soil-based studies on PS to address issues on Fe(III) and metal speciation and competition in the rhizosphere, and the efficacy of different sources of soil-metals for solubilization of metals and as sites of adsorption for PS. In the absence of this research, we propose to further investigate these issues through a general consideration of the chemistry involved and by using conceptual modeling.

5. GENERAL METAL–PHYTOSIDEROPHORE BIOGEOCHEMISTRY CONSIDERATIONS

5.1. Metal-binding properties of phytosiderophores

A preliminary set of binding constants for MA with Fe(III), Fe(II), Cu(II) and Zn(II) was determined by Sugiura *et al.* (1981). Later, Murakami *et al.* (1989) conducted a more extensive study using MA, DMA and epiHMA purified from root washings of Fe-deficient barley and wheat. The MAs contain three strongly acidic carboxyl functional groups; the pK_a of the 4″ carboxyl is about 3.2, while those of the 1 and 4′ carboxyls are $\leq \sim 2.7$ (Sugiura *et al.*, 1981; Murakami *et al.*, 1989). The amine groups are considerably weaker acids: the pK_a of the ring N ranges from about 7.1 to 8.3, while that of the chain N ranges from ~9.6 to 10.0 (Murakami *et al.*, 1989). Thus, while the MAs have a formal charge of −3 at extremely high pH, at circumneutral pH values specific to Fe nutrition, they are more typically zwitterions.

Murakami *et al.* (1989) determined binding constants for Fe(III), Fe(II), Cu(II), Zn(II), Ca(II), Mn(II) and Ni(II) with MA, DMA and epiHMA. The metal-binding characteristics for the three MAs are quite similar, and the representative stability constants for DMA are presented in Table 1.

Table 1
Stability constants for DMA (25°C, I = 0.1 M) as reported by Murakami *et al.* (1989)

Metal	$\log K_{11}$	$\log K_{11\text{-}1}$
Fe(III)	18.4	16.3
Ca(II)	3.3	—
Mn(II)	8.3	—
Fe(II)	10.5	—
Ni(II)	14.8	—
Cu(II)	18.7	—
Zn(II)	12.4	—

With the exception of Fe(III), all of the metals tested exhibited only 1:1 binding with DMA; thus, the reaction schemes appear to be quite straightforward for these metals. With Fe (III), however, Murakami *et al.* (1989) found that the following reaction predominates at pH values much above 3.0:

$$Fe^{3+} + L \rightarrow FeL(H_{-1})^{-} + H^{+} \qquad (1)$$

They attributed the shedding of an extra proton upon binding with Fe(III) to dissociation of the hydroxyl group on the 3″ carbon. Stoichiometrically, this reaction is indistinguishable from

$$Fe^{3+} + L \rightarrow FeLOH^{-} + H^{+} \qquad (2)$$

We have used the latter convention when describing the relevant stability constant ($\beta_{11\text{-}1}$).

The 11-1 complex formed between Fe(III) and the MAs is important for several reasons. First, it is by far the dominant complex at pH values near 7.0, and as we will show in the next section, this increases the effective binding constant considerably above that prediced on just the 1:1 complex. Second, it seems to have been overlooked in at least one study of comparative binding strength. von Wirén *et al.* (1999) compared the binding strengths of DMA and the closely related intracellular Fe chelator, NA. Among other findings, they concluded that NA would outcompete DMA for Fe(III) at physiological pH values, but this was predicated on simulations where only the 1:1 complexes were considered. However, speciation modeling by Reichman and Parker (2002) showed that the reverse is true: if the Fe(III)DMA$(H_{-1})^{-}$ complex is included in calculations, and if DMA and NA compete for Fe(III), almost 100% of Fe is complexed to DMA at pH values relevant to the cytoplasm, xylem and phloem. Had the relative binding strengths been properly evaluated, von Wirén *et al.* (1999) would have probably reached a very different conclusion about the ability of NA to "strip" Fe(III) from DMA in the cytoplasm and; thus, about NA's exact role in long-distance Fe transport.

The binding constants proposed by Murakami *et al.* (1989) for MA, DMA and epiHMA provide an excellent starting point, and the values seem plausible and reasonable. For example, the relative binding strength of Mn(II) < Fe(II) < Ni(II) < Cu(II) > Zn(II) follows exactly the well-known Irving–Williams order (Morel and Hering, 1993). However, some qualifiers seem warranted. The Murakami *et al.* (1989) values were determined using "purified" root washings, yet no quantitative information about the purity of the compounds studied was provided. Additionally, only a single titration at a 1:1 metal/ligand ratio was conducted for each combination, all at a single ionic strength of 0.1 M. Finally, there is no published stability constant for Cd(II), and although we might expect the binding strength to be similar to that for Zn(II), the ecotoxicological concerns surrounding Cd seem to justify a critical and empirical determination.

5.2. Use of conditional constants to understand the metal-binding chemistry of PS in the rhizosphere

The overall complexity of metal–ligand chemistry means that it can often be difficult to draw meaningful comparisons solely by examining the 1:1 stability constants (K_{ML}). As we have shown above, this problem is well exemplified in the case of Fe(III) binding by MAs, where an $ML(H_{-1})$ complex (stoichiometrically equivalent to ML(OH)) predominates at all relevant pH values, and the concentration of ML is rather negligible. While simulations using a geochemical modeling program such as GEOCHEM-PC (Parker *et al.*, 1995) can answer questions on metal–ligand binding, more facile evaluations can be made using *conditional* or *effective* binding constants. These conditional constants can readily be used to evaluate:

(a) which of two ligands will "outcompete" the other for a given metal; and
(b) which of two competing metals will form complexes with a limited quantity of a given ligand.

We also believe that the derivation and use of effective and conditional constants can be a useful tool to assist in understanding the important factors in metal–ligand competition and speciation.

In addition to the commonly used thermodynamic stability constants (formulated using ionic activities) and concentration-based constants (${}^{C}K$ values), an operationally defined set of stability constants can also be developed, usually referred to as *effective* constants (Morel and Hering, 1993). These constants, while derived from thermodynamic values, are no longer general, but instead, quite specific to the assumed conditions (ionic strength *and* pH) used in their calculation. Let us consider the case where, in addition to ML, a protonated MLH complex also forms. In addition, we will assume that the complexing ligand, L, is a weak acid so that in addition to the free ligand, species such as HL can also exist. Thus, a generalized effective constant can be defined as

$$K^{\mathrm{eff}} = \frac{\left[\sum \mathrm{ML}\right]}{\left[\mathrm{M}^{2+}\right]\left[\sum \mathrm{H}_x\mathrm{L}\right]} \qquad (3)$$

where Σ ML is the sum of the metal-ligand complexes (ML + MLH) and, thus, $\Sigma\ H_x L$ represents L + HL. Thus,

$$K^{\mathrm{eff}} = \frac{[\mathrm{ML}]+[\mathrm{MLH}]}{[\mathrm{M}^{2+}]([\mathrm{L}]+[\mathrm{HL}])} \qquad (4)$$

If the relevant equilibrium quotients are written for the formation of ML, MLH and HL, then rearranged so that the left-hand terms are [ML], [MLH] and [HL] respectively, and then substituted back into Eq. (4), we get

$$K^{\text{eff}} = \frac{(\beta_{\text{ML}}[\text{M}^{2+}][\text{L}])+(\beta_{\text{MLH}}[\text{M}^{2+}][\text{L}][\text{H}^+])}{[\text{M}^{2+}]([\text{L}]+\beta_{\text{HL}}[\text{H}^+][\text{L}])} \tag{5}$$

Cancellation leads to

$$K^{\text{eff}} = \frac{\beta_{\text{ML}} + \beta_{\text{MLH}}[\text{H}^+]}{1 + \beta_{\text{HL}}\,[\text{H}^+]} \tag{6}$$

If polyprotic species of MLH or HL exist, then Eq. (6) can be extended to a more general form, i.e. (Morel and Hering, 1993)

$$K^{\text{eff}} = \frac{\beta_{\text{ML}} + \beta_{\text{MLH}}[\text{H}^+] + \beta_{\text{MLH}_2}[\text{H}^+]^2...}{1 + \beta_{\text{HL}}\,[\text{H}^+] + \beta_{\text{H}_2\text{L}}[\text{H}^+]^2...} \tag{7}$$

As indicated by the ellipses, Eq. (7) can be expanded to other exponential terms in $[\text{H}^+]$ as needed. If the metal and ligand tend to form mixed complexes of the type ML(OH)_x, then the analogous expression can similarly be written as

$$K^{\text{eff}} = \frac{\beta_{\text{ML}} + \beta_{\text{MLOH}}/[\text{H}^+] + \beta_{\text{ML(OH)}_2}/[\text{H}^+]^2....}{1 + \beta_{\text{HL}}\,[\text{H}^+] + \beta_{\text{H}_2\text{L}}[\text{H}^+]^2....} \tag{8}$$

Finally, for metals that are hydrolytic, especially at alkaline pH values, we can devise a more comprehensive version of Eq. (3). This might best be termed a *conditional* constant (Stumm and Morgan, 1996). Thus

$$K^{\text{cond}} = \frac{\left[\sum \text{ML}\right]}{\left[\sum \text{MOH}\right]\left[\sum \text{H}_x\,\text{L}\right]} \tag{9}$$

where Σ MOH denotes the sum of the species M^{n+}, MOH, M(OH)_2, M(OH)_3, etc. By substituting in the appropriate hydrolysis expressions for Σ MOH, and using the same process as before, we obtain

$$K^{\text{cond}} = \frac{\beta_{\text{ML}}+\beta_{\text{MLOH}}/[\text{H}^+]+\beta_{\text{ML(OH)}_2}/[\text{H}^+]^2....}{\left(1+\beta_{\text{MOH}}/[\text{H}^+]+\beta_{\text{M(OH)}_2}/[\text{H}^+]^2....\right)\left(1+\beta_{\text{HL}}\,[\text{H}^+]+\beta_{\text{H}_2\text{L}}[\text{H}^+]^2....\right)} \tag{10}$$

Perhaps an easier way to visualize Eq. (10) is

$$K^{\text{cond}} = \frac{\beta_{\text{ML}} + \text{MCT}}{(\text{MHT})(\text{LAT})} \tag{11}$$

where MCT is the mixed complex term, MHT the metal hydrolysis term and LAT the ligand acidity term. For the specific pH of interest, these three terms can be applied to the original stability constant for the 1:1 metal–ligand complex. Thus, from Eq. (11) we can see that for a given pH and ionic strength, the apparent affinity of a ligand for a given metal can be:

(a) *increased* if complexes such as MLH or MLOH dominate over the 1:1 complex;
(b) *decreased* by the tendency for the metal to hydrolyze (competition from OH^-); or,
(c) *decreased* by the weakly acidic nature of the ligand (competition from H^+)

Table 2 illustrates how these adjustments influence the value of the conditional constants for complexation of DMA by Fe^{3+}, Zn^{2+}, Cu^{2+} and Ni^{2+}. With Fe^{3+}, the predominance of FeDMA(H_{-1}) leads to an effective stability constant that is $\sim 10^{4.9}$ greater than the stability constant for FeDMA formation ($10^{23.3}$ vs. $10^{18.4}$). In comparison, the other three metals form only 1:1 complexes with DMA, hence there is no adjustment for the MCT. The second adjustment accounts for competition from hydroxyl ligands. Thus, because of the highly hydrolytic character of Fe(III), the conditional constant for Fe(III)DMA is lowered by about $10^{7.6}$ compared with the other metals being adjusted by a factor of 2.1 or less. Finally, adjusting for the acidity of the ligand at the pH of interest results in a lowering of K^{cond} value by $10^{4.3}$ for all metals because of the predominance of H_2DMA^- and $HDMA^{2-}$ at pH 7.0. Inspection of Table 2 reveals that

Table 2
Contributions of various correction terms to the conditional constants (pH 7.0, $I = 0.1$ M) for Fe(III), Zn, Cu and Ni binding by 2′-deoxymugineic acid (Murakami *et al.*, 1989)

Metal	log K_{11}	MCT	MHT	LAT × 10^4	log K^{cond}
Fe^{3+}	18.4	1.8×10^{23}	4.6×10^7	1.9×10^4	11.3
Zn^{2+}	12.8	—	1.1	1.9×10^4	8.5
Cu^{2+}	18.7	—	2.1	1.9×10^4	14.1
Ni^{2+}	14.8	—	1.0	1.9×10^4	10.5

the operationally relevant binding constant at pH 7.0 is about four to seven orders of magnitude lower than the simple 1:1 constant for all the four metals.

Interpretation of Table 2 suggests that at circumneutral pH, Cu dominates over Fe(III) for binding to DMA. In comparison, Zn, and to a lesser extent, Ni, are bound less strongly by DMA than Fe(III) at pH 7. However, by pH 8, Ni and Fe(III) have almost identical K^{cond} (Fe(III) = 12.1, Ni = 12.3), and further increases in pH increase the dominance of Ni over Fe(III). Thus, at pH values common to alkaline soils, both Cu and Ni could displace Fe(III) from DMA; through a comparison of the K^{cond} values, induced Fe deficiency could be a component of phytotoxicities of these metals in the Poaceae. However, the K^{cond} also suggests that Cu–DMA (and Ni–DMA) complexes might aid Cu (and Ni) nutrition at lower metal availabilities. The weak binding of Zn relative to Fe(III) lessens the possibility that PS can significantly enhance Zn acquisition in neutral soils that contain both labile Fe and Zn, despite some of the solution culture research outlined in Section 3.

The same calculation of K^{cond} values can be used advantageously to assess the relative binding strength of PS and other ligands for Fe(III). For example, in Table 3 the relative binding of Fe(III) by DMA and by desferrioxamine B (DFOB), a microbial siderophore, can be assessed. Thus, although DFOB is a more efficient chelator of Fe(III) at circumneutral pH values, use of conditional constants shows that the difference is 5.5, rather than 12 orders of magnitude, as suggested by the 1:1 binding constants. This is in agreement with research suggesting that microbial siderophores can supply Fe to grasses but much less effectively than Fe-DMA i.e. PS do not appear to efficiently strip Fe(III) from siderophores (Yehuda *et al.*, 1996). Indeed, maize and oats supplied with ferrioxamine B (FOB) absorbed more Fe under nonaxenic conditions than under axenic conditions (Crowley *et al.*, 1992), suggesting that biodegradation of siderophores is a prerequisite step for complexation of Fe(III) by DMA. Thus, instead of being a ready source of Fe(III) for plants, microbial siderophores are likely to strip Fe(III) from DMA in the rhizosphere, and if present at high enough concentrations, microbial siderophores could cause short-term losses of available Fe in the rhizosphere system (see review in Parker *et al.*, 2005).

Table 3
Contributions of various correction terms to the conditional constants (pH 7.0, I = 0.1 M) for Fe(III) binding by 2′-deoxymugineic acid (DMA) and desferrioxamine B (DFOB) (Murakami *et al.*, 1989; Parker *et al.*, 1995)

Ligand	log K_{11}	MCT	MHT	LAT	log K^{cond}
DMA	18.4	1.8×10^{23}	4.6×10^{7}	1.9×10^{4}	11.3
DFOB	30.6	1.0×10^{26}	4.6×10^{7}	1.3×10^{6}	16.8

While inferences drawn from conditional constants allow for facile investigations of metal-ligand equilibria, to assess the overall significance of PS in the rhizosphere we need to be able to address a more complex scenario. For a more accurate reflection of competitive metal-ligand binding scenarios in the rhizosphere, there is a need to include the soil solid phase as a sparingly soluble source of metals. An approach to this problem is outlined in the section below.

5.3. Modeling phytosiderophore-induced metal mobilization from the soil solid-phase, and implications for plant nutrition

The insolubility of Fe, Zn, Cu, Ni and Cd in soil effectively constitutes a competing ligand from which PS may or may not be able to strip metals. We ran simulations using GEOCHEM-PC (Parker *et al.*, 1995) and imposed solid-phase pools with relevant solubilities (see below for explanation). GEOCHEM-PC has a conveniently editable database to which the necessary ligands and mixed–solid phases were added. We were thus able to evaluate the issue of metal–metal competition for PS in a realistic geochemical setting.

The affinity with which soil solids bind transition metals has been modeled in several ways. The simplest way, best exemplified in Lindsay's (1979) textbook, is to treat all metal solubilities as idealized solid phases. For example, the solubility of Fe(III) can be modeled using known oxyhydroxide phases such as goethite or lepidocrocite, or as an amorphous solid. The assumed log *K* for this reaction can be varied from 0.02 (goethite) to −3.54 (amorphous $Fe(OH)_3$) (Lindsay, 1979). We used a log *K* of −2.7 in our modeling work as a value typically found for soil solutions (Lindsay, 1991). In all cases, the calculated Fe^{3+} activity declines three log units for every unit increase in pH, reflecting the stoichiometry of

$$Fe^{3+} + 3H_2O \rightleftharpoons Fe(OH)_{3(s)} + 3H^+ \tag{12}$$

For other metals, such as Cd, Zn, Cu, and Ni, no simple solid with properties simulating metal solubility in soils exists. Lindsay (1979) previously advocated the concept of a fictitious solid phase called "soil-Cu." There are a number of theoretical and semi-theoretical models that have been used to describe (ad)sorption of transition metals onto reactive surfaces (Fe, Mn or Al oxides; soil organic matter). While probably more correct in a mechanistic sense than the solubility relations discussed below, these models have not proven to be particularly useful with intact soils because they contain a very complex assemblage of colloidal surfaces. Moreover, they do not seem to adequately predict increases in metal solubility with increases in total soil metal burden. This has led an increasing number of researchers to develop purely empirical models that describe trace-metal solubility as a function of simple soil parameters such as pH, organic matter content, and total metal content (e.g. McBride *et al.*, 1997; Gray *et al.*,

1999). These models are generated using multiple linear regression methods on large data sets, and yield empirical models of the general form

$$\log (M^{n+}) = a + b\,\text{pH} + c \log M_T + d \log \text{OM} \tag{13}$$

where the *a* term is an empirical constant (intercept), and the other three coefficients reflect the dependence of solubility on pH, total soil metal content and organic matter (or carbon), respectively. The *b* and *d* coefficients are negative, reflecting the suppression of metal solubilities with increasing pH or organic matter content, while the coefficient *c* is universally positive. Virtually all of the reported models include the first three terms, but the coefficient for organic matter is often statistically insignificant and the last term is thus sometimes omitted.

These purely empirical models can often describe metal solubilities with reasonable accuracy. For example, Sauvé *et al.* (2000) measured free Cd^{2+} activities in 64 soils with diverse Cd levels, pH coefficients, and OM contents. A three-term version of Eq. (13) (the OM term was nonsignificant) could explain 70% of the observed variation in Cd^{2+} activities. When Gray *et al.* (1999) combined their data set with another developed by McBride *et al.* (1997), they were able to explain 81% of the variation in (Cd^{2+}) using a full, four-term version of the model (Eq. (13)). We note that the pH coefficients for these data sets rarely approach the value of -2 suggested by Lindsays' (1979), or by limited studies of metal solubility in calcareous soils (e.g. Elfalaky *et al.*, 1991).

The value of *b* – the slope of the metal solubility as a function of pH – is often found to be between -0.5 and -0.7 for Cd^{2+} and Zn^{2+} (McBride *et al.*, 1997; Salam and Helmke, 1998; Gray *et al.*, 1999; Sauvé *et al.*, 2000). If we take -0.67 as a reasonable value for Cd^{2+}, then metal solubility will show a 2:3 dependence on soil solution pH. This can be handled by having GEOCHEM-PC consider the formation of a *mixed solid phase* with a "mythical soil ligand," the "MSL":

$$3Cd^{2+} + 4MSL^{-} + 2H_2O \rightarrow Cd_3MSL_4(OH)_{2(s)} + 2H^{+} \tag{14}$$

There is a stability constant, $K_{34\text{-}2}$, that describes the equilibrium quotient for this reaction. Taking logarithms of that quotient and rearranging yields

$$\log (Cd^{2+}) = -1/3 \log K_{34\text{-}2} - 4/3 \log (MSL^{-}) - 2/3 \log \text{pH} \tag{15}$$

Eq. (15) now possesses the desired 2:3 dependency of (Cd^{2+}) on pH. The empirically derived *a* term (intercept) in Eq. 13 can be incorporated by choosing an appropriate value for $K_{34\text{-}2}$. For example, with the "combined" data set of Gray *et al.* (1999), the *a* term is -3.73 after conversion from $\mu g\ L^{-1}$ to molarity; thus, a value of 11.2 would be employed for $\log K_{34\text{-}2}$ in GEOCHEM-PC.

Finally, the c and d terms in Eq. (13) that account for the effects of metal loading and organic matter, respectively, can be incorporated into the GEOCHEM-PC simulations by adjusting the activity of free MSL^- in Eq. (15) for each specific soil. Again, looking at the combined data set in Gray *et al.* (1999), c was found to be 0.85 and d to be -0.45, when Cd_T and OM were in mg/kg and g C/kg, respectively. Thus, for a soil containing 10 mg kg^{-1} of Cd and 10 mg kg^{-1} organic C, the sum of the two terms in Eq. (13) would be 0.40. The concentration of total MSL would thus be adjusted so that $(MSL^-) = 0.50$ M, and Eq. (15) would be treated by GEOCHEM-PC as

$$\log(Cd^{2+}) = -3.73 + 0.40 - 2/3 \log pH \tag{16}$$

For other soils with different Cd and OM contents, the activity of the MSL^- can be adjusted so that the middle term on the right-hand side of Eq. (16) appropriately shifts the solubility up or down. Similarly, the value of log $K_{34\text{-}2}$ can be adjusted in the GEOCHEM-PC database to employ different versions of the empirical solubility relation shown in Eq. (15).

We employed the same approach to accommodate other pH-dependent solubilities for Zn, Cu and Ni with important parameters in Table 4. When modeling such solid-phase solubilities using GEOCHEM-PC, an "excess" concentration of each metal is imposed so that the solid-phase reservoir is "infinite." Similarly, the activity of the MSL^- that is imposed to adjust solubility is large and invariant. Its high value (see above) will not interfere with our simulations because ionic strength can be fixed at an appropriate value by GEOCHEM-PC. In our study, we assumed a more realistic ionic strength of 0.03 M and soil moisture was set at 25% (w/w).

Table 4
Important parameters used to develop the mythical soil ligand (MSL) for input into Geochem-PC (Parker *et al.*, 1995)

Metal	Coefficient				pH–solubility dependence	log K	Reference
	a	b	c	d			
Cd	4.31	0.60	0.85	0.45	~2:3	11.2	Gray *et al.* (1999)
Zn	4.44	0.71	0.68	n.a.	~2:3	10.8	McBride *et al.* (1997)
Cu	1.37	0.21	0.93	0.21	~1:5	32.2	Sauvé *et al.* (2000)
Ni	7.02	1.05	1.21	0.85	~1:1	0.8	Sauvé *et al.* (2000)

Notes: The empirical solubility model used was: $\log(M^{n+}) = a + b\,pH + c \log M_T + d \log OM$, where a is an empirical constant (intercept) and the other three coefficients reflect the dependence of solubility on pH, total soil metal content and organic matter (or carbon), respectively.

For metal–DMA solubility, we used stability constants determined by Murakami *et al.* (1989) (Table 1). No stability constants have been determined for Cd; hence, the log *K* for Cd was assumed to be the same as Zn. As discussed in Section 2.3, the better quantitative studies and more accurate modeling suggest that concentrations of DMA in the rhizosphere are most likely in the μM range. Thus, we focused our simulations here on DMA concentrations of 25 μM.

To test the ability of Fe(III) to compete for DMA with other metals, two-metal scenarios were modeled. The findings from this suggest that while Fe(III) is able to effectively compete with Zn, Cd, Cu or Ni at low pH and complex the majority of the DMA present, at higher pHs, the other metals predominate DMA complexation (see Fig. 3). Further GEOCHEM-PC simulations with 25 μM concentrations of DMA, Fe(III), and Cd/Zn/Cu/Ni, and with neither precipitation nor "mythical soil ligand," were of the same general shape as those in Fig. 1 but had

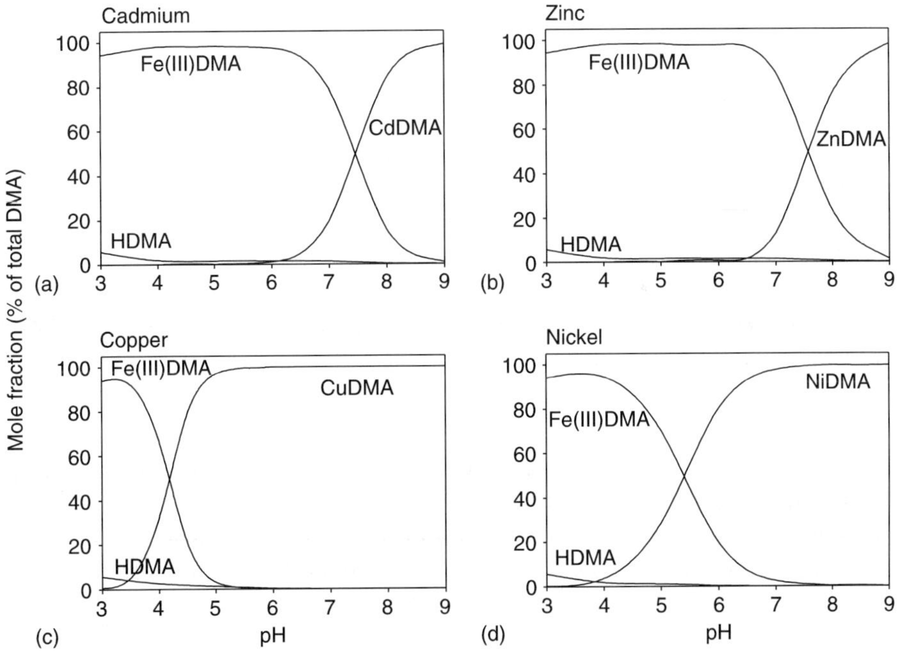

Fig. 3. Distribution of metal-2′-deoxymugineic acid (DMA) in solution as a function of pH. Metal binding by the soil solid phase is simultaneously considered using empirically derived solubilities. For Cd, Zn, Cu, and Ni the empirical solubility model used was: $\log(M^{n+}) = a + b\,\mathrm{pH} + c \log M_T + d \log \mathrm{OM}$, where *a* is an empirical constant (intercept) and the other three coefficients reflect the dependence of solubility on pH, total soil metal content and organic matter (or carbon), respectively. $Fe(OH)_3$ had an assumed solubility of $\log K = -2.7$ (Lindsay, 1979). The total soil metal concentration was 10 mg kg^{-1}, organic C was 10 g kg^{-1} and DMA was 25 μM.

different crossover points. For example, Cu and Fe(III), were at equimolar concentrations at a pH of ~3.5 (cf. ~4.5 in Fig. 3) and Fe(III) and Cd at ~8.5 (cf. ~7.5). This indicates the importance of the assumed solubility of the soil solid-phase for influencing the speciation of PS in the rhizosphere. These findings have a number of possible implications for bioavailability of metals to plants.

Regardless of whether field-grown plants excrete PS in response to deficiencies of micronutrients other than Fe, metal-binding by PS may have implications for solubilization and enhanced uptake of metals from contaminated soils, and hence, the potential to increase food-chain transfer hazards. Cd is highly toxic to humans and animals, and is thus is of particular concern. While the mechanisms utilized by plants for Cd uptake are poorly understood (Welch and Norvell, 1999), recent research has at least implicated PS in the plant uptake of Cd. Cd uptake has been demonstrated to be greater in Fe-efficient vs. Fe-inefficient species and cultivars, presumably reflecting greater PS production by the former (Mench and Fargues, 1994; Awad and Römheld, 2000; Römheld and Awad, 2000). In contrast, although PS solubilized Cd from $Cd_3(PO_4)_2$, Cd uptake in wheat and barley in solution culture was best described by the free metal ion activity of Cd in solution (Shenker *et al.*, 2001). However, this anomaly may be explained by the absence of any diffusive limits to transport to the root in well-mixed nutrient solutions.

The affinity of DMA for transition metals other than Fe may also have implications for the bioavailability of those metals (Cu, Ni, Zn) for which phytotoxicity is the principal concern when they are present at excessive levels in soil. A comparison of metal uptake in Fe-efficient and Fe-inefficient oats grown in a contaminated sludge demonstrated more Zn uptake in the Fe-efficient (and presumably high PS-producing) cultivar (Mench and Fargues, 1994). Preculturing of wheat and sorghum in Fe-deficient cultures resulted in an increased uptake of Zn and Ni when the plants were transferred to soils contaminated with either sewage sludge or individual metal (Awad and Römheld, 2000; Römheld and Awad, 2000). The increase in metal uptake in wheat was more pronounced than in sorghum, which tallies with the lower Fe-efficiency and lower levels of PS release found in sorghum (Römheld and Awad, 2000).

The binding of metals such as Cu and Ni by DMA could interfere with the ability of PS to provide adequate Fe to meet plant nutritional needs. Uptake of Fe complexed with PS in barley was reduced by excess Cu and Zn in nutrient solution (Ma *et al.*, 1993). McBride (2001) found an inverse relationship between Fe and Cu in the shoots of maize grown in a number of Cu-contaminated peat soils which could indicate Cu competition with Fe for binding with PS. In addition, Kukier and Chaney (2000) have postulated that the distinct chlorotic banding in Ni-intoxicated oats is related to the diurnal pattern of PS release. That is, Ni may have a negative impact on Fe-PS chelation and uptake, but the severity of this effect will be at a minimum during the daily peak release of PS, resulting in a narrow band of greener leaf tissue produced during that time.

At alkaline pH values, where plants are most likely to be Fe deficient and hence to produce significant quantities of PS, Fe(III) is unable to compete with common metals in the rhizosphere for binding to the DMA complex (Fig. 1). This is likely to be a problem even in noncontaminated soils. So, how are the Poaceae able to utilize DMA for Fe(III) uptake if Fe(III) is a weak binder of DMA at relevant pH values? The diurnal pattern of PS release may be important for providing localized high concentrations of PS in the rhizosphere, whereby undersaturation of the PS complex by other metals provides sites for Fe(III) binding. In addition, organic ligands with comparatively low affinities for Cu, Ni, Cd and Zn e.g. microbial siderophores, humic acids, and organic acids, may be able to strip Fe from the soil and provide a source of Fe for binding with PS.

6. CONCLUSIONS

While PS concentrations in the rhizosphere are in a continual state of flux through time and space and only appear to be present at μM concentrations, they have the potential to have a significant impact on environmental processes and human health. Phytosiderophores are not highly selective for Fe, and other transition metals, such as Cd, Zn, Cu and Ni, are able to effectively compete with Fe for chelation by PS at pHs found in the rhizosphere. In addition, the available evidence suggests that PS are capable solubilizers of a range of transition metals from the soil solid-phase and organic matter in general. However, while the evidence points to PS as efficient solubilizers of transition metals, we do not yet conclusively know if metal–PS complexes, besides Fe(III)-PS, are absorbed by plant roots in significant quantities. Thus, the efficient solubilization of transition metals may only have implications via the production of a soluble pool for uptake rather than via direct uptake of metal–ligand complexes. Therefore, the evidence suggests that PS could provide a soluble pool of metals for uptake with a number of potential implications including:

(1) increasing the bioavailability of trace metals where deficiency is the primary concern (particularly Cu and Ni, and to a lesser extent, Zn),
(2) increasing the bioavailability of metals where phytotoxicity is the main concern (particularly Cu and Ni, and to a lesser extent, Zn),
(3) the potential for food-chain transfer hazards for metals such as Cd where toxicity to humans and animals is the chief concern, and,
(4) competition between Fe(III) and other metals for binding to PS complexes may reduce solubilized Fe(III), and hence, has the potential to induce Fe deficiency (especially Cu but also Ni, Cd and Zn).

However, while we have been able to review the literature and undertake modeling to give probable scenarios for the biogeochemical interactions of metal–PS complexes in the rhizosphere, there is still a need for validation with empirical data. Indeed, our review of the literature has demonstrated the importance of minimizing confounding factors if unambiguous answers are to be found, and we look forward to studies in our laboratory and others that are able to give more definitive answers to questions on metal–PS speciation in the rhizosphere.

REFERENCES

Alvarez Fernandez, A., Garate, A., Lucena, J.J., 1997. Interaction of iron chelates with several soil materials and with a soil standard. J. Plant Nutr. 20, 559–572.

Awad, F., Römheld, V., 2000. Mobilization of heavy metals from contaminated calcareous soils by plant born, microbial and synthetic chelators and their uptake by wheat plants. J. Plant Nutr. 23, 1847–1855.

Awad, F., Römheld, V., Marschner, H., 1988. Mobilization of ferric iron from a calcareous soil by plant- borne chelators. J. Plant Nutr. 11, 701–713.

Bernards, M.L., Jolley, V.D., Stevens, W.B., Hergert, G.W., 2002. Phytosiderophore release from nodal, primary, and complete root systems in maize. Plant Soil 241, 105–113.

Bertrand, I., Hinsinger, P., 2000. Dissolution of iron oxyhydroxide in the rhizosphere of various crop species. J. Plant Nutr. 23, 1559–1577.

Cakmak, I., Erenoglu, B., Gulut, K.Y., Derici, R., Römheld, V., 1998. Light-mediated release of phytosiderophores in wheat and barley under iron or zinc deficiency. Plant Soil 202, 309–315.

Cakmak, I., Ozturk, L., Karanlik, S., Marschner, H., Ekiz, H., 1996a. Zinc-efficient wild grasses enhance release of phytosiderophores under zinc deficiency. J. Plant Nutr. 19, 551–563.

Cakmak, I., Sari, N., Marschner, H., Ekiz, H., Kalayci, M., Yilmaz, A., Braun, H.J., 1996b. Phytosiderophore release in bread and durum wheat genotypes differing in zinc efficiency. Plant Soil 180, 183–189.

Cesco, S., Nikolic, M., Römheld, V., Varanini, Z., Pinton, R., 2002. Uptake of Fe-59 from soluble Fe-59-humate complexes by cucumber and barley plants. Plant Soil 241, 121–128.

Cesco, S., Romheld, V., Varanini, Z., Pinton, R., 2000. Solubilization of iron by water-extractable humic substances. J. Plant Nut. Soil Sci. 163, 285–290.

Chaney, R.L., Brown, J.C., Tiffin, L.O., 1972. Obligatory reduction of ferric chelates in iron uptake by soybeans. Plant Physiol. 50, 208–213.

Clarkson, D.T., Sanderson, J., 1978. Sites of absorption and translocation of iron in barley roots. Plant Physiol. 61, 731–736.

Crowley, D.E., Gries, D., 1994. Modeling of iron availability in the plant rhizosphere, In: Manthey, J.A., Crowley, D.E., Luster, D.G., (Eds.), Biochemistry of Metal Micronutrients in the Rhizosphere. Lewis Publishers, Boca Raton, pp. 199–223.

Crowley, D.E., Romheld, V., Marschner, H., Szaniszlo, P.J., 1992. Root-microbial effects on plant iron uptake from siderophores and phytosiderophores. Plant Soil 142, 1–7.

Curie, C., Panaviene, Z., Loulergue, C., Dellaporta, S.L., Briat, J.F., Walker, E.L., 2001. Maize *yellow stripe1* encodes a membrane protein directly involved in Fe(III) uptake. Nature 409, 346–349.

Elfalaky, A.A., Aboulroos, S.A., Lindsay, W.L., 1991. Measurement of cadmium activities in slightly acidic to alkaline soils. Soil Sci. Soc. Am. J. 55, 974–979.

Figueira, A., Kido, E.A., Almeida, R.S., 2001. Identifying sugarcane expressed sequences associated with nutrient transporters and peptide metal chelators. Genet. Mol. Biol. 24, 207–220.

Fushiya, S., Takahashi, K., Nakatsuyama, S., Sato, Y., Nozoe, S., Takagi, S., 1982. Co-occurrence of nicotianamine and avenic acids in *Avena sativa* and *Oryza sativa*. Phytochemistry 21, 1907–1908.

Gerke, J., 1997. Aluminum and iron(III) species in the soil solution including organic complexes with citrate and humic substances. Z. Pflanz. Bodenkunde. 160, 427–432.
Gerke, J., 2000. Mathematical modelling of iron uptake by graminaceous species as affected by iron forms in soil and phytosiderophore efflux. J. Plant Nutr. 23, 1579–1587.
Gray, C.W., McLaren, R.G., Roberts, A.H.C., Condron, L.M., 1999. Solubility, sorption and desorption of native and added cadmium in relation to properties of soils in New Zealand. Eur. J. Soil Sci. 50, 127–137.
Gries, D., Brunn, S., Crowley, D.E., Parker, D.R., 1995. Phytosiderophore release in relation to micronutrient metal deficiencies in barley. Plant Soil 172, 299–308.
Gries, D., Klatt, S., Runge, M., 1998. Copper-deficiency-induced phytosiderophore release in the calcicole grass *Hordelymus europaeus*. New Phytol. 140, 95–101.
Gries, D., Runge, M., 1992. The ecological significance of iron mobilization in wild grasses. J. Plant Nutr. 15, 1727–1737.
Grusak, M.A., Pearson, J.N., Marentes, E., 1999. The physiology of micronutrient homeostasis in field crops. Field Crop. Res. 60, 41–56.
Hiradate, S., Inoue, K., 1998. Interaction of mugineic acid with iron (hydr)oxides: sulfate and phosphate influences. Soil Sci. Soc. Am. J. 62, 159–165.
Inoue, K., Hiradate, S., Takagi, S., 1993. Interaction of mugineic acid with synthetically produced iron oxides. Soil Sci. Soc. Am. J. 57, 1254–1260.
Jones, D.L., 1998. Organic acids in the rhizosphere—a critical review. Plant Soil 205, 25–44.
Kawai, S., Takagi, S.I., Sato, Y., 1988. Mugineic acid-family phytosiderophores in root-secretions of barley, corn and sorghum varieties. J. Plant Nutr. 11, 633–642.
Kukier, U., Chaney, R.L., 2000. Remediating Ni-phytotoxicity of contaminated quarry muck soil using limestone and hydrous iron oxide. Can. J. Soil Sci. 80, 581–593.
Lindsay, W.L., 1979. Chemical Equilibria in Soils. Wiley, New York.
Lindsay, W.L., 1991. Inorganic equilibria affecting micronutrients in soils, In: Mortvedt, J.J., Cox, F.R., Shuman, L.M., Welch, R.M., (Eds.), Micronutrients in Agriculture, second ed. Soil Science Society of America, Madison, pp. 113–144,
Ma, J.F., 1993. Inhibition of mugineic acid ferric complex uptake in barley by copper, zinc, and cobalt. Physiol. Plant. 89, 331–334.
Ma, J.F., Kusano, G., Kimura, S., Nomoto, K., 1993. Specific recognition of mugineic acid ferric complex by barley roots. Phytochemistry 34, 599–603.
Ma, J.F., Nomoto, K., 1996. Effective regulation of iron acquisition in graminaceous plants. The role of mugineic acids as phytosiderophores. Physiol. Plant. 97, 609–617.
Marschner, H., Römheld, V., Kissel, M., 1987. Localization of phytosiderophore release and of iron uptake along intact barley roots. Physiol. Plant 71, 157–162.
McBride, M., Sauve, S., Hendershot, W., 1997. Solubility control of Cu, Zn, Cd and Pb in contaminated soils. Eur. J. Soil Sci. 48, 337–346.
McBride, M.B., 2001. Cupric ion activity in peat soil as a toxicity indicator for maize. J. Environ. Qual. 30, 78–84.
Mench, M.J., Fargues, S., 1994. Metal uptake by iron-efficient and inefficient oats. Plant Soil 165, 227–233.
Morel, F.M.M., Hering, J.G., 1993. Principles amd Application of Aquatic Chemistry. Wiley, New York.
Mori, S., 1994. Mechanisms of iron acquisition by graminaceous (Strategy II) plants, In: Crowley, D.E., Luster, D.G., (Eds.), Biochemistry of Metal Micronutrients in the Rhizosphere, CRC Press, Boca Raton, pp. 225–249,
Murakami, T., Ise, K., Hayakawa, M., Kamei, S., Takagi, S.I., 1989. Stabilities of metal-complexes of mugineic acids and their specific affinities for iron(III). Chem. Lett. 12, 2137–2140.

Noma, M., Noguchi, M., 1976. Occurrence of nicotianamine in higher-plants. Phytochemistry 15, 1701–1702.

Nomoto, K., Sugiura, Y., Takagi, S.I., 1987. Mugineic acids, studies on phytosiderophores, In: Winkelmann, G., van der Helm, D., Neilands, J.B. (Eds.), Iron Transport in Microbes, Plants and Animals, VCH Publishers, New York, pp. 401–425,

Nomoto, K., Yoshioka, H., Arima, M., Fushiya, S., Takagi, S., Takemoto, T., 1981. Structure of 2'-deoxymugineic acid, a novel amino-acid possessing an iron-chelating activity. Chimia 35, 249–250.

Pandeya, S.B., Singh, A.K., Dhar, P., 1998. Influence of fulvic acid on transport of iron in soils and uptake by paddy seedlings. Plant Soil 198, 117–125.

Parker, D.R., Chaney, R.L., Norvell, W.A., 1995. GEOCHEM-PC: a chemical speciation program for IBM and compatible personal computers, In: Loeppert, R.H., Schwab, A.P., Goldberg, S., (Eds.), SSSA Spec. Pub. No. 38. American Society of Agronomy, Soil Science Society of America, Madison, WI, pp. 253–269.

Parker, D.R., Reichman, S.M., Crowley, D.E., 2005. Metal chelation in the rhizosphere, In: Wright, S.F., Zabeb, R.W. (Eds.), Root Soil Management: Interactions Between Roots and the Soil, Monograph No. 48. American Society of Agronomy, Soil Science Society of America, Madison, WI (in press).

Pedler, J.F., Parker, D.R., Crowley, D.E., 2000. Zinc deficiency-induced phytosiderophore release by the Triticaceae is not consistently expressed in solution culture. Planta 211, 120–126.

Reichman, S.M., Parker., D.R., 2002. Revisiting the metal-binding chemistry of nicotianamine and 2′-deoxymugineic acid. Implications for iron nutrition in Strategy II plants. Plant Physiol. 129, 1435–1438.

Rengel, Z., 1999. Zinc deficiency in wheat genotypes grown in conventional and chelator-buffered nutrient solutions. Plant Sci. 143, 221–230.

Römheld, V., 1991. The role of phytosiderophores in acquisition of iron and other micronutrients in graminaceous species – an ecological approach. Plant Soil 130, 127–134.

Römheld, V., Awad, F., 2000. Significance of root exudates in acquisition of heavy metals from a contaminated calcareous soil by graminaceous species. J. Plant Nutr. 23, 1857–1866.

Römheld, V., Marschner, H., 1986. Evidence for a specific uptake system for iron phytosiderophores in roots of grasses. Plant Physiol. 80, 175–180.

Rovira, A.D., 1969a. Diffusion of carbon compounds away from wheat roots. Aust. J. Biol. Sci. 22, 1287–1290.

Rovira, A.D., 1969b. Plant root exudates. Bot. Rev. 35, 35–58.

Salam, A.K., Helmke, P.A., 1998. The pH dependence of free ionic activities and total dissolved concentrations of copper and cadmium in soil solution. Geoderma 83, 281–291.

Sauvé, S., Norvell, W.A., McBride, M., Hendershot, W., 2000. Speciation and complexation of cadmium in extracted soil solutions. Env. Sci. Tech. 34, 291–296.

Scholz, G., Becker, R., Pich, A., Stephan, U.W., 1992. Nicotianamine–a common constituent of strategy-I and strategy-II of iron acquisition by plants–a review. J. Plant Nutr. 15, 1647–1665.

Shenker, M., Fan, T.W.M., Crowley, D.E., 2001. Phytosiderophores influence on cadmium uptake by wheat and barley plants. J. Environ. Qual. 30, 2091–2098.

Shi, W.M., Chino, M., Yousef, R.A., Mori, S., Takagi, S., 1988. The occurrence of mugineic acid in the rhizosphere soil of barley plant. Soil Sci. Plant Nutr. 34, 585–592.

Singh, K., Chino, M., Nisizawa, N.K., Goto, S., Nakanishi, T., Takagi, S., Mori, S., 1992. Iron extraction efficacy of plant borne mugineic acid family phytosiderophores in Indian calcareous soils. J. Plant Nutr. 15, 1625–1645.

Singh, K., Sasakuma, T., Bughio, N., Takahashi, M., Nakanishi, H., Yoshimura, E., Nishizawa, N.K., Mori, S., 2000. Ability of ancestral wheat species to secrete mugineic acid family phytosiderophores in response to iron deficiency. J. Plant Nutr. 23, 1973–1981.

Stumm, W., Morgan, J.J., 1996. Aquatic chemistry: chemical equilibria and rates in natural waters. John Wiley and Sons, New York.

Sugiura, Y., Nomoto, K., 1984. Phytosiderophores structures and properties of mugineic acids and their metal complexes. Struct. Bond. 58, 107–135.

Sugiura, Y., Tanaka, H., Mino, Y., Ishida, T., Ota, N., Inoue, M., Nomoto, K., Yoshioka, H., Takemoto, T., 1981. Structure, propeties and transport mechanism of iron(III) complex of mugineic acid, a possible phytosiderophore. J. Am. Chem. Soc. 103, 6979–6982.

Takagi, S., 1976. Naturally occurring iron-chelating compounds in oat- and rice-root washings. i. Activity measurement and preliminary characterization. Soil Sci. Plant Nutr. 22, 423–433.

Takagi, S., Nomoto, K., Takemoto, T., 1984. Physiological aspect of mugineic acid, a possible phytosiderophore of graminaceous plants. J. Plant Nutr. 7, 469–477.

Takagi, S.I., 1993. Production of phytosiderophores, In: Barton, L.L., Hemming, B.C. (Eds.), Iron Chelation in Plants and Soil Microorganisms, Academic Press, San Diego, pp. 111–131,

Takagi, S.I., Kamei, S., Yu, M.H., 1988. Efficiency of iron extraction from soil by mugineic acid family phytosiderophores. J. Plant Nutr. 11, 643–651.

Takemoto, T., Nomoto, K., Fushiya, S., Ouchi, R., Kusano, G., Hikino, H., Takagi, S., Matsuura, Y., Kakudo, M., 1978. Structure of mugineic acid, a new amino acid possessing an iron-chelating activity from roots washings of water-cultured *Hordeum vulgare* L. Proc. Jap. Acad. Ser. B. Phys. Biol. Sci. 54, 469–473.

Treeby, M., Marschner, H., Römheld, V., 1989. Mobilization of iron and other micronutrient cations from a calcareous soil by plant-borne, microbial, and synthetic metal chelators. Plant Soil 114, 217–226.

von Wirén, N., Klair, S., Bansal, S., Briat, J.F., Khodr, H., Shioiri, T., Leigh, R.A., Hider, R.C., 1999. Nicotianamine chelates both Fe(III) and Fe (II). Implications for metal transport in plants. Plant Physiol. 119, 1107–1114.

von Wirén, N., Marschner, H., Römheld, V., 1996. Roots of iron-efficient maize also absorb phytosiderophore-chelated zinc. Plant Physiol. 111, 1119–1125.

von Wirén, N., Mori, S., Marschner, H., Römheld, V., 1994. Iron inefficiency in maize mutant ys1 (*Zea mays* L. cv yellow-stripe) is caused by a defect in uptake of iron phytosiderophores. Plant Physiol. 106, 71–77.

Walter, A., Pich, A., Scholz, G., Marschner, H., Römheld, V., 1995. Effects of iron nutritional status and time of day on concentrations of phytosiderophores and nicotianamine in different root and shoot zones of barley. J. Plant Nutr. 18, 1577–1593.

Walter, A., Römheld, V., Marshner, H., Mori, S., 1994. Is the release of phytosiderophores in zinc-deficient plants a response to impaired iron utilization? Physiol. plant 92, 493–500.

Weavind, T.E.F., Hodgson, J.F., 1971. Iron absorption by wheat roots as a function of distance from the root tip. Plant Soil 34, 697–705.

Welch, R.M., 1995. Micronutrient nutrition of plants. Crit. Rev. Plant Sci. 14, 49–82.

Welch, R.M., Norvell, W.A., 1999. Mechanisms of cadmium uptake, translocation and deposition in plants, In: McLaughlin, M. J., Singh, B. R., (Eds.), Cadmium in Soils and Plants. Kluwer Academic Publishers, Dordrecht, pp. 125—150.

Welch, R.M., Norvell, W.A., Schaefer, S.C., Shaff, J.E., Kochian, L.V., 1993. Induction of iron(III) and copper(II) reduction in pea (*Pisum sativum* L) roots by Fe and Cu status: does the root-cell plasmalemma Fe(III)-chelate reductase perform a general role in regulating cation uptake. Planta 190, 555–561.

Yehuda, Z., Shenker, M., Hadar, Y., Chen, Y.N., 2000. Remedy of chlorosis induced by iron deficiency in plants with the fungal siderophore rhizoferrin. J. Plant Nutr. 23, 1991–2006.

Yehuda, Z., Shenker, M., Römheld, V., Marschner, H., Hadar, Y., Chen, Y.N., 1996. The role of ligand exchange in the uptake of iron from microbial siderophores by gramineous plants. Plant Physiol. 112, 1273–1280.

Zhang, F., Römheld, V., Marschner, H., 1989. Effect of zinc deficiency in wheat on the release of zinc and iron mobilizing root exudates. Z. Pflanz. Bodenkunde. 152, 205–210.

Zhang, F.S., Treeby, M., Römheld, V., Marschner, H., 1991. Mobilization of iron by phytosiderophores as affected by other micronutrients. Plant Soil 130, 173–178.

Biogeochemistry of Trace Elements in the Rhizosphere
P.M. Huang and G.R. Gobran (Editors)

Chapter 5

Effects of organic ligands on the adsorption of trace elements onto metal oxides and organo–mineral complexes

A. Violante, M. Ricciardella, M. Pigna, and R. Capasso

Dipartimento di Scienze del Suolo, della Pianta e dell' Ambiente, Via Università 100, 80055 Portici (Napoli), Italy

ABSTRACT

Biomolecules released by plants and microorganisms have an important effect on the mobility of trace elements in soil environments. We have found that in the presence of low-molecular-mass organic ligands (e.g. oxalate (OX) and citrate), the adsorption of heavy metals on short-range-ordered oxides increased with an increase in the initial organic ligand/Me molar ratio (r_L) up to 10, whereas on well-crystallized Al- or Fe-oxides, the adsorption initially increased up to 2.5 (goethite) or 5 (bayerite) and then decreased. The r_L value, the nature of the organic ligands and heavy metals, and the surface properties of the sorbents are critical in determining whether metal adsorption is enhanced or inhibited. Organic ligands inhibit the adsorption of trace elements in anionic form (arsenate (As), arsenite, and selenite). The adsorption of heavy metals and metalloids on variable-charge minerals is influenced not only by the presence, nature, and concentration of organic ligands but also by the sequence of addition of trace elements on the sorbents.

Trace elements in cationic and anionic form show a different adsorption capacity onto metal oxides, organic matter, and organomineral complexes. Heavy metals also compete for sorption sites on these soil components. However, we found that Cu inhibits the sorption of Zn more on ferrihydrite than on an organomineral complex or a humic acid-like material.

1. INTRODUCTION

In the rhizosphere, a wide variety and considerable amounts of organic compounds are released by plants or microorganisms, which are present in significantly greater amounts than in the bulk soil. High- and low-molecular-mass substances, such as mucilages, polysaccharides, proteins, carbohydrates, phenolics, phytosiderophores, amino acids, and organic acids, are released at the soil–root interface (Marschner, 1995; Jones, 1998; Huang and Germida, 2002).

Excretion products of roots include a variety of low-molecular-mass organic acids (LMMOAs) such as acetic, butyric, oxalic, malic, propionic, succinic, citric, tartaric, and fumaric acid. The relative abundance of LMMOAs in the rhizosphere is both species and cultivar dependent (Marschner, 1995).

Studies have shown that rhizosphere soil exhibits differences in weathering, physical characteristics, and mineralogy compared with bulk soil. Larger amounts of Al and Fe were extracted from rhizosphere than from nonrhizosphere clay samples (Sarkar *et al.*, 1979).

Organic ligands interact with Fe and Al released from primary and secondary minerals, promoting the formation of short-range-ordered precipitates and organomineral complexes. Biomineralization of noncrystalline Al-oxides common in the cells of mature root bodies has been described earlier by April and Keller (1990a, b). This behavior is similar to that ascertained in the weathering interface between lichens and rocks, leading to the first stages in the formation of soil (Wilson, 1995). Lichen weathering results in poorly ordered secondary products, and although these may become better crystallized with time, this has rarely been observed. Ferrihydrite is likely to be the main component of the short-range-ordered Fe-oxyhydroxides pool. Goethite was detected in association with hematite in the *Sterocaulon vesuvianum*–Vesuvius volcanic rock-weathering interface (Adamo and Violante, 2000).

It is likely that at the rock–lichen interface as well as at the soil–root interface, Fe and Al mobilized through weathering of silicates are primarily bound by biomolecules in organic complexes (Robert and Berthelin, 1986; Wilson, 1995; Jones, 1998; Violante and Gianfreda, 2000; Huang and Germida, 2002). Organics retained on the surfaces or present in the network of short-range-ordered precipitates strongly inhibit the crystallization of metal hydroxides or oxyhydroxides (Huang and Violante, 1986; Violante *et al.*, 2002c).

Crystalline, noncrystalline, or short-range-ordered Fe, Al and Mn hydroxides and oxyhydroxides as well as some short-range-ordered aluminosilicates (allophane and imogolite) react readily with inorganic anions, cations, and organic molecules of LMMOAs and biopolymers (e.g. enzymes, toxins, DNA, RNA, and polysaccharides). Organic compounds, present at the soil–root interface, compete for common sites on metal oxides and influence the adsorption of other organic and inorganic anions and cations, both nutrients or pollutants (Nagarajah *et al.*, 1970; Lopez-Hernandez *et al.*, 1986; Xu *et al.*, 1988; Inskeep, 1989; Marschner, 1995; Liu *et al.*, 1999, 2001; Huang *et al.*, 2002; Violante *et al.*, 2002a, b).

The effect of LMMOAs on the adsorption/desorption of pollutants on soil components has received some attention only in the last two decades (*inter alias*, Balistrieri and Chao, 1987; Dynes and Huang, 1997; Jackson, 1998; Gobran *et al.*, 2000; Liu *et al.,* 2001; Huang and Germida, 2002; Violante *et al.*, 2002a, b, c; Violante *et al.*, 2003).

The role of organic (and inorganic) ligands on the adsorption of heavy metals in cationic form on the surfaces of variable-charge minerals is still obscure. It has been demonstrated that organic ligands may either hinder or promote the sorption of metal cations, depending, respectively, on whether the metal complexes they form remain in solution or are themselves sorbed by the mineral, and on whether sorbed complexes are bound more strongly or less strongly than the free cations (Bar-Josef, 1991; Chairidchai and Ritchie, 1992; Jackson, 1998).

Recently, we have carried out studies on the effect of LMMOAs on the adsorption of selected heavy metals and metalloids onto/from metal oxides, variable-charge soils, and organomineral complexes. The aim of this work is to present some of our significant findings on the effect of biomolecules, usually present in the rhizosphere, on the mobility of trace elements in soil environments. We also compare our results with those reported in the literature.

2. CHEMISTRY AND MINERALOGY OF METAL OXIDES

2.1. Metal oxides in soil environments

Oxides of Fe, Al, and Mn commonly occur in soils, particularly those in advanced stages of weathering (Hsu, 1989; Cornell and Schwertmann, 1996; Kampf *et al.*, 2000; Churchman, 2000; Violante *et al.*, 2002c).

Al-hydroxides crystallize in three polymorphs: gibbsite, bayerite, and nordstrandite. Gibbsite is a common product of tropical and subtropical weathering. It is one of the major minerals in many Oxisols and a minor mineral component of many Ultisols. The apparent rarity of nordstrandite and bayerite in nature reflects difficulties in identifying them due to their low concentrations and masking by the presence of gibbsite. Gibbsite often shows hexagonal plates, but elongated hexagonal rods have also been observed. Bayerite crystals have a triangular or barrel-shaped morphology, whereas nordstrandite has been found in nature as crystals radiating into solution cavities and as small pellets in alkaline environments (see Hsu, 1989; Violante *et al.*, 2002c and references therein).

The two crystalline Al-oxyhydroxide polymorphs that exist in natural environments are boehmite and diaspore. Gibbsite, boehmite, and diaspore are the three hydrates of Al, which are the main constituents of bauxites and laterites.

Short-range-ordered or noncrystalline Al precipitation products are ubiquitous in soil environments (Parfitt, 1980; Kampf *et al.*, 2000). They dominate the chemical reactions in soils because of their extremely small particle sizes and highly reactive surfaces. In pure form they are not stable, but in the presence of chelating anions they may remain unchanged indefinitely. Short-range-ordered Al precipitates and soluble OH-Al species often coat crystalline minerals in soils or may be interlayered into the interlamellar spaces of vermiculites or smectites, altering the surface properties of these phyllosilicates. OH-Al interlayered vermiculites and smectites are particularly abundant in Ultisols and Alfisols (Barnhisel and Bertsch, 1989).

The major Fe oxides and oxyhydroxides may exist as crystalline minerals (hematite, goethite, lepidocrocite, maghemite, and magnetite), short-range-ordered

(6- or 2-line ferrihydrite), or noncrystalline precipitates, which are partly present as coatings on clay minerals and humic substances (Cornell and Schwertmann, 1996).

Goethite and hematite are the most abundant crystalline Fe-oxides in soils. Hematite forms from ferrihydrite through a solid-state reaction, whereas goethite forms from ferrihydrite via a dissolution–precipitation process. Hematite tends to form under higher temperatures in dry situations and with low organic matter contents, whereas goethite formation is promoted by low temperatures, high soil moisture, and relatively high organic matter contents. Hematite is usually in the form of hexagonal plates, but other morphologies have been described, such as rods and granular ellipsoids. The basic morphologies of goethite are acicular needles.

Ferrihydrite covers a range of poorly ordered compounds, whose degree of ordering depends on the rate of hydrolysis and the time of aging. It is mostly found in soils, especially podsols (B horizons), Andisols, and placic horizons. Its formation is favored in the presence of high concentrations of silicates, phosphate, organic matter, and some cations, including Al and Mn, which inhibit the formation of crystalline Fe oxides. Ferrihydrite forms very small particles, 4–6 nm in size and more or less spherical in shape.

Many studies have demonstrated that at certain critical Fe/Al ratios, Fe and Al ions may coprecipitate, forming short-range-ordered mixed oxides (aluminous ferrihydrites) of different chemical composition, size, nature, solubility, and reactivity (Cornell and Schwertmann, 1996; Colombo and Violante, 1996).

Lepidocrocite is much less common than hematite and goethite, but it is not rare. Iron coatings around rice roots, formed of goethite and lepidocrocite, have been identified. Maghemite, magnetite, schwertmannite, and akaganeite are other Fe-oxides present in soil environments, which form under specific conditions (Cornell and Schwertmann, 1996).

2.2. Surface chemistry of metal oxides

The nature of the surfaces of pure oxides in contact with aqueous solutions is pH-dependent (Hsu, 1989; McBride, 1994; Goldberg *et al.*, 1996; Cornell and Schwertmann, 1996). On a hydroxylated or hydrated surface, positive or negative charge is developed by adsorption or desorption of protons (H^+) or hydroxyl ions (OH^-), or dissociation of surface species.

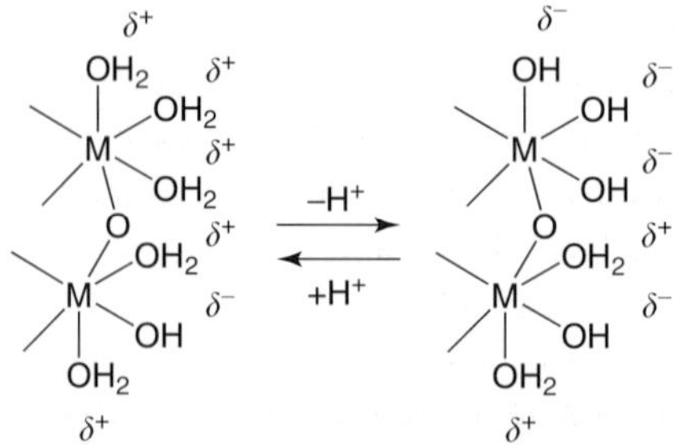

The pH at which the net variable charge on the surface is zero is called the point of zero charge (PZC). Because of its dependency on pH, the surface charge of the metal oxides is called "variable charge." The reported PZC of Fe-oxides usually ranges from 7.0 to 9.5, whereas that of Al-oxides ranges from pH 8 to 9.2 (Hsu, 1989; Sparks, 1995; Cornell and Schwertmann, 1996; Bigham *et al.*, 2002).

Crystalline metal oxides usually have specific surface of 15–70 $m^2\ g^{-1}$, whereas short-range-ordered precipitates may have specific surface even greater than 200–300 $m^2\ g^{-1}$ (Hsu, 1989; Cornell and Schwertmann, 1996). Multi-functional organic compounds associated with Fe and Al play an important role in both structural distortion of metal precipitates and promotion of aggregation of the reaction products. Probably, structural distortion and aggregation occur simultaneously, but aggregation may be more prominent when the amount of certain ligands in the solid is sufficiently high (Violante and Huang, 1992; Violante *et al.*, 2002c; Huang *et al.*, 2002). Cornell and Schwertmann (1996) reported that the specific surface of natural ferrihydrite decreases with increasing content of natural organic carbon and increases after H_2O_2 treatment.

Inorganic and organic ligands with a strong affinity for Fe, Al, and Mn (e.g. phosphate, inositolphosphate, oxalate (OX), citrate, tartrate (TR), malate (MAL), fulvate, and humate) are strongly adsorbed on the surfaces of metal oxides through a ligand-exchange mechanism, forming inner-sphere complexes. Other anions (e.g. chloride, sulfate, arsenite, chromate, acetate, benzoate, and formate) are moderately or weakly adsorbed on oxide surfaces. Adsorption of anions via ligand exchange results in a shift in the PZC of the oxide to a higher acid value (Goldberg *et al.*, 1996).

Metal oxides selectively adsorb divalent cations even at solution pH values lower than the PZC of metal oxides. The mechanism of metal ion association with hydrous-oxide surfaces involves an ion-exchange process in which the adsorbed cations replace bound protons. Specifically, adsorbed cations raise the value of PZC of oxides. pH affects adsorption of metal cations, either by changing the number of sites available for adsorption or by changing the concentration of the cation species (Me^{2+}, $MeOH^+$, $Me(OH)_2$) that are preferentially adsorbed (Jackson, 1998).

3. MATERIALS AND METHODS

3.1. Synthetic metal oxides

Gibbsite was prepared by precipitating 0.1 mol L^{-1} $Al(NO_3)_3$ by NaOH 0.5 mol L^{-1} up to pH 7.0. The suspension was aged for 7 days at room temperature and for a further 60 days at 40°C. Bayerite was prepared by precipitating 0.1 mol L^{-1} $Al(NO_3)_3$ by NaOH 0.5 mol L^{-1} up to pH 10.0. The suspension was aged for 60 days at room temperature. Goethite was prepared by slowly adding 200 ml of 2.5 mol L^{-1} NaOH to 50 g of $Fe(NO_3)_3 \cdot 9H_2O$ solubilized in 825 mL of deionized water. The final suspension was aged for 6 days at 60°C.

All the suspensions were washed free of salts by dialysis, freeze-dried, lightly ground, and passed through a 0.16-mm sieve (Liu *et al.*, 2001).

Under transmission electron microscopy (TEM) observations, goethite consisted of acicular crystals. Gibbsite was mainly in the form of hexagonal plates, but also showed the presence of noncrystalline materials under TEM observations. Bayerite showed the presence of triangular particles and some short-range-ordered materials.

Goethite had a surface area, determined by using the retention method of ethylene glycol monoethyl ether (EGME; Eltanawy and Arnold, 1973) of 80 m^2 g^{-1} and a point of zero salt effect (PZSE) of 8.2. Gibbsite and bayerite had surface areas (EGME) of 120 and 135 m^2 g^{-1}, respectively, and a PZSE of 8.5–8.9. The PZSE of the oxides was determined by the method of Sakurai *et al.* (1988).

Ferrihydrite and mixed Fe–Al oxides were prepared according to the method described by Colombo and Violante (1996). Stock solutions of 0.1 mol L^{-1} $Al(NO_3)_3$ and 0.1 mol L^{-1} $Fe(NO_3)_3$ were mixed in different proportions to synthesize samples having initial Fe/Al molar ratios (R) of 1, 2, 4, 10, and ∞ (no Al present). The solutions (henceforth referred to as R1, R2, R4, R10, and R∞) were potentiometrically titrated to pH 5.5 by adding 0.5 mol L^{-1} NaOH. After 7 days of aging at room temperature, the suspensions were dialyzed with deionized water and freeze-dried. A noncrystalline Al precipitation product (R0) was obtained by precipitating 0.1 mol L^{-1} $Al(NO)_3$ to pH 7.0 by adding 0.5 mol L^{-1} NaOH; the precipitate was immediately washed, dialyzed, and freeze-dried. The X-ray diffraction (XRD) patterns of the samples R1–R∞ showed the characteristic broad peaks of the 6-line ferrihydrite (Cornell and Schwertmann, 1996), whereas the XRD trace of R0 showed the presence of very poorly crystalline gibbsite. However, electron microscopy of the latter sample showed morphologically ill-defined materials characteristic of noncrystalline precipitates, and very distorted hexagonal crystals of gibbsite.

3.2. Preparation of a humic acid-like polymer and an organo–mineral complex

A humic acid-like polymer, called polymerin (POL), was extracted from olive oil mill waste waters and prepared and characterized according to the procedure determined by Capasso *et al.* (2002). POL chemical composition is as follows: carbohydrates 52.40%, proteins 10.40%, melanin 26.14%, metals 11.06%. The relative molecular weight of POL ranged from 3500 to 20,000.

Aliquots of 1 g of POL were dissolved in 500 ml of 0.1 mol L^{-1} $FeCl_3$. The mixture was titrated using an automatic titrator (Vit 90, Abu 93 Triburette) with 1 mol L^{-1} NaOH to reach pH 6.0. The sample obtained by precipitation was washed and then dialyzed with distilled water to reach a conductivity of 2 μS. This complex was noncrystalline to XRD and showed an elemental C and Fe content, respectively, of 4.35 and 48.0% (ww) (Capasso *et al.*, 2004). The non-permeated fraction was lyophilized, producing $Fe(OH)_x$–POL.

3.3. Soil sample

The <2 mm fraction of an Andisol (a 2C horizon of a Typic Hapludand) containing 40% of allophanic material, from Roccamonfina Volcano (Italy), was used in this work. The general descriptions and selected chemical, physico-chemical, and mineralogical properties of this soil are given by Vacca *et al.* (2003).

3.4. Adsorption of heavy metals onto metal oxides

A total of 20 mg of each R0–R∞ sample (Table 1) was equilibrated at 20°C with 50 mL of 0.02 mol L^{-1} KCl solution to produce a constant ionic strength. The sorption experiments of heavy metals (Cu, Pb, Zn, or Co) were carried out by adding solutions containing fixed quantities of elements as - $MeCl_2$. The amount of each element in the solutions was 50 mmol kg^{-1}. The samples were maintained at a constant pH (3.0–8.0) kept for 4 h by adding 0.1 or 0.01 mol L^{-1} HCl or NaOH continuously. Competitive adsorption experiments between Zn and Cu were carried out at pH 6.2. Cu and Zn were added as a mixture to ferrihydrite (R∞), keeping Cu concentration constant (50 mmol kg^{-1}) in the presence of different amounts of Zn to produce an initial Zn/Cu molar ratio of 1, 2, 3, 4 or 8. The experiments were carried out at pH 6.2 because previous

Table 1
Surface area and PZC of metal oxides

Samples	Surface area ($m^2\ g^{-1}$)	PZC
Noncrystalline Al-hydroxide (R0)[a]	234	8.5
Fe–Al oxide (R1)[a]	275	n.d.
Fe–Al oxide (R2)[a]	275	n.d.
Fe–Al oxide (R4)[a]	280	n.d.
Fe–Al oxide (R10)[a]	285	n.d.
Ferrihydrite (R∞)[a]	198	7.3
Goethite	80	8.2
Bayerite	105	8.9
Gibbsite	120	8.5

[a] R0, R1, R2, R4, R10, and R∞ indicate the initial Fe/Al molar ratio.
The sample R0 showed also the presence of poorly crystalline gibbsite; the R1–R10 samples showed the presence of (aluminous) ferrihydrite.

experiments showed that after adding 50 mmol kg^{-1} of Cu or Zn, Cu adsorption was complete, whereas only about 50% of Zn was adsorbed on ferrihydrite.

The adsorption experiments of Cu onto bayerite and goethite in the absence or presence of oxalate were carried out by adding solutions containing suitable quantities of Cu (40 mmol kg^{-1}) to 100 mg of each sorbent in a volume of 9.7 mL (0.02 mol L^{-1}). The samples were kept for 4 h at a constant pH (3.0–7.0) and the final suspensions were brought at 10 mL. The final suspensions were centrifuged at 10,000 **g**, and the heavy-element concentrations in the supernatants were determined by atomic absorption spectrophotometer (AAS) (Perkin–Elmer Analyst 700). The adsorbed amounts of Cu and Zn were calculated by the difference between the amount of elements recovered in the supernatants and the amount initially added.

3.5. Adsorption of Pb in the presence of organic ligands

The adsorption experiments of Pb in the absence or presence of OX or TR were carried out at pH 4.0 by solutions containing fixed quantities of Pb and suitable quantities of OX or TR as a mixture adding to 20 mg of different metal oxides (R0–R∞; Table 1). The amount of Pb in the solutions was 117 mmol kg^{-1} with organic ligand/Pb molar ratios (r_L) ranging from 0 to 7. Some experiments were carried out at pH 4.0 by changing the order of addition of Pb (117 mmol kg^{-1}) and TR (TR/Pb molar ratio of 4) to ferrihydrite (R∞) as follows:

(1) Pb and TR were added together (Pb + TR system);
(2) Pb was added 30 min before TR (Pb before TR system);
(3) TR was added 30 min before Pb (TR before Pb system).

The final suspensions were centrifuged at 10,000 **g** and Pb concentration in the supernatants was determined as described for Cu and Zn.

3.6. Adsorption of As in the absence or presence of organic ligands

Adsorption of As on selected sorbents (ferrihydrite, noncrystalline Al-hydroxide, and Andisol) was determined at pH values ranging from 4.0 to 8.0. Suitable amounts of As were added to an Andisol sample at pH 4.5 as a mixture with LMMOAs (OX, MAL, or citrate) or phosphate in order to have a maximum surface coverage of As (as determined by adsorption isotherms) and an initial organic ligand or phosphate/As molar ratios ranging from 0 to 5.

Other experiments were carried out on the adsorption of As in the absence or presence of MAL on different sorbents (goethite, ferrihydrite, noncrystalline Al hydroxide, and Andisol). Suitable amounts of As (90–100% of surface coverage of As for each sorbent) were added to each sorbent in the presence of increasing concentrations of MAL to have initial MAL/As molar ratios ranging from 0 to 5. After 24 h of reaction, the final suspensions were centrifuged at 10,000 **g** for 20 min and filtered through a 0.22-μm filter.

Some experiments were carried out at pH 5.0 using ferrihydrite as sorbent by adding MAL 24 h before As (MAL before As) or As and MAL as a mixture (MAL + As) at MAL/As molar ratios ranging from 0 to 1. The suspensions were maintained and stirred for 48 h.

As present in the final solutions was determined by ion chromatography, using a Dionex DX-300 Ion Chromatograph (Dionex Co, Sunnyvale, CA), an IonPac AS11 column (4.0 mm), an eluent of 0.05 mol L^{-1} NaOH at a flow rate of 2 ml min^{-1}, and a CD20 Conductivity Detector combined with autosuppression.

3.7. Adsorption of heavy metals and metalloids onto POL and $Fe(OH)_x$–POL

One hundred milligram of sorbent (POL or $Fe(OH)_x$–POL), previously dried at 100°C for 1 h, was equilibrated at 20°C with 20 mL of 0.03 mol L^{-1} KNO_3. Solutions of 0.1 mol L^{-1} individual heavy metals and metalloids (Cu, Zn, or As[V]), adjusted previously to pH 4, were added to each sorbent to give an initial concentration ranging from 0.50 to 7.50 mol L^{-1}. The pH of each suspension was kept constant for 24 h by addition of 0.1 or 0.01 mol L^{-1} HCl or KOH. The final suspensions were centrifuged at 10,000 **g** for 30 min, then ultrafiltered through an ultrafiltration cell (200 mL) equipped with a magnetic stirrer and a membrane with a cutoff of 1000 Da. The final concentration of the metals was determined in the permeated solution by AAS.

4. RESULTS

4.1. Adsorption of heavy metals and metalloids onto Fe and/or Al oxides

Adsorption of trace elements in cationic form is pH-dependent and is characterized by a pH range, where the amount of a heavy metal that is bound to a sorbent increases abruptly to nearly 100%, known as "adsorption edge." Fig. 1 shows the adsorption of Pb, Cu, Zn, and Co at different pH values onto selected short-range-ordered oxides (a noncrystalline Al precipitation product [R0], ferrihydrite [R∞], and mixed Fe–Al oxides formed at different initial Fe/Al molar ratios [R1–R10]) (Table 1), whereas Fig. 2 shows the adsorption of Cu onto well-crystallized metal oxides, goethite, and bayerite. In the region in which adsorption increases rapidly, the species $MeOH^{+}$ and $Me(OH)_2^{\circ}$ of each metal were negligible.

The adsorption selectivity sequence for heavy metals among the precipitates was Pb > Cu > Zn > Co (Fig. 1). The same sequence was found for gibbsite, bayerite, and goethite (data not shown). Selectivity is used in the sense that adsorption at a lower pH is more selective (Kinniburgh and Jackson, 1976). However, some authors found a different sequence on different crystalline or short-range-ordered Fe- or Al-oxides. The relative affinities of metals for ferrihydrite (Kinniburgh and Jackson, 1976, 1981; Pickering, 1979) and hematite (Forbes *et al.*, 1976; McKenzie, 1980) have been reported as Pb > Cu > Zn > Co, but as Cu > Pb > Zn > Co for goethite and crystalline and noncrystalline

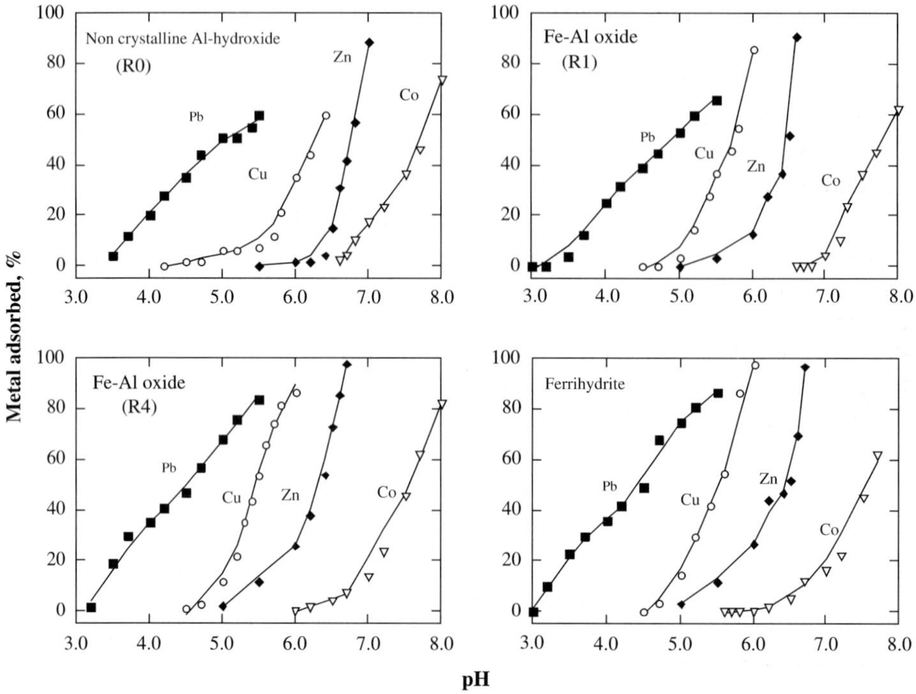

Fig. 1. Adsorption of Pb, Cu, Zn, and Co (%) on noncrystalline Al-hydroxide (R0), mixed Fe–Al oxides (R1–R4) and ferrihydrite (R∞) at different pH values (50 mmol of each metal per kg of oxide). Initial Fe/Al molar ratio of 0 (R0), 1 (R1), 4 (R4), and ∞ (R∞).

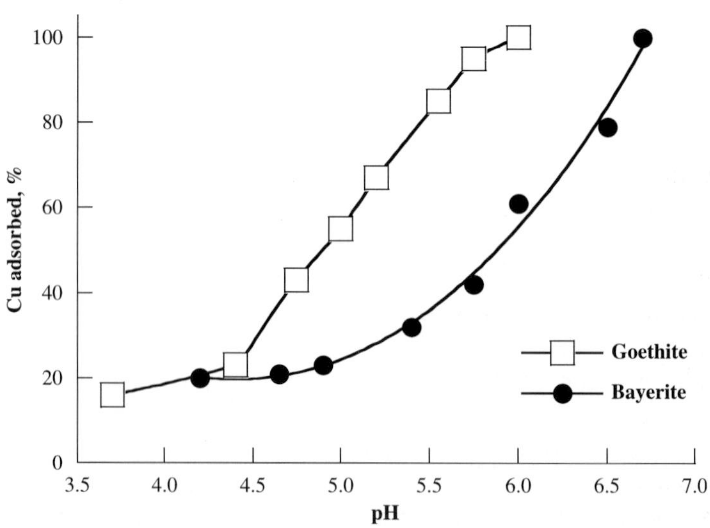

Fig. 2. Percentage of Cu adsorbed on goethite and bayerite (40 mmol of Cu added per kg of oxide) as a function of pH (3.5–7.0).

Al-hydroxides (Kinniburgh and Jackson, 1976, 1981; Jackson, 1998). For Mn-oxides and -hydroxides, the selective sequence of these heavy metals has usually been found to be Pb > Cu > Co > Zn (Mc Kenzie, 1989; Jackson, 1998). A body of evidence accumulated by different workers (Jackson, 1998 and references therein) has established that Mn-oxides are usually more effective sorbents than Fe- and Al-oxides.

A measure of the relative affinity of the heavy elements for an oxide is given by the pH at which 50% of the original cation is adsorbed, termed pH_{50} (Kinniburgh and Jackson, 1976; Violante *et al.*, 2003). The pH_{50} values of Pb, Cu, Zn and Co adsorbed on the noncrystalline Al-oxide, ferrihydrite and mixed Fe–Al oxides (R1–R10) were different from sample to sample (Fig. 1; Table 2). It can be seen from the data presented in Table 2 that the pH_{50} value of the samples R0–R4 decreased with increase in Fe content of the Fe–Al oxides. For example, the pH_{50} value of Pb decreased from 5.01 for R0 to 4.35 for R∞. In other words, the ΔpH_{50} (R0–R∞) for Pb was 0.66. The ΔpH_{50} for Cu, Zn and Co was 0.76, 0.34, and 0.30, respectively. The mixed Fe–Al oxides clearly showed different sorption affinity, particularly for metals (e.g. Pb and Cu), which are more selectively adsorbed. It appears evident that the greater the selectivity of a heavy metal for the gels, the greater will be the ΔpH_{50} from the noncrystalline Al-hydroxide (R0) to ferrihydrite (R∞).

Table 2
Value of pH_{50}[a] for Pb, Cu, Zn and Co adsorbed onto selected crystalline and short-range-ordered metal oxides

Samples	Pb	Cu	Zn	Co
Noncrystalline				
Al-hydroxide (R0)[b]	5.01	6.24	6.73	7.77
Fe–Al oxide (R1)[b]	4.91	5.59	6.46	7.73
Fe–Al oxide (R2)[b]	4.79	5.56	6.42	7.59
Fe–Al oxide (R4)[b]	4.49	5.47	6.41	7.47
Fe–Al oxide (R10)[b]	4.35	5.35	6.38	7.47
Ferrihydrite (R∞)[b]	4.35	5.29	6.39	7.47
Goethite	n.d.	4.95	n.d.	n.d.
Bayerite	n.d.	5.85	n.d.	n.d.

[a] pH_{50} indicates the pH at which 50% of the original cation was adsorbed. [b] R0, R1, R2, R4, R10, and R∞ indicate the initial Fe/Al molar ratio.
The sample R0 showed the presence of poorly crystalline gibbsite; the R1–R10 samples showed the presence of (aluminous) ferrihydrite.

The greater affinity of heavy metals for Fe-oxides than for Al-oxides has been reported (Jackson, 1998 and references therein). The PZC of Fe-oxides is lower than that of Al-oxides (Table 1; Cornell and Schwertmann, 1996; Violante *et al.*, 2002c). At a given pH, Fe–Al oxides containing greater amounts of Al must have a more positive surface charge (and a greater PZC) than Fe–Al oxides richer in Fe. Consequently, trace elements were adsorbed more easily on the Fe–Al oxides containing more Fe. However, it is important to note that despite the increased affinity of Pb, Cu, Zn, and Co for mixed Fe–Al oxides with increasing amounts of Fe coprecipitated with Al, the adsorption was not linearly related to Fe content. For example, the pH_{50} values of Zn and Co were practically constant for R4, R10, and R∞ (Table 2). Conversely, the pH_{50} for Pb and Cu, which showed greater affinity for the oxides, significantly decreased from R0 to R4 and then remained practically constant for R10 and R∞ (Table 2). These results indicate that the mixed Fe–Al oxides were not simple mixtures of different amounts of Fe- and Al-oxides, but oxides with different mineralogy, chemical composition, and surface properties (Colombo and Violante, 1996; Violante *et al.*, 2003; Tables 1 and 2), and therefore reactive toward cations and anions (Fig. 2; Table 2). Furthermore, the sorption capacity of an oxide is a function of its crystallinity, particle size, and specific surface, as well as elemental composition (Colombo and Violante, 1996; Jackson, 1998; Violante and Pigna, 2002). Recently, Harvey and Rhue (2003) showed that phosphate adsorption onto noncrystalline Fe–Al mixed hydroxides decreased with the Fe/Al ratio and pH. The adsorption of phosphate was largely controlled by Fe and was more sensitive to pH change between 4 and 6 than pure Fe- or Al-hydroxides.

The pH_{50} value of Cu was 4.95 for goethite and 5.85 for bayerite (Table 2; Fig. 2). The pH_{50} values found for goethite and bayerite were lower than those obtained for ferrihydrite (R∞) or the noncrystalline Al precipitation product (R0), which were 5.29 and 6.24, respectively (Fig. 1; Table 2). This behavior is probably due to differences in adsorption conditions in the experiments carried out using R0–R∞ oxides or goethite and gibbsite.

In contrast to the trace elements in cationic form, the adsorption of trace elements in anionic form (e.g. As, selenite, molybdate, chromate) usually decreases with increase in pH owing to a decrease in the positive charge of the sorbent at higher pH values (Fig. 3). However, some ligands (e.g. arsenite and selenite) may be adsorbed more easily at high pH values because they form weak acids at low pH values and may consequently only be dissociated in alkaline environments (Sparks, 1995; Goldberg *et al.*, 1996).

The behavior of As adsorbed on to metal oxides and variable-charge soils is usually similar to that of phosphate (Smith *et al.*, 1998; Liu *et al.*, 2001). Arsenate, arsenite, selenite, molybdate, and chromate adsorb on variable-charge minerals at pH values greater than the PZC of the sorbents, or where the surface is negatively charged (Balistrieri and Chao, 1987; Sparks, 1995).

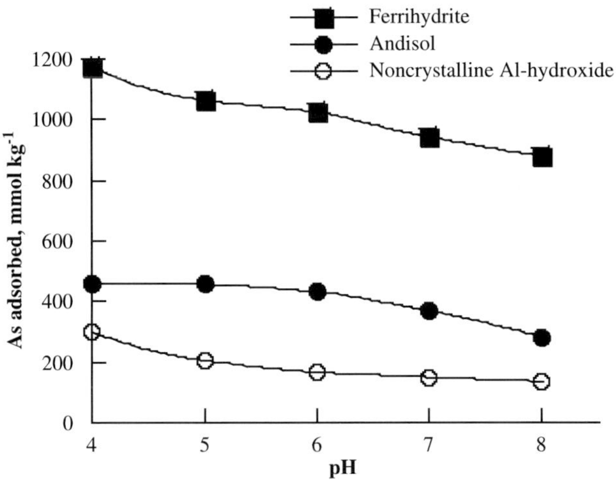

Fig. 3. Arsenate (As) adsorption (mmol kg^{-1}) on ferrihydrite (■), Andisol (●) and noncrystalline Al-hydroxide (O) as a function of pH (4.0–8.0).

4.2. Adsorption of heavy metals and metalloids onto organic matter and organomineral complexes

Trace elements in cationic and anionic forms show a different adsorption capacity on metal oxides, organic matter, and organo mineral complexes. The isotherms of Cu, Zn, and As absorbed at pH 4.0 onto POL (extracted from olive oil mill waste waters), a Fe $(OH)_x$–POL complex, and ferrihydrite are shown in Fig. 4.

The amounts of As adsorbed on POL were lower than those of Cu and Zn because anions are not easily adsorbed on organic, negatively charged sorbents. However, Thanabalasingan and Pickering (1986) showed that adsorption of As on humic acids is pH-dependent, the maximum adsorption being around pH 5.0–5.5.

The amounts of As adsorbed on the Fe $(OH)_x$–POL complex were not only much greater than on POL (Fig. 4B) but were also much higher than those of Cu (Fig. 4A) and Zn adsorbed on the organo–mineral complex. More As was sorbed on ferrihydrite than on the $Fe(OH)_x$–POL complex (Fig. 4B), evidently because the organic molecules partially prevented As adsorption (as discussed below). On ferrihydrite, the difference between the amounts of As and those of Cu and Zn adsorbed was still much greater than on $Fe(OH)_x$–POL and POL (data not shown).

4.3. Competitive adsorption of heavy metals onto soil components

Heavy metals compete for sorption sites on soil components. Reports on competition in the adsorption of two or more heavy metals on variable-charge minerals and humic substances in soils are rare. Benjamin and Leckie (1981) found a nearly complete lack of competition among Cd, Cu, Zn and Pb added to a poorly crystalline Fe- oxyhydroxide. Probably, the lack of competition among the heavy metals could be attributed to low surface coverage effect. Saha *et al.*

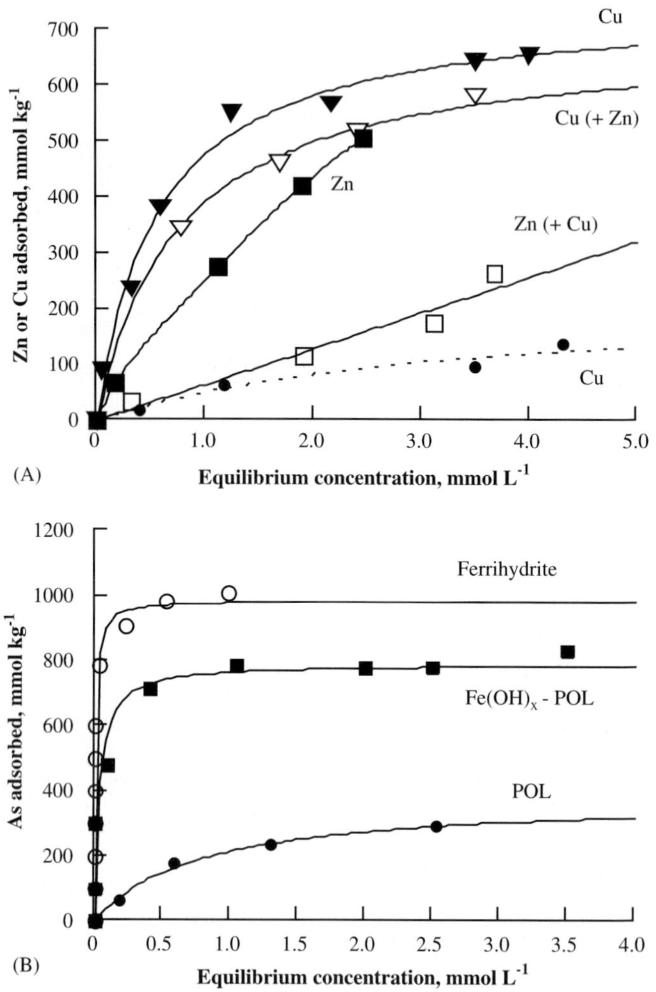

Fig. 4. (A) Adsorption of Cu and Zn at pH 4.0 on a humic acid-like polymer (POL) added alone (▼, Cu; ■, Zn) or as a mixture (▽Cu adsorbed in the presence of Zn [Cu (+ Zn)]; □, Zn adsorbed in the presence of Cu [Zn (+ Cu)]. Dashed curve indicates the adsorption of Cu on the $Fe(OH)_x$–POL complex. (B) Adsorption of arsenate (As) at pH 4.0 onto ferrihydrite, $Fe(OH)_x$–POL complex and humic acid-like polymer (POL).

(2002) demonstrated that Cd, Zn, and Pb competed with each other for adsorption sites on hydroxyaluminum- and hydroxyaluminosilicate–montmorillonite complexes.

Recently, we carried out experiments on the competitive adsorption of Cu and Zn on ferrihydrite, the humic acid-like sample (POL), and the Fe $(OH)_x$–POL complex. Fig. 5 shows the adsorption of Zn on ferrihydrite in the presence of Cu (50 mmol added per kg of sorbent) at different initial Zn/Cu molar ratios at pH 6.2.

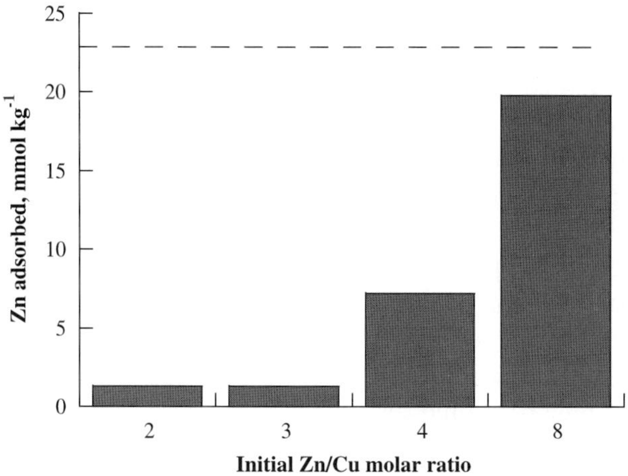

Fig. 5. Adsorption of Zn at pH 6.2 on ferrihydrite (R∞) in the presence of Cu (50 mmol kg^{-1}) at different initial Zn/Cu molar ratio. Dashed line indicates that the amounts of Cu adsorbed did not change in the presence of increasing Zn concentrations.

The adsorption of Cu was not affected by the presence of Zn, even at a Zn/Cu molar ratio of 8.0 (Fig. 5, dashed line); conversely, Zn adsorption was strongly prevented by Cu. When 400 mmol kg^{-1} of Zn (Zn/Cu molar ratio of 8.0) was added to ferrihydrite, only 20 mmol kg^{-1} of Zn was adsorbed, a quantity even lower than that fixed (23 mmol kg^{-1}) when only 50 mmol kg^{-1} of Zn was added in the absence of Cu. These findings strengthen the observation that Cu has a greater affinity for the surfaces of ferrihydrite and thus inhibits the sorption of Zn on common sites.

The duration also affected the adsorption of Zn in the presence of Cu. Zn adsorption increased from 4 to 20 mmol kg^{-1} when Cu was added (initial Zn/Cu molar ratio of 2) from 1 to 336 h after Zn addition (data not shown; Sparks, 1999; Violante *et al.*, 2003).

On POL, Cu was adsorbed more selectively than Zn (Fig. 4A). In the presence of equimolar concentrations of Zn, the amounts of Cu adsorbed were only slightly reduced when compared to the amounts adsorbed in the absence of Zn. Conversely, the presence of Cu strongly reduced Zn adsorption. A similar trend was observed using a Fe $(OH)_x$–POL complex as sorbent, but the quantities of Cu and Zn sorbed on this complex were much lower than those fixed on the humic acid-like sample (Fig. 4A). However, some evidence seems to demonstrate that Cu prevents Zn adsorption more strongly on ferrihydrite than on the organomineral complex [$Fe(OH)_x$–POL] or the humic acid-like material (POL).

4.4. Effect of organic ligands on the adsorption of trace elements

LMMOAs, mainly those with a high chelating power toward Fe, Al or Mn, play an important role in the adsorption, desorption, and mobilization of trace elements because they form soluble complexes with trace elements in cationic form and then modify their mobility in natural environments.

We have studied the influence of increasing concentrations of OX or TR on the adsorption of Pb or Cu at pH 4.0–4.5 on noncrystalline Al precipitation products (R0), ferrihydrite (R∞), mixed Fe–Al gels (R1–R10), goethite, and bayerite, at organic ligand/Pb or Cu molar ratios (r_L) ranging from 0.1 to 15.0 (Table 1; Figs. 6 and 7).

Lead adsorption at pH 4.0 in the absence of organic ligands varied from 20% on the noncrystalline Al-oxide to 33–36% on Fe–Al oxides (R4–R10) or ferrihydrite (R∞) of the amount initially added (117 mmol Pb per kg). In the presence of OX or TR, the quantities of Pb adsorbed on the oxides usually increased with an increase in r_L (Fig. 6). Some experiments revealed that adsorption of Pb on

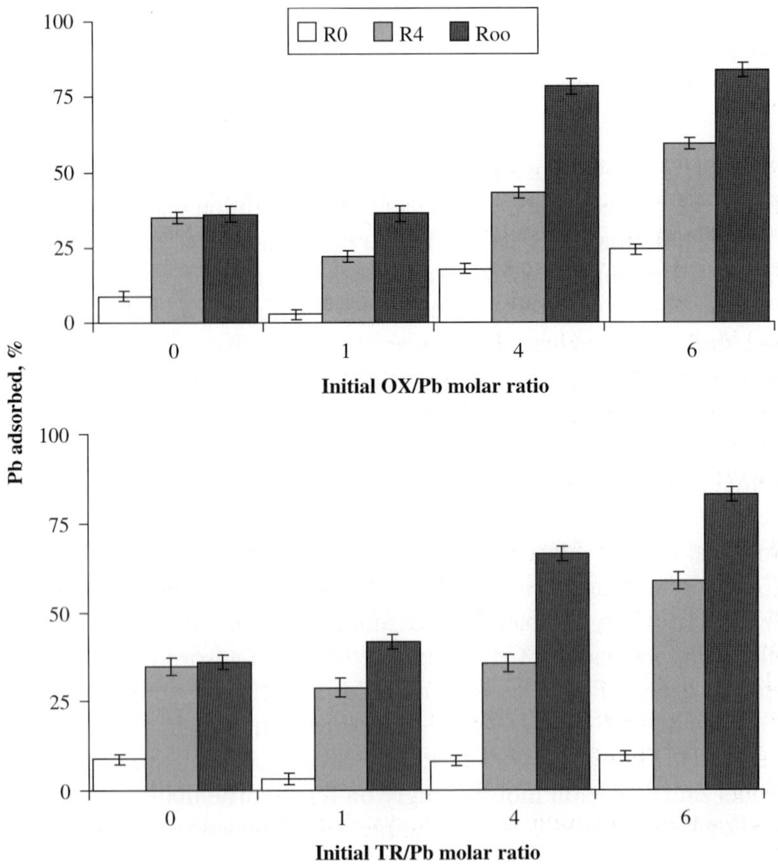

Fig. 6. Effect of increasing concentrations of oxalate (OX) or tartrate (TR) on Pb adsorption (%) on Fe–Al oxides at pH 4.0. R0, R4 and R∞ indicate samples formed at an initial Fe/Al molar ratio of 0, 4, and ∞ (Table 1); 117 mmol kg^{-1} of Pb was initially added on each sample.

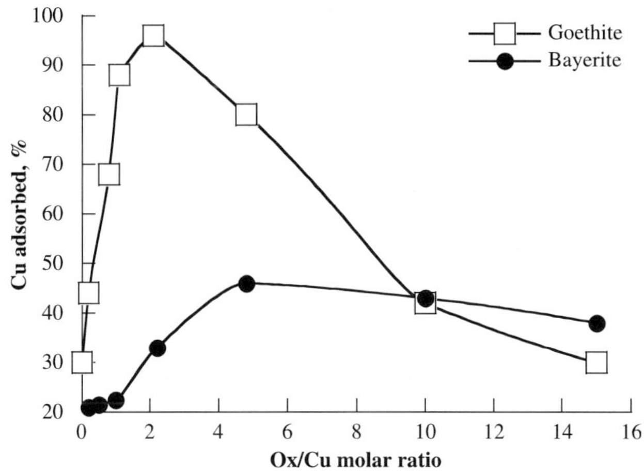

Fig. 7. Amounts (%) of Cu adsorbed at pH 4.5 on goethite and bayerite in the presence of increasing oxalate concentration (OX/Cu molar ratio ranging from 0 to 15); 40 mmol kg^{-1} of Cu was added to the sorbents.

R4–R∞ increased by increasing r_L up to 10 (data not shown). Chelating organic ligands may enhance metal adsorption on oxides of Fe, Al, Mn and Si by forming stable surface–metal–ligand complexes (McBride, 1989; Jackson, 1998; Zhou *et al.*, 1999). The adsorption increase of Pb was particularly high on the Fe–Al oxides richer in Fe (R4–R10) and ferrihydrite. This behavior may be due to the fact that OX and TR have higher complexation constants with Fe than with Al^{3+}. For example, the stability constant (log *K*) of OX–Fe(III) complex with stoichiometry of the complex metal:ligand ratio 1:1 is 7.7, whereas that of the complex OX-Al is 6.1 (Jones, 1998; Kurek, 2003).

Experiments carried out on the adsorption of Cu on well-crystallized oxides showed a different behavior. Fig. 7 shows the amounts of Cu adsorbed on goethite and bayerite at pH 4.5 in the presence of increasing concentrations of OX. In the absence of OX, 31% of added Cu (40 mmol kg^{-1}) was adsorbed on goethite and 23% on bayerite. Cu adsorption on goethite strongly increased by about three times by increasing r_L from 0 to 2.5 and then rapidly decreased at greater r_L values. Adsorption of Cu on bayerite strongly increased by increasing OX concentrations. The maximum Cu adsorption occurred at $r_L = 5$ (47% of added Cu), but at $r_L > 5$ Cu adsorption did not decrease as rapidly as observed for goethite, probably owing to the presence of small amounts of short-range-ordered materials (poorly crystalline boehmite and noncrystalline materials), as discussed below. The adsorption of Cu on kaolinite (data not shown) also showed a small increase by increasing r_L from 0 to 0.1 and a sharp decrease at $r_L > 0.5$ (Zhou *et al.*, 1999). This may be due to the fact that kaolinite did not adsorb anionic metal complexes to any significant extent. Farrah and Pickering (1976a, b), Pickering (1979), and

Sakurai and Huang (1995) also found that OX strongly reduced Cu and Cd adsorption on kaolinite and montmorillonite. Recently, Wu *et al.* (2003) found that Pb adsorbed onto montmorillonite, goethite, and humic acid decreased with increasing concentrations of EDTA and citric acid.

Many factors, such as pH, surface properties of the sorbents, surface charge of the adsorbents, the number of sites available for sorption, and the nature and charge of Me-L species in solution influence heavy metal adsorption on metal oxides, allophane, and phyllosilicates in the presence of increasing concentrations of chelating organic ligands (Kinniburgh *et al.*, 1976, 1981; McBride, 1989; Goldberg *et al.*, 1996; Jackson, 1998; Violante *et al.*, 2003).

According to many authors (see McBride, 1989; Jackson, 1998 and references therein), some organic ligands enhance metal adsorption on oxides of Al, Fe, and Si by forming stable surface–metal–ligand complexes. The increased negative charge brought to the surface by LMMOAs must also be considered to promote additional metal adsorption (Goldberg *et al.*, 1996). However, at relatively high concentrations of organic ligands, opposite phenomena could occur. In fact, large amounts of ligands may either occupy many adsorption sites on the clay minerals, which are then not available for heavy metals, or may favor metal desorption owing to the formation of soluble metal complexes. McBride (1985a, b, 1989) demonstrated that high levels of inorganic and organic ligands adsorbed onto a noncrystalline Al-hydroxide and allophane reduced Cu adsorption, owing to the blocking of surface sites. Many studies have also reported that the addition of LMMOAs suppresses the metal adsorption (Goldberg *et al.*, 1996; Jackson, 1998; Sakurai and Huang, 1995). Chairidchai and Ritchie (1992) found that the effectiveness of an organic ligand in influencing heavy metal adsorption on soil minerals or soils is affected by the PZC of a sorbent, the pH of the soil solution, and the quantity of complex formed. When the pH is above the PZC of a sorbent, the organic ligand decreases the heavy metal adsorption, but the reverse occurs when the pH is below the PZC of a sorbent. Certainly, the ligand/metal ratio, the nature of organic ligands and heavy metals, and the surface properties of the sorbents seem to be critical in determining whether metal adsorption at surfaces is enhanced or inhibited, because large amounts of chelating ligands in solution shift the equilibrium in favor of soluble metal complexes.

The increased adsorption of Pb on poorly crystalline sorbents (Fig. 6) due to increasing TR or OX/Pb molar ratio up to 10 must be attributed to the large surface area of the oxides (Table 1), which allows the adsorption of great amounts of PbLx complexes; vice versa, on goethite and bayerite, the decrease in the adsorption of Cu, respectively, at OX/Cu molar ratio > 2.5 or 5 is due to the lower surface area of these crystalline metal oxides (Table 1).

The presence of organic and inorganic ligands that strongly interact with variable-charge minerals and soils also affects the adsorption of trace elements in anionic form through competition for available binding sites. The competition depends on the affinity of the anions for the surfaces of the sorbents as well as the nature and surface properties of the minerals and soils. The reduction of surface charge of the sorbents after adsorption of an organic or inorganic ligand certainly helps in preventing the adsorption of trace elements in anionic form (Barrow, 1992).

We have demonstrated that organic ligands and phosphate inhibit As adsorption on metal oxides and variable-charge soils. Fig. 8 shows the influence of increasing concentrations of selected organic ligands and phosphate on the adsorption of As onto an Andisol containing 40% of allophanic materials. Phosphate prevented As adsorption more than citrate, MAL, and OX (in the order cited).

Balistrieri and Chao (1987) suggested that for a given anion concentration ratio, the competition sequence with selenite on goethite is phosphate>silicate>citrate>molybdate>bicarbonate/carbonate>OX>fluoride>sulfate. More recently, Dynes and Huang (1997) showed that the ability of 12 LMMOAs to inhibit selenite sorption on poorly crystalline Al-hydroxides was in the order OX>MAL>citrate>succinate>glycolate>aspartate>salycilate>p-hydroxybenzoate>glycine = formiate = acetate. Generally, the larger the stability constant of the Al–organic solution complexes ($K_{\text{Al-L}}$), the more effective the organic acid was in competing with selenite for the adsorption sites of the Al hydroxides. However, some of the organic acids competed less successfully than expected based on their $K_{\text{Al-L}}$ values. This is attributed to the stereochemical and electrostatic effects originating from both the surface of the Al hydroxides and the

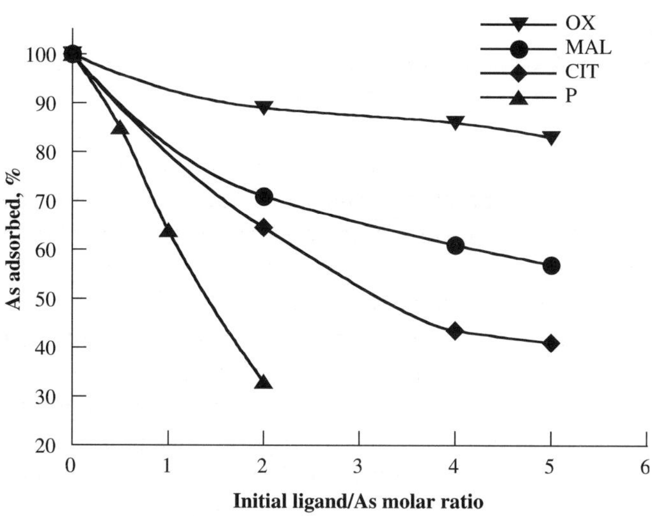

Fig. 8. Influence of organic [oxalate (OX), malate (MAL), and citrate (CIT)] and inorganic (phosphate) (P) ligands on the adsorption of arsenate on Andisol at pH 4.5; 416 mmol kg^{-1} of arsenate (As) was added to the sorbents.

organic acids, which lowered the ability of some organic acids to compete with selenite for the sorption sites of the Al-hydroxides.

The efficiency of organic anions in preventing As adsorption was also influenced by the nature and surface properties of variable-charge minerals and soils. Fig. 9 shows the adsorption of As at pH 4.0 and in the presence of increasing concentrations of malic acid (MAL/As molar ratio ranging from 0 to 5) on different metal oxides of Fe (goethite), Al (noncrystalline Al precipitation product), and on an Andisol. MAL either did not inhibit or, very poorly, prevented As adsorption on goethite, even at an initial MAL/As molar ratio of 5. In contrast, the organic ligand strongly inhibited As adsorption on noncrystalline Al hydroxide and Andisol, despite the fact that MAL forms complexes more stable with Fe than with Al (Jones, 1998). Previous studies have demonstrated that As is adsorbed more strongly on metal oxides of Mn, Fe or Ti than on metal oxides of Al (Violante and Pigna, 2002); consequently, MAL ligands prevented As adsorption on Al-hydroxide or allophanic material to a greater extent than on goethite. The great influence of chelating organic ligands in reducing As adsorption onto a noncrystalline Al-hydroxide and allophane in the Andisol may also be due to partial dissolution of Al from the sorbents (Figs. 8 and 9).

Grafe *et al.* (2001) found that As adsorption on goethite was reduced by humic and fulvic acid, but not by citric acid, whereas As adsorption was decreased by all the three organic acids between pH 3.0 and 8.0 in the order of citrate > fulvic acid > humic acid.

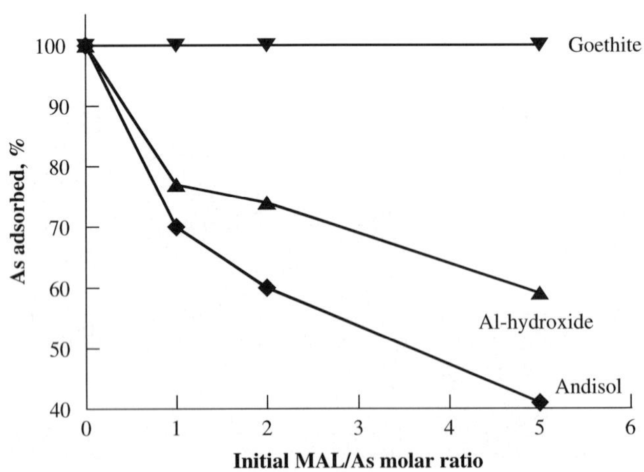

Fig. 9. Amounts (%) of arsenate (As) adsorbed onto goethite, noncrystalline Al-hydroxide, and Andisol at pH 4.0 in the presence of increasing concentrations of malate (MAL) (150, 150, 416 mmol As was initially added per kg of goethite, noncrystalline Al-hydroxide, and Andisol, respectively).

4.5. Adsorption of trace elements in the presence of organic ligands as affected by the sequence of their addition on the sorbents

The adsorption of trace elements on variable-charge minerals is influenced not only by the presence and concentration of organic ligands but also by the sequence of addition of heavy metals or metalloids and organic ligands to the sorbents.

It has been ascertained that larger amounts of Pb were adsorbed when TR was added before Pb, usually according to the following sequence: TR before Pb > Pb before TR > Pb + TR > Pb (Fig. 10). Similar findings were obtained using other sorbents e.g. mixed Fe–Al oxides and noncrystalline Al-hydroxide (Violante *et al.*, 2003). Certainly, by changing the order of addition of TR and Pb to metal oxides, different species were adsorbed on the surfaces of the sorbents. The species (Pb, TR, $PbTR_x$, etc.), initially present in solution, showed different affinities for the surfaces of the sorbents. However, it is particularly difficult to determine the species that may form in solution and may be adsorbed on the surfaces of an oxide with different affinity when a metal ion and an organic ligand are present in the system. The surface complexation models, which describe the formation of charge, potential, and the adsorption of different MeLx species at the oxide–water interface are the most popular (Goldberg *et al.*, 1996). We may only hypothesize that in the TR before Pb systems, the increased negative charge brought to the surfaces by previous adsorption of TR anions promoted additional Pb adsorption on the surfaces of the variable-charge mineral. Furthermore, other Pb ions could be sorbed by complexation with TR ions previously adsorbed on the surfaces of the oxides. In the Pb before TR systems, the previous adsorption

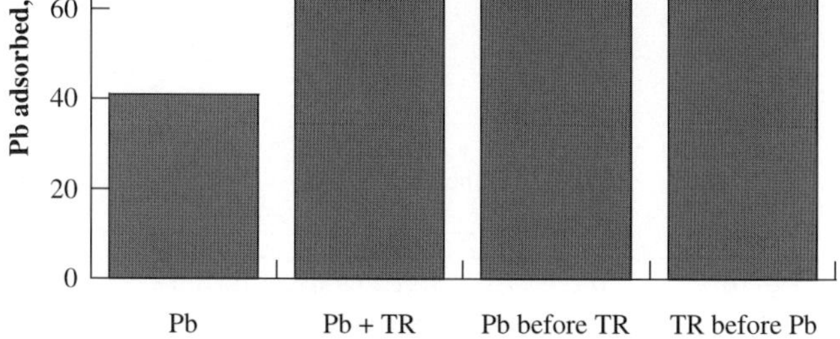

Fig. 10. Effect of the sequence of addition of Pb and tartrate (TR) on the adsorption of Pb (117 mmol kg^{-1}) on ferrihydrite at pH 4.0, when Pb was added alone (Pb), with tartrate (Pb + TR), before (Pb before TR) or after tartrate (TR before Pb).

of Pb could increase the PZC of the metal oxides, promoting further adsorption of some $PbTR_x$ complexes formed after TR addition.

The order of addition of an organic ligand and a toxic element in anionic form to a sorbent strongly affected the adsorption of the pollutant as well. The adsorption of As on ferrihydrite at pH 4.0 when As and MAL were added as a mixture (MAL+As) or when MAL was added before As (MAL before As) (initial MAL/As molar ratio ranging from 0 to 1) is shown in Fig. 11. At an initial MAL/As molar ratio of 1, the inhibition of MAL was about 7% when the ligands were added together and 22% when MAL was added before As. When As was added before MAL (As before MAL), the inhibition of MAL to prevent As adsorption was negligible (data not shown).

These results have important implications in soil environments, mainly in rhizospheric soils, where root exudates and trace elements may compete with each other in different combinations of interactions with the sorbents.

5. CONCLUDING REMARKS

Heavy metals and metalloids are selectively adsorbed on variable-charge minerals (e.g., Al, Fe and Mn oxides), which occur in soils in advanced stages of weathering and in rhizosphere soils. Weathering induced by LMMOAs and other biomolecules facilitates the release of Fe and Al from primary and secondary minerals, promoting the formation of Fe and Al oxides, usually of short-range order.

The adsorption of toxic elements in cationic form on metal oxides may be enhanced or inhibited in the presence of strongly chelating organic ligands, but

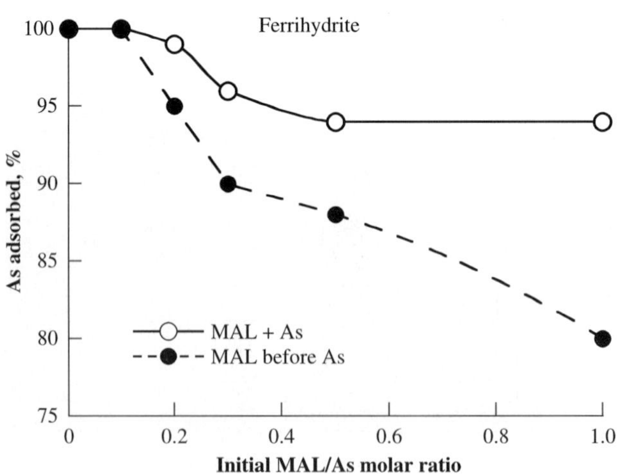

Fig. 11. Effect of the sequence of addition of arsenate (As) and malate (MAL) on the adsorption of As on ferrihydrite at pH 5.0, when As was added as a mixture (MAL/As molar ratio ranging from 0 to 1) with MAL (○, MAL + As) or after MAL (●, MAL before As).

the evidence seems to demonstrate that the ligand/metal molar ratio, pH, crystallinity, surface properties, chemical composition of metal oxides, and nature of the organic ligands play a great role in the adsorption of heavy metals. LMMOAs usually decrease the adsorption of toxic elements in anionic form and promote their desorption, but the nature of the sorbents also plays a determining role in the adsorption/desorption processes.

We have demonstrated that the adsorption of heavy metals and metalloids is affected not only by the presence, nature, and concentration of organic ligands but also by the sequence of reaction of pollutants and organics with the sorbents.

In the rhizosphere, the presence of biomolecules continuously released by plants as root exudates or microorganisms play a crucial role in the dynamics and bioavailability of metals and metalloids for plants and biota. The influence of biomolecules as well as of biotic and abiotic components on trace elements transformation and mobility deserves to be studied with particular attention.

ACKNOWLEDGMENTS

This work was supported in part by Ministero dell' Università e della Ricerca Scientifica e Tecnologica (MURST), Programmi di Ricerca Scientifica di interesse nazionale (PRIN 2002). Contribution No. 72 from Dipartimento di Scienze del Suolo, della Pianta e dell'Ambiente (DiSSPA).

REFERENCES

Adamo, P., Violante, P., 2000. Weathering of rocks and neogenesis of minerals associated with lichen activity. Appl. Clay Sci. 16, 229–256.

April, R., Keller, D., 1990a. Interactions between minerals and roots in forest soils. In: Farmer, V.C., Tardy, Y. (Eds.), Proceedings of Ninth International Clay Conference, Sci. Geol. Mem., pp. 85–89.

April, R., Keller, D., 1990b. Mineralogy of the rhizosphere in forest soils of the eastern United States. Biogeochemistry 9, 1–12.

Balistrieri, L.S., Chao, T.T., 1987. Selenium adsorption by goethite. Soil Sci. Soc. Am. J. 51, 1145–1151.

Barnhisel, R.I., Bertsch, P.M., 1989. Chlorite and hydroxy-interlayered vermiculite and smectite. In: Dixon, J.B., Weed, S.B. (Eds.), Minerals in Soil Environments. Soil Sci. Soc. of Am., Madison, WI, pp. 729–788.

Barrow, N.J., 1992. The effect of time on the competition between anions for sorption. Soil Sci. J. 43, 424–428.

Bar-Yosef, B., 1991. Root excretions and their environmental effects. In: Waisal, Y., Eshel, A., Kafkafi, U. (Eds.), Influence on availability of phosphorus. In: The Plant Root: The Hidden Half. Marcell Dekker Inc, New York, pp. 529–557.

Benjamin, M.M., Leckie, J.O., 1981. Multiple-site adsorption of Cd, Cu, Zn, and Pb on amorphous iron oxyhydroxide. J. Colloid Interface Sci. 79, 209–221.

Bigham, J.M., Fitzpatrick, R.W., Schulze, D.G., 2002. Iron Oxides. In:. Dixon, J.B, Schulze, D.G. (Eds.), Soil Mineralogy with Environmental Applications, Soil Sci. Soc. Am., Book Series 7, Madison, WI, pp. 323–366.

Capasso, R., De Martino, A., Arienzo, M., 2002. Recovery and characterization of the metal polymeric organic fraction (polymerin) from olive oil mill waste waters. J. Agric. Food Chem. 50, 2846–2855.

Capasso, R., Pigna, M., De Martino, A., Pucci, M., Sannino, F., Violante A., 2004. Potential remediation of waters contaminated with Cr(III), Cu and Zn by sorption on the organic polymeric fraction of olive mill wastewater (polymerin) and its derivatives. Envir. Sci. Tech., 38, 5170–5176.

Chairidchai, P., Ritchie, G.S.P., 1992. The effect of pH on zinc adsorption by a lateritic soil in the presence of citrate and oxalate. J. Soil Sci. 43, 713–728.

Churchman, G.J., 2000. The alteration and formation of soil minerals by weathering. In: Sumner M. E., (Ed.), Handbook of Soil Science, CRC Press, Boca Raton, FL, pp. F3–F76.

Colombo, C., Violante, A., 1996. Effect of time and temperature on the chemical composition and crystallization of mixed iron and aluminum species. Clays Clay Miner. 45, 113–120.

Cornell, R.M., Schwertmann, U., 1996. The Iron Oxides. Structure, Properties, Reactions and Uses. VCH Publishes, New York, USA.

Dynes, J.J., Huang, P.M., 1997. Influence of organic acids on selenite sorption by poorly ordered aluminum hydroxides. Soil Sci. Soc. Am. J. 61, 772–783.

Eltanawy, I.M., Arnold, P.W., 1973. Reappraisal of ethylene glycol monoethyl ether (EGME) method for surface area estimation of clays. J. Soil Sci. 24, 232–238.

Farrah, H., Pickering, W.F., 1976a. The sorption of copper species by clays. I Kaolinite. Aust. J. Chem. 29, 1167–1176.

Farrah, H., Pickering, W.F., 1976b. The sorption of copper species by clays. II Illite and montmorillonite. Aust. J. Chem. 29, 1177–1184.

Forbes, E.A., Posner, A.M., Quirk, J.F., 1976. The specific adsorption of divalent Cd, Co, Cu, Pb, and Zn on goethite. J. Soil Sci. 27, 154–166.

Gobran, G.R., Wenzel, W.W., Lombi, E., 2000. Trace Elements in the Rhizosphere. CRC Press, Boca Raton, FL.

Goldberg, S., Davis, J.A., Hem, J.D., 1996. The surface chemistry of aluminum oxides and hydroxides. In: G. Sposito (Ed.), The Environmental Chemistry of Aluminum, second ed. Lewis Publishers, Boca Raton, FL, pp. 271–331.

Grafe, M., Eick, M.J., Grossl, P.R., 2001. Adsorption of arsenate (V) and arsenite (III) on goethite in the presence and absence of dissolved organic carbon. Soil Sci. Soc. Am. J. 65, 1680–1687.

Harvey, O.R., Rhue, R.D., 2003. The effect of Fe:Al ratio in synthetic amorphous mixed metal hydr(oxides) on P sorption. 2003 ASA Meetings, Denver, CO.

Hsu, P.H., 1989. Aluminum hydroxides and oxyhydroxides. In: Dixon, J.B., Weed, S.B. (Eds.), Minerals in Soil Environments, second ed. Soil Sci. Soc. Am., Book Series 1, Madison, WI, pp. 331–378.

Huang, P.M., Bollag, J.-M., Senesi, N., 2002. Interactions between Soil Particles and Microorganisms: Impact on the Terrestrial Ecosystem. John Wiley & Sons, New York.

Huang, P.M., Germida, J.J., 2002. Chemical and Biochemical Processes in the Rhizosphere: Metal Pollutants. In: Huang, P.M., Bollag, J.-M., Senesi, N. (Eds.), Interactions between Soil Particles and Microorganisms: Impact on the Terrestrial Ecosystem. John Wiley & Sons, New York, pp. 381–438.

Huang, P.M., Violante, A., 1986. Influence of organic acids on crystallization and surface properties of precipitation products of aluminium. In: Huang, P.M., Schnitzer, M. (Eds.), Interactions of Soil Minerals with Natural Organics and Microbes. Special Publication 17, Soil Science Society of America, Madison, WI, pp. 159–221.

Huang, P.M., Wang, M.K., Kampf, N., Schulze, D.G., 2002. Aluminum Hydroxides. In: Dixon, J.B., Schulze, D.G. (Eds.), Soil Mineralogy with Environmental Applications, Soil Science Society of America, Book Series 7, Madison, WI, pp. 261–289.

Jackson, T.A., 1998. The Biogeochemical and Ecological Significance of Interactions between Colloidal Minerals and Trace Elements. In: Parker, A., Rae, J.E. (Eds.), Environmental Interactions of Clays. Springer-Verlag, Berlin, pp. 93–205.

Jones, D.L., 1998. Organic acids in the rhizosphere: a critical review. Plant Soil 205, 25–44.

Kampf, N., Scheinost, A.C., Schultze, D.G., 2000. Oxides Minerals. In: Sumner, M.E. (Ed.), Handbook of Soil Science, CRC Press, Boca Raton, FL, F125–F168.

Kinniburgh, D.G., Jackson, M.L., 1976. Adsorption of alkaline earth, transition and heavy metal cations by hydrous oxides gels of iron and aluminum. Soil Sci. Soc. Am. J. 40, 796–799.

Kinniburgh, D.G., Jackson, M.L., 1981. Cation adsorption by hydrous metal oxides and clay. In: Anderson, M.A., Rubin, A.J. (Eds.), Adsorption of Inorganics at Solid–Liquid Interfaces. Ann. Arbor. Science Publishers, Ann Arbor, pp. 91–160.

Kurek, E., 2002. Microbial mobilization of metals from soil minerals under aerobic conditions. In: Huang, P.M., Bollag, J.-M., Senesi, N. (Eds.), Interactions between Soil Particles and Microorganisms: Impact on the Terrestrial Ecosystem. John Wiley & Sons, pp. 189–225.

Inskeep, W.P., 1989. Adsorption of sulfate by kaolinite and amorphous iron oxide in the presence of organic ligands. J. Environ. Qual. 18, 379–385.

Liu, F., He., Z., Colombo, C., Violante, A., 1999. Competitive adsorption of sulfate and oxalate on goethite in the absence or presence of phosphate. Soil Sci. 164, 180–189.

Liu, F., De Cristofaro, A., Violante, A., 2001. Effect of pH phosphate and oxalate on the adsorption/desorption of arsenate on/from goethite. Soil Sci. 166, 197–208.

Lopez-Hernandez, D., Siegert, G., Rodriguez, J.V., 1986. Competitive adsorption of phosphate with malate and oxalate by tropical soils. Soil Sci. Soc. Am. J. 50, 1460–1462.

Marschner, H., 1995. Mineral Nutrition of Higher Plants. Academic Press, London.

McBride, M.B., 1985a. Influence of glycine on Cu^{2+}adsorption by microcrystalline gibbsite and boehmite. Clays Clay Miner. 33, 397–402.

McBride, M.B., 1985b. Sorption of copper(II) on aluminum hydroxide as affected by phosphate. Soil Sci. Soc. Am. J. 49, 843–846.

McBride, M.B., 1989. Reactions controlling heavy metal solubility in soils. Adv. Soil Sci 10, 1–56.

McBride, M.B., 1994. Environmental Chemistry of Soils, Oxford University Press, New York.

McKenzie, R.M., 1980. The adsorption of lead and other heavy metals on oxides of manganese and iron. Aust. J. Soil Res. 21, 505–513.

McKenzie, R.M., 1981. The surface charge on manganese dioxides. Aust. J. Soil Res. 19, 41–50.

McKenzie, R.M., 1989. Manganese oxides and hydroxides. In: Dixon, J.B., Weed, S.B. (Eds.), Minerals in Soil Environments, second ed. Soil Sci. Soc. Am., Book Series 1, Madison, WI, pp. 439–465.

Nagarajah S., Posner, A.M., Quirk, J.P., 1970. Competitive adsorption of phosphate with polygalacturonate and other organic anions on kaolinite and oxide surface. Nature (London). 228, 83–84.

Parfitt, R.L., 1980. Chemical properties of variable charge soils. In: Theng, B.K.G. (Ed.), Soils with Variable Charge. New Zealand Society of Soil Science, Lower Hutt, NZ, pp. 167–194.

Pickering, W.F., 1979. Copper retention by soil/sediment components. In: Nriagu, J.O. (Ed.), Copper in the Environment. I. Ecological Cycling. Wiley, New York, pp. 217–253.

Robert, M., Berthelin, J., 1986. Role of biological and biochemical factors in soil mineral weathering. In: Huang, P.M., Schnitzer, M. (Eds.), Interactions of Soil Minerals with Natural Organics and Microbes. Soil Sci. Soc. of Am., Special Publication 17, Madison, WI, pp. 453–495.

Saha, U.K., Taniguchim, S., Sakurai, K., 2002. Simultaneous adsorption of cadmium, zinc, and lead on hydroxyaluminum- and hydroxyaluminosilicate-montmorillonite complexes. Soil Sci. Soc. Am. J. 66, 117–128.

Sakurai, K., Huang, P.M., 1995. Cadmium adsorption on the hydroxyaluminum-montmorillonite complex as influenced by oxalate. In: Huang, P.M., Berthelin, J., Bollag, J.-M., McGill, W.B.,

Page, A.L. (Eds.), Environmental Impact of Soil Component Interactions: Natural and Anthropogenic Organics. CRC Press, Boca Raton, FL, pp. 39–46.

Sakurai, K., Ohdate, Y., Kyuma, K., 1988. Comparison of salt titration and potentiometric titration methods for the determination of zero point of charge. Soil Sci. Plant Nutr. 34, 171–182.

Sarkar, A.N., Jenkin, R.G., Wyn Jones, R.G., 1979. Modification to mechanical and mineralogical composition of soil within the rhizosphere. In: Harley, J.L., Scott Russell, R. (Eds.), The Soil–Root Interface. Academic Press, London, pp. 125–136.

Smith, E., Naidu, R., Alston, A.M., 1998. Arsenic in the Soil Environment: A Review. Adv. Agron. 64, 149–195.

Sparks, D.L., 1995. Environmental Soil Chemistry. Academic Press, San Diego.

Sparks, D.L., 1999. Kinetic and mechanisms of chemical reactions at the soil mineral/water interface. In: Sparks, D.L. (Ed.), Soil Physical Chemistry, second ed. CRC Boca Raton, FL, pp. 135–191.

Thanabalasingan, P., Pickering, W.F., 1986. Arsenic sorption by humic acids. Envir. Pollut. 12, 233–246.

Vacca, A., Adamo, P., Pigna, M., Violante, P., 2003. Properties and classification of selected soils from the Roccamonfina volcano, Central-Southern Italy. Soil Sci. Soc. Am. J. 67, 198–207.

Violante, A., Gianfreda, L., 2000. Role of biomolecules in the formation of variable-charge minerals and organo-mineral complexes and their reactivity with plant nutrients and organics in soil. In: Bollag, J.-M., Stotzky, G. (Eds.), Soil Biochemistry, vol. 10, Marcell Dekker, New York, pp. 207–270.

Violante, A., Huang, P.M., 1992. Effect of tartaric acid and pH on the nature and physicochemical properties of short-range ordered aluminum precipitation products. Clays Clay Miner. 40, 462–469.

Violante, A., Huang, P.M., Bollag, J.-M., Gianfreda, L. 2002a. Soil Mineral–Organic Matter–Microorganism Interactions and Ecosystem Health: Dynamics, Mobility and Transformations of Pollutants and Nutrients. Developments in Soil Science 28A. Elsevier, Amsterdam.

Violante, A., Huang, P.M., Bollag, J.-M., Gianfreda, L., 2002b. Soil Mineral–Organic Matter–Microorganism Interactions and Ecosystem Health: Ecological Significance of the Interactions Among Clay Minerals, Organic Matter and Soil Biota. Developments in Soil Science 28B. Elsevier, Amsterdam.

Violante, A., Krishnamurti, G. S. R., Huang, P.M., 2002c. Impact of organic substances on the formation of metal oxides in soil environments. In: Huang, P.M., Bollag, J.-M., Senesi, N. (Eds.), Interactions Between Soil Particles and Microorganism: Impact on the Terrestrial Ecosystem. John Wiley & Sons, pp. 133–188.

Violante, A., Pigna, M., 2002. Competitive Sorption of Arsenate and Phosphate on Different Clay Minerals and Soils. Soil Sci. Soc. Am. J. 66, 1788–1796.

Violante, A., Ricciardella, M., Pigna, M., 2003. Adsorption of heavy metals on mixed Fe-Al oxides in the absence or presence of organic ligands. Water, Air, and Soil Pollution 145, 289–306, 2003.

Wilson, M.J., 1995. Interactions between Lichens and Rocks: A Review. Cryptogamic Botany 5, 299–305.

Wu, Z., Gu, Z., Wang, X., Evans, L., Guo, H., 2003. Effects of organic acids on adsorption of lead onto montmorillonite, goethite and humic acid. Environ. Poll. 121, 469–475.

Xu, H., Allard, B., Grimvall, A., 1988. Influece of pH and organic substance on the adsorption of As(V) on geologic materials. Water, Air, Soil Pollution 40, 293–305.

Zhou, D., De Cristofaro, A., He, J.Z., Violante, A., 1999. Effect of oxalate on adsorption of copper on goethite, bayerite and kaolinite. In: Kodama, H., Mermut, A.R., Torrance, J.K. (Eds.), Clays for our Future. Proceedings 11th International Clay Conference, Ottawa, Canada, 1977, pp. 523–529.

Biogeochemistry of Trace Elements in the Rhizosphere
P.M. Huang and G.R. Gobran (Editors)

Chapter 6

Kinetics of cadmium desorption from iron oxides formed under the influence of citrate

C. Liu[a] and P.M. Huang[b]

[a] Kuo Testing Labs, Inc., 337 South 1st Avenue, Othello, WA 99344, USA

[b] Department of Soil Science, University of Saskatchewan, 51 Campus Drive, Saskatoon SK S7N 5A8, Canada
E-mail: huangp@sask.usask.ca

ABSTRACT

Research on cadmium (Cd) dynamics in the environment has received increasing attention. However, little is known about desorption kinetics of Cd from short-range-ordered mineral colloids formed under the influence of various ionic environments. The present study examines the desorption kinetics of Cd following its adsorption on iron oxides formed under the influence of citric acid which is common in soil and water. Iron oxides were formed at pH 6.00 ± 0.05 and 25°C at an initial Fe(II) concentration of 10^{-2} M in the presence of citrate ligand at initial citrate/Fe(II) molar ratios (MRs) of 0, 0.001, 0.01 and 0.1. The kinetics of Cd desorption by 0.01 M KNO_3, KCl, K-acetate, and K-citrate (adjusted to pH 5.0) from the iron oxides, following 1-day Cd adsorption at the initial Cd concentration of 50 μM, was investigated at 20°C from 5 min to 7 days. The fast (5 min to 2 h) and slow (2–24 h) processes of Cd desorption from the iron oxides can be satisfactorily described by the overall parabolic diffusion equation. The amount of Cd released and the reaction rate varied greatly with surface properties of the iron oxides formed under the influence of citrate. The amount of Cd released by all the extractants from the iron oxides was in the sequence of the initial citrate/Fe(II) MR of 0.1 > 0.01 > 0 > 0.001, which is in accord with the specific surface, microporosity, citrate content in the precipitates, degree of disorder of iron oxides, and amount of Cd adsorbed. The rate of Cd release by different extractants for the fast reaction was in the order chloride > citrate > acetate > nitrate, which was generally consistent with the sequence of the stability constants of Cd-extractant ligand complexes. The rate of Cd release by different extractants for the slow reaction was in the order citrate > acetate ≥ chloride > nitrate, indicating that a longer induction period was required for an extractant with a larger molecule to desorb the Cd adsorbed on the micropore surface. The research findings are of fundamental significance in understanding the role of the nature and properties of iron oxides formed under the influence of organic acids such as citric acid in influencing Cd dynamics in soil and related environments.

1. INTRODUCTION

Iron oxides are the most abundant metallic oxides in soils (Schwertmann and Taylor, 1989; Cornell and Schwertmann, 1996; Bigham *et al.*, 2002). Iron oxides usually form via solution from Fe^{2+} ions released from Fe(II)-bearing silicates and sulfide minerals on weathering (Oades, 1963; Schwertmann and Taylor, 1989; Cornell and Schwertmann, 1996). Once formed in soil and other natural environments, the mineral phase, composition, and distribution of iron oxides can be continually modified by the alteration of their environments (Schwertmann and Taylor, 1989). Therefore, the formation and transformation of pedogenic iron oxide mineral phases depend on the pedo-environmental conditions under which they have formed.

The soil rhizosphere is the first few millimeters of soil in contact with the roots (Huang and Germida, 2002; Jauert *et al.*, 2002). Organic ligands excreted by plant roots, and pH changes in the rhizosphere are two major factors that may modify the phytoavailability of heavy metals such as Cd (Collins *et al.*, 2003), through a number of chemical processes. Citric acid, a low-molecular-weight organic acid, is commonly present in soil rhizosphere. Its concentration ranges from 10^{-5} to 10^{-3} M (Robert and Berthelin, 1986). The previous studies show that the presence of citric acid during the formation of iron oxides significantly modifies their surface properties: surface structure, chemical composition, fine-scale surface morphology, specific surface area, surface porosity, and surface charge (Liu and Huang, 1999a, b).

Iron oxides have long been recognized as playing a vital role in controlling the fate of heavy metals in soils and sediments (Schwertmann and Taylor, 1989; Cornell and Schwertmann, 1996; Liu, 1999; Appel and Ma, 2002; Glover *et al.*, 2002) owing to the very reactive $-OH$ and $-OH_2$ functional groups exposed on their surface. It is usually accepted that trace metal concentrations in soil solution are primarily controlled by sorption and desorption reactions at the particle–water interface. While numerous studies have been conducted to understand adsorption of these metals on soil minerals (Grundl and Sparks, 1999; Selim and Sparks, 2001), less is known about desorption processes, especially their kinetics (Krishnamurti *et al.*, 1997; Onyatta and Huang, 2003). Furthermore, little is known about desorption of heavy metals from short-range-ordered mineral colloids formed under the influence of various ionic environments.

Research on cadmium pollution has received increasing international attention inasmuch as cadmium is toxic and can cause severe human health problems such as kidney disorder and *itai–itai* disease (Webb, 1979; Alloways, 1995). The influence of residence time and organic acids on Cd desorption from pure goethite has been studied (Glover *et al.*, 2002). However, pure iron oxides rarely exist in natural environments. The desorption kinetics of Cd from iron oxides, which are formed under the influence of organic acids such as citric acid, a common organic ligand in soil rhizosphere, still remains to be uncovered.

The objective of the present study is to examine the desorption kinetics of Cd following its adsorption on iron oxides. To simulate the effects of organic ligands in soil rhizosphere environment and chloride-bearing fertilizer, Cd desorption caused by citrate, acetate and chloride was investigated.

2. METHODS

2.1. Preparation and characteristics of iron oxide samples

Iron oxides were formed at pH 6.00 ± 0.05 and 25°C at an initial Fe(II) concentration of 10^{-2} M in the presence of citrate ligands at initial citrate/Fe(II) molar ratios (MRs) of 0, 0.001, 0.01 and 0.1. Citric acid concentration ranged from 10^{-5} to 10^{-3} M. The synthesis procedure is the same as the method reported before (Liu and Huang, 1999a). The sample was freeze-dried at −40°C. The minerals of iron oxides were identified by using X-ray diffraction (XRD) powder on a Philips (Model PW 1031) X-ray diffractometer using Mn-filtered FeKα radiation at 35 kV and 16 mA, and by using infrared (IR) spectrometry on a Perkin-Elmer infrared spectrophotometer (Model 983) using the KBr pellet technique (1 mg of sample mixed with 250 mg of KBr).

2.2. Determination of surface properties of iron oxides

The specific surface area of iron oxides was measured by using a multiple point BET-N_2 adsorption isotherm obtained with a Quantachrome Autosorb-1 apparatus (Quantachrome Corp., Syosset, NY). Prior to N_2 adsorption, about 100 mg samples were outgassed for 24 h at 10 mtorr. During N_2 adsorption the solids were thermostated in liquid N_2 (77–78 K). The pore-specific surface area of the iron oxides was determined from the 93-point N_2 adsorption isotherms using the t-plot method of de Boer, and the average pore diameter was estimated using the Kelvin equation (assuming cylindrical pores) (Gregg and Sing, 1982) with the Quantachrome Autosorb-1. The point of zero salt effect (PZSE) was determined in 0.01, 0.1 and 1 M NaCl solutions by the potentiometric titration method on a Metrohm titroprocessor (682 model). The organic carbon content of the iron oxides was determined with a Leco CR12 C analyzer (Leco Corp., St. Joseph, MI) at 1123 K (Wang and Anderson, 1998).

The basic characteristics of the iron oxides formed at various initial citrate/Fe(II) MRs are given in Table 1.

2.3. Kinetic experiments

The kinetics of Cd desorption from iron oxides, which formed at initial citrate/Fe(II) MRs of 0, 0.001, 0.01 and 0.1, following 1-d Cd adsorption at an

Table 1
Characteristics of iron oxide formed at various initial citrate/Fe(II) MRs

	Citrate/Fe MR			
	0	0.001	0.01	0.1
Dominant mineral[a]	G, M, L	L	L	noncrystalline
Specific surface area (m^2 g^{-1})	135.2 ± 0.5	102.8 ± 0.3	189.3 ± 0.4	209.6 ± 1.4
Micropore area (m^2 g^{-1})	0	0	22.2 ± 0.1	136.1 ± 0.9
Average pore diameter (nm)	10.2 ± 0.1	18.4 ± 0.2	5.4 ± 0.0	2.6 ± 0.0
PZSE	7.0 ± 0.2	6.0 ± 0.0	5.6 ± 0.1	3.9 ± 0.1
Organic C (g kg^{-1})	0.2 ± 0.0	1.1 ± 0.1	6.2 ± 0.4	55.7 ± 0.9

[a] G, goethite; M, maghemite; L, lepidocrocite.

initial Cd concentration of 50 μM at 20°C was investigated using the conventional batch method in the present study.

Fifty milligrams of freeze-dried iron oxides was dispersed in 35 mL deionized distilled water by ultrasonification (Sonifier, Model 350, Danbury, CT) at 150 W for 2 min. The suspension was adjusted to pH 5.0 with 0.01 M HNO_3 or 0.01 M KOH, and diluted to 40 mL. An aliquot of 40 mL of 100 μM Cd stock solution (pH 5.0) containing $Cd(NO_3)_2$ and KNO_3 was added to each flask, which contained the dispersed iron oxides, to obtain the initial concentrations of Cd and KNO_3 of 50 μM and 0.01 M, respectively. The final volume of the suspension was 80 mL. The suspension was shaken for 24 h at 20°C by placing the flask into a shaker with constant-temperature water bath. The suspensions at the end of adsorption period were filtered through a 0.1-μm Millipore membrane. The Cd concentration of the filtered solution was determined by a graphite furnace atomic absorption spectrometer (GFAAS) (Model 3100, Perkin-Elmer Corp., Norwalk, CT). The amount of Cd adsorbed by the iron oxides was based on the difference between the initial and final Cd concentrations.

The solids on the filter membrane were washed with deionized distilled water three times under vacuum. Then, the solids with membrane were transferred to a 250 mL flask and an 80 mL extractant was added into the flask. The extractants included 0.01 M (pH 5.0) KNO_3, KCl, K-citrate and K-acetate. The flasks were shaken at 20°C for 5 min to 7 days. The entire suspensions at the end of each reaction period were filtered through a 0.1-μm Millipore membrane. The Cd in the solutions was determined by the GFAAS. The experiment was carried out in duplicate.

3. RESULTS AND DISCUSSION

3.1. Adsorption of Cd by the iron oxides formed under the influence of citrate

The adsorption of Cd under the condition of the present study reached equilibrium at the end of a 24-h reaction period (data not shown). The amounts of Cd adsorbed by the iron oxides formed at various initial citrate/Fe(II) molar ratios are given in Fig. 1. The sequence of Cd adsorption by the iron oxides formed at different citrate/Fe(II) MRs was in the order of $0.1 > 0.01 > 0 > 0.001$. This sequence in the amounts of Cd adsorbed by the iron oxides was generally in accord with the increase in the N_2-BET-specific surface area of the iron oxides (Fig. 2). The iron oxides with higher specific surface would adsorb more Cd. In the absence of citrate ligands, a mixture of goethite, maghemite, and lepidocrocite was formed. The presence of citrate ligands at the initial citrate/Fe(II) MR of 0.001 improved the crystallization of lepidocrocite at the expense of goethite and maghemite, and decreased the specific surface area of the precipitation products (Table 1). This is because the Fe-citrate complexation possibly influences oxygen coordination and plays a positive role in the way the double rows of $[Fe(O,OH)_6]$ octahedra of lepidocrocite are linked during crystallization of the precipitation products through catalytic processes, as reported in a previous study (Liu and Huang, 1999a). Although lepidocrocite was still formed at an initial citrate/Fe(II) MR of 0.01, the degree of crystallinity of lepidocrocite was reduced (Liu and Huang, 1999a) and the surface area was increased (Table 1). The atomic force micrograph showed that some amorphous iron oxide also formed at this citrate/Fe(II) MR (Liu and Huang, 1999b). As the initial citrate/Fe(II) MR was further increased to 0.1, only X-ray noncrystalline iron oxide was formed, resulting in the largest specific surface area.

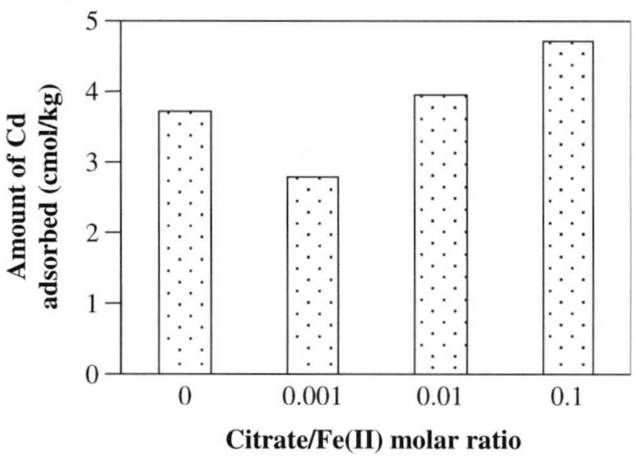

Fig. 1. Cadmium adsorbed by the Fe oxides formed at various initial citrate/Fe(II) MRs after a 24-h reaction period at 20°C.

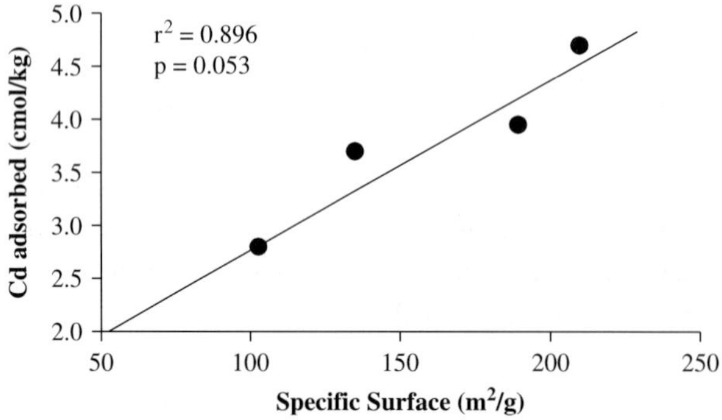

Fig. 2. The relation of the amount of Cd adsorbed and the specific surface area of the Fe oxides.

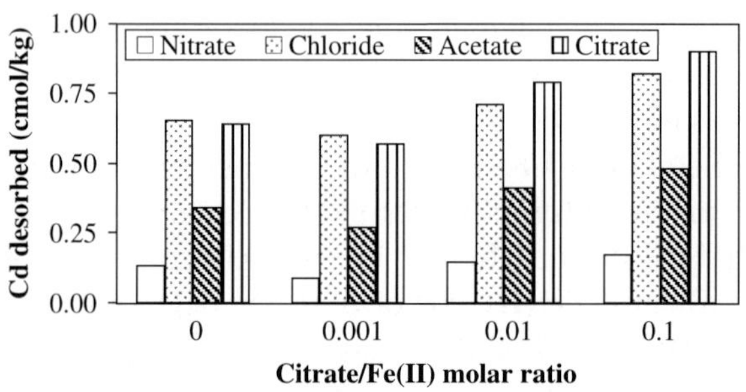

Fig. 3. Amount of Cd desorbed from the Fe oxide formed at various initial citrate/Fe(II) MRs by different extractants at pH 5.0 at the end of a 7-day reaction period.

3.2. Amounts of Cd desorbed from the iron oxides

The amounts of Cd released from the iron oxides at the end of a 7-day reaction period are presented in Fig. 3. The sequence of the amount of Cd desorbed was in accord with that of the amount of the Cd preadsorbed. For all the extractants studied, the amount of Cd released from the iron oxides formed at various initial citrate/Fe(II) MRs was in the order of $0.1 > 0.01 > 0 > 0.001$. The Cd released from the surface of the iron oxides by various extractants may proceed by (1) complexation of the ligands, i.e. nitrate, chloride, acetate, or citrate, with Cd; and (2) cation-exchange reaction of K with Cd. However, the exchange of Cd by K cannot explain the difference in Cd desorption from the iron oxides formed at various citrate/Fe(II) MRs, since the same concentration of K solution was used in all the systems studied. The complexation of these inorganic and organic ligands with Cd^{2+} on the surface should be the dominant process in releasing Cd into the solution phase. At any citrate/Fe(II) MR, the sequence of amounts of the Cd released by these ligands (Fig. 3) is generally in accord with that of the stability constants of Cd complexes with these ligands (log K_1 = 1.3 and log K_2 = 0.85 for Cd–NO_3 complexes, log K_1 = 1.52 and log K_2 = 0.94 for Cd–acetate complexes, log K_1 = 2.0 and log K_2 = 5.32 for Cd–Cl complexes, and log K_1 = 3.74 and log K_2 = 4.12 for Cd–citrate complexes at 25°C and ionic strength of 0.1 (Academic Software, 1993)). The iron oxide with a higher specific surface area adsorbed more Cd (Fig. 1) as discussed above. It would have more Cd ions exposed on the surface, and is thus more accessible to the complexing ligands. Therefore, for any extractants studied, the amounts of Cd released followed the sequence of 0.1 > 0.01 > 0 > 0.001.

The percentage of the amount of Cd desorbed by any extractant over the amount of Cd preadsorbed was not significantly different for different iron oxides (Fig. 4). However, in each iron oxide system, the different extractants led to substantially different percentages of the amount of Cd desorbed. Chloride and citrate resulted in the release of 17–21% of preadsorbed Cd. About 9–10% of Cd was desorbed by acetate. Only about 4% of preadsorbed Cd was released by nitrate. The percentage of Cd released from the surface of iron oxides is also consistent with the stability constants of the complexes of Cd with those ligands. Only relatively small portions of preadsorbed Cd were released from the iron oxides in all the systems, indicating that Cd is strongly bound by iron oxides and by citrate for iron oxides formed at the initial citrate/Fe(II) MR of 0.1. Iron oxides, as great sinks to retain Cd, play a vital role in controlling Cd mobility and bioavailability in natural environments.

3.3. Kinetics of Cd desorption by different extractants

The Cd remaining in the iron oxides decreased with increase in reaction time. For example, the time function of the Cd remaining in the iron oxides formed at the initial citrate/Fe(II) MRs of 0.001 and 0.1 is illustrated in Fig. 5. The data show that the desorption of Cd from the iron oxides approached equilibrium

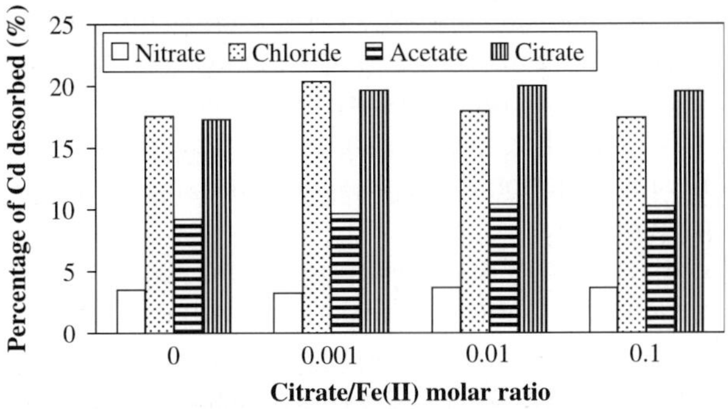

Fig. 4. Percentage of Cd released at the end of a 7-day reaction period from Cd adsorbed by the Fe oxides formed at various initial citrate/Fe(II) MRs.

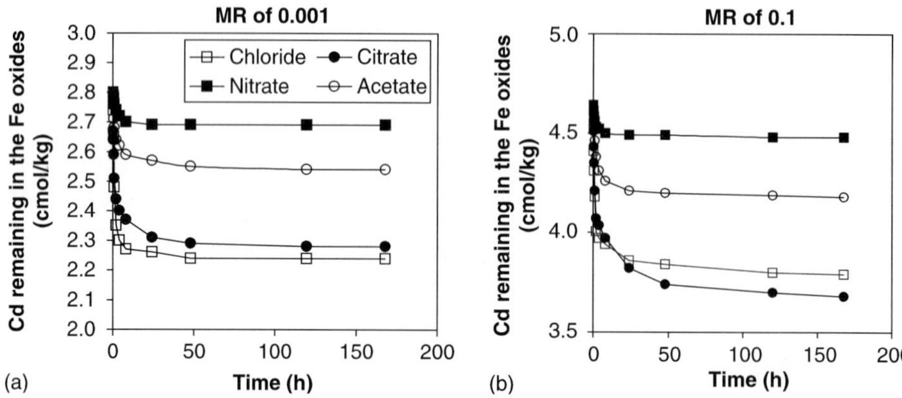

Fig. 5. Effects of various extractants on the time function of the Cd remaining in the Fe oxides (cmol kg^{-1}) formed at the initial citrate/Fe(II) MRs of 0.001 and 0.1.

after a 48-h reaction period. The kinetic and empirical equations, including the zero-, first-, and second-order-rate equation, and the overall parabolic diffusion equation, the modified Freundlich equation, and the Elovich equation were used to fit Cd desorption data at the end of reaction times from 5 min to 2 h and from 2–24 h. The degree of fit of the rate equations to the data was examined using the correlation coefficient (r^2), probability (p), and standard error (SE) of linear regression analysis. As an example, the fitting degree of various equations to the data of the fast reaction (5 min to 2 h) of Cd desorption from the iron oxide formed at the initial citrate/Fe(II) MR of 0.1 by chloride and citrate is given in Table 2. A similar trend was observed in other iron oxide systems. The data show that both fast and slow reactions of the Cd desorption from the iron oxides can be satisfactorily described by the overall parabolic diffusion equation as illustrated in Fig. 6. The parabolic diffusion equation is often used to indicate that diffusion-controlled phenomena are rate-limiting (Sparks, 1999). Several scientists applied the overall parabolic equation to study the kinetics of sorption of nutrients and pollutants by soil components (Jardine and Sparks, 1984; Simard *et al.*, 1992; Mengel and Uhlenbecker, 1993; Dang *et al.*, 1994; Krishnamurti *et al.*, 1997; Ma and Liu, 1997; Liu *et al.*, 2001). Cd desorption from the iron oxides is well described by the parabolic diffusion equation in the present study, suggesting that its rate is limited by diffusion processes.

The rate coefficients of Cd desorption from the iron oxides by various extractants are presented in Table 3. The data show that in all of the systems studied, the rate coefficient of the fast reaction was much greater than that of the slow reaction. The rate of Cd release by different extractants for the fast reaction from all the systems studied was in the order $Cl^- >$ citrate $>$ acetate $> NO_3^-$ (Table 3), which was generally consistent with the stability constants of Cd-extractant ligand

Table 2
The degree of fitting of various equations to the fast reaction (5 min to 2 h) of Cd desorption from the Fe oxide formed at the initial citrate/Fe(II) MR of 0.1 by chloride and citrate

Equation	Chloride			Citrate		
	r^2	p	SE[a]	r^2	p	SE
0-order	0.963	1.52×10^{-4}	0.17	0.919	8.43×10^{-3}	0.12
1st-order	0.852	2.01×10^{-3}	0.21	0.878	1.05×10^{-2}	0.14
2nd-order	0.830	6.15×10^{-3}	0.34	0.853	4.36×10^{-2}	0.26
Overall parabolic diffusion	0.992	7.26×10^{-5}	0.06	0.974	6.42×10^{-4}	0.04
Elovich	0.983	4.59×10^{-4}	0.07	0.934	7.89×10^{-3}	0.04
Freundlich	0.984	4.45×10^{-4}	0.10	0.966	8.45×10^{-4}	0.06

[a] Standard error (cmol kg^{-1}).

complexes. The rate of Cd release by different extractants for the slow reaction from all the systems followed the order citrate > acetate = Cl^- > NO_3^-. The fast reaction of Cd desorption could predominantly occur on the mesopore surface of the iron oxides, while the slow reaction of Cd desorption could predominantly take place on the micropore surface of the iron oxides. Therefore, a longer induction period for the desorption reaction occurring on the micropore surface is required for extractants with larger molecules, such as citrate and acetate. This may explain the observation that the rate of the slow reaction of Cd desorption by citrate and acetate with a larger molecular size was much greater than that of the slow reaction of Cd desorption by chloride and nitrate with a smaller molecular size. Although the stability constants of complexes of acetate with Cd ions are lower than those of chloride with Cd ions (Academic Software, 1993; Herbelin and Westall, 1994), the rate coefficients of the slow reaction of Cd released by acetate from the iron oxides formed at initial citrate/Fe(II) MRs of 0.001, 0.01 and 0.1 were higher than those released by chloride from the same iron oxides.

In the chloride and nitrate extractant systems, the iron oxides formed at the initial citrate/Fe(II) MRs of 0, 0.001, and 0.01 had similar rate fast Cd desorption reaction rates (Table 3). However, the iron oxide formed at the MR of 0.1 had a significantly slower rate of fast Cd desorption reaction (Table 3), although it had the greatest specific surface area (Table 1). Unlike the iron oxides formed at the initial citrate/Fe(II) MRs of 0, 0.001 and 0.01, the surface area of the iron oxide formed at the MR of 0.1 is predominantly composed of micropore surface instead of mesopore surface (Liu and Huang, 1999a). This further confirmed that the fast reaction of Cd desorption could predominantly occur on the mesopore surface of the iron oxides.

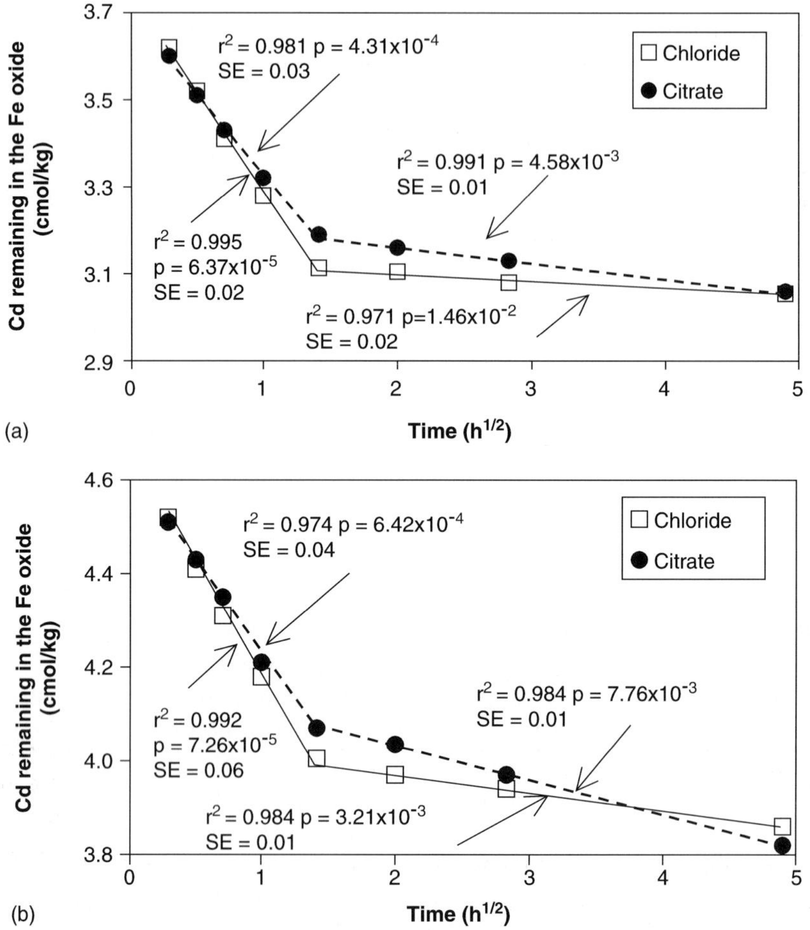

Fig. 6. Plotting of Cd desorption by chloride or citrate from Fe oxides formed at initial citrate/Fe(II) MRs of (a) 0 and (b) 0.1 based on overall parabolic diffusion equation.

However, this trend is not very clear in the citrate and acetate extractant systems. As indicated by the total C content in the iron oxide samples (Table 1), there was a significant amount of citrate coprecipitated with iron oxides formed at the initial citrate/Fe(II) MRs of 0.01 and 0.1. The coprecipitated citrate with iron oxides could have different effects on the Cd desorption induced by inorganic and organic extractants, since the stability constants for the Cd– and Fe–ligand complexes vary greatly.

In all the extractant systems, the rate of slow reaction of Cd desorption from the iron oxides formed at various initial citrate/Fe(II) MRs followed the order $0.1 > 0.01 > 0 > 0.001$ (Table 3), also indicating that the slow reaction of Cd desorption could predominantly occur on the micropore surface of the iron oxides. This is consistent with the microporosity of the iron oxides formed (Table 1).

Table 3
Rate coefficients (cmol kg^{-1} $h^{-1/2}$) of Cd desorption in overall diffusion equation[a]

Reaction	Initial citrate/Fe(II) MR			
	0	0.001	0.01	0.1
Chloride				
Fast	0.452	0.450	0.459	0.438
Slow	0.017	0.006	0.025	0.046
Nitrate				
Fast	0.062	0.061	0.061	0.046
Slow	0.009	0.004	0.013	0.021
Citrate				
Fast	0.366	0.308	0.405	0.399
Slow	0.038	0.036	0.048	0.070
Acetate				
Fast	0.165	0.153	0.172	0.169
Slow	0.017	0.016	0.039	0.051

[a] $LSD_{0.05} = 0.007$ for the fast reaction and $LSD_{0.05} = 0.002$ for the slow reaction.

The data in the present study show that compared with nitrate and acetate, citrate and chloride can result in the release of a greater amount of Cd from the iron oxides through a fast reaction process. Citrate is one of the most abundant low-molecular-weight organic acids in soil rhizosphere. To reduce food chain contamination through Cd uptake by plants, selection of plant cultivars that excrete different types and amounts of organic acids merits attention in cropping systems. Chloride is commonly present in irrigation water and chloride-bearing fertilizers. Recent field experiments have shown that high levels of chloride in irrigation waters can increase soil Cd uptake by crops (Chien *et al.*, 2003). The potential increase of Cd bioavailability by application of chloride-bearing fertilizers and irrigation water containing substantial amounts of chloride also deserves close attention.

4. CONCLUSIONS AND FUTURE PROSPECTS

The bioavailability and toxicity of Cd in soil and related environments are governed by the speciation and dynamics of Cd. Therefore, desorption kinetics of Cd from soils and sediments has received increasing attention. The data obtained in the present study revealed that rate of desorption of Cd adsorbed on iron oxides formed under the influence of organic acids such as citric acid was greatly influenced by the increase of their degree of crystal disorder, specific surface, and microporosity.

The desorption kinetics can be divided into the fast and slow reactions. For the fast reaction, the rate of Cd desorption is generally in accord with the stability constants of Cd-extractant ligand complexes: chloride > citrate > acetate > nitrate. For the slow reaction, the rate of Cd desorption is apparently influenced by the size of extractant molecules: citrate > acetate ≥ chloride > nitrate. A longer induction period for an extractant with a larger molecule is required for Cd desorption from the micropore surface. Therefore, stability constants of Cd-extractant ligand complexes, and steric factor of the molecular size of the extractants in relation to the pore size of the oxides merit attention in understanding the kinetics of Cd desorption.

The research findings obtained in the present study are of fundamental significance in understanding the transformation and mobility of Cd in soil, especially in the rhizosphere, as influenced by soluble salts and organic acids common in environment. Further microscopic and spectroscopic studies are needed to confirm the Cd transformation mechanisms. The effect of chemical aging of Cd during adsorption processes should also be investigated in the future to understand the long-term transformation of Cd in soil and related environments.

ACKNOWLEDGMENTS

This study was supported by Discovery Grant GP2383 – Huang of the Natural Sciences and Engineering Research Council of Canada.

REFERENCES

Academic Software, 1993. IUPAC Stability Constants Database. Version 1.02. Academic Software. W. Yorks, UK.

Alloways, B.J., 1995. Heavy Metals in Soils, Blackie Academic & Professional, London, UK.

Appel, C., Ma, L., 2002. Concentration, pH, and surface charge effects on cadmium and lead sorption in three tropical soils. J. Envir. Qual. 31, 581–589.

Bigham, J.M., Fitzpatrick, R.W., Schulze, D.G., 2002. Iron oxides. In: Dixon, J.B., Schulze, D.G., (Eds.), Soil Mineralogy with Environmental Applications. SSSA Book Series: 7. Soil Sci. Soc. Am., Madison, WI. pp. 323–366.

Chien, S.H., Carmona, G., Prochnow, L.I., Austin, E.R., 2003. Cadmium availability from granulated and bulk-blended phosphate-potassium fertilizers. J. Envir. Qual. 32, 1911–1914.

Collins, R.N., Merrington, G., McLaughlin, M.J., Morel, J.-L., 2003. Organic ligand and pH effects on isotopically exchangeable cadmium in polluted soils. Soil Sci. Soc. Am. J. 67, 112–121.

Cornell, R.M., Schwertmann, U., 1996. The Iron Oxides. Structure, Properties, Reactions, Occurrence, and Uses, VCH, Weinheim, FRG.

Dang, Y.P., Dalal, R.C., Edwards, D.G., Tiller, K.G., 1994. Kinetics of zinc desorption from Vertisols. Soil Sci. Soc. Am. J. 58, 1392–1399.

Glover II, L.J., Eick, M.J., Brady, P.V., 2002. Desorption kinetics of Cd^{2+} and Pb^{2+} from goethite. Influence of time and organic acids. Soil Sci. Soc. A. J. 66, 797–804.

Gregg, S.J., Sing, K.S.W., 1982. Adsorption, Surface Area and Porosity, second ed., Academic Press, London.

Grundl, T.J., Sparks, D.L., 1999. Kinetics and mechanisms of reactions at the mineral–water interface: an overview. In: Sparks, D.L., Grundl, T.J., (Eds.), Mineral–Water Interfacial Reactions: Kinetics and Mechanisms. ACS Symp. Series 715. Am. Chem. Soc., Washington, DC, pp. 2–11.

Herbelin, A.L., Westall, J.C., 1994. FITEQL. A Computer Program for Determination of Chemical Equilibrium Constants from Experimental Data. Version 3.1. Department of Chemistry, Oregon State University, Or.

Huang, P.M., Germida, J.J., 2002. Chemical and biological processes in the rhizosphere: Metal pollutants. In: Huang, P.M., Bollag, J.-M., Senesi, N., (Eds.), Interactions of Soil Particles with Microorganisms and the Impact on the Terrestrial Ecosystem. IUPAC Series on Analytical and Physical Chemistry of Environmental Systems, Vol. 8. Wiley, Chichester, UK, pp. 381–438.

Jardine, P.M., Sparks, D.L., 1984. Potassium–calcium exchange in a multireactive soil system. I. Kinetics. Soil Sci. Soc. Am. J. 48, 39–45.

Jauert, P., Schumacher, T.E., Boe, A., Reese, R.N., 2002. Rhizosphere acidification and cadmium uptake by strawberry clover. J. Envir. Qual. 31, 627–633.

Krishnamurti, G.S.R., Cieslinski, G., Huang, P.M., Van Rees, K.C.J., 1997. Kinetics of cadmium release from soils as influenced by organic acids: Implication in cadmium availability. J. Envir. Qual. 26, 271–277.

Liu, C., 1999. Surface Chemistry of iron oxide minerals formed in different ionic environments. Ph.D. Thesis, University of Saskatchewan, Saskatoon, Saskatchewan, Canada.

Liu, C., Frenkel, A.I., Vairavamurthy, A., Huang, P.M., 2001. Sorption of cadmium on humic acid: Mechanistic and kinetic studies with atomic force microscopy and X-ray absorption fine structure spectroscopy. Can. J. Soil Sci. 81, 337–348.

Liu, C., Huang, P.M., 1999a. Properties of iron oxides formed at various citrate concentrations. In: Kodama, K., Mermut, A.R., Torrance, J.K., (Eds.), Clays for Our Future. The Proceedings of the 11th International Clay Conference, Ottawa, Canada. The International Association for the Study of Clays. Ottawa, Canada, pp. 513–522.

Liu, C., Huang, P.M., 1999b. Atomic force microscopy and surface characteristics of iron oxides formed in citrate solutions. Soil Sci. Soc. Am. J. 63, 65–72.

Ma, Y.B., Liu, J.F., 1997. Adsorption kinetics of zinc in a calcareous soil as affected by pH and temperature. Commun. Soil Sci. Plant Anal. 28, 1117–1126.

Mengel, K., Uhlenbecker, K., 1993. Determination of available interlayer potassium and its uptake by ryegrass. Soil Sci. Soc. Am. J. 57, 761–766.

Oades, J.M., 1963. The nature and distribution of iron compounds in soils. Soil Fert. 27, 69–80.

Onyatta, J.O., Huang, P.M., 2003. Kinetics of cadmium release from selected tropical soils from Kenya by low-molecular-weight organic acids. Soil Sci. 168, 234–252.

Robert, M., Berthelin, J., 1986. Role of biological and biochemical factors in soil mineral weathering. In: Huang, P.M., Schnitzer, M. (Eds.), Interactions of Soil Minerals with Natural Organics and Microbes. SSSA Special Publication 17, Soil Sci. Soc. of Am., Madison, WI, pp. 453–496.

Schwertmann, U., Taylor, R.M.,1989. Iron oxides. In: Dixon, J.B., Weed S.B. (Eds.), Minerals in Soil Environments. 2nd Edition, Book Ser. 1, Soil Science Society of America, Madison, WI, pp. 379–438.

Selim, H.M., Sparks, D.L., 2001. Heavy Metals Release in Soils. Lewis Publishers, Boca Raton, FL.

Simard, R.R., De Kimpe, C.R., Zizka, J., 1992. Release of potassium and magnesium from soil fractions. Soil Sci. Soc. Am. J. 56, 1421–1428.

Sparks, D.L., 1999. Kinetics of sorption/release reactions on natural particles. In: Huang, P.M., Senesi, N., Buffle, J., (Eds.), Structure and Surface Reactions of Soil Particles. IUPAC Series on Analytical and Physical Chemistry of the Environmental Systems. Wiley. Chrichester, UK, pp. 413–448.

Wang, D., Anderson, D.W., 1998. Direct measurement of organic carbon content in soils by the Leco 12 Carbon Analyzer. Commun. Soil Sci. Plant Anal. 29, 15–21.

Webb, M., 1979. The Chemistry, Biochemistry, and Biology of Cadmium. Elsevier, Amsterdam, The Netherlands.

Biogeochemistry of Trace Elements in the Rhizosphere
P.M. Huang and G.R. Gobran (Editors)

Chapter 7

Biogeochemistry of soil cadmium and the impact on terrestrial food chain contamination

G.S.R. Krishnamurti[a], D.F.E. McArthur[a], M.K. Wang[b], L.M. Kozak[a], and P.M. Huang[a]

[a]Department of Soil Science, University of Saskatchewan, 51 Campus Drive, Saskatoon SK S7N 5A8, Canada
E-mail: huangp@sask.usask.ca

[b]Department of Agricultural Chemistry, National Taiwan University, Taipei, Taiwan, Republic of China 10764

ABSTRACT

Cadmium has been recognized to be a highly toxic element, but it was not until recently that concern has been expressed about the possible effects on human health of long-term exposure to low concentrations of this element. The discovery that Cd pollution from a basic metal mining operation could cause serious illness and possibly death has led to public anxiety as well as medical interest. There are many reviews on the chemistry, biochemistry, and biology of soil cadmium. However, to assess the impact of Cd pollution on food chain contamination and ecosystem health, major biogeochemical pathways, particularly those in the rhizosphere, have to be elucidated, and gaps in our knowledge must be identified for future research planning. This chapter addresses the biogeochemistry of soil Cd and the impact of Cd pollution on terrestrial food chain contamination through the rhizosphere. Knowledge of the total content of Cd does not imply comprehensive knowledge of its chemical behavior; rather, it is the chemical speciation of Cd that influences Cd's chemical reactivity, mobility, bioavailability, and toxicity in the ecosystem. Therefore, it is essential to investigate the chemical speciation of Cd in soils and sediments, and especially in the rhizosphere, which is the bottleneck of terrestrial food chain contamination. The fractionation of metal–organic complex-bound Cd species is a recent innovation in sequential extraction schemes. The metal–organic complex-bound Cd is the most common among the particulate-bound Cd species of surface soils in temperate and tropical regions. Cadmium present as metal–organic complex-bound species is especially enriched in the rhizosphere soils after application of phosphate fertilizer. The importance of metal–organic complex-bound Cd fractions in assessing phytoavailability of soil Cd has been demonstrated, and thus merits attention.

1. INTRODUCTION

Cadmium (Cd) is a relatively rare element and is usually present in the natural terrestrial environments at low concentrations. In the absence of contributions from anthropogenic activities, the base levels of the trace elements are controlled by the original lithology and weathering effects. Cadmium was recognized to be a highly toxic element, but it was not until recently that concern has been expressed about the possible effects on human health of long-term exposure to low concentrations of this element. The discovery that Cd pollution from a basic metal mining operation could cause serious illness and possibly death has led to public anxiety (Kobayashi and Hagino, 1965) as well as medical interest, owing to reports from Japan of "Ouch-Ouch" (Itai-Itai in Japanese) disease (Emmerson, 1970). Although industrial operations are major sources of Cd pollution, many countries now show concern that disposal of metal-rich sewage sludge and application of Cd-rich phosphate fertilizers to agricultural fields may render crops a health hazard if consumed by humans and animals. Nevertheless, the increasing awareness of the potential hazards of Cd contamination should not obscure the fact that Cd is present in natural ecosystems and is present in all living organisms.

There are many reviews on the chemistry, biochemistry, biology, and transport of soil Cd (Page and Bingham, 1973; Fleischer *et al.*, 1974; Webb, 1979; Foerstner, 1991; Ross, 1994; Alloway, 1995; McLaughlin and Singh, 1999; Adriano, 2001). However, to assess the impact of Cd pollution on food chain contamination and ecosystem health, major biogeochemical pathways, especially those in the rhizosphere, have to be elucidated, and gaps in our knowledge must be identified for future research planning. This chapter addresses the biogeochemistry of soil Cd and the impact of Cd pollution on terrestrial food chain contamination through the rhizosphere.

2. GEOCHEMISTRY OF CADMIUM

Significant aspects of the geochemistry of Cd with regard to soil–plant–human relationships are its low concentration in the Earth's crust, its estimated average crustal abundance being 0.15 mg kg^{-1} (Weast, 1979), and its impact on ecosystem health (Huang *et al.*, 1998). The Cd concentrations in the geosphere range from 0.005 mg kg^{-1} in schists to 219 mg kg^{-1} in black shales (Alloway, 1995). Consideration of crustal abundance can be confusing because the Earth's crust, at least the top 16 km, comprises 95% igneous and metamorphic rocks and 5% sedimentary rocks. Of the sedimentary rocks, 80% are shales, 15% sandstones, and 5% limestones (Mitchell, 1964). Sedimentary rocks are much more important soil parent materials since they overlie most igneous formations and account for 75% of the outcrops on the Earth's surface (Ross, 1994).

2.1. Igneous and metamorphic rocks

Page and Bingham (1973) reported that soils derived from igneous rocks have Cd contents of 0.1–0.3 mg kg^{-1} and those from metamorphic rocks have Cd contents of contain 0.1–1.0 mg kg^{-1}. The Cd contents of selected types of rocks and ores are presented in Table 1. The low contents of Cd in igneous rocks account for its low crustal abundance. Cadmium has the same valency and similar ionic radius as Ca (Cd^{2+}, 0.97 Å; Ca^{2+}, 0.99 Å), but does not commonly substitute for

Table 1
Cadmium content of selected types of rocks and ores[a]

	Range (mg kg^{-1})	Mean (mg kg^{-1})
Igneous rocks		
Rhyolites	0.03–0.57	0.23
Granites	0.01–1.60	0.20
Basalts	0.01–1.00	0.13
Pitchstone	0.05–0.34	0.17
Schists	0.01–0.87	0.02
Metamorphic rocks		
Gneisses	0.01–0.26	0.04
Eclogite	0.04–0.26	0.11
Sedimentary rocks		
Shales	0.02–11.00	1.40
Bentonite	0.03–11.00	1.40
Marine shales	0.30–219.0	15.50
Sandstones	0.01–0.65	< 0.03
Lacustrine sediments	5.0–19.0	11.00
Manganese nodules	3.0–21.2	8.00
Phosphorites	10–980	25.00
Sulfide ores		Median
Sphalerite	< 50,000	3000
Galena	< 5000	10.0
Tetrahedrite-tennartite	< 2400	600.0
Metacinnabar	< 117,000	10.0

[a] Data from Gulbransden (1966), Waketa and Schmitt (1970), Williams and David (1973), Fleischer *et al.* (1974), Holmes (1976), Rose *et al.* (1979), Adriano (2001), Page *et al.* (1987) and Thornton (1992).

Ca in minerals such as plagioclase or calcite. This is because Cd has a greater tendency to form covalent bonds, whereas those formed by Ca are predominantly ionic (Krauskopf, 1967), accounting for the low concentration of Cd in igneous and metamorphic rocks.

The main minerals in igneous rocks, which have been reported to contain significant amounts of Cd, are biotite (4.8; Waketa and Schmitt, 1970) and riebeckite (5.8 mg kg^{-1}; Holmes, 1976). The most abundant source of Cd is in zinc sulfide minerals such as sphalerite and wurtzite, and mercury sulfide minerals such as metacinnabar, where Cd can occur at concentrations up to 50,000 and 117,000 mg kg^{-1} of the minerals, respectively (Fleischer *et al.*, 1974).

Cd is unique among metals currently used in industry in that there are no ore bodies exploited solely for its production; however, it is normally obtained as a by-product from smelting and purification of other metal ores (Page and Bingham, 1973). Consequently, smelting operations involving metal sulfides and other types of ores give rise to significant amounts of local Cd pollution in the environment (Peterson and Alloway, 1979).

2.2. Sedimentary rocks

In contrast to the low concentrations in igneous rocks, a wider range of Cd concentrations is found among various types of sedimentary rocks (Table 1). Sandstones and limestones contain relatively low concentrations of Cd; conversely, organic-rich lacustrine sediments, black shales, and phosphorites contain high concentrations of Cd and other trace elements. The high concentration of heavy metals in the latter rock types is a result of their relatively high organic matter content; a direct effect is through the formation of metal–organic complexes or indirectly because of the organic matter acting as a substrate for microbiological activity (Peterson and Alloway, 1979). Black shales are recognized as being relatively enriched with Cd, as indicated by the reports of high contents of Cd in black Pierre shales of the late Cretaceous age (11 mg kg^{-1}; Tourtelot *et al.*, 1964), in Permian Marl Slates (73.5 mg kg^{-1}; Holmes, 1976), and in marine black shales of the Carboniferous age (up to 219 mg kg^{-1}; Holmes, 1976). Even though the black shales comprise only a very small proportion of the total outcrop area on the Earth's surface, they are locally important and lead to Cd pollution in soils derived from these parent materials. Marples and Thornton (1980) reported a concentration of up to 20 mg kg^{-1} in topsoils derived from carboniferous black shales.

Phosphorites are reported to have high concentrations of Cd up to 980 mg kg^{-1} (Table 1). Even though on a world-area basis, outcrops of phosphorites are of little significance, the environmental impact of Cd in these rocks is likely to be much greater than that in black shales because phosphorites are the major source of phosphate for agricultural fertilizer manufacture. Some of the phosphate fertilizers are reported to contain 140 mg Cd kg^{-1} (Idaho monoammonium phosphate fertilizer; Krishnamurti *et al.*, 1996) and 170 mg Cd kg^{-1} (superphosphate fertilizer; Peterson and Alloway, 1979).

3. SOURCES OF CADMIUM IN SOIL ENVIRONMENTS

3.1. Background level of cadmium in soils

Concentrations of Cd in uncontaminated soils (to 100 cm depth) are closely related to their parent materials, typical concentrations being around 0.2 mg kg^{-1} (Nriagu, 1980). The highest Cd values reported for soils derived from black shales include < 22 mg kg^{-1} on outcrops on the Monterey shale in California, USA (Page *et al.*, 1987), < 24 mg kg^{-1} in soils on carboniferous black shale in Derbyshire, UK (Marples and Thornton, 1980), and < 11 mg kg^{-1} in alluvial soils in an area of black shale and black slate outcrops in South Korea (Kim and Thornton, 1993). There are many areas, however, where natural Cd levels are augmented through variations in farming practices and by its release from industry as effluents.

Values for total Cd concentrations in unpolluted soils vary from < 0.01 to 2.50 mg kg^{-1}; Sillanpaa and Jansson (1992) reported an average Cd concentration around 0.05 mg kg^{-1} for 3500 soils collected from 30 countries around the world. Jensen and Bro-Rasmussen (1992) reported that the Cd concentrations in European soils ranged from 0.06 mg kg^{-1} in Finland to 0.50 mg kg^{-1} in the UK, depending on the nature of the parent material. The Cd concentration in the soils can be as low as 0.03 mg kg^{-1} or as high as 10 mg kg^{-1}. Holmgren *et al.* (1993) reported an average total Cd concentration value of 0.265 mg kg^{-1} for the 3045 soils in the USA. Roberts *et al.* (1994) recently examined the Cd status of unpolluted, non-agricultural soils in New Zealand and reported that Cd concentrations in the top 7.5 cm surface horizons ranged from 0.02 to 0.77 mg kg^{-1}, with an average value of 0.20 mg kg^{-1}.

3.2. Impact of farming practices

Cadmium entering the soil constitutes a relatively lasting form of pollution, since uptake by plants can continue long after the source of pollution ceases to exist. Most of the Cd applied to surface soils is adsorbed onto organic matter and mineral colloids, and owing to its insolubility, remains within the upper 0–15 cm of the soil profile or plough layer (Williams and David, 1976; Peterson and Alloway, 1979; Mulla *et al.*, 1980; Rothbaum *et al.*, 1986; McGrath, 1987; Huang and Germida, 2002). It was also shown that plants took up more than 90% of the Cd from the surface horizons (Christensen and Tjell, 1983).

The concentrations of impurities in manufactured nitrogenous or potash fertilizers are generally low, ranging from 0.1 to 2.0 mg kg^{-1} (McLaughlin *et al.*, 1996). The ranges of Cd concentrations present in farmyard manure, household and municipal solid waste, compost, sewage sludge, and phosphate fertilizers are shown in Table 2. All of these soil amendments add Cd to agricultural soils, but their impacts on the Cd levels in agricultural soils vary.

Farmyard manure is commonly applied to many agricultural soils. The Cd in farmyard manure comes from vegetation that is fed to the farm animals: the Cd in the vegetation has been taken out of the soil by the plant. However, since most of the Cd ingested with the vegetation remains in the manure (Chaney *et al.*, 1999), the concentration of Cd in manure is generally higher than the concentration of Cd in the vegetation consumed by the animals. Because the Cd in the farmyard manure comes indirectly from the soil in the first place, the overall net effect of manure application on total soil Cd is relatively insignificant. However, localized increases in soil Cd concentration can occur with repeated applications of manure onto fields that were not the original source of all of the Cd in the manure.

Household and municipal solid waste compost also contains Cd that had its origins in agricultural soil. Composts consist of food wastes that have been microbially degraded. The food wastes contain trace amounts of Cd, most of which originated from the soil (Adriano, 2001). The degradation process results in compost with a much higher Cd concentration than the food wastes that went into it. Just like the farmyard manure, repeated applications of compost onto fields that were not the source of all of the Cd in the compost will result in localized increases in soil Cd concentration. However, the agricultural soils affected by compost application are in the minority, since compost is applied mainly to soils around urban areas.

Sewage sludge is the solid residue formed during the treatment of sewage waste. The treatment process involves a digestion step, during which the Cd in the sewage waste is concentrated. The Cd in sewage sludge comes from both industrial sewage and domestic sewage. The Cd concentrations in sewage sludge vary from 1 to 3650 mg kg^{-1} in the dry matter (Peterson and Alloway, 1979). The application of sewage sludge heavily contaminated with Cd, may lead to very high Cd concentrations of the order of 26–159 mg kg^{-1} in the soils (Pike *et al.*,

Table 2
Major agricultural soil amendment sources of Cd (McArthur, 2001)

	Cd concentration (mg kg^{-1})	Reference
Farmyard manure	0.3–1.8	McGrath (1984); Kabata-Pendias and Pendias (1992)
Compost	0.26–11.7	Alloway and Steinnes (1999)
Sewage sludge	< 1–3650	Alloway (1990)
Phosphorus fertilizers	0.2–345	Alloway and Steinnes (1999)

1975; Chumbley and Unwin, 1982; Alloway *et al.*, 1990; Dudka and Chlopecka, 1990; Jackson and Alloway, 1991a, b).

Heavy metals are considered to be the main factor limiting the agricultural use of sewage sludge, because relatively large amounts of sludge have to be applied to meet the nitrogen requirements of the crop (Chaney and Hornick, 1977). Organically bound Cd from sewage sludge was reported to be more readily adsorbed and translocated within the plants than similar amounts of the element in ionic form (Peterson and Alloway, 1979). On the contrary, Street *et al.* (1977) observed that Cd added alone to soil as sulfate was more readily taken up by plants than Cd added in sewage sludge. This may be attributable to the difference in the form of organically bound Cd, soil properties, and plant species.

In the past, most of the sewage sludge was disposed of at sea or in landfills. However, since sewage sludge can be a useful source of nutrients, such as P and N, and contain colloidal organic matter, which can have a beneficial effect on soil structure, there has been a trend in recent years to apply sewage sludge to agricultural land (Alloway and Steinnes, 1999). Because of the relatively high concentrations of Cd in some sewage sludge (Table 2), regular applications of sewage sludge can result in large increases in the Cd concentrations of agricultural soils. However, the agricultural soils affected by sewage sludge are in the minority, since the sludge is applied mainly to soils around urban areas (Brown and Jacobson, 1987).

While both compost and sewage sludge applications to agricultural soil are fairly localized, application of Cd-contaminated phosphate fertilizer to the soil is not. Every year in major agricultural areas around the world, farmers routinely apply Cd-contaminated phosphate fertilizer (Table 2) to their soils. The Cd found in phosphate fertilizers has its origin in the rock phosphate from which the fertilizer is made. The rock phosphates have a wide range of Cd concentrations, ranging between 1 and 156 mg kg^{-1}, depending on the source (Williams and David, 1973; Stenstrom and Vahter, 1974; Reuss *et al.*, 1978; Freedman and Hutchinson, 1981; Smilde and van Luit, 1983; Syers *et al.*, 1986; Bramley, 1990; Singh, 1991). Fertilizers made from magmatic phosphates contain low or even negligible concentrations of Cd, whereas those made from sedimentary phosphates contain high concentrations of Cd. The mg of Cd kg^{-1} of P in these rocks ranges from 1 to at least 570 (McLaughlin *et al.*, 1996). The mg of Cd kg^{-1} of P does not change when the rock is processed into superphosphate fertilizer. Although it drops by 30–35% when the ore is processed into higher analysis phosphate fertilizers (Potash & Phosphate Institute/Potash & Phosphate Institute of Canada, Foundation for Agronomic Research, 1998), there can still be as much as 532 mg Cd kg^{-1} of P in the processed fertilizers.

Continuous application of phosphate fertilizers to agricultural soils has caused up to a 14-fold increase in Cd concentration in the surface soil (Williams and David, 1976; Mulla *et al.*, 1980; Mortvedt, 1987). The mean input of Cd to agricultural soils is estimated to vary from 3.5 to 5 g ha^{-1} yr^{-1} (Hutton and

Symon, 1986; Kloke *et al.*, 1984; Rothbaum *et al.*, 1986), however, the local inputs may be as high as 80 g ha^{-1} yr^{-1} (Jackson and Alloway, 1992). In a long-term field experiment at Rothamstead Experimental Station, UK, large annual applications of farmyard manure were found to be a more significant source of Cd than the combined inputs from phosphatic fertilizers (Jones *et al.*, 1987). By contrast, Tiller *et al.* (2000) reported that in Australia, the main source of Cd contamination in soils was widespread fertilization by phosphate fertilizers.

In the Canadian Prairies, the total soil Cd, Cd availability index (CAI), and total soil organic C of cultivated soils were found to be significantly lower than those of adjacent virgin soils that had the same parent material, whereas an opposite trend was observed for the soil pH and the aromaticity of the organic C (McArthur *et al.*, 2001). The reduced CAI in the cultivated soils was related to the increase in both the soil pH and the aromaticity of the organic C. A significant positive correlation was found between the fraction of total Cd removed from the soil after long-term cultivation (31–94 y) and the corresponding fraction of organic C removed. Further, the low frequency of application of phosphate fertilizers and low Cd concentration in the phosphate fertilizers applied in the Canadian Prairies in the past also partially accounted for the decrease of total soil Cd and CAI after long-term cultivation.

McArthur (2001) developed a model to predict the 100-year change in the profile distribution of Cd in the Orthic Chernozemic soils studied on the basis of a durum wheat system under annual cropping, and applications of low, medium, and high Cd-bearing phosphate fertilizers. The model predicts that with the application of any of the three phosphate fertilizers, there will be a statistically significant increase in both the concentration and mass of A horizon Cd (Table 3). The level of significance of the increase in Cd concentration due to applications of Florida fertilizer, which contained 10 mg Cd kg^{-1}, is relatively low at $\alpha = 0.10$. But the predicted 100-year increases in A-horizon soil Cd concentration and mass as a result of applying phosphate fertilizers made from Togo (60 mg Cd kg^{-1}) and Idaho (144 mg Cd kg^{-1}) phosphate ores are very large and highly significant. If phosphate fertilizer made from Idaho ore is applied, both the concentration and the total mass of A-horizon Cd are predicted to be about 3.5 times higher in 2097. The 100-year Cd concentration resulting from the use of phosphate fertilizer made from Idaho phosphate ore could reach 1.18 mg kg^{-1} soil (Table 3), which is in the commonly accepted critical region of 1.0–3.0 mg Cd kg^{-1} soil (Tiller *et al.*, 2000). Typically, when the concentration of soil Cd enters this region, the crops grown in the soil have unacceptably high levels of Cd (Tiller *et al.*, 2000).

3.3. Industrial pollution and atmospheric deposition

Cadmium enrichment of soil can also be associated with industrial pollution (Little and Martin, 1972). Topsoils contaminated by mine spoil showed a

Table 3
Predicted 100 year change in the profile distribution of Cd of the Orthic Chernozemic soils studied based on a durum wheat system under annual cropping and applications of low, medium, and high Cd-bearing phosphate fertilizers (McArthur, 2001)

Fertilizer ore source	Year	Cd concentration (mg kg^{-1})		Cd mass (mg m^{-2})	
		A horizon	B horizon	A horizon	B horizon
	1997	0.332	0.202	43	59
Florida[a]	2097	0.39[b]	0.21[c]	50[d]	54[c]
Togo[e]	2097	0.68[f]	0.21[c]	88[f]	54[c]
Idaho[g]	2097	1.18[h]	0.21[c]	152[h]	54[c]

[a] Fertilizer Cd concentration of 10 mg kg^{-1} (Alloway and Steinnes, 1999). [b] Significantly different from the 1997 value at $\alpha = 0.10$. [c] No significant difference from the 1997 value at $\alpha = 0.10$. [d] Significantly different from the 1997 value at $\alpha = 0.05$. [e] Fertilizer Cd concentraiton of 60 mg kg^{-1} (Potash & Phosphate Institute/ Potash & Phosphate Institute of Canada, Foundation for Agronomic Research, 1998). [f] Significantly different from the Idaho, Florida, and 1997 values at $\alpha = 0.001$. [g] Fertilizer Cd concentration of 144 mg kg^{-1} (Krish-namurti *et al.*, 1996). [h] Significantly different from the Togo, Florida, and 1997 values at $\alpha = 0.001$.

wide range of Cd concentrations ranging from 540 to 1700 mg kg^{-1} (Bauchauer, 1973; Davies and Roberts, 1978).

Industrial effluents are commonly discharged to surface water drainage systems after clarification in tailing ponds. Recent investigations have revealed very high concentrations of Cd, up to 150 (Macklin and Klimek, 1992) and 140 mg kg^{-1} (Helios Rybicka, 1992, 1993) in the overbank and bottom sediments of Upper Vistula and Przemsza rivers in Poland. The maximum concentrations of Cd in these sediments are particularly high and approach one of the highest recorded in Europe. The range of Cd concentration in the metal-processing areas in Poland is between 6 and 270 mg kg^{-1} (Kabata-Pendias and Pendias, 1992; Verner *et al.*, 1994). Progressive pollution of both the rivers and their tributaries is a real danger and may lead to an environmental catastrophe (Helios Rybicka, 1996).

Cadmium deposited from the atmosphere onto soil can come from a variety of natural and anthropogenic sources (Alloway and Steinnes, 1999; McArthur, 2001). It is estimated that volcanoes, the dominant natural source of atmospheric Cd, emit 520 mg of Cd into the atmosphere annually. The dominant anthropogenic source of emission of Cd into the atmosphere is primary non-ferrous metal production, which accounts for an estimated 4721 mg of Cd emitted into the atmosphere annually. The amount of Cd emitted into the atmosphere from all anthropogenic sources is estimated to be about one order of magnitude greater than that emitted from all natural sources.

4. CHEMICAL SPECIATION OF SOIL CADMIUM

The ecological consequences of heavy metal pollution of soils are largely related to heavy metal solubility and mobility within the soil profile. These interrelated factors determine the leaching and availability of heavy metals to microbes, plants, and ultimately, to humans. However, leaching of heavy metals through the soil profile onto the groundwater, even in the highly polluted soils treated with sewage sludge, does not occur to any appreciable extent (Emmerich *et al.*, 1982a, b). Movement of heavy metals within the soil profile will be mainly in solution phase. Soil chemical reactions controlling mobility of heavy metals and their subsequent uptake by the plants can be broadly classified as adsorption/desorption or dissolution/precipitation, which are, in turn, controlled by the concentration and ionic species of Cd in solution, and nature of solid surface phases present. Adsorption/ desorption has provided a useful means of describing heavy metal reactions in soils, but the boundary between adsorption and precipitation is sensitive to small changes in soil components and soil conditions (Brummer *et al.*, 1983). The important role played by soluble metal–organic complexes in ionic speciation and their application to soil chemistry (Hodgson *et al.*, 1965) have improved our understanding of heavy metal reactions in soils through the use of advanced computer models, such as GEOCHEM (Parker *et al.*, 1987; Parker, 1991), SOILCHEM (Sposito and Coves, 1988) and MINTEQL (Allison *et al.*, 1991), for solution speciation.

Soil biological processes that can be considered sensitive to heavy metals are mineralization of N and P, cellulose degradation, and possibly, N_2 fixation, although there is little evidence to date to suggest that soil biological processes in most heavily polluted soils are affected by heavy metals (Domsch, 1984). However, adverse impacts of Cd can occur, depending on the loading and mobility of Cd in soil (McGrath, 1999). The latter depends on the soil's physicochemical properties.

While classifying soil types, determination of the total content of selected elements often suffices. However, the identification of particular species present tends to be far more informative than the mere calculation of total elemental percentages. For agricultural and environmental purposes, the amount of nutrient "available" to plants or the amount of pollutant easily "mobile" becomes more important since this fraction, even though it represents only a small fraction of the total content, greatly influences plant growth and the quality of groundwaters. The generalized schematic cycle of Cd in agrosystems can be represented as shown in Fig. 1.

Many studies (e.g. Sauerbeck and Rietz, 1983) have indicated the existence of different binding forms and a strongly pH-dependent solubility effect of the trace elements. Plant uptake of a trace element occasionally correlates well with the amount of metal extracted by a specific chemical reagent. Unfortunately, the nature of the most appropriate reagent seems to vary with the type of plant being chosen, the element being considered, and the soil type. The diverse response

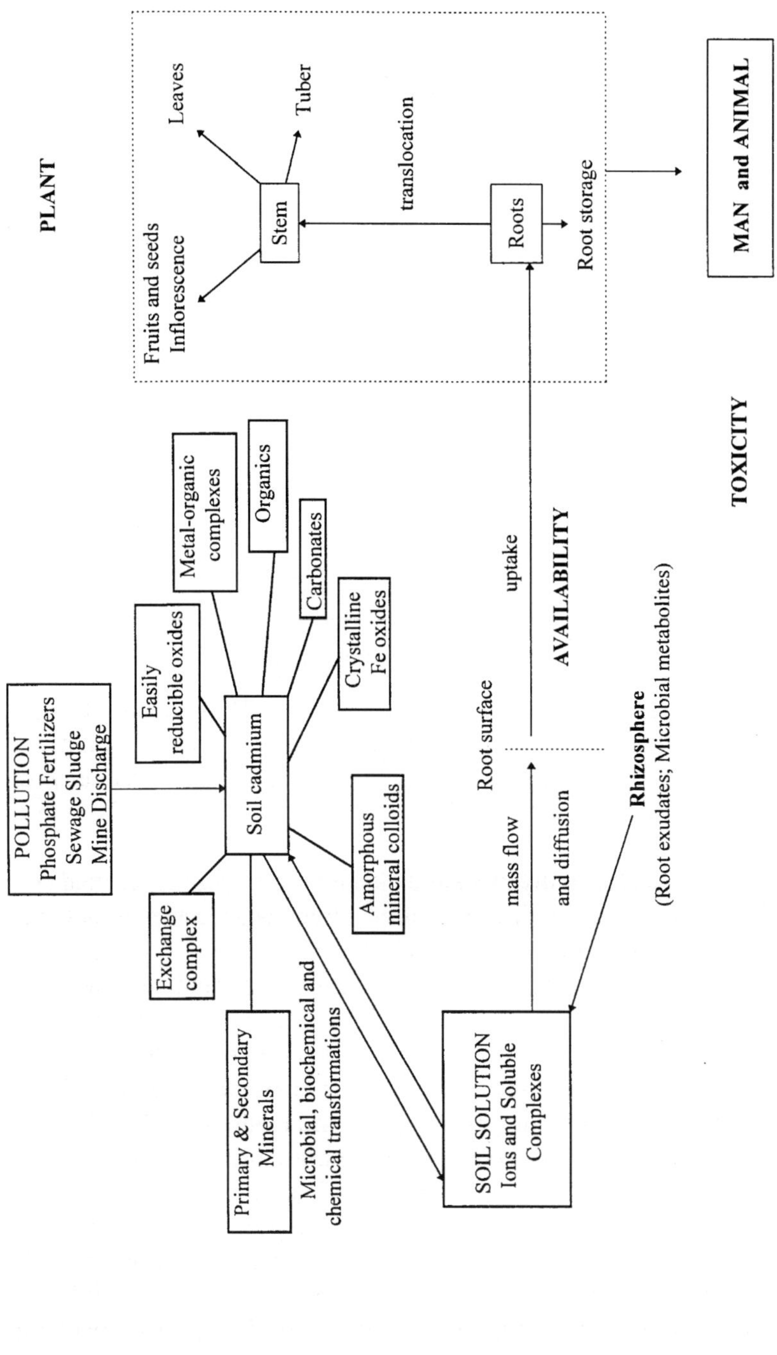

Fig. 1. Cadmium in the soil–plant–human/animal system.

arises, in part, from the fact that the trace element of soil is bound, in different ways, to individual components, so that chemicals differ in their ability to effect the release. In addition, the fraction present in the associated soil solution can also possess chemical forms that are not readily transformed into the root systems and translocated into other parts of the plant (Pickering, 1986). It can therefore be argued that elemental analysis should be supplemented by a "speciation scheme" that facilitates (i) assessment of the chemical form in the aqueous phase (solution speciation) and (ii) identification of bonding modes or component associations (soil particulate-bound speciation).

It is possible to summarize the principal forms of mobile and mobilizable toxic elements in soil (Berrow and Burridge, 1980) as follows:

(i) Soil solution: ionic, molecular, and chelated forms.
(ii) Exchange interface: readily exchangeable ions in inorganic and organic fractions.
(iii) Adsorption complex: more firmly bound ions.
(iv) Incorporated in precipitated Fe, Mn, and Al oxides. and
(v) Crystal lattices of secondary minerals (illustrated in Fig. 1).

In order to quantify these fractions and to assess the mobility and bioavailability of different forms, scientists have attempted to extract the different fractions using a range of chemical extractants. Soil Cd speciation is fundamental to understanding the mobility and bioavailability of different chemical forms of soil Cd.

4.1. Solution speciation

Cadmium speciation in the soil solution may well play a role in its bioavailability. An element's bioavailability is reported to be a function of at least three parameters (Brummer, 1986):

(i) the total amount of potentially available elements (the quantity factor),
(ii) the concentration or activity and ionic ratios of elements in the soil solution (the intensity factor), and
(iii) the rate elements transfer from solid to liquid phases and to plant roots (reaction kinetics).

The initial chemical form of the metal, and the environmental and edaphic conditions, such as pH, redox status, and soil organic matter content, influence the fate of toxic metals in soils. Metal speciation in soils can be broadly discussed via the following approaches:

(1) Computer-based modeling – based on geochemical principles, using either the equilibrium constants or using Gibbs free energy values. Both the approaches are subject to the conditions of equilibrium and mass balance.

(2) Soil environmental constraints – based on the understanding of the soil processes and conditions that control the formation and transformation of metal species.
(3) Chemical analysis using extractants – based on the reactivity of the extractants with the metal under study.

These different approaches, individually, provide an idea of the processes involved in the retention and mobility of toxic metals. However, there are no published examples of soil studies that have utilized a combination of these approaches to understand the fate of metals in soil ecosystems. Numerous studies have used chemical extractants to quantify different metal fractions, but only a limited few have attempted to characterize metal phytoavailability by correlating soil-extractable metal fractions with plant metal uptake (e.g. Lake *et al.*, 1984; Krishnamurti *et al.*, 1995a).

4.1.1. Geochemical theory

Using basic chemical theory, Sposito (1986) and Morgan (1987) have outlined the principles for elucidating metal speciation in the natural environment. Sposito and Mattigod (1980) and Sposito and Coves (1988) have developed computer-based geochemical modeling approaches specifically for soils.

Two different thermodynamic approaches can be used to calculate metal ion speciation, either by using equilibrium constants or by using Gibbs free energy values, which are subject to conditions of equilibrium and mass balance. Even though the same approach of thermodynamic calculations can be applied to both inorganic and organic metal complexes, the structure and binding mechanism of the soil metal–organic complexes is not clearly understood at the present time, and hence, we do not have reliable thermodynamic data for realistic organic species.

4.1.1.1. Equilibrium constant approach. The equilibrium equation of a two-component system can be expressed as

$$\mathrm{M}_a \cdot \mathrm{L}_{b(\mathrm{s})} = \mathrm{aM}^{m+}_{(\mathrm{aq})} + \mathrm{bL}^{n-}_{(\mathrm{aq})} \tag{1}$$

where M is the trace metal, L the ligand, m and n the valencies of the metal and ligand, respectively, and a and b the stoichiometric constants.

The equilibrium constant for this reaction (K) can be written as

$$K = [(\mathrm{M}^{m+})^a\,(\mathrm{L}^{n-})^b]/[(\mathrm{M}_a \cdot \mathrm{L}_{b(\mathrm{s})}] \tag{2}$$

or in its log form as

$$\log(\mathrm{M}^{m+}) = (1/a)\,[\log(\mathrm{M}_a \cdot \mathrm{L}_{b(\mathrm{s})}) - b\log(\mathrm{L}^{n-}) + \log K] \tag{3}$$

Among the two factors that regulate the activity of the metal cation in the soil solution, viz., the activities of $M_a \cdot L_b$ and L^{n-}, the contribution of L^{n-} is more significant, as the ligand L^{n-} can participate in a variety of sorption and complexation reactions in the soil system. The dominant species of the metal that can form in a given soil environmental condition can be assessed using activity graphs. These graphs plot the log $[(M_a{\cdot}L_{b(s)})/(M^{m+})]$ against any soil characteristic, such as pH or pCO_2. The methods to obtain such information have been discussed in detail by Sposito (1983).

4.1.1.2. Gibbs free energy approach. The change in the Gibbs free energy for selected processes was also used to assess the stability of compounds and the dissolution sequence. For an equation such as Eq. (1) above, the free energy of the reaction is calculated as

$$\Delta G^{o}_{r} = \Delta G^{o}_{f}\{\text{products}\} - \Delta G^{o}_{f}[\text{reactants}] \tag{4}$$

If the result is found to be negative, then the reactants are unstable and the reaction will proceed spontaneously. On the other hand, if the result is positive, the reaction is less likely to occur without other additions to the system. The relation between the reaction equilibrium constant (K^{o}) and the change in free energy of reaction (ΔG^{o}_{r}) is given by

$$\Delta G^{o}_{r} = -\text{RT} \ln K^{o} \tag{5}$$

At 25° C, Eq. (5) reduces to

$$\log K^{o} = -\,0.733\,\Delta G^{o}_{r} \tag{6}$$

This equation is useful for calculating the equilibrium constants for chemical reactions for which thermodynamic data are difficult to measure by conventional methods. Lindsay (1979) used this method to calculate the log K^{o} values of an extremely large number of trace metal reactions in soils.

Computer-based chemical equilibrium models of natural systems, such as soils, have undergone a great deal of development in recent years and have become useful tools for studying water quality criteria (Jenne, 1979). The widely used programs, viz., GEOCHEM (Sposito and Mattigod, 1980) used exclusively for soil systems and MINEQL (Westall *et al.*, 1976; James and Parks, 1978) applied for the study of water bodies, are progeny of the program REDEQL2 (Morel and Morgan, 1972; McDuff and Morel, 1973). Details of REDEQL2 program were discussed by Legatt (1977) and by Ingle *et al.* (1978). Some typical applications of GEOCHEM include (Sposito, 1983)

(i) Prediction of the concentration of inorganic and organic complexes of a metal cation in a soil solution;

(ii) Calculation of the concentration of a particular chemical form of a nutrient element in a solution bathing plant roots so as to correlate that form to nutrient uptake;
(iii) Prediction of the chemical fate of pollutant metal added to a soil solution of known characteristics; and
(iv) Estimation of the effect of changing pH, ionic strength, redox potential, water content or the concentration of some constituent on the solubility of a chemical element of interest in soil solution.

GEOCHEM has more than twice as much thermodynamic data as REDEQL2. It differs from REDEQL2 principally in using thermodynamic data that have been selected especially for soil systems, in containing a method for describing the cation exchange, and in employing different subroutine for correcting thermodynamic equilibrium constants for the effect of non-zero ionic strength. MINEQL differs from REDEQL2 in the subroutines that describe solid phases and the adsorption phenomena.

In recent years, several computer-based models have become widely available and are used successfully, e.g. SOILCHEM (the updated version of GEOCHEM; Sposito and Coves, 1988), HYDRAQL (Papelis *et al.*, 1988), ECOSAT (Keizer, 1991), and MINTEQA2 (Allison *et al.*, 1991).

4.1.2. Soil environmental constraints

An increasing number of studies have focused on modeling the impact of anthropogenic metal inputs such as sewage sludge (Mattigod and Sposito, 1979), geothermal brines (Sposito *et al.*, 1979), acid rain (Sposito *et al.*, 1980), and coal gasification (Ireland *et al.*, 1982) into soil systems. Elemental speciation was predicted by GEOCHEM. These studies provided valuable insights into the chemistry of speciation of metals in soil systems, and a basis for predicting the behavior of the elements in contaminated soils.

It is also generally recognized that metal ion availability and uptake by plants is controlled by several factors, such as pH, ionic strength, redox potential, composition of solution, ionic size, and valence (Frausto da Silva and Williams, 1976). Therefore, elemental speciation is a major factor controlling the availability and uptake of various non-essential elements by plants (Sposito and Bingham, 1981). Speciation of 10 metals and 13 ligands in saturation extracts of soils was computed by GEOCHEM, and the uptake of Cd by sweet corn grown on soils subjected to known additions of Cd was studied (Sposito and Bingham, 1981). The results (Fig. 2) showed that Cd uptake by sweet corn was highly correlated with the concentration of $CdCl^+$ in soil solution and not with Cd^{2+} concentration in soil solution. The authors concluded that reduced charge on Cd through complex formation appeared to enhance Cd uptake.

The speciation of Cd in soil solutions of soils from Northern France that had been contaminated by effluent from Pb–Zn metallurgical plant was calculated

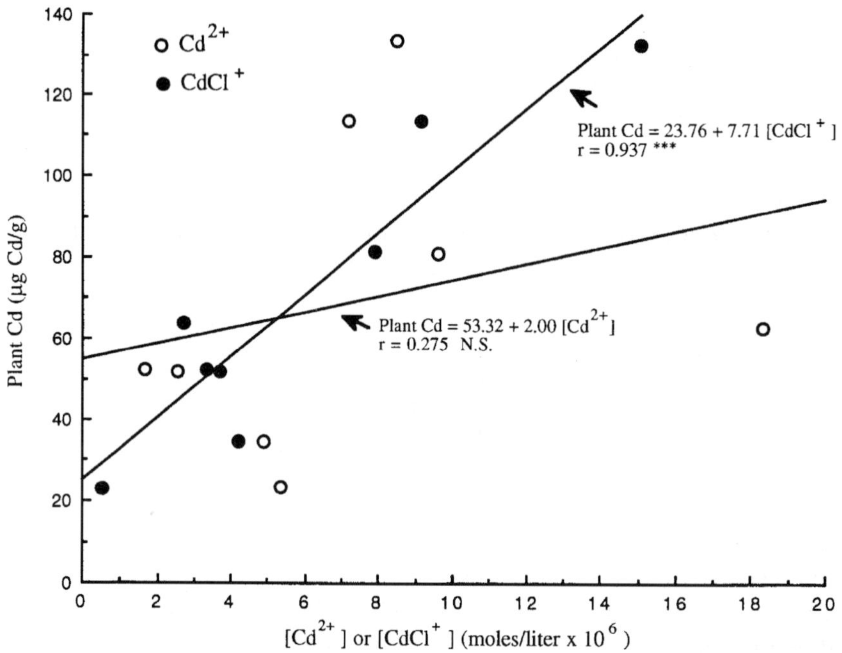

Fig. 2. Cadmium uptake by corn shoots against molar concentration of $CdCl^+$ and $Cd2^+$ in saturation extracts of the soils (redrawn from Sposito and Bingham, 1981).

using the program SOILCHEM (Charlatchka *et al.*, 1997). The database, Geodata, was complemented with the stability constants of Cd-butyrate, Cd-propionate (Sillen and Martell, 1971), and Cd-fulvates (Lamy *et al.*, 1994). The free species, Cd^{2+}, accounted for around 60%, and the inorganic species of $CdCl^+$ and $CdHCO_3^+$ accounted for around 40%. The organic species increased from $< 1\%$ to 20% with increased flooding and a concomitant decrease in the concentration of Cd^{2+} species (Charlatchka *et al.*, 1997). The input data consisted of the measured total concentration of Cd, pH, and the thermodynamic constants from the database. The ionic strength was calculated by the program as well as activity coefficients (Sposito, 1981; Sposito and Cove, 1988).

Few studies report actual measurement of Cd^{2+} speciation in soil solutions. Many authors report Cd^{2+} speciation based on the measured total dissolved Cd concentration and computation of free Cd^{2+} using chemical equilibrium models. The accuracy of the speciation depends on the correctness of Cd–complex stability constants (Turner, 1995). For example, the proportion of free Cd (as Cd^{2+}) in solution varies from 16 to 82% depending on the log K values for Cd–(DOM) dissolved organic matter complexes (Table 4). The stability constants of Cd-humics are particularly uncertain. The reported high values of pCd^{2+} ranged

Table 4
Speciation of Cd in soil solution using different log *K* values for Cd–DOM complexes (from Krishnamurti and Naidu, 2003)

$\log K_{Cd\text{-}DOM}$	Cd species in soil solution ($\mu g\ L^{-1}$)		
	Cd^{2+}	Cd–DOM	Cd–inorganic[a]
4.10 (Schnitzer and Hansen, 1970)	418.80 (81.8)[b]	43.50 (8.5)	49.65 (9.7)
5.30 (Almas *et al.*, 2000)	185.85 (36.3)	304.10 (59.4)	22.05 (4.3)
5.80 (Alberts and Geisy, 1983)	81.90 (16.0)	420.50 (82.1)	9.60 (1.9)

Speciation arrived at using MINTEQA2 computer model.
[a] Cd–inorganic ligand complexes such as $CdCl^+$, $CdSO_4$, and $CdNO_3^+$. [b] values in parentheses are the percentage distribution of the species.

between 8 and 5, and the proportion of Cd–organic complexes was usually reported as negligible (Hirsch and Banin, 1990; Jopony and Young, 1994; Temminghoff *et al.*, 1995; Candelaria and Chang, 1997; Elzinga *et al.*, 1999). Nevertheless, the association of Cd with organics in soil solution is not insignificant (del Castilho *et al.*, 1993; Naidu and Harter, 1998). Several investigations have shown that soluble organics increase the Cd concentration in soil solutions (Dunnivant *et al.*, 1992; McBride *et al.*, 1997; Krishnamurti *et al.*, 1997c; Naidu and Harter, 1998).

The geochemical equilibrium speciation model MINTEQA2 (Allison *et al.*, 1991) is now widely used to derive metal speciation in soil solutions. The database includes over 900 dissolved species, including 13 species of trace metal complexes with dissolved organic matter (Allison and Brown, 1995). In this model, dissolved organic matter is treated as a complex material consisting of various types of monoprotic acid sites, and was assumed to be normally distributed with respect to the log *K* values for protons and metals (Dobbs *et al.*, 1989). The stability constants of metal–organic complexes obtained using Scatchard plot approach has been the method of choice in most of the recent studies on metal binding by organic acids (Fitch and Stevenson, 1983, 1984). Krishnamurti and Naidu (2003) used the stability constants of the metal-DOM complexes, as reported by Stevenson and Fitch (1986), in the database of the MINTEQA2 model, for arriving at the speciation of Cd in soil solutions of a few typical soils from South Australia, which were incubated with 10 mg Cd Kg^{-1} soil. The Cd–DOM complexes were found to be the most dominant species, with small amounts (average $< 5\%$) of free Cd^{2+} species (Table 5). Recent reports (Almas *et al.*, 2000; Sauve *et al.*, 2000; Krishnamurti *et al.*, 2004) have also shown the Cd–organic complexes as the dominant Cd species in most of the soil solutions tested, using differential pulse anodic stripping voltammetry for measuring Cd^{2+} activity.

Table 5
Speciation of Cd in soil solution of incubated soils of south Australia (from Krishnamurti and Naidu, 2003)

Soil	Cd in soil solution (μg L^{-1})	Cd species in soil solution (μg L^{+1})			
		Cd^{2+}	Cd–DOM	$CdCl^{+}$	$CdSO_4$
Kapinnie	512.0	81.90 (16.0)[a]	420.50 (82.1)	5.29 (1.0)	3.21 (0.6)
Cookes plain	115.0	5.30 (4.6)	109.90 (95.0)	0.30 (0.3)	0.10 (0.1)
Bute	106.0	3.04 (2.8)	102.81 (97.0)	0.06 (0.1)	0.07 (0.1)
Nangari	115.0	1.31 (1.1)	113.52 (98.7)	0.06 (0.1)	0.11 (0.1)
Freeling	400.0	0.54 (0.1)	399.02 (99.7)	0.11 (0.1)	0.34 (0.1)
Border town	321.5	0.63 (0.2)	320.86 (99.8)	< 0.01	< 0.01
Pinnaroo	69.95	3.77 (5.4)	65.92 (94.2)	0.19 (0.3)	0.07 (0.1)

Speciation arrived at using MINTEQA2 computer model; species $\geq 0.1\%$ of the total Cd are shown in the table.
[a] values in parentheses are the percentage distribution of the species.

In recent years, the understanding of colloid surfaces and soil constituents has increased tremendously. Surface coordinating functional groups on particulate inorganic and humic materials are viewed as complexant ligands (Stumm, 1992). The surface complex models (SCM) are now finding increased application in the fields of pollutant-retention behavior (Zachara *et al.*, 1992), the chemistry of plant nutrient retention (Goldberg and Traina, 1987), and transport of pollutants by colloids (Goldberg, 1992). The elegance of the surface complexation approach lies in the fact that it can be incorporated into the thermodynamic speciation models used for soluble complexes. The SCMs that are commonly in use are the diffuse double-layer model (DDLM) (Huang and Stumm, 1973; Dzombek and Morel, 1990), the constant capacitance model (CCM) (Stumm *et al.*, 1970, 1976, 1980; Schindler *et al.*, 1976), the triple-layer model (TLM) (Davis *et al.*, 1978; Davis and Leckie, 1978, 1980; Hayes and Leckie, 1987; Hayes *et al.*, 1988), and the 1 pK basic Stern model (Bolt and van Riemsdijk, 1982; Van Riemsdijk *et al.*, 1986, 1987). The application of many of the commonly used computer models in the determination of speciation in solution phase has been discussed exhaustively by Lumsdon and Evans (1995).

4.2. Solid-phase cadmium speciation

The distribution of trace elements among soil components such as organic matter or hydrous metal oxides is important for assessing the potential of soil to supply sufficient micronutrients for plant growth or to contain toxic quantities of trace metals, and for determination of amelioration procedures for soils at risk of causing trace metal contamination. Metal cations may be soluble, readily exchangeable, complexed with organic matter or hydrous oxides, substituted in stoichiometric compounds, or occluded in mineral structures. Delineating the speciation of metals in soils is considered essential for understanding the mobility, bioavailability, and toxicity of the metals, and for developing useful environmental guidelines for potential toxic hazards (Davies, 1980, 1992). Trace elements in soils are normally found in many different physicochemical forms (Lake *et al.*, 1984; Tessier and Campbell, 1988) and may be associated with various soil components, such as clay minerals, metal hydrous oxides, carbonates, and soil organic matter. The nature of this association is referred to here as "speciation". Speciation may also refer to the type of bonding between an element and other solid components. For example, an element in ionic form may bind to colloidal-size fractions of soils, such as clay minerals or organic matter by coulombic forces, whereas covalent bonds may be formed with surface ligands on hydrous oxide surfaces (Fendorff *et al.*, 1994; Manceau *et al.*, 1995). Ligands can form either inner- or outer-sphere complexes with cations on an adsorbent (Ritchie and Sposito, 1995). Factors affecting distribution of an element among different forms include pH, ionic strength of the solution, the solid and solution components and their relative concentration and affinities for an element, and time (Jones and Jarvis, 1981; Tiller, 1983; McBride, 1989, 1991; Alloway, 1990; Foerstner, 1991; Ritchie and Sposito, 1995).

4.2.1. Single reagent extraction

Single chemical extractants are generally used to determine "available" amounts of soil metals and usually aim to extract the water-soluble, easily exchangeable, metals bound to organic matter or hydrous oxides, and occluded in mineral structures.

4.2.1.1. Water-soluble and exchangeable. The use of electrolytes, such as $CaCl_2$, $MgCl_2$, NH_4Cl, KNO_3, $Mg(NO_3)_2$, and $Ca(NO_3)_2$ as extractants promotes displacement of metal ions held by electrostatic attraction to negative sites on particle surfaces (exchangeable). Usually, 1 M solutions are employed as extractants (M $MgCl_2$ – Tessier *et al.*, 1979; Hoffman and Fletcher, 1979; Shuman, 1979; Maher, 1984; Hickey and Kitrick, 1984; Nielsen *et al.*, 1986; Elliot *et al.*, 1990; M KNO_3 – Stover *et al.*, 1976; Silviera and Sommers, 1977; Schalscha *et al.*, 1980, 1982; Miller and McFee, 1983; M $Mg(NO_3)_2$ – Shuman, 1983; Krishnamurti *et al.*, 1995a). More dilute solutions have also been proposed (0.5 M KNO_3 – Emmerich *et al.*, 1982a, b; Sposito *et al.*, 1982; 0.05 M $CaCl_2$ – MacLaren and Crawford, 1973; Shuman, 1979; Iyengar *et al.*, 1981; 0.5 M $CaCl_2$ – MacLaren *et al.*, 1986; 0.5 M $Ca(NO_3)_2$ – Tiller *et al.*, 1972; Miller *et al.*, 1986a; 0.5 M $MgCl_2$ – Gibbs, 1973, 1977; 0.25 M $Ca(NO_3)_2$ – Miller *et al.*, 1986b), since they more closely resemble the electrolyte concentrations that can occur in natural systems. Use of nitrate salts was preferred to chloride salts on the grounds that the chloride ion has a specific complexing effect (Tiller *et al.*, 1972; Krishnamurti *et al.*, 1995a).

The use of M NH_4OAc was also advocated as an effective reagent for the extraction of exchangeable phase (Jackson, 1958; Gupta and Chen, 1975; Salomons and Foerstner, 1980, 1984; Schoer and Eggersgluess, 1982; Kersten and Foerstner, 1986; Rule and Alden, 1992). However, acetate ion has a complexing effect, particularly with heavy metal ions.

4.2.1.2. Specifically sorbed carbonate-bound. Significant concentrations of trace elements can be associated with sediment carbonates. A mixture of M NaOAc with HOAc at pH 5 has been shown to extract $> 99\%$ of total carbonate present in the soils (Jackson, 1958). This method was used as a specific extractant for the determination of carbonate-bound heavy metals in soils (Tessier *et al.*, 1979; Foerstner *et al.*, 1981; Harrison *et al.*, 1981; Robbins *et al.*, 1984; Hickey and Kittrick, 1984; Krishnamurti *et al.*, 1995a; Krishnamurti and Naidu, 2000). Other reagents that were also used include 2.5% HOAc (MacLaren and Crawford, 1973; Gupta and Chen, 1975; Garcia-Mirgaya *et al.*, 1981) and 0.05 M Na_2-EDTA (Sposito *et al.*, 1982).

4.2.1.3. Bound to hydrous oxides of Fe and Mn. Fe and Mn oxides, present in soils as nodules, concretions, matrix components, and cement between particles or as coatings on particles, are excellent scavengers of trace metals (Jenne, 1968). A mixture of sodium dithionite, sodium citrate, and sodium bicarbonate buffered at pH 7.3 was suggested as a suitable reagent for determining the total free iron contents (Mehra and Jackson, 1960), and has been used widely in soil/sediment studies (Tessier *et al.*, 1979). This reagent dissolves both the crystalline and amorphous oxyhydroxides. However, the dithionite salt contaminated with metal impurities, such as Zn and metal ions, may be lost from solution through the formation of metal sulfides (Tessier *et al.*, 1979; Shuman, 1982).

Hydroxylamine hydrochloride, dissolved in acetic or nitric acid, was shown to selectively extract Mn oxyhydroxides and Mn oxides, as well as amorphous Fe oxides (Chester and Hughes, 1967; Chao, 1972; Chao and Zhou, 1983). This reagent was used for the extraction of trace metals associated with hydrous oxides of Fe and Mn in soils (Gupta and Chen, 1975; Stover *et al.*, 1976; Tessier *et al.*, 1979; Harrison *et al.*, 1981; Miller and McFee, 1983; Miller *et al.*, 1986a, b; Rule and Alden, 1992; Krishnamurti *et al.*, 1995a; Krishnamurti and Naidu, 2000).

For total recovery of amorphous Fe oxides, treatment of samples with acidified ammonium oxalate (pH 3) in the absence of the catalyzing effect of light has been proposed (LeRiche and Wier, 1963; Schwertmann, 1964; McKeague and Day, 1966). This reagent has been extensively used for selectively extracting trace elements associated with amorphous Fe oxides (Shuman, 1979; Salomons and Foerstner, 1980, 1984; Kersten and Foerstner, 1986; Krishnamurti *et al.*, 1995a; Krishnamurti and Naidu, 2000).

4.2.1.4. Organically bound metals. Hydrogen peroxide is the commonly used extractant (Gupta and Chen, 1975; Tessier *et al.*, 1979; Hickey and Kitrick, 1984; Gibson and Farmer, 1986; Rule and Alden, 1992; Krishnamurti *et al.*, 1995a; Krishnamurti and Naidu, 2000), even though it dissolves Mn oxides (Shuman, 1979, 1983; Keller and Vedy, 1994). However, Orsini and Bermond (1994) found that the kinetics of the destruction of organic matter was slower and took almost 24 h. Other reagents used include NaOCl buffered at pH 9.5 (Gibbs, 1973; Hoffmann and Fletcher, 1981; Shuman, 1983) and 0.1 M sodium or potassium pyrophosphate (Schalscha *et al.*, 1982; Miller and McFee, 1983; MacLaren *et al.*, 1986). However, 0.1 M sodium or potassium pyrophosphate was shown to extract the metal–organic complexes selectively (McKeague, 1967; Bascomb, 1968), and may not completely extract the organic-bound Cd. Even the C removal with NaOCl is said to be higher than that achieved by H_2O_2 (Lavkulich and Wiens, 1970); the associated precipitation of released metal ions in the alkaline medium and the possible alteration of mineral constituents are major disadvantages.

The use of various extractants in the selective extraction of trace metals associated with different fractions of sediments, soils, sewage sludges, and sludge-amended soils has been comprehensively discussed (Pickering, 1986; Beckett, 1989).

4.2.2. Sequential extraction

Sequential chemical extraction involves treatment of a sample of soil or sediment with a series of reagents in order to partition the trace element content. The principal advantage claimed for sequential extraction over the use of single extractants is that the phase specificity is improved. This technique has been used to determine the chemical forms of trace elements in soils, sediments, and suspended solids in natural waters, and is based theoretically and experimentally on more than 100 years of research (Jackson, 1985). A basic requirement of any extraction procedure should be the ability of the extractant to dissolve a specific component of a soil or sediment (Chao, 1984). Many different methods have been employed to fractionate trace elements, and these have proved useful for metal speciation (Jones and Hao, 1993). Reviews of the fractionation methods used to determine the chemical forms of trace elements in soils and sediments (e.g. Pickering, 1981, 1986; Ross, 1994; Sheppard and Stevenson, 1997), in geochemical exploration (Chao, 1984), and in natural waters (Florence and Batley, 1977) are available. A few commonly used fractionation schemes for delineating different forms of trace elements are given in Table 6.

Gupta and Chen (1975) were the first to use a sequential extraction procedure for the speciation of Cd in sediments. The scheme delineated Cd as

Table 6
Extractants used in a few typical sequential extraction schemes for delineating different forms of cadmium in soils

Species	Tessier *et al.* (1979)	Salomons and Foerstner (1980)	Shuman (1985)	Krishnamurti *et al.* (1995a)
Exchangeable	[I] M $MgCl_2$, pH 7[a]	[I] M NH_4OAc, pH 7	[I] M $Mg(NO_3)_2$, pH 7	[I] M $Mg(NO_3)_2$, pH 7
Carbonates	[II] M NaOAc, pH 5	[II] M NaOAc, pH 5		[II] M NaOAc, pH 5
Easily reducible metal oxides		[III] 0.1 M $NH_2OH{\cdot}HCl$ in 0.01 M HNO_3	[III] 0.1 M $NH_2OH \cdot HCl$ (pH 2)	[IV] 0.1 M $NH_2OH \cdot HCl$ in 0.01 M HNO_3
Organic matter	[IV] 30% H_2O_2, pH 2, 85°C; extract with 3.2 M NH_4OAc in 20%HNO_3	[V] 30% H_2O_2, pH 2, 85°C; extract with M NH_4OAc in 6% HNO_3	[II] 0.7 M NaOCl, pH 8.5	[V] 30% H_2O_2, pH 2, 85°C extract with M $Mg(NO_3)_2$ in 20%HNO_3
Fe and Mn oxides	[III] 0.04M $NH_2OH \cdot HCl$ in 25% HOAc			
Amorphous oxides		[IV] 0.1 M ammonium oxalate/ oxalic acid (pH 3)	[IV] 0.2 M ammonium oxalate/ oxalic acid (pH 3, dark)	[VI] 0.2 M ammonium oxalate/ oxalic acid (pH 3, dark)
Crystalline Fe oxides			[V] 0.2 M ammoniam oxalate/ oxalic acid (pH 3) in 0.1 M ascorbic acid, 95°C	[VII] 0.2 M ammoniam oxalate/ oxalic acid (pH 3) in 0.1 M ascorbic acid 95°C
Metal–organic complexes				[III] 0.1 M Na pyrophosphate pH 10)
Residual	[V] HF-$HClO_4$ digestion	[VI] Hot HNO_3	[VI] HF-HCl–HNO_3 digestion	[VIII] HF-$HClO_4$ digestion

[a] The Roman numerals refer to the order of each stage in the different sequential extraction schemes.

exchangeable (M NH_4OAc), carbonate-bound (M HOAc), easily reducible metal oxide-bound (0.1 M $NH_2OH{\cdot}HCl$ in 0.01 M HNO_3), organic-bound (30% H_2O_2, NH_4OAc), iron oxide-bound (0.04 M $NH_2OH{\cdot}HCl$ in 25% HOAc) and residual (HF, $HClO_4$). However, the procedure used most extensively, with minor modifications, by various environmental researchers for the speciation of particulate-bound heavy metals in soils and sediments is that of Tessier *et al.* (1979). This procedure delineates the metal species sequentially as exchangeable, carbonate-bound, Fe and Mn oxide-bound, organically bound, and residual.

Shuman (1982, 1985), working on the speciation of Cu, Mn, Fe and Zn, modified the sequence of the extraction as exchangeable, oxidizable (organically bound), easily reducible (Mn oxides-bound), moderately reducible (amorphous Fe oxides-bound), strongly reducible (crystalline Fe oxides-bound), and residual. The low percentages of the oxidizable fractions obtained following the sequential extraction scheme of Shuman (1985) as compared to those obtained following the scheme of Tessier *et al.* (1979) were attributed to the lack of selectivity of the method used (Charlatchka *et al.*, 1997).

The differentiation of the metal–organic-complex-bound Cd species as distinct from the other organically bound species was the innovation in the selective sequential extraction scheme suggested by Krishnamurti *et al.* (1995a). This scheme proportionates the particulate-bound Cd species in soils as exchangeable, carbonate-bound, metal–organic-complex-bound, easily reducible metal oxide-bound, organic-bound, amorphous mineral colloid-bound, crystalline Fe oxide-bound, and residual. The metal–organic-complex-bound Cd species were selectively extracted using 0.1 M pyrophosphate as the extractant in the sequential extraction scheme:

Fraction 1. Exchangeable: M Mg $(NO_3)_2$ at pH 7 (soil/reagent = 1:10); shake for 4 h.

Fraction 2. Carbonate-bound: M NaOAc at pH 5 (soil/reagent = 1: 25); shake for 6 h.

Fraction 3. Metal–organic complex-bound: 0.1 M $Na_4P_2O_7$ (pH 10) (soil/reagent = 1:30); shake for 20 h.

Fraction 4. Easily reducible metal oxide-bound: 0.1 M $NH_2OH{\cdot}HCl$ in 0.01 M HNO_3 (soil/reagent = 1:20); shake for 30 min.

Fraction 5. Organically bound: 5 mL 30% H_2O_2 (pH 2) and 3 mL 0.02 M HNO_3, heat for 2 h at 85°C; add 3 mL 30% H_2O_2 (pH 2), heat for 2 h at 85°C; cool and add 10 mL 1 M Mg $(NO_3)_2$ in 20% HNO_3; shake for 30 min.

Fraction 6. Amorphous mineral colloid-bound: 0.2 M $(NH_4)_2C_2O_4$ (pH 3) (soil/reagent = 1:10); shake for 4 h (in the dark).

Fraction 7. Crystalline Fe oxide-bound: 0.2 M $(NH_4)_2C_2O_4$ (pH 3) in 0.1 M ascorbic acid (soil/reagent = 1:25); heat for 30 min at 95°C.

Fraction 8. Residual: HF/$HClO_4$ acid digestion.

Krishnamurti *et al.* (1995a) determined the distribution of particulate-bound Cd species in selected typical soils of southern Saskatchewan, Canada following the schemes of Tessier *et al.* (1979) and the scheme modified by them. They found that Cd in these soils was predominantly bound to the metal–organic complexes, accounting for, on average, 40% of the total Cd present in the soils; whereas, on an average, 36% of Cd in the soils was observed to be in the Fe, Mn oxide-bound form, following the fractionation scheme of Tessier *et al.* (1979). The average percent distribution of solid-bound Cd species in surface horizons of the temperate soils of Southern Saskatchewan, Canada studied is in the order metal–organic complex-bound > carbonate-bound > residual > organic-bound > crystalline Fe oxide-bound > easily reducible metal oxide-bound > amorphous mineral colloid-bound; that of the tropical soils of Kenya (Onyatta and Huang, 1999) is in the order metal-organic complex-bound > residual > crystalline Fe oxide-bound > organic-bound > amorphous mineral colloid-bound > easily reducible metal oxide-bound > carbonate-bound (not detectable) (Fig. 3). The metal–organic complex-bound Cd in the surface horizons of the temperate soils accounts for 31–55%, with an average of 40%, of the total Cd present in the soils. The metal–organic complex-bound Cd is also generally the highest among the particulate-bound Cd species of the tropical soils, accounting for 25–46%, with an average of 37%, of the total Cd in the soils. However, in certain tropical soils, residual or crystalline Fe oxide-bound Cd is

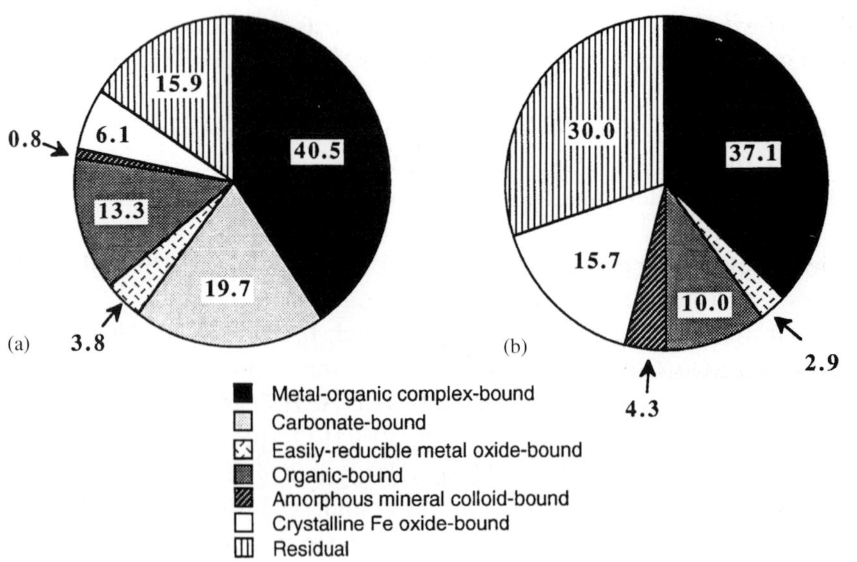

Fig. 3. Comparison of particulate-bound Cd species of temperate and tropical soils. (a) temperate soils (Krishnamurti *et al.*, 1995a) and (b) tropical soils (Onyatta and Huang, 1999).

predominant (Onyatta and Huang, 1999). The importance of metal–organic complex-bound Cd species to the bioavailability of Cd, and the nature of bonding of Cd sites in the temperate soils was also reported (Krishnamurti *et al.*, 1995a, 1997a, b).

Krishnamurti and Naidu (2000) have recently modified the sequential extraction scheme of Krishnamurti *et al.* (1995a), sub-fractionating the metal–organic complex-bound Cd into Cd-humic and Cd-fulvic species. The Cd in the surface soils of South Australia, developed under a Mediterranean climate, was observed to be predominantly present in residual fraction, accounting for on average 33.4% of the total Cd in the soils. Cd present as metal–organic complex-bound species was predominantly bound to the fulvic acid fraction of the organic matter, accounting for on an average 72.9% of the species (Krishnamurti and Naidu, 2000). The importance of Cd associated with the exchangeable and fulvic acid binding sites in assessing the plant-available Cd was also reported.

Compared with bulk soils, solid-phase speciation of Cd differs substantially in phosphate fertilizer-treated rhizosphere soils (Krishnamurti *et al.*, 1996). The amounts of carbonate-bound Cd and metal–organic complex-bound Cd species of the rhizosphere soils at 2-week plant growth stage, particularly in the soils treated with Idaho phosphate fertilizer, are appreciably higher than those of the corresponding bulk soils (Fig. 4). In comparison to the corresponding bulk soils, the amount of carbonate-bound Cd species of the rhizosphere soils increased by 15–18% in the control soils and by 79–92% in the soils treated with Idaho phosphate fertilizer, whereas the metal-organic complex-bound Cd species increased by 4–7% in the control soils and by 2–3 times in the soils treated with Idaho phosphate fertilizer. The increase in the carbonate-bound Cd

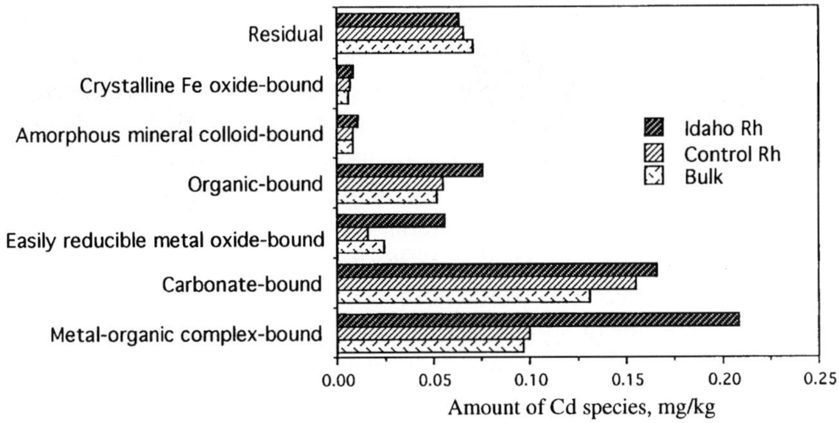

Fig. 4. Distribution of particulate-bound Cd species in the bulk and rhizosphere (Rh) soils collected at 2-week plant growth stage (redrawn from Krishnamurti *et al.*, 1996).

species in the rhizosphere soils is attributed to the increased amounts of carbonate, a product of plant respiration, present at the soil–root interface. Xian and Shokohifard (1989) suggested that exudation of H_2CO_3 by roots may help to solubilize metal carbonates and make them more bioavailable. The prolific root growth of the plants, particularly in the soils treated with Idaho phosphate fertilizer, might have resulted in the secretion of increased amounts of low-molecular-weight organic acids (LMWOAs) into the rhizosphere, which resulted in high amounts of metal–organic-complex-bound Cd species by chelation. Appreciable amounts of LMWOAs were detected in the root exudates of durum wheat, with the actual amount dependent on the cultivar (Cieslinski *et al.*, 1994; Szmigielska *et al.*, 1995). The sustaining Cd release from the soils by LMWOAs was shown to have the same trend as that of Cd accumulated in the plant (Krishnamurti *et al.*, 1997c).

4.2.3. Non-destructive analysis

The experimental detection and quantification of surface species on *in situ* soil particles and other natural colloids is a difficult area of research because of sample heterogeneity, low surface concentrations, and the necessity of investigating the solid adsorbents in the presence of water. Unambiguous information can be obtained only with *in situ* surface spectroscopy, such as X-ray photoelectron (XPS), extended X-ray absorption fine structure (EXAFS), X-ray absorption near edge structure (XANES), inelastic electron tunnelling (IETS), and electron energy loss (EEL) spectroscopies. Recent advances in the development of non-invasive, *in situ* spectroscopic scanned-probe and microscopic techniques have been applied successfully to study mineral particles in aqueous suspensions (Hawthorne, 1988; Hochella and White, 1990; Bertsch and Hunter, 1998).

Among the various spectroscopic methods, the most notable one is synchrotron-based X-ray techniques that are revolutionizing soil and environmental chemistry research (Manceau *et al.*, 1992). The most utilized synchrotron-based X-ray technique in soil and environmental chemistry to date has been X-ray absorption spectroscopy. The extended X-ray absorption fine structure spectroscopy (EXAFS) provides specific information on the local environment of the absorber, including coordination number, identity, and distances to nearest and sometimes next nearest neighboring atoms (Fendorff *et al.*, 1994; Schulze and Bertsch, 1995; Fendorff and Sparks, 1996; Bertsch and Hunter, 1998; Sparks, 2000). Application of EXAFS would advance our understanding of the nature of solid-bound Cd species of soils as the methods develop further and can operate at the trace levels of Cd present in soils. However, sampling techniques required for these methods often annihilate or change irreversibly the surface species of interest. Molecular-level information about the mechanisms, orientation or dynamics of surface species inferred from data obtained by these methods may bear little resemblance to the chemical mechanisms operating in a natural soil–water system (Ritchie and Sposito, 1995). Therefore, improvement of sampling techniques merits close attention in uncovering the *in situ* Cd species at the molecular level.

5. CADMIUM RHIZOSPHERE CHEMISTRY AND PLANT UPTAKE

It has been shown repeatedly that an increase in soil Cd content results in an increased plant uptake of the metal. This has been demonstrated for soils with naturally elevated Cd levels (Lund *et al.*, 1981), soils contaminated by non-ferrous metal mining (Alloway *et al.*, 1988), and soils that have received Cd via sewage sludge application (Davis and Coker, 1980). It is this basic relationship that makes the soil–crop pathway of human exposure susceptible to increased levels of soil Cd.

Although much effort has been spent testing different soil extraction techniques for characterizing metal phytoavailability, Sharma and Shupe (1977) found a surprisingly good relation (r = 0.883 for Cd) between total metal concentrations in the soil and total metal concentrations in plants. The statistical relationship between total soil and plant metal concentration may be fortuitous, since the total amount of Cd in soil is seldom indicative of its effect on Cd accumulation in plants (Cottenie *et al.*, 1983). More attention should be paid to Cd speciation in relation to Cd bioavailability, especially in the rhizosphere.

5.1. Rhizosphere chemistry of cadmium

The term "rhizosphere" was first used by Hiltner (1904) but has since been modified and redefined. It is a narrow zone of soil influenced by the root and exudates. The extent of the rhizosphere may vary with soil type, plant species, age, and many other factors (Curl and Truelove, 1986), but it is usually considered to extend from the root surface out into the soil for a few millimeters. More intense microbial activity and larger microbial populations occur in this zone than in bulk soil, in response to the release by roots of large amounts of organic compounds. Upto 18% of the C assimilated through photosynthesis can be released from roots. Microbial populations in the rhizosphere can be 10–100 times larger than the populations in bulk soil (Sposito and Reginato, 1992). The rhizosphere is bathed in root exudates and microbial metabolites. Thus, the chemistry and biology of Cd in the rhizosphere differs significantly from its chemistry and biology in bulk soil.

A series of complexation reactions in the soil solution affect Cd transformation in the rhizosphere. Complexation reactions of Cd with ligands in the soil solution are significant in determining the chemical behavior and toxicity of Cd in the rhizosphere. In view of the occurrence of organic and inorganic ligands in the rhizosphere (McLaughlin *et al.*, 1998; Huang and Germida, 2002) and the stability constants of the complexes of Cd with these ligands (NIST, 1997), a large fraction of soluble Cd ions in the soil solution may actually be complexed with a series of organic and inorganic ligands commonly present in the rhizosphere.

The study of metal speciation in the soil solution has been encouraged by the free metal hypothesis in environmental toxicology (Lund, 1990). This hypothesis states that the toxicity and bioavailability of a metal is related to the activity of the free aqua ion. Although this hypothesis is gaining popularity in studies of soil–plant relations (Parker *et al.*, 1995), some evidence is now emerging that the free metal ion hypothesis may not be valid in all situations (Tessier and Turner, 1995). Therefore, the role of metal–organic and metal–inorganic complexes in metal uptake merits attention.

Exudates of various kinds isolated from axenically grown plants have been shown to complex metals (Morel *et al.*, 1986; Gries *et al.*, 1995). Hamon *et al.* (1995) reported that Cd was complexed in the soil solutions of the rhizosphere after radish growth (Table 7). Seasonal changes in the concentration of metals in the rhizosphere are related to the presence of complexing agents of biological origin (Linehan *et al.*, 1989). Krishnamurti *et al.* (1996) reported variations in pH and the CAI of the bulk and rhizosphere soils collected after 2 and 7 weeks of crop growth (Table 8). At the 2-week growth stage, the pH of the rhizosphere soil was lower than that of the corresponding bulk soil, and the CAI

Table 7
Percentages of total Cd or Zn in soil solution present as the free ion in relation to days of growth of radish plants. Note the soil solution at day 0 is assumed to be equivalent to bulk soil solution and that at 30 days rhizosphere soil solution as a result of root exploration throughout the pot (Hamon *et al.*, 1995)

Days of radish growth	Cd^{2+}(%)
0	95
14	15
16	11
18	15
20	25
22	31
24	50
26	36
28	28
30	28

Table 8
The influence of application of Idaho phosphate fertilizer on pH and cadmium availability index (CAI)[a] of the bulk and rhizosphere soils collected at 2- and 7-week crop growth stages. Modified from Krishnamurti *et al.* (1996)

Soil and cultivar		Bulk soil	Rhizosphere soil				LSD[b]	
			Control[c]		Idaho[d]		0.01	0.05
			2-week	7-week	2-week	7-week		
Luseland soil								
Kyle	pH	7.95	7.75	7.90	7.38	7.90	0.11	0.08
	CAI	87	97	88	152	88	85	63
Arcola	pH	7.90	7.80	7.85	7.38	7.85	0.11	0.08
	CAI	87	102	87	208	89	85	63
Jedbergh soil								
Kyle	pH	8.15	7.95	8.10	7.38	8.10	0.14	0.10
	CAI	9	12	9	80	10	27	19
Arcola	pH	8.10	7.90	8.05	7.70	8.05	0.14	0.10
	CAI	9	16	9	84	8	27	19

[a] CAI = cadmium availability index (in $\mu g\ kg^{-1}$) determined by 1 mol L^{21} NH_4Cl extraction method of Krishnamurti *et al.* (1995b). [b] LSD = least significant difference at $p = 0.01$ and $p = 0.05$. [c] Without application of Idaho phosphate fertilizer. [d] With application of Idaho phosphate fertilizer.

values, which were determined by the method of Krishnamurti *et al.* (1995b), were higher in the rhizosphere soil, indicating that more Cd was complexed with the LMWOAs at the soil–root interface. Compared with the bulk soils, the CAI values were 2–9 times higher in the rhizosphere of field plots fertilized with Idaho monoammonium phosphate fertilizer at the 2-week growth stage, which was attributed to the combined effects of the Cd introduced into the rhizosphere from the fertilizer, and of complexation reactions of phosphate and LMWOAs with soil Cd. At the 7-week plant growth stage, such differences were not observed. Appreciable amounts of LMWOAs were detected in the root exudates in the rhizosphere soils collected at the 2-week plant growth stage. The high value of CAI observed in the rhizosphere soils at the 2-week growth stage was attributed to the result of complexation of the particulate-bound Cd with solution LMWOAs at the soil–root interface. The enhanced root growth of the plants, particularly in the soils treated with Idaho phosphate fertilizer, might have resulted in high amounts of metal–organic complex-bound Cd species in the soil solution by complexation. The prolific plant and microbial activity with the application of phosphate fertilizer is expected to result in increased amounts of LMWOAs in solution at the soil–root interface. Therefore, a larger fraction of Cd will be in a complexed and usually soluble form in the rhizosphere than in the bulk soils.

Both the amounts and proportion of organic compounds of root exudates vary substantially with plant species and cultivars. Further, the same plant cultivar grown in different soils shows variations with respect to the kind and amount of LMWOAs present in the rhizosphere (Table 9). More recent data show that the kind and amount of root exudates vary with the level of Cd in contaminated soils (Chou *et al.*, 2003; Wang *et al.*, 2003), as illustrated in Table 10. Further, the amount of LMWOAs correlates significantly with the Cd concentration in the rhizosphere soils (Table 11). Cadmium speciation in relation to its bioavailability in the rhizosphere merits increased attention.

Table 9
Amount of low-molecular-weight organic acids (μg kg^{-1} dry soil) in rhizosphere soil of durum wheat cv. Kyle grown in three different soils (Szmigielska *et al.*, 1997)

	Soil		
Acid	Yorkton	Sutherland	Waitville
Malonic	99a	56a	68a
Succinic	22a	35476c	10826b
Fumaric	12a	150b	71ab
Malic	45a	898c	370b
Tartaric	ND[a]	665b	214a
trans-Aconitic	ND	13a	3a
Citric	ND	195b	81a
Acetic	865a	29245c	12240b
Propionic	ND	499a	ND[a]
Butyric	NC	7604b	2127a
Total	1043a	74801c	26000b

[a] ND = not detected.
Means within the same row having the same letter are not significantly different ($p \leq 0.05$).

Table 10
Species and amounts of LMWOAs in tobacco rhizosphere soils treated with various Cd concentrations (Chou *et al.*, 2003)

Treatment	LMWOA (μmol kg^{-1} soil)					
	acetic	lactic	glycolic	maleic	succinic	total
10[a]	260.8d	154.3c	797.3c	503.5b	173.7c	1889.6
5	146.4bc	145.3c	671.3ab	484.5ab	131.6ab	1579.1
1	113.5b	113.1b	594.6a	429.0a	114.6a	1364.8
Control	59.9a	66.9a	488.5a	384.1a	98.1a	1097.5
LSD 0.05	37.3	27.0	148.3	103.7	30.4	

[a] Cd concentration (mg Cd kg^{-1} soil).

Table 11
Relationship between LMWOAs and Cd concentration in tobacco rhizosphere soils (Chou *et al.*, 2003)

	Acetic	Lactic	Glycolic	Maleic	Succinic
r	0.98	0.90	0.97	0.92	0.99
P	0.02	0.10	0.03	0.07	0.01

5.2. Cadmium uptake by plants

Values for the uptake of Cd in the harvested crop tend to be low when compared to the total mass of Cd in the plough layer (McGrath, 1987). The apparent increase in the Cd uptake in wheat grain in long-term experimental plots was attributed to improved yields (Jones and Johnstone, 1989). However, the main source of Cd to a plant is the soil itself. The deposition of atmospheric Cd can also play an important role at sites where the concentration of Cd in the atmosphere is high (Dollard and Davies, 1989). Plants accumulate Cd at different rates, and the final concentration of Cd in plant tissues differs between species growing concurrently on the same soil (Dowdy and Larsen, 1975; Hansen and Tjell, 1983; Keefer *et al.*, 1986; Kim *et al.*, 1988).

In addition to the inter- and intraspecies variations in Cd concentrations in plants, marked differences also occur in the accumulations of Cd between various plant parts, in the order, roots > stems and leaves >> grain and tubers (Table 12). High Cd concentrations in root tissues have been reported in a number of studies of a variety of plant species (Mitchell *et al.*, 1978; Kraffczyck *et al.*, 1984; Florijn and Van Beusichem, 1993; Cieslinski *et al.*, 1996). However, root concentrations

Table 12
Cadmium concentrations (μg Cd kg^{-1} of dry weight) in various parts of 7-week-old plants of durum wheat and flax grown in Waitville and Regina soils of Southern Saskatchewan, Canada (from Cieslinski *et al.*, 1996)

Cultivar	Roots		Leaves		Stems		Heads	
	Waitville	Regina	Waitville	Regina	Waitville	Regina	Waitville	Regina
Durum wheat								
Kyle	495.9	822.1	402.0	823.6	76.6	535.8	16.2	43.4
Sceptre	470.4	675.6	613.2	684.4	15.2	639.8	22.9	55.7
DT 627	415.0	340.6	244.3	343.1	0.3	159.1	27.9	56.5
DT 623	343.5	363.7	6.5	394.9	0	235.3	0	55.5
Flax								
AC Emerson	400.3	541.3	810.0	1402.0	407.3	1090.0	na[a]	na
Flanders	499.0	483.3	569.7	1390.0	307.8	289.5	na	na
YSED 2	461.0	925.1	476.0	2011.7	541.8	1355.8	na	na

[a] na = not applicable.

usually do not correspond with Cd concentrations in shoots and leaves (Mitchell *et al.*, 1978; Mench and Martin, 1991; Cieslinski *et al.*, 1998).

A considerable number of soil variables, such as heavy metal content, pH, nature and quantity of sorption sites, temperature, and metal speciation in solution are considered to be important factors affecting Cd concentrations in crops. Those have been demonstrated to influence heavy metal bioavailability (Tyler and McBride, 1982; Brummer, 1986; Adriano, 2001; Sommers *et al.*, 1987; Cabrera *et al.*, 1988; Alloway, 1990). The sorptive properties of inorganic (e.g. Al, Fe and Mn oxides) and organic colloidal fractions (e.g. humic and fulvic acids), solution ligands, pH, ionic strength, and metal– and mineral–organic complexes determine the partitioning of the heavy metal between the solid and liquid phases. Krishnamurti and Naidu (2003) have shown that the soil solution concentrations of Cd in a few typical soils of South Australia can be predicted significantly using a multivariate model, with pH and the amounts of Cd associated with organic binding sites in the equation (Table 13).

5.2.1. Natural soils

The Cd content of a range of plant species collected from unpolluted soils is given in Table 14. Cadmium concentrations in plants were in the range of 0.012–6.6 mg kg^{-1} (dry weight), with leafy vegetables accumulating higher

Table 13
Regression analysis of Cd concentration in soil solution (Cd_s) with pH and solid-phase particulate-bound Cd fractions (from Krishnamurti and Naidu, 2003)

Regression equation	R^2 value	p value
$\log(Cd_s) = 4.778 - \underset{(0.02)}{0.132}\,pH + \underset{(0.03)}{1.323}\log(\text{organic Cd}) - \underset{(0.03)}{3.765}\log(\text{fulivic complex Cd}) - \underset{(0.04)}{0.691}\log(\text{humic complex Cd})$	0.992	0.02

The solid-phase fractionation of Cd in the soils incubated with 10 mg Cd kg^{-1} soil was arrived at following the procedure outlined by Krishnamurti and Naidu (2000). The values in parentheses shown under the different parameters in the multivariate model are the level of significance of respective parameters.

Table 14
Cadmium concentration in selected plants grown on unpolluted and contaminated soils

Type of plant	Cadmium concentration (mg kg^{-1} dry weight)		
	Unpolluted soils	Industrial discharge	Sewage sludge application
Grasses	0.03–0.3 (Fleischer *et al.*, 1974)	< 50 (Burkitt *et al.*, 1972)	30.0 (Boswell, 1975)
Cabbage	0.2 (Chaney and Giardano, 1977)		37.2 (Furr *et al.*, 1976)
Spinach	3.6 (Chaney and Giardano, 1977)		161.0 (Chaney and Giardano, 1977)
Lettuce	1.1–6.6 (Alloway, 1986)	1.2–28.8 (Davies and White, 1981)	2.9–58.3 (Alloway, 1986)
		1.9–24.2 (van Driel *et al.*, 1987)	3.5–8.4 (King, 1986)
Carrot	0.9 (Chaney and Giardano, 1977)	8.0 (Lagerwerff and Brower, 1974)	16.0 (Chaney and Giardano, 1977)
Potato	0.05–0.3 (Fleischer *et al.*, 1974)	14–17.0 (Kobayashi, 1972)	2.0 (Furr *et al.*, 1976)
Rice	0.029 (Masironi *et al.*, 1977)		0.2 (Chaney and Giardano, 1977)
Wheat	0.012–0.036 (Williams and David, 1973)		0.1–14 (Bingham *et al.*, 1975)

amounts of Cd (0.2–6.6 mg kg^{-1}). Much lower concentrations of Cd were reported in grasses (0.03–0.3 mg kg^{-1}) and cereal grains (0.01–0.04 mg kg^{-1}). Plants accumulate Cd at different rates, and the final concentration of Cd in plant tissues differs between species growing concurrently on the same soil (Fig. 5).

5.2.2. Contaminated soils

Cd entering the soil constitutes a more lasting form of pollution, since its uptake by plants can continue long after the direct source of pollution has ceased. The most important sources of Cd contamination in soils are phosphate fertilizers, sewage sludge amendments, and industrial discharge. The concentration of Cd in the plant species grown on contaminated soils can be very high, depending on the nature and the proximity of the source of contamination. Some of the typical values reported are given in Table 14. The accumulation of Cd by different species of plants growing on sewage-amended soils showed a trend (Dowdy and Larsen, 1975; Keefer *et al.*, 1986; Kim *et al.*, 1988) similar to that in uncontaminated soils (Fig. 5). Sauerbeck (1991), summarizing the results of a number of long-term field experiments with sewage sludge treatment in Germany, found that many dicotyledonous crop plants, such as spinach and lettuce, absorbed more Cd than did monocotyledonous crop plants, such as oats and wheat.

5.3. Cadmium transport within plant and cadmium phytotoxicity

Cd is readily translocated to the plant tops after absorption through the roots (Chaney and Giordano, 1977). MacLean (1976) showed that Cd was present in higher concentrations in the roots than in other parts of crops such as oats,

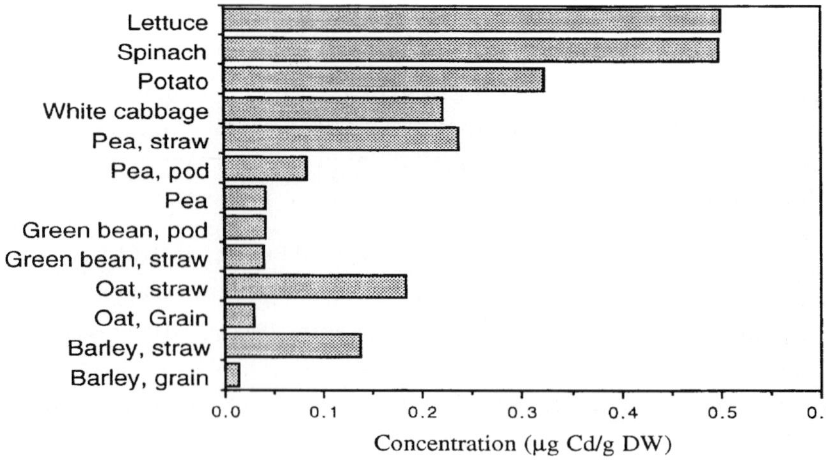

Fig. 5. Concentrations of cadmium in plants grown under identical conditions. (Redrawn from Hansen and Tjell, 1983).

soybean, maize, and tomato; these crops are not grown for the consumption of the roots. However, in lettuce, carrot, and potato, Cd contents were highest in leaves (MacLean, 1976). The Cd concentrations in various plant parts grown on soils treated with different Cd concentrations are generally in the order new roots > old roots > rhizomes > stem–leaf stalks > stem leaf-blades > reproductive organs (Balsberg, 1982). The Cd content of wheat grown in Skane, Sweden had an average of 73–100 μg Cd kg^{-1} d.w. in comparison with an average value of 29 μg kg^{-1} d.w. in wheat grown in central Sweden; concurrently, the wells in the soil landscape of Skane, polluted with industrial effluents, have an average level of 400 μg Cd L^{-1} (Selinas *et al.*, 1996).

The speciation of Cd in the tissue of plants is an important factor in determining its accumulation in human body. Cd has been found to bind to cytoplasmic proteins that usually contain cysteine, collectively called phytochelatins. Some plants appear to be able to produce metal-binding polypeptides to sequester metals and render them unavailable for transport within the plant (Steffens, 1990). Elevated concentrations of Cd in plant tissues may trigger the formation of phytochelatins (Grill *et al.*, 1987; Scheller *et al.*, 1987; Salt *et al.*, 1989). However, the exact mechanism of the triggering of phytochelatins by heavy metals such as Cd is still not known.

Terrestrial plants accumulate Cd, but the rate of accumulation is higher under experimental conditions, where Cd is available in solution, than it is in plants grown in soil, when part of the Cd is bound and less available. Stunted growth and toxic signs on leaves of lettuce, cabbage, carrot and radish plants were reported (Alloway *et al.*, 1990), but only at the highest concentrations of Cd tested, which resulted in a Cd content of around 20 mg kg^{-1} in the upper parts of the plants. Cd reduced transpiration and photosynthesis at concentrations of 100–200 mg Cd L^{-1} (Bazzaz *et al.*, 1974).

5.4. CAI

Many extractants were recommended for use as bioavailable indices based on significant correlation between quantities of metal extracted from the soils by the extractants and metal uptake by plants. The most commonly used extractants were ammonium bicarbonate-diethyl triamine penta acetic acid (AB-DTPA) (Soltanpour and Schwab, 1977; Norwell, 1984) and ammonium acetate acetic acid-ethylene diamine tetra acetic acid (AAAc-EDTA) (Lakanen and Ervio, 1971; Sillanpaa and Jansson, 1992). AB-DTPA was used successfully as an extractant for characterizing the bioavailability of both native soil metals as well as metals added to soils in sewage sludges. Some authors also reported insignificant relationships between the AB-DTPA-extracted metals and test plant metal concentrations (e.g. Haq and Miller, 1972; Rappaport *et al.*, 1988). O'Connor (1988) has subsequently identified a whole series of "misuses" of the DTPA test, which probably account for failure of the test. The chelate-based extractants tend to extract significantly higher amounts of trace elements and may not necessarily reflect the plant-available content in the soils. Extraction of trace metals from soils by unbuffered salt solutions could be a fast and simple way of evaluating their bioavailability to plants (Jackson and Alloway, 1991a; Lebourg *et al.*, 1996), e.g. 0.1 M $NaNO_3$ (VSBo, 1986); 0.05 M $CaCl_2$ (Sauerbeck and Styperek, 1984; Andrewes *et al.*, 1996); 0.01 M $CaCl_2$ (Houba *et al.*, 1990; Whitten and Ritchie, 1991; Andrewes *et al.*, 1996); 1 M NH_4NO_3 (Symeonides and McRae, 1977; DIN, 1995); and 1 M NH_4Cl (Krishnamurti *et al.*, 1995b). However, in some cases, these extractants failed, and causes could not be explained satisfactorily (Jackson and Alloway, 1991a, b; Gupta and Aten, 1993; Singh *et al.*, 1995).

Krishnamurti *et al.* (1995b) have compared the suitabilities of different extractants, as proposed earlier, for estimating the bioavailability index of the soils. They have shown that the 1 M NH_4Cl-extractable Cd correlated at least one order of magnitude more significantly with the Cd content of the grain of durum wheat cultivars than other extractants (Table 15). Studies on the soils from New Zealand (Gray *et al.*, 1999) and South Australia (Krishnamurti *et al.*, 2000) have also validated the use of 1 M NH_4Cl as a better extractant than other extractants studied for estimating the phytoavailability index of the soils (Table 15). The stability constant of the Cd–chloride complex (Smith and Martell, 1976) is of a similar order of magnitude as that of the Cd–organic acid complexes of many of the LMWOAs (Smith and Martell, 1977) present in the soil rhizosphere. This may account for the very high correlation observed between the 1 M NH_4Cl-extractable soil Cd and the

Table 15
Mean cadmium concentrations extracted from soils (mg Cd/kg soil) using different extractants and their regression coefficients (r^2) with the phytoavailable cadmium content in durum wheat

Extractant	Saskatchewan[a]		New Zealand[b]		South Australia[c]	
	Mean	r^2	Mean	r^2	Mean	r^2
ABDTPA	105.1	0.59**				
0.05 M $CaCl_2$	7.3	0.84**	58.2	0.51*	46.3	0.58**
M NH_4NO_3	3.4	0.56*	18.9	0.54*	12.1	0.71**
M NH_4OAc	23.9	0.68**	55.7	0.52*		
EDTA			156.4	0.43*	102.2	0.43*
M NH_4Cl	60.3	0.88***	121.9	0.59**	75.8	0.86***

[a] From Krishnamurti *et al.*(1995b). [b] from Gray *et al.* (1999). [c] From Krishnamurti *et al.* (2000).
Level of significance: * < 0.01, ** < 0.001; *** < 0.0001.

uptake of Cd by durum wheat in Canada, New Zealand, and Australia (Krishnamurti *et al.*, 1995b, 2000; Gray *et al.*, 1999). The formation of metal–organic complexes in cytoplasm and the subsequent transport of the complexes to the vacuoles has been proposed to be the general mechanism for the accumulation of heavy metal ions in plant vacuoles (Woolhouse, 1983; Krotz *et al.*, 1989; Rauser, 1990; Vogeli-Lange and Wagner, 1990). The mechanisms of metal–organic complexes in influencing the bioavailability of Cd are yet to be unravelled.

The phytoavailability of a metal ion depends on the form of the metal, soil properties, the plant species, and management practices. It is extremely difficult to assess the value of the very large number of studies that have reported metal extractants and plant availability of metals on different soils with or without the application of contaminants, such as sewage sludges. Comparisons between different studies are virtually impossible. Furthermore, only a few studies report metal availability tests, determined from chemical extractions, for native species under real field conditions (e.g. Gough *et al.*, 1980; Krishnamurti *et al.*, 1995b).

5.5. Cadmium speciation and availability

Morrison *et al.* (1989) stated: "research has so far failed to provide a realistic estimate of the toxic or available fraction in soils by chemical tests." Elaborate sequential chemical extraction schemes, designed to release a particular geochemical fraction, have frequently been used. However, little progress has been made in identifying the particular species of the element that may contribute to its bioavailability.

The speciation of metal cations governs their availability to plants and their potential to contaminate waterways and the ecosystem (Bernhard *et al.*, 1986). Available forms of metal ions are not necessarily associated with one particular chemical species or specific soil component. Hence, in order to predict the availability of metal cations, we either have to establish the species involved and develop the methods that specifically determine those forms only (solution and particulate-bound speciation), or develop an empirical relationship between the accepted diagnostic measure of the metal and plant uptake i.e. intrinsic availability index.

Both speciation in solution and in solid phase can affect plant uptake of metals. The amount of trace metal in soluble or exchangeable form can qualitatively indicate the easy uptake of metal cations by plants. At lower pH values, organic matter appears to be the only solid-phase component capable of retaining the trace metal cations and decreasing the soluble and exchangeable forms (Gerritse and Van Driel, 1984; Mann and Ritchie, 1993). The uptake of metals by plants is a function of their content and speciation in solution and the ability of the solid phase to replenish the solution metals. With respect to potentially toxic metal cations such as Cd, it would appear that the free ion and monomeric hydrolysis species are the major toxic chemical forms (Cabrera *et al.*, 1988). The free ion, Cd^{2+}, appears to be the form of Cd that is readily taken up by the plants, whereas $CdCl^{+}$ is taken up more slowly, and Cd-humate is not adsorbed (Bingham *et al.*, 1984, 1986; Cabrera *et al.*, 1988). Thus, it was argued that complexing of Cd with both inorganic and organic ligands decreases its toxicity (Chen and Stevenson, 1986). On the contrary, analysis of toxicity data obtained from growth studies with a soil alga (*Chlorococcum* sp.) indicated that Cd–citrate, as well as the Cd–DOM complexes, are bioavailable and contributed toward the toxicity to alga (Table 16, Krishnamurti *et al.*, 2004). These data contradict the long-held notion that Cd–DOM complexes are not bioavailable to soil biota, although they may increase the mobility of Cd.

Voluminous data are available on the distribution of different species of Cd in soils around the world, following many of the available sequential extraction schemes. However, little is known about the specific species of Cd important in its bioavailability. An attempt in this direction was made by Krishnamurti *et al.* (1995a), who carried out multiple regression analysis between the CAI, as measured by AB-DTPA-extractable Cd (Soltanpour and Schwab, 1977) and different forms of particulate-bound Cd. The importance of metal–organic complex-bound Cd species in the bioavailability of Cd, and the nature of the bonding sites of Cd were also studied in detail using multiple regression analysis and differential Fourier Transform Infrared Spectroscopy (FTIR) analysis (Krishnamurti *et al.*, 1995a, 1997b).

The beta coefficients (standard regression coefficients, Snedecor and Cochran, 1980) of the different species obtained from the multiple regression analysis of the data were in the order Fe, Mn oxide-bound $\cong$ organic-bound $>$ carbonate-bound (following the scheme of Tessier *et al.*, 1979), indicating the importance of both Fe, Mn oxide-bound and organic-bound Cd species in estimating the CAI of the soils (Eq. (1), Table 17). In the scheme of Krishnamurti

Table 16
Influence of Cd and Cd species, as determined by MINTEQA2 computer model on the growth of alga (*Chlorococcum* sp.) in soil pore water (Krishnamurti *et al.*, 2004)

Total Cd ($\mu g\ L^{-1}$)	Cd species ($\mu g\ L^{-1}$) Cd^{2+}	Cd-DOM	Inhibition of of algal growth (%)
115	5	109	11
302	31	269	50
479	50	427	79

DOM = dissolved organic matter.

Table 17

Data on the multiple regression analysis for the relationship of the availability index (CAI)[a], the particulate-bound Cd species and the components extracted by 0.1 M $Na_4P_2O_7$ (from Krishnamurti *et al.*, 1995a)

Regression equation with "beta" coefficients in parentheses[b]	Multiple correlation coefficient[a]
Tessier et al. (1979) scheme:	
Equation (1)	
$CAI = 0.0009 + 0.5728\ X_a + 1.7980\ X_b - 0.2976\ X_c + 0.5851\ X_d$	0.955
(0.4550) (0.4851) (−0.0175) (0.2206)	(9.8×10^{-6})
Krishnamurti et al. (1995a) scheme:	
Equation (2)	
$CAI = 0.0033 + 0.9420\ X_1 + 0.4311\ X_2 + 0.3199\ X_3 + 0.8782\ X_4 - 0.7059\ X_5 - 0.9054\ X_6$	0.968
(0.7817) (0.1651) (0.1405) (0.1198) (−0.1199) (-4.8×10^{-6})	(6.2×10^{-5})
Equation (3)	
$X_1 = 3.502 \times 10^{-2} + 8.815 \times 10^{-4}\ X_7 + 2.122 \times 10^{-3}\ X_8 + 4.497 \times 10^{-5}\ X_9 - 6.150 \times 10^{-5}\ X_{10}$	0.745
(0.5860) (0.0141) (0.3728) (− 0.4096)	(4.7×10^{-2})
Equation (4)	
$CAI = -3.071 \times 10^{-2} + 1.098\ X_1 - 1.648 \times 10^{-4}\ X_7 + 3.089 \times 10^{-2}\ X_8 + 8.379 \times 10^{-6}\ X_9$	0.932
(0.9111) (− 0.0909) (0.1703) (0.0576)	(8.5×10^{-5})

[a] CAI: cadmium available index determined as ABDTPA-extractable Cd following the method of Soltanpour and Schwab (1977). [b] Values in parenthesis are levels of significance; X_a, X_b, X_c and X_d, are the Fe, Mn oxide-bound Cd, organic-bound Cd, exchangeable Cd and carbonate-bound CD, respectively, following the Tessier *et al.* (1979) scheme, and X_1, X_2, X_3, X_4, X_5 and X_6 are the metal-organic complex-bound Cd, organics-bound Cd, carbonate-bound Cd, crystalline Fe oxide-bound Cd, easily reducible metal oxide-bound Cd, and amorphous mineral colloid-bound Cd, respectively, and X_7, X_8, X_9 and X_{10} are the Mn, organic C, Al and Fe-extracted by 0.1 M $Na_4P_2O_7$, following the Krishnamurti *et al.* (1995a) scheme.

et al. (1995a), the Fe, Mn oxide-bound and organic-bound Cd species were subfractionated into five distinct species viz., metal–organic complex-bound, easily reducible metal oxide-bound, organic-bound, amorphous mineral colloid-bound, and crystalline Fe oxide-bound Cd species. The beta coefficient of metal–organic complex-bound Cd species was at least 4–5 times higher than any of the other species, clearly bringing out the importance of metal–organic complex-bound Cd species in estimating the CAI (Eq. (2), Table 17). Even though Cd in the metal–organic complexes was associated strongly with Mn (Eq. (3) in Table 17), the importance of the Al–organic complex-bound Cd in influencing the bioavailability of Cd in the soils studied was clearly brought out by their beta coefficients in the multiple regression analysis (Eq. (4) in Table 17). Based on the differential FTIR spectra of the metal–organic complexes extracted by the 0.1 M sodium pyrophosphate extractant used in the speciation scheme, Krishnamurti *et al.* (1997a) showed that Cd in the soils was apparently bonded at the COO– of carboxyl and the OH of the phenolic groups and the OH of the Fe, Al and Mn in the metal–organic complexes. Recently, Krishnamurti and Naidu (2000) have shown the importance of Cd associated with exchangeable sites and fulvic acid binding sites in metal–organic complexes in assessing plant-available Cd in the soils of South Australia (Table 18).

6. CADMIUM CONTAMINATION IN THE TERRESTRIAL FOOD CHAIN AND HUMAN HEALTH

Approximately 85% of the Cd released into the environment results from anthropogenic emissions, mainly from smelting and refining of non-ferrous metals, fossil fuel combustion, and municipal waste incineration (IPCS, 1992). A direct relationship between geochemistry and human health is highly plausible, potentially exciting, repeatedly tantalizing, but rarely proven because the number of variables is legion. Human experiments are difficult to arrange and conduct, and epidemiological evidence seldom proves causality (Crounce *et al.*, 1983). Cadmium is considered to be a toxic trace element. Its known geochemical implications for human health include: (i) bone and renal disease in populations exposed to contaminated drinking water; (ii) lung and renal dysfunction in industrial workers exposed to air-borne Cd; and (iii) possible linkages to hypertension (Schroeder, 1965a, b; Celebresi *et al.*, 1980). Medical interest in Cd was sparked by reports from Japan of "Ouch-Ouch" (Itai-Itai in Japanese) disease, which is characterized by bone pain, multiple bone fractures, and renal loss of protein and Ca due to bone resorption. Death resulted from kidney failure, and the autopsies showed high Cd tissue concentrations. The Cd was traced to rice and soybean grown in local soils contaminated by a Pb/Zn mining operation. The persons most affected were those living where soil Cd levels were the highest and who had been the most stressed from malnutrition, multiple pregnancies, or the

Table 18
Data on the multiple regression analysis between plant-available Cd[a] and the particulate-bound Cd species of the soils[b] (Krishnamurti and Naidu, 2000)

Simple correlation analysis:

	Metal–organic complex-bound	Exchangeable	Detrital	Easily reducible metal oxide-bound	Specifically adsorbed	Organic -bound
Plant-available Cd	0.761	0.735	0.446	− 0.239	− 0.219	− 0.453
	(6.5×10^{-3})	(1×10^{-2})	(1.7×10^{-1})	(4.8×10^{-1})	(5.2×10^{-1})	(1.6×10^{-1})
	Metal–fulvic acid-bound	Metal–humic acid-bound				
	0.824	0.168				
	(1.8×10^{-3})	(6.2×10^{-1})				

Multiple regression analysis:

Plant-available Cd = 0.0004 + 3.5676 exchangeable Cd + 2.6500 metal-fulvic acid-bound Cd	$R^2 = 0.915$
	$(p = 0.0001)$
Plant-available Cd = 0.0645 + 2.8734 exchangeable Cd + 2.5746 metal-fulvic acid-bound Cd – 0.0079 pH	$R^2 = 0.919$
	$(P = 0.0003)$

The values in parentheses under the correlation coefficients are the levels of significance.
[a] Cd concentration in leaf and stem of the plant. [b] Following the method of Krishnamurti and Naidu (2000).

menopause (Waldbodt, 1978; Lauwerys, 1979). The daily Cd intake was up to 300 μg or more compared with an average 50 μg day^{-1} in other parts of the world (Prasad, 1978). Interpretation of the consequences of borderline high Cd intake in humans is quite complex. Modest epidemiological evidence suggests a relationship between Cd and renal cancer, supported by several types of animal and *in vitro* studies related to carcinogenesis (Stout and Rawson, 1981; Degreave, 1981). Suggestions that toxic elements in small amounts may be related to behavioral and learning disorders in children include reports of increased levels of toxic elements in the hair of affected individuals (Capell *et al.*, 1981; Thatcher *et al.*, 1982).

Chronic intoxication by Cd was first described by Friberg in Sweden, and was subsequently confirmed by a number of workers in other countries (see the reviews by Friberg *et al.*, 1986; Bernard and Lauwerys, 1986; IPCS, 1992; Bernard *et al.*, 1992). The studies confirmed that the earliest manifestation of Cd nephrotoxicity is renal dysfunction, leading to increased proteinuria. Cd nephropathy is irreversible and may accelerate the age-related decline of renal function. Depending on the indicator of renal dysfunction used, critical concentrations of Cd in urine vary between 5 and 10 μg g^{-1} creatinine, and the corresponding concentrations in the renal cortex range from 140 to 180 mg kg^{-1} (Bernard *et al.*, 1992).

In recent years, the availability of regional geochemical data for many of the developed countries has demonstrated that in addition to pollution caused by human activities, large areas have high concentrations of heavy metals, which occur naturally (Appleton, 1992). Certain associations between environmental geochemistry, diet, and degenerative disease have also been suggested (Martyn *et al.*, 1989). Attempts to link the occurrence of degenerative diseases such as cancer and heart disease to diet in developed countries may have been hampered partly because of the lack of knowledge of dietary components, and partly because the effect of trace element status on humans is masked by the use of food from different areas. By contrast, people in developing countries who live on subsistence agriculture obtain much of their food from local sources and hence, problems of pollution and increasing urbanization may be particularly intense and more easily identifiable. Further, trace element toxicities may be much more critical for human and animal health in developing countries than in developed countries (Plant *et al.*, 1996). Studies of the relationship between environmental geochemistry and health are, therefore, likely to be of more immediate value in developing than in developed countries, although the results could have worldwide significance.

The Cd pollution caused by anthropogenic sources represents a potential source of exposure for future generations in both developing and developed countries through food chain contamination. In the terrestrial ecosystem, the rhizosphere is the bottleneck of food chain contamination. Plant–microbe interactions affect physicochemical reactions in the rhizosphere. Physicochemical properties that can be different in the rhizosphere as compared to the bulk soil include pH, concentration of complexing biomolecules, redox potential, ionic strength, moisture, and nutrient status. The total rhizosphere environment is governed by root–microbe–soil interactions (Lynch, 1990). It is difficult to separate the impacts of microbial activity on the chemistry of Cd from those of plant root activity in the rhizosphere. Microorganisms can act in a way similar to plant roots. They can accumulate Cd through adsorption and uptake. They can also mobilize Cd through the action of microbial excretions. Bacteria are especially abundant in the rhizosphere (Lynch, 1990) and have a large capacity to sorb metals because of their high surface area-to-volume ratio (Beveridge, 1988). Most of the information on the uptake of Cd by microorganisms has been obtained *in vitro*. The effect of the free-living rhizosphere biota on Cd uptake by plants remains to be established and the effect of the rhizosphere biota is even less understood. Furthermore, the effect of the nature and properties of soil particles on plant–microbe interactions, and the impact on Cd uptake and subsequent food chain contamination remain obscure.

7. CONCLUSIONS AND FUTURE PROSPECTS

Knowledge of the total content of Cd does not indicate comprehensive information on its chemical behavior. It is the chemical speciation of Cd that influences Cd chemical reactivity, mobility, bioavailability, and toxicity in the ecosystem. Therefore, it is essential to investigate chemical speciation of Cd in soils and sediments, and especially in the rhizosphere, which is the bottleneck of terrestrial food chain contamination.

The differentiation of the metal–organic complex-bound Cd species from other metal oxide-bound and organically bound Cd species is a recent innovation in the sequential extraction scheme. This scheme proportionates the particulate-bound Cd species in soils as exchangeable, carbonate-bound, metal–organic complex-bound, easily reducible metal oxide-bound, organic-bound, amorphous mineral colloid-bound, crystalline Fe oxide-bound, and residual. The metal–organic complex-bound Cd is generally the highest among the particulate-bound Cd species of surface soils in temperate and tropical regions. Cadmium present as metal–organic complex-bound species is especially enriched in the rhizosphere soils after application of phosphate fertilizer. Its importance in assessing phytoavailability of soil Cd has been demonstrated and thus merits attention.

In the rhizosphere, the kinds and concentrations of substrates are different from those in the bulk soil, because of root exudation. This leads to colonization by different populations of bacteria, fungi, protozoa, and nematodes. Plant–microbe interactions, in turn, affect physicochemical reactions in the rhizosphere. The total rhizosphere environment is governed by an interacting trinity of the soil, the plant, and the organism associated with the root. The reactions and processes of Cd in the rhizosphere can only be interpreted satisfactorily with interdisciplinary approaches. Much of the research on physicochemical reactions of Cd in soils has used well-defined model systems that simulate bulk soil characteristics. The impacts of physicochemical–biological interfacial reactions in the rhizosphere on Cd uptake, food chain contamination, and ecosystem health merit increasing attention.

It is advantageous to combine theroretical (thermodynamic) calculations, sequential extraction schemes, and spectroscopies (e.g., synchrotron-based methods, e.g. EXAFS and XANES) to advance our understanding of Cd speciation in the rhizosphere which is bathed in root exudates and microbial metabolites. This information is fundamental in order to advance the frontiers of knowledge on the biogeochemistry of soil Cd at the molecular level.

ACKNOWLEDGMENTS

This study was supported by Discovery Grant 2383 of the Natural Sciences and Engineering Research Council of Canada and Project NSC 92-2811-B-002-075 of the Republic of China (Taiwan).

REFERENCES

Adriano, D.C., 2001. Trace Elements in Terrestrial Environments: Biogeochemistry, Bioavailability, and Risks of Metals. Second ed. Springer-Verlag, New York.

Alberts, J.J., Geisy, J.P., 1983. Conditional stability constants of trace metals and naturally occurring humic materials: their application in equilibrium models and verification with field data. In: Christman, R.F., Gjessing, E.T. (Eds.), Aquatic and Terrestrial Humic Material. Ann Arbor Science, Ann Arbor, MI, pp. 333–348.

Allison, J.D., Broen, D.S., Nova-Gradac, K.J., 1991. MINTEQA2/PRODEFA2: A Geochemical Assessment Model for Environmental Systems. Version 3.11 (EPA/600/3-91/021). USEPA, Athens, GA.

Allison, J.D., Brown, D.S., 1995. MINTEQA2/PRODEFA2: a geochemical speciation model and interactive preprocessor. In: Loeppert, R.H., Schwab, A.P., Goldberg, S. (Eds.), Chemical Equilibrium and Reaction Models. SSSA, Madison, WI, pp. 241–252.

Alloway, B.J., 1986. Cadmium and lead in soils and vegetables: investigation of the factors controlling concentrations in soil solutions and the tissues of crop plants. DOE Project PECD 7/8/05.

Alloway, B.J., 1990. Soil processes and the behavior of metals. In: Alloway, B.J. (Ed.), Heavy Metals in Soils. Blackie Academic & Professional, Glasgow, pp. 7–28.

Alloway, B.J., 1995. Cadmium. In: Alloway, B.J. (Ed.), Heavy Metals in Soils. second ed. Blackie Academic & Professional, Glasgow, pp. 122–151.

Alloway, B.J., Jackson, J.P., Morgan, H., 1990. The accumulation of cadmium by vegetables grown on soils contaminated with a variety of sources. Sci. Total Environ. 91, 223–236.

Alloway, B.J., Steinnes, E., 1999. Anthropogenic additions of cadmium to soils. In: McLaughlin, M.J., Singh, B.R. (Eds.), Cadmium in Soils and Plants. Developments in Plant and Soil Sciences, vol. 85. Kluwer Academic Publishers, Dordrecht, Holland, pp. 97–123.

Alloway, B.J., Thornton, I., Smart, G.A., Sherlock, J.C., Quinn, M.J., 1988. Metal availability. Sci. Total Environ. 75, 41–69.

Almas, A.R., McBride, M.B., Singh, B.R., 2000. Solubility and lability of cadmium and zinc in two soils treated with organic matter. Soil Sci. 165, 250–259.

Andrewes, P., Town, R.M., Hedley, M.J., Loganathan, P., 1996. Measurement of plant available cadmium in New Zealand soils. Aust. J. Soil Res. 34, 441–452.

Appleton, D.J., 1992. Review of the use of geochemical maps for identifying areas where mineral deficiencies or Excesses may Affect Cattle Productivity in Tropical Agriculture. British Geological Survey Technical Report WC/92/24, The Geological Society, London.

Balsberg, A.M., 1982. Seasonal changes in concentration and distribution of supplied cadmium in a Fillipendula ulmaria ecosystem. Oikios 38, 91–98.

Bascomb, C.L., 1968. Distribution of pyrophosphate-extractable iron and organic carbon in soils of various groups. J. Soil Sci. 19, 251–258.

Bauchauer, M.J., 1973. Contamination of soil and vegetation near a zinc smelter by zinc, cadmium, copper and lead. Environ. Sci. Technol. 7, 1331–1335.

Bazzaz, F.A., Carlson, R.W., Rolfe, G.L., 1974. The effects of heavy metals on plants. part 1. Inhibition of gas exchange in sun flower by Pb, Cd, Ni and Ti. Environ. Pollut. 7A, 241–246.

Beckett, P.H.T., 1989. The use of extractants in studies on trace metals in soils, sewage sludges, and sludge-treated soils. Adv. Soil Sci. 9, 143–176.

Bernard, A., Lauwerys, R., 1986. Effects of Cd exposure in humans. In: Foulkes, E.C. (Ed.), Handbook of Experimental Pharmacology. Springer, Berlin, pp. 135–177.

Bernard, A., Roels, H., Buchet, J.P., Cardenas, A., Lawerys, R., 1992. Assessment of the causality of the cadmium–protein relationships in the urine of the general population with reference to the Cadmibel study. In: Norberg, G.F., Herber, R.F.M., Alessio, L. (Eds.), Cadmium in the Human Environment: Toxicity and Carcinogenicity. IARC, Lyon, pp. 341–346.

Bernhard, M., Brinkmann, F.E., Sadler, P.J., (Eds.), 1986. The Importance of Chemical Speciation in Environmental Processes. Springer, Berlin.

Bertsch, P.M., Hunter, D.B., 1998. Elucidating fundamental mechanisms in soil and environmental chemistry: the role of advanced analytical, spectroscopic, and microscopic methods. In: Huang, P.M., Sparks, D.L., Boyd, S.A. (Eds.), Future Prospects for Soil Chemistry. SSSA Special Publication No. 55, Soil Sci.Soc. Am. Madison, WI, pp. 103–122.

Berrow, M.J., Burridge, J.C., 1980. Trace element levels in soil: effects of sewage sludge. In: Inorganic Pollution and Agriculture. MAFF Reference Book No. 326. HMSO, London, pp. 159–190.

Beveridge, T.J., 1988. The bacterial surface: general consideration towards design and function. Can. J. Microbiol. 34, 363–372.

Bingham, F.T., Page, A.L., Mahler, R.J., Ganje, T.J., 1975. Growth and cadmium accumulation of plants grown on a soil treated with a cadmium-enriched sewage sludge. J. Environ. Qual. 4, 207–211.

Bingham, F.T., Sposito, G., Strong, J.E., 1984. The effect of chloride on the availability of cadmium. J. Environ. Qual. 13, 71–74.

Bingham, F.T., Sposito, G., Strong, J.E., 1986. The effect of sulfate on the availability of cadmium. Soil Sci. 141, 172–175.

Bolt, G.H., Van Riemsdijk, W.H., 1982. Ion adsorption on inorganic variable charge constituents. In: Bolt, G.H. (Ed.), Soil Chemistry. B. Physico-Chemical Models. second ed. Elsevier, Amsterdam, pp. 495–502.

Boswell, F.C., 1975. Municipal sewage sludge and selected element applications to soil: effect on soil and fescue. J. Environ. Qual. 4, 267–273.

Bramley, R.G.V., 1990. Cadmium in New Zealand agriculture. NZ. J. Agric. Res. 33, 505–519.

Brown, L.R., Jacobson, J., 1987. Assessing the future of urbanization. In: State of the World. World Watch Institute. Norton and Company, New York, pp. 38–56.

Brummer, G.W., 1986. Heavy metal species, mobility and availability. In: Bernhard, M., Brinkman, F.E., Sadler, P.J. (Eds.), The Importance of Chemical Speciation in Environmental Processes. Springer, Berlin, pp. 169–192.

Brummer, G., Tiller, K.G., Herms, U., Clayton, P.M., 1983. Adsorption/desorption and/or precipitation dissolution processes of zinc in soils. Geoderma 31, 337–354.

Burkitt, A., Lester, P., Nickless, G., 1972. Distribution of heavy metals in the vicinity of an industrial complex. Nature 238, 327–328.

Cabrera, D., Young, S.D., Rowell, D.L., 1988. The toxicity of cadmium to barley plants as affected by complex formation with humic acid. Plant Soil 105, 195–204.

Candelaria, L.M., Chang, A.C., 1997. Cadmium activities, solution speciation, and solid-phase distribution of cadmium in cadmium nitrate- and sewage sludge-treated soil systems. Soil Sci. 162, 722–732.

Capell, I.D., Pinnock, M.H., Dorrell, H.M., Williams, D.C., Grant, E.C.G., 1981. Comparison of concentrations of some trace, bulk and toxic metals in the hair of normal and dyslexic children. Clin. Chem. 27, 879–881.

Celebresi, E.J., Moore, G.S., Tuthill, R.W., Siewger, T.L., 1980. Drinking water and cardiovascular disease. J. Environ. Path. Toxicol. 4, 1–326.

Chaney, R.L., Giordano, P.M., 1977. Microelements as related to plant deficiencies and toxicities. In: Elliot, L.F., Stevenson, F.J. (Eds.), Soils for the Management of Organic Wastes and Waste Waters. SSSA. Madison, WI, pp. 235–279.

Chaney, R.L., Hornick, S.B., 1977. Accumulation and effects of cadmium on crops. In: Proceedings First International Cadmium Conference San Fransisco, Metal Bull. Ltd., Worcester, London, pp. 125–140.

Chaney, R.L., Ryan, J.A., Li, Y.,-M. Brown, S.L., 1999. Soil cadmium as a threat to human health. In: McLaughlin, M.J., Singh, B.R. (Eds.), Cadmium in Soils and Plants. Developments in Plant and Soil Sciences, vol. 85. Kluwer Academic Publishers, Dordrecht, Holland, pp. 219–256.

Chao, T.T., 1972. Selective dissolution of manganese oxides from soils and sediments with acidified hydroxylamine. Soil Sci. Soc. Am. Proc. 36, 764–768.

Chao, T.T., 1984. Use of partial dissolution techniques in geochemical exploration. J. Geochem. Explor. 20, 101–135.

Chao, T.T., Zhou, L., 1983. Extraction techniques for selective dissolution of amorphous iron oxides from soils and sediments. Soil Sci. Soc. Am. J. 47, 225–232.

Charlatchka, R., Cambier, P., Bourgeois, S., 1997. Mobilization of trace metals in contaminated soils under anaerobic conditions. In: Prost, R., (Ed.), Contaminated Soils. Proceedings of Third International Conference on Biogeochemistry Trace Elements, Paris (France), May 15–19, 1995, INRA, Paris, pp. 159–174.

Chen, Y., Stevenson, F.J., 1986. Soil organic matter interactions with trace elements. In: Chen, Y., Avnimelech, Y. (Eds.), The Role of Organic Matter in Modern Agriculture. Kluwer Academic, Dordrecht, pp. 73–116.

Chester, R., Hughes, M.J., 1967. A chemical technique for separation of ferro-manganese minerals, carbonate minerals and adsorbed trace elements from pelagic sediments. Chem Geol. 2, 249–262.

Chou, S.Y., Chiu, C.Y., Chen, C.T., Zhoung, S.Y., Wang, M.K., 2003. Influence of cadmium treatment on low molecular weight organic acids exudation in tobacco rhizosphere soils. Soil Environ. 6, 233–240.

Christensen, T., Tjell, J., 1983. Interpretation of experimental results on cadmium crop uptake from sewage sludge amended soil. In: L'Hermite, P., Ott, H. (Eds.), Processing and Use of Sewage Sludge. D. Reidel Publishing Company, Dordrecht, pp. 358–370.

Chumbley, C.J., Unwin, R.J., 1982. Cadmium and lead content of vegetables grown on land with a history of sewage sludge application. Environ. Pollut. 4B, 231–237.

Cieslincki, G., Van Rees, K.C.J., Huang, P.M., Kozak, L.M., Rostad, H.P.W., Knott, D.R., 1996. Cadmium uptake and bioaccumulation in selected cultivars of durum wheat and flax as affected by soil type. Plant Soil 182, 115–124.

Cieslinski, G., Van Rees, K.C.J., Szmigielska, A.M., Huang, P.M., 1994. Plant growing conditions for collection of low-molecular-weight organic acids from root exudates. In: Agronomy Abstracts. ASA, Madison, WI, p. 264.

Cieslincki, G., Van Rees, K.C.J., Szmigielska, A.M., Krishnamurti, G.S.R., Huang, P.M., 1998. Low-molecular-weight organic acids in rhizosphere soils of durum wheat and their effect on cadmium bioaccumulation. Plant Soil 203, 109–117.

Cottennie, A., Verloo, M., Kiekens, L., Camerlynck, R., Vleghe, G., Dhaese, A. 1983. Essential and non-Essential Trace Elements in the System Soil–Water-Plant. Lab. Analyt. Agrochem. State Univ. Ghent, Belgium.

Crounce, R.G., Pories, W.J., Bray, J.T., Mauger, R.L., 1983. Geochemistry and man: health and disease. In: Thornton, I., (Ed.), Applied Environmental Geochemistry. Academic Press, New York, pp. 267–333.

Curl, E.A., Truelove, B., 1986. The Rhizosphere. Springer, Berlin.

Davies, B.E., 1980. Trace element pollution. In: Davies, B.E., (Ed.), Applied Soil Trace Elements. Wiley, Chichester, pp. 287–351.

Davies, B.E., 1992. Trace elements in the environment: retrospect and prospect. In: Adriano, D.C. (Ed.), Biogeochemistry of Trace Metals. Lewis Publishers, Boca Raton, FL, pp. 1–17.

Davies, B.E., Roberts, L.J., 1978. The distribution of heavy metals in contaminated soils in northeast Clyde, Wales. Water Air Soil Pollut. 9, 507–508.

Davies, B.E., White, H.M., 1981. Trace elements in vegetables grown on soils contaminated by base metal mining. J. Plant Nutr. 3, 387–396.

Davis, J.A., James, R.O., Leckie, J.O., 1978. Surface ionization and complexation at the oxide/water interface. I. Computation of electrical double layer properties in simple electrolytes. J. Colloid Interface Sci. 63, 480–499.

Davis, J.A., Leckie, J.O., 1978. Surface ionization and complexation at the oxide/water interface. II. Surface properties of amorphous iron oxyhydroxide and adsorption of metal ions. J. Colloid Interface Sci. 67, 90–107.

Davis, J.A., Leckie, J.O., 1980. Surface ionization and complexation at the oxide/water interface. III. Adsorption of anions. J. Colloid Interface Sci. 74, 32–43.

Davis, R.D., Coker, E.G., 1980. Cadmium in Agriculture, with Special Reference to the Utilization of Sewage Sludge on Land. Technical Report No. TR/139. Water Resource Centre, Medmenham, UK.

Degreave, N., 1981. Carcinogenic, teratogenic and mutagenic effects of cadmium. Mutat. Res. 86, 115–135.

Del Castilho, P., Chardon, W.J., Salomos, W., 1993. Influence of cattle-manure slurry application on the solubility of cadmium, copper, and zinc in a manured acidic, loamy-sand soil. J. Environ. Qual. 22, 689–697.

DIN (Deutsches Instutut fur Normung), 1995. Bodenbeschaffenheit. Extraktion von Spurenelemente mit Ammonium-nitratlosung. Vornorm DIN V19730. Boden-Chemische Bodenuntersuchungsverfahren, Berlin, Germany.

Dobbs, J.C., Susetyo, W., Knight, F.E., Castles, M.A., Carriera, L.A., Azarraga, L.V., 1989. A novel approach to metal–humic complex studies by lanthanide ion probe spectroscopy. Anal. Chem. 61, 483–488.

Dollard, G.J., Davies, T.J., 1989. A Study of the Contribution of Airborne Cadmium to the Cadmium Burdens of Several Vegetable Species. United Kingdom Atomic Energy Authority, Didkot, UK.

Domsch, K.H., 1984. Effects of pesticides and heavy metals on biological processes in soil. Plant Soil 76, 367–378.

Dowdy, R.H., Larsen, W.E., 1975. The availability of sludge-borne metals to various vegetable crops. J. Environ. Qual. 4, 278–282.

Dudka, S., Chlopecka, A., 1990. Effect of solid-phase speciation on metal availability and phytoavailability in sludge-amended soil. Water Air Soil Pollut. 51, 153–160.

Dunnivant, F.M., Jardine, P.M., Taylor, D.L., McCharty, J.F., 1992. Cotransportation of cadmium by hexachlorobiphenyl of dissolved organic carbon through columns containing aquifer material. Environ. Sci. Technol. 26, 360–368.

Dzombek, D.A., Morel, F.M.M., 1990. Surface Complexation Modelling: Hydrous Ferric Oxide. Wiley, New York.

Elliot, H.A., Dempsey, B.A., Maille, M.J., 1990. Content and fractionation of heavy metals in water treatment sludges. J. Environ. Qual. 19, 330–334.

Elzinga, E.J., van Grissen, J.J.M., Swartjes, F.A., 1999. General purpose Freundlich isotherms for cadmium, copper and zinc in soils. Eur. J. Soil Sci. 50, 139–149.

Emmerich, W.E., Lund, L.J., Page, A.L., Chang, A.C., 1982a. Movement of heavy metals in sewage-sludge treated soils. J. Environ. Qual. 11, 174–178.

Emmerich, W.E., Lund, L.J., Page, A.L., Chang, A.C., 1982b. Solid phase forms of heavy metals in sewage sludge-treated soils. J. Environ. Qual. 11, 178–181.

Emmerson, B.T., 1970. "Ouch-Ouch" disease: the Osteomalacia of cadmium nephropathy. Ann. Int. Med. 73, 854–855.

Fendorff, S.E., Sparks, D.L., 1996. X-ray absorption fine structure spectroscopy. In: Sparks, D.L. (Ed.), Methods of Soil Analysis. Part 3. Chemical Methods. Soil Science Society of America, Madison, WI, pp. 377–416.

Fendorff, S.E., Sparks, D.L., Lamble, G.M., Kelley, M.J., 1994. Application of X-ray absorption fine structure spectroscopy to soils. Soil Sci. Soc. Am. J. 58, 1583–1595.

Fitch, A., Stevenson, F.J., 1983. Stability constants of metal-organic matter complexes: theoretical aspects and mathematical models. In: Augustithis, S.S. (Ed.), The Significance of Trace Elements in Solving Paragenetic Problems and Controversies. Theoprastus Publications S.A., Athens, Greece, pp. 645–649.

Fitch, A., Stevenson, F.J., 1984. Comparison of models for determining stability constants of metal complexes with humic substances. Soil Sci. Soc. Am. J. 48, 1044–1049.

Fleischer, M., Sarofim, A.F., Fassett, D.W., Hammond, P., Shacklette, H.T., Nisbet, I.C.T., Epstein, S., 1974. Environmental impact of cadmium. A review by the Panel on Hazardous Trace Substances. Environ. Health Prospect. 7, 253–323.

Florence, T.M., Batley, G.E., 1977. Determination of chemical forms of trace metals in natural waters with special reference to copper, lead, cadmium and zinc. Talanta 24, 151–158.

Florijn, P.J., Van Beusichem, M.L., 1993. Uptake and distribution of cadmium in maize inbred lines. Plant Soil 150, 25–32.

Foerstner, U., 1991. Soil pollution phenomena-mobility of heavy metals in contaminted soil. In: Bolt, G.H., Hayes, M.H.B., McBride, M.B. (Eds.), Interactions at the Soil Colloid–Soil Solution Interface. Kluwer Academic Publishers, Dordrecht, pp. 543–582.

Foerstner, U., Calmano, K., Conrad, H., Jaksch, H., Schimkus, C., Schoer, J., 1981. Chemical speciation of heavy metals in solid waste materials (sewage sludge, mining wastes, dredged materials, polluted sediments) by sequential extraction. Proceedings of International Conference on Heavy Metals Environment, WHO/EED, pp. 698–704.

Frausto da Silva, J.J.R., Williams, R.J.P., 1976. The uptake of elements by biological systems. Struct Bond 29, 67–121.

Freedman, B., Hutchinson, T.C., 1981. Sources of metal and elemental contamination of terrestrial environments. In: Lepp, N.W. (Ed.), Effect of Heavy Metal Pollution on Plants vol. 2. Metals in the Environment. Applied Science Publishers, London, pp. 35–94.

Friberg, L., Elinder, C.G., Kjellstrom, T., Nordberg, G.F., 1986. Cadmium and Health: a Toxicology and Epedimeology Appraisal vol. II. Effects and Response. CRC Press, Boca Raton, FL.

Furr, A.K., Kelly, W.C., Bache, C.A., Gutenmann, W.H., Lisk, D.J., 1976. Multielement absorption by crops grown in pots on municipal sewage sludge-amended soil. J. Agric. Food Chem. 24, 889–892.

Garcia-Mirgaya, J., Castro, S., Paolini, J., 1981. Lead and zinc levels and chemical fractionation in roadside soils of Caracas, Venezuela. Water Air Soil Pollut. 15, 285–297.

Gerritse, R.G., Van Driel, W., 1984. The relationship between adsorption of trace metals, organic matter and pH in temperate soils. J. Environ. Qual. 13, 197–204.

Gibbs, R.J., 1973. Mechanisms of trace metals transport in rivers. Science 180, 171–173.

Gibbs, R.J., 1977. Transport phases of transition metals in Amajon and Yukon rivers. Geol. Soc. Am. Bull. 88: 829–843.

Gibson, J.J., Farmer, J.G., 1986. Multistep sequential chemical extraction of heavy metals from urban soils. Environ. Pollut. B 11, 117–135.

Goldberg, S., 1992. Use of surface complexation models in soil chemical systems. Adv. Agron. 47, 223–329.

Goldberg, S., Traina, S.J., 1987. Chemical modelling of anion competition on oxides using the constance capacitance model-mixed-ligand approach. Soil Sci. Soc. Am. J. 51, 929–932.

Gough, L.P., McNeal, J.M., Severson, R.C., 1980. Predicting native plant copper, iron, manganese and zinc levels using DTPA and EDTA soil extractants. Northern Great Plains Soil Sci. Soc. Am. J. 44, 1030–1036.

Gray, C.W., McLaren, R.G., Roberts, A.H.C., Condron, L.M., 1999. Cadmium phytoavailability in some New Zealand soils. Austr J. Soil Res. 37, 461–477.

Gries, D., Brunor, S., Crowley, D.E., Parker, D.R., 1995. Phytosiderophore release in relation to micronutrient metal deficiency in barley. Plant Soil 172, 299.

Grill, E., Winnacker, E.-L., Zenk, M.H., 1987. Phytochelatins, a class of heavy metal-binding peptides from plants, are functionally analogous to metallotheonines. Proc. Natl. Acad. Sci. USA. 84, 439–443.

Gulbransden, R.A., 1966. Chemical composition of phosphorites of phosphoria formation. Geochim. Cosmochim. Acta 30, 769–778.

Gupta, S.K., Aten, C., 1993. Comparison and evaluation of extraction media and their suitability in a simple model to predict the biological relevance of heavy metal concentrations in contaminated soils. Int. J. Environ. Anal. Chem. 51, 25–46.

Gupta, S.K., Chen, K.Y., 1975. Partitioning of trace elements in selective chemical fractions of nearshore sediments. Environ. Lett. 10, 129–158.

Hamon, R.E., Lorenz, S.E., Holm, P.E., Christensen, T.H., McGrath, S.P., 1995. Changes in trace metal species and other components of the rhizosphere during growth of radish. Plant Cell Environ. 18, 749–756.

Hansen, J.A., Tjell, J.C., 1983. Sludge application to land-overview of Cd problem. In: Davis, R.D., Hucker, G., L'Hermite, P. (Eds.). Environmental Effects of Organic and Inorganic Contaminants in Sewage Sludge. Reidel, Dordrecht, pp. 91–113.

Haq, A.U., Miller, M.H., 1972. Prediction of available soil Zn, Cu and Mn using chemical extractants. Agron. J. 64, 779–782.

Harrison, R.M., Laxe, D.P.H., Wilson, S.J., 1981. Chemical associations of lead, cadmium, copper and zinc in street dusts and road side soils. Environ. Sci. Technol. 15, 1378–1383.

Hawthorne, F.C., 1988. Spectroscopic Methods in Mineralogy and Geology. Mineralogical Society of America, Washington, DC.

Hayes, K.F., Leckie, J.O., 1987. Modelling ionic strength effects on cation adsorption at hydrous oxide/solution interfaces. J. Colloid Interface Sci. 115, 564–572.

Hayes, K.F., Papelis, C., Leckie, J.O., 1988. Modelling ionic strength effects on anion adsorption at hydrous oxide/solution interfaces. J. Colloid Interface Sci. 125, 717–726.

Helios Rybicka, E., 1992. Heavy metal partitioning in polluted river and sea sediments: clay minerals effects. Mineralagia Petrografia Acta XXXV-A, 297–305.

Helios Rybicka, E., 1993. Phase-specific bonding of heavy metals in sediments of the Vistula river, Poland. Appl. Geochem. 2, 45–48.

Helios Rybicka, E., 1996. Environmental impact of mining and smelting industries in Poland. In: Appleton, J.D., Fuge, R., McCall, G.J.H., (Eds.), Environmental Geochemistry and Health. Geological Society Spl. Pub. No. 113, The Geological Society, London, pp. 183–193.

Hickey, M.G., Kittrick, J.A., 1984. Chemical partitioning of Cd, Cu, Ni and Zn in soils and sediments containing high levels of heavy metals. J. Environ. Qual. 13, 372–376.

Hiltner, L., 1904. Uber neuere Erfahrungen und Probleme auf dem Gebiet der Bodenbakteriologie und unter besonderer Berucksichtigung und Brache, Arb Dtsch. Landwirt. Ges. 98, 59.

Hirsch, D., Banin, A.J., 1990. Cadmium speciation in soil solutions. J. Environ. Qual. 19, 366–372.

Hochella, M.F., White, A.F., (Eds), 1990. Mineral–Water Interface Geochemistry. Mineralogical Society of America, Washington, DC.

Hodgson, J.F., Geering, H.R., Norvell, W.A., 1965. Micronutrient cation complexes in soil solution: Partition between complexed and uncomplexed forms by solvent extraction. Soil Sci. Soc. Am. Proc. 29, 665–669.

Hoffman, S.J., Fletcher, W.K., 1979. Extraction of Cu, Zn, Mo, Fe and Mn from soils and sediments using a sequential procedure. In: Watterson, J.R., Theobald, P.K. (Eds.), Geochemical Exploration. Association of Exploration Geochemists, Rexdale, Ontario, Canada, pp. 289–299.

Hoffman, S.J., Fletcher, W.K., 1981. Detailed lake sediment geochemistry of anamolous lakes on the Nechako Plateau, central British Colombia – comparison of trace metal distributions in Capoose and Fish Lakes. J. Geochem. Explor. 14, 221–244.

Holmes, R., 1976. The regional distribution of cadmium in England and Wales. Ph.D. Thesis, University of London, London, UK.

Holmgren, C.G.S., Meyer, M.W., Chaney, R.L., Daniels, R.B., 1993. Cadmium, lead, zinc, copper, and nickel in agricultural soils of the United States of America. J. Environ. Qual. 22, 335–348.

Houba, V.J.G., Novozamsky, I., Lexmond, T.M., van der Lee, J.J., 1990. Applicability of 0.01 M $CaCl_2$ as a single extraction solution for the assessment of nutrient status of soils and other diagnostic purposes. Commun. Soil Sci. Plant Anal. 21, 2281–2290.

Huang, C.P., Stumm, W., 1973. Specific adsorption of cations on hydrous γ-aluminum oxide. J. Colloid Interface Sci. 43, 409–417.

Huang, P.M., Adriano, D.C., Logan, T.J., Checkai, R.T., 1998. Soil Chemistry and Ecosystem Health. SSSA Special Publication 52. Soil Science Society of America, Madison, WI.

Huang, P.M., Germida, J.J., 2002. Chemical and biological processes in the rhizosphere: metal pollutants. In: Huang, P.M., Bollag, J.-M., Senesi, N. (Eds.), Interactions between Soil Particles and Microorganisms. Impact on the Terrestrial Ecosystem. IUPAC Series on Analytical and Physical Chemistry of Environmental Systems, vol. 8 Wiley Chichester, UK, pp. 381–438.

Hutton, M., Symon, C., 1986. The quantities of cadmium, lead, mercury, and arsenic entering the U.K. environment from human activities. Sci. Total Environ. 57, 129–150.

Ingle, S.E., Schuldt, M.D., Schults, D.W., 1978. A User's Guide for REDEQL-EPA. US EPA Report EPA-600/3-78-024, Corvallis, ORE.

IPCS, 1992. Environment Health Criteria 135. Cadmium – Environmental Aspects. WHO, Geneva.

Ireland, R.R., McConachie, W.A., Stuermer, D.H., Wang, F.T., Koszykowski, R.F., 1982. The analysis of inorganic constituents in the ground water at an underground coal gassification site. In: Filby, R.H., (Ed.), Atomic and Nuclear Methods in Fossil Energy Research (Proceedings of the Topical Conference Am. Nuclear Soc., Am. Chem. Soc. Dec. 1980, Mayaguez, Puerto Rico). Plenum Press, New York, pp. 175–188.

Iyengar, S.S., Martens, D.C., Miller, W.P., 1981. Distribution and plant availability of soil zinc fractions. Soil Sci. Soc. Am. J. 45, 735–739.

Jackson, A.P., Alloway, B.J., 1991a. The transfer of cadmium from sewage sludge amended soils into the edible componernts of food crops. Water Air Soil Pollut. 57, 883–891.

Jackson, A.P., Alloway, B.J., 1991b. The bioavailability of cadmium to lettuce and cabbage in soils previously treated with sewage sludges. Plant Soil 132, 179–186.

Jackson, A.P., Alloway, B.J., 1992. The transfer of cadmium from agricultural soils to the human food chain. In: Adriano, D.C. (Ed.), Biogeochemistry of Trace Elements. Lewis Publishers, Boca Raton, FL, pp. 109–158.

Jackson, M.L., 1958. Soil Chemical Analysis. Elsevier, Englewood Cliffs, NJ.

Jackson, M.L., 1985. Soil Chemical Analysis – An Advanced Course, second ed. Published by the author, Department of Soil Science, University of Wisconsin, Madison, WI.

James, R.O., Parks, G.A., 1978. Application of the Computer Program MINEQL to Solution of Problems in Surface Chemistry. Technical Note, Stanford University, Stanford, CA.

Jenne, E.A., 1968. Controls on Mn, Fe, Co, Ni, Cu and Zn concentrations in soils and waters: the dominant role of hydrous Mn and Fe oxides. In: Trace Inorganics in Water. Adv. in Chemistry Series, 73. Americal Chemical Society, Washington, DC, pp. 337–387.

Jenne, E.A., 1979. Chemical Modelling in Aqueous Systems. ACS Symposium Series No. 93. American Chemical Society, Washington, DC.

Jensen, A., Bro-Rasmussen, F., 1992. Environmental cadmium in Europe. Rev. Environ. Contam. Toxicol. 125, 101–181.

Jones, J.M., Hao, J., 1993. Ombrotrophic peat as a medium for historical monitoring of heavy metal pollution. Environ. Geochem. Health 15, 67–74.

Jones, L.H.P., Jarvis, S.C., 1981. The fate of heavy metals. In: Greenland, D.J., Hayes, M.H.B., (Eds.), The Chemistry of Soil Processes. Wiley, New York, pp. 593–620.

Jones, K.C., Johnstone, A.E., 1989. Cadmium in cereal grain and herbage from long-term experimental plots in Rothamstead, UK. Environ. Pollut. 57, 199–216.

Jones, K.C., Symon, C.J., Johnstone, A.E., 1987. Retrospective analysis of an archived soil collection. II. Cadmium. Sci. Total Environ. 67, 75–89.

Jopony, M., Young, S.D., 1994. The solid-solution equilibria of lead and cadmium in polluted soils. Eur. J. Soil Sci. 45, 59–70.

Kabata-Pendias, A., Pendias, H., 1992. Trace Elements in Soils and Plants, second ed. CRC Press, Boca Raton, FL.

Keefer, R.F., Singh, R.N., Horvath, D.J., 1986. Chemical composition of vegetables grown on agricultural soil amended with sewage sludges. J. Environ. Qual. 15, 146–152.

Keizer, M.G., 1991. ECOSAT: A Computer Program for the Calculation of Speciation in Soil–Water Systems. Department of Soil Science and Plant Nutrition, Agricultural University, Wageningen, The Netherlands.

Keller, C., Vedy, J.-C., 1994. Heavy metals in the environment: distribution of copper and cadmium fractions in two forest soils. J. Environ. Qual. 23, 987–999.

Kersten, M., Foerstner, U., 1986. Chemical fractionation of heavy metals in anoxic estuarine coastal sediments. Water Sci. Technol. 18, 121–130.

Kim, S.J., Chang, A.C., Page, A.L., Warneke, J.E., 1988. Relative concentrations of cadmium and zinc in tissue of selected food plants grown on sludge-treated soils. J. Environ. Qual. 17, 569–573.

Kim, S.J., Thornton, I., 1993. Influence of uraniferous black shales on cadmium, molybdenum, and selenium in soils and crop plants in the Deog-Pyoung area of Korea. Env. Geochem. Health 15, 119–133.

King, L.D., 1986. Agricultural Use of Municipal and Industrial Sludges in Southern United States. Southern Cooperative Series Bulletin 314. North Carolina State University, Raleigh, p. 52.

Kloke, A., Sauerbeck, D.R., Vetter, H., 1984. The contamination of plants and soils with heavy metals and the transport of metals in terrestrial food chains. In: Nriagu, J.O. (Ed.), Changing Metal Cycles and Human Health. Springer, Berlin, pp. 113–141.

Kobayashi, J., 1972. Air and water pollution by cadmium, lead, and zinc attributed to the largest zinc refinery in Japan. In: Hemphill, D.D. (Ed.), Trace substances in Environmental Health. No.5. University of Missouri, Colombia, pp. 117–128.

Kobayashi, J. Hagino, N., 1965. Progress Report WP00359. Okayama University, Okayama, Japan.

Kraffczyck, I., Trolldenier, G., Beringer, H., 1984. Soluble root exudates of maize: influence on potassium supply and rhizosphere microorgansms. Soil Biol. Biochem. 16, 315–322.

Krauskopf, K.B., 1967. Introduction to Geochemistry. McGraw-Hill, New York.

Krishnamurti, G.S.R., Cieslinski, G., Huang, P.M., Van Rees, K.C.J., 1997c. Kinetics of cadmium release from soils as influenced by organic acids: implication in cadmium availability. J. Environ. Qual. 26, 271–277.

Krishnamurti, G.S.R., Huang, P.M., Van Rees, K.C.J., 1996. Studies on soil rhizosphere: speciation and availability of cadmium. Chem. Spec. Bioavailab. 8, 23–28.

Krishnamurti, G.S.R., Huang, P.M., Van Rees, K.C.J., Kozak, L.M., Rostad, H.P.W., 1995a. Speciation of particulate-bound cadmium of soils and its bioavailability. Analyst 120, 659–665.

Krishnamurti, G.S.R., Huang, P.M., Van Rees, K.C.J., Kozak, L.M., Rostad, H.P.W., 1995b. A new soil test method for the determination of plant available cadmium in soils. Commun. Soil Sci. Plant Anal. 26, 2857–2867.

Krishnamurti, G.S.R., Huang, P.M., Van Rees, K.C.J., Kozak, L.M., Rostad, H.P.W., 1997a. Differential FTIR study of pyrophosphate extractable material of soils: implication in cadmium-bonding sites and availability. In: Prost, R. (Ed.), Contaminated Soils (Proceedings of the 3rd International Conference Biogeochem. Trace Elements, May 15–19, 1995, Paris, France). INRA, Paris. pp. 1–10.

Krishnamurti, G.S.R., Huang, P.M., Van Rees, K.C.J., Kozak, L.M., Rostad, H.P.W., 1997b. Distribution of cadmium in selected soil profiles of Sakatchewan, Canada; speciation and availability. Can. J. Soil Sci. 77, 613–619.

Krishnamurti, G.S.R., Megharaj, M., Naidu, R., 2004. Bioavailability of cadmium-organic complexes to soil alga – an exception to the free ion model. J. Agri. Food. Chem. 52, 3894–3899.

Krishnamurti, G.S.R., Naidu, R., 2000. Speciation and phytoavailability of cadmium in selected surface soils of South Australia. Aust. J. Soil Res. 38, 991–1004.

Krishnamurti, G.S.R., Naidu, R., 2003. Solid-solution equilibria of cadmium in soils. Geoderma 113, 17–30.

Krishnamurti, G.S.R., Smith, L.H., Naidu, R., 2000. Method for measuring plant-available cadmium in soils. Aust. J. Soil Res. 38, 823–836.

Krotz, R.M., Evangelou, B.P., Wagner, G.J., 1989. Relationship between cadmium, zinc, cadmium-peptide, and organic acid in tobacco suspension cells. Plant Physiol. 91, 780–787.

Lagerwerff, J.V., Brower, D.L., 1974. Effect of a smelter on the agricultural conditions in the surrounding environment. In: Hemphill, D.D. (Ed.), Trace Substances in Environmental Health. No.8. University of Missouri, Colombia. pp. 203–212.
Lakanen, E., Ervio, R., 1971. A comparison of eight extractants for the determination of plant available micronutrients in soils. Acta Agral. Fenn. 123, 223–232.
Lake, D.L., Kirk, P.W.W., Lester, J.N., 1984. Fractionation, characterization, and speciation of heavy metals in sewage sludge and sludge-amended soils: a review. J. Environ. Qual. 13, 175–183.
Lamy, I., Cambier, P., Bourgeois, B., 1994. Lead and cadmium complexation with soluble organic carbon and speciation in alkaline soil leachates. Env. Geochem. Health 16, 1–16.
Lauwerys, R., 1979. Cadmium in man. In: Webb, M. (Ed.), The Chemistry, Biochemistry and Biology of Cadmium. Elsevier/North Holland Biomedical Press, Amsterdam, pp. 433–455.
Lavkulich, L.M., Wiens, J.H., 1970. Comparison of organic matter destruction by hydrogen peroxide and sodium hypochlorite and its effect on selected mineral constituents. Soil Sci. Soc. Am. Proc. 34, 755–758.
Lebourg, A., Sterckeman, T., Ciesielski, H., Proix, N., 1996. Interet de differents reactifs d'extraction chimique pour l'evaluation de la biodisponibilite des metaux en traces du sol. Agronomie 16, 201–215.
Legatt, D.J., 1977. Machine computation of equilibrium concentrations – some practical considerations. Talanta 24, 535–542.
LeRiche, H.H., Wier, A.H., 1963. A method for studying trace elements in soil fractions. J. Soil Sci. 14, 225–235.
Lindsay, W.L., 1979. Chemical Equilibria in Soils. Wiley, New York.
Linehan, D.J., Sinclair, A.H., Mitchell, M.C., 1989. Seasonal changes in Cu, Mn, Zn, and Co concentrations in soil in the root zone of barley (*Hordeum vulgure* L.). J. Soil Sci. 40, 103.
Little, P., Martin, M.H., 1972. A survey of zinc, lead and cadmium in soil and natural vegetation around smelting complex. Environ. Pollut. 3, 241–254.
Lumsdon, D.G., Evans, L.J., 1995. Predicting chemical speciation and computer simulation. In: Ure, A.M., Davidson, C.M. (Eds.), Chemical Speciation in the Environment. Blackie Academic & Professional, London, pp. 86–134.
Lund, L.J., Betty, E.E., Page, A.L., Elliott, R.A., 1981. Occurrence of naturally high cadmium levels in soils and its accumulation by vegetation. J. Environ. Qual. 10, 551–556.
Lund, W., 1990. Speciation analysis-why and how? Fresenius J. Anal. Chem. 337, 557.
Lynch, J.M., 1990. Introduction: some consequences of microbial rhizosphere competence for plant and soil. In: Lynch, J.M. (Ed.), The Rhizosphere. Wiley, Chichester, pp. 1–10.
Macklin, M.G., Klimek, K., 1992. Dispersal, storage and transformation of metal-contaminated alluvium in the upper Vistula basin, south-west Poland. Appl. Geogr. 12, 7–30.
MacLaren, R.G., Crawford, D.V., 1973. Studies on soil copper. I. The fractionation of Cu in soils. J. Soil Sci. 24, 172–181.
MacLaren, R.G., Lawson, D.M., Swift, R.S., 1986. The forms of cobalt in some Scottish soils as determined by extraction and isotope exchange. J. Soil Sci. 37, 223–234.
Maclean, A.J., 1976. Cadmium in different plant species and its availability in soils as influenced by organic matter and additions of lime, P, Cd and Zn. Can. J. Soil Sci. 56, 129–138.
Maher, W.A., 1984. Evaluation of a sequential extraction scheme to study associations of trace elements in estuarine and oceanic sediments. Bull. Environ. Contam. Toxicol. 32, 339–344.
Manceau, A., Charlet, M.C., Boisset, M.C., Didier, B., Spadini, L., 1992. Sorption and speciation of heavy metals on hydrous Fe and Mn oxides. From microscopic to macroscopic. Appl. Clay Sci. 7, 201–233.
Manceau, A.M., Harge, J.C., Sarret, G., Hazelamnn, J.L., Boisset, M.C., Mench, M., Cambier, P.H., Prost, R., 1995. Direct determination of heavy metal speciation in soils by EXAFS

spectroscopy. In: Prost, R. (Ed.), Contaminated Soils. Proceedings of Third International Conference, Biogeochemistry of Trace Elements (May 15–19, 1993, Paris), INRA, Paris, pp. 99–120.

Mann, S.S., Ritchie, G.S.P., 1993. The influence of pH on the forms of cadmium in four western Australian soils. Aust. J. Soil Res. 31, 255–270.

Marples, A.E., Thornton, I., 1980. The distribution of cadmium derived from geochemical and industrial sources in agricultural and pasture herbage in parts of Britain. In: Cadmium (Proceedings of 2nd International Cadmium Conference, Cannes, 1979). Metal Bulletin Ltd., Worcester, London, pp. 74–79.

Martyn, C.N., Barker, D.J.P., Osmond, C., Harris, E.C., Edwardson, J.A., Lacey, R.F., 1989. Geographical relation between Alzheimer's disease and Al in drinking water. Lancet 1, 59–62.

Masironi, R., Koirtyohann, S.R., Pierce, J.O., 1977. Zinc, copper, Cadmium and chromium in polished and unpolished rice. J. Sci. Tot. Environ. 7, 27–43.

Mattigod, S.V., Sposito, G., 1979. Chemical modelling of trace metal equilibria in contaminated soil solutions using the computer program GEOCHEM. In: Jenne, E.A. (Ed.), Chemical Modelling in Aqueous Systems. Speciation, Sorption, Solubility and Kinetics. ACS Symp. Ser. No. 93. ACS, Washington, DC, pp. 837–850.

McArthur, D.F.E., 2001. Impact of Long-term Cultivation on the Status of Cadmium in Chernozemic Soils. Ph.D. Thesis, The University of Saskatchewan, Saskatoon, SK Canada.

McArthur, D.F.E., Huang, P.M., Kozak, L.M., 2001. Impact of long-term cultivation on the status of organic matter and cadmium in soil. Can. J. Soil Sci. 81, 349–355.

McBride, M.B., 1989. Reactions controlling heavy metal solubility in soils. Adv. Soil Sci. 10, 1–56.

McBride, M.B., 1991. Processes of Heavy and Transition Metal Sorption by Soil Minerals. Kluwer Academic Publishers, Dordrecht.

McBride, M.B., Sauve, S., Hendershot, W., 1997. Solubility control of Cu, Zn, Cd, and Pb in contaminated soils. Eur. J. Soil Sci. 48, 337–346.

McDuff, R.E., Morel, F.M.M., 1973. Description and Use of the Chemical Equilibrium Program REDEQL2. Tech. Report. EQ-73-02. California Institute of Technology, Pasadena, CA.

McGrath, S.P., 1984. Metal concentration in sludges and soil from a long-term field trial. J. Agric. Sci. Camb. 103, 25–35.

McGrath, S.P., 1987. Long term studies of metal transfer following application of sewage sludge. In: Coughetry, P.J., Martin, M.H., Unsworth, M.H., (Eds.), Pollutant Transport and Fate in Ecosystems. Spl. Publ. No. 6. British Ecological Society, Backwell Scientific, Oxford, pp. 301–317.

McGrath, S.P., 1996. Application of single and sequential extraction procedures to polluted and unpolluted soils. Sci. Total Environ. 178, 38–44.

McGrath, S.P., 1999. Adverse effects of cadmium on soil microflora and fauna. In: McLaughlin, M.J., Singh, B.R., (Eds.), Cadmium in Soils and Plants. Kluwer Academic Publishers, Dordrecht, pp. 199–218.

McKeague, J.A., 1967. An evaluation of 0.1 M pyrophosphate and pyrophosphate-dithionite in comparison with oxalate as extractants of the accumulation products in Podzols and some other soils. Can. J. Soil Sci. 47, 95–102.

McKeague, J.A., Day, J.H., 1966. Dithionite and oxalate extractable Fe and Al as aids in differentiating various classes of soils. Can. J. Soil Sci. 46, 13–22.

McLaughlin, M.J., Tiller, K.G., Naidu, R., Stevens, D.P., 1996. Review: the behaviour and environmental impact of contaminants in fertilizers. Aust. J. Soil Res. 34, 1–54.

McLaughlin, M.J., Smolders, E., Merckx, R., 1998. Soil–root interface: physicochemical processes. In: Huang, P.M., Adriano, D.C., Logan, T.J., Checkai, R.T., (Eds.), Soil Chemistry and Ecosystem Health SSSA Special Publication 52. Soil Science Society of America, Madison, WI, pp. 233–277.

McLaughlin, M.J., Singh, B.R., 1999. Cadmium in Soils and Plants. Kluwer Academic Publishers, Dordrecht, Holland.

Mehra, O.P., Jackson, M.L., 1960. Iron oxide removal from soils and clays by a dithionite-citrate system buffered with sodium bicarbonate. Clay. Clay Miner. 7, 317–327.

Mench, M., Martin, E., 1991. Mobilization of cadmium and other metals from two soils by root exudates of *Zea mays* L., *Nicotiana tabacum* L., and *Nicoiana rustica* L. Plant Soil 132, 187–196.

Miller, W.P., McFee, W.W., 1983. Distribution of Cd, Zn, Cu and Pd in soils of industrial north western Indiana. J. Environ. Qual. 12, 29–33.

Miller, W.P., Martens, D.C., Zelazny, L.W., 1986a. Effect of sequence in extraction of trace metals from soils. Soil Sci. Soc. Am. J. 50, 558–560.

Miller, W.P., Martens, D.C., Zelazny, L.W., Kornegay, E.T., 1986b. Forms of solid phase copper in copper-enriched swine manure. J. Environ. Qual. 15, 69–72.

Mitchell, G.A., Bingham, F.T., Page, A.L., 1978. Yield and metal composition of lettuce and wheat grown on soils amended with sewage sludge enriched with cadmium, copper, nickel and zinc. J. Environ. Qual. 7, 165–171.

Mitchell, R.L., 1964. Trace elements in soils. In: Bear, F.E. (Ed.), Chemistry of the Soil. Reinhold, New York, pp. 320–368.

Morel, F., Morgan, J.J., 1972. A numerical method for computing equilibria in aqueous chemical systems. Environ. Sci. Technol. 6, 58–67.

Morel, J.L., Mench, M., Guckert, A., 1986. Measurement of Pb^{2+}, Cu^{2+}, and Cd^{2+} binding with mucilage exudates from maize (*Zea mays* L.) roots. Biol. Fertil. Soils 2, 29–34.

Morgan, J.J., 1987. General affinity concepts, equilibria and kinetics in aqueous metal chemistry. In: Patterson, J.W., Passino, R. (Eds.), Metal Speciation, Separation and Recovery. Lewis Publishers, Michigan, pp. 27–61.

Morrison, G.M.P., Batley, G.E., Florence, T.M., 1989. Metal speciation and toxicity. Chem. Brit. 1989, 791–796.

Mortvedt, J.J., 1987. Cadmium levels in soils and plants from some long-term soil fertility experiments in the United States of America. J. Environ. Qual. 16, 137–142.

Mulla, D.J., Page, A.L., Ganje, T.J., 1980. Cadmium accumulations and bioavailability in soils from long-term phosphorus fertilization. J. Environ. Qual. 9, 408–412.

Naidu, R., Harter, R.D., 1998. Effect of different organic ligands on cadmium sorption by and extractability from soils. Soil Sci. Soc. Am. J. 62, 644–650.

Nielsen, D., Hoyt, P.B., MacKenzie, A.F., 1986. Distribution of soil Zn fractions in British Colombia interior orchard soils. Can. J. Soil Sci. 66, 445–454.

NIST, 1997. Standard Reference Database 46. Critically Selected Stability Constants of Metal Complexes, Version 5.0, National Institute of Standards and Technology, US Department of Commerce, Gaitherburg, MD.

Norwell, W.A., 1984. Comparison of chelating agents as extractants for metals in diverse soil materials. Soil Sci. Soc. Am. J. 48, 1285–1292.

Nriagu, J.O., 1980. Human influence on the global cadmium cycle. In: Nriagu, J.O. (Ed.), Cadmium in the Environment. Part 1. Ecological Cycling. Wiley-Interscience, New York, pp. 2–12.

O'Connor, G.A., 1988. Use and misuse of the DTPA soil test. J. Environ. Qual. 17, 715–718.

Onyatta, J.O., Huang, P.M., 1999. Cadmium speciation and bioavailability index for selected Kenyan soils. Geoderma 91, 87–101.

Orsini, L., Bermond, A.P., 1994. Copper biodisponibility in calcareous soil samples. Part 1. Chemical fractionation of copper. Environ. Technol. 15, 695–700.

Page, A.L., Bingham, F.T., 1973. Cadmium residues in the environment. Residue Rev. 48, 1–43.

Page, A.L., Chang, A.C., Mohamed El-Amamy, 1987. Cadmium levels in soils and crops in the United States. In: Hutchinson, T.C., Meema, K.M., (Eds.), Lead, Mercury, Cadmium and Arsenic in the Environment. SCOPE 31. Wiley, Chichester, pp. 119–146.

Papelis, C., Hayes, K.F., Leckie, J.O., 1988. HYDRAQL: a program for the computation of chemical equilibrium composition of aqueous batch systems including surface complexation modelling of ion adsorption at the oxide/solution interface. Technical Report 306. Department of Civil Engineering, Stanford University, CA.

Parker, D.R., 1991. GEOCHEM-PC. Version 2.0. A Computer Program for the Calculation of Chemical Equilibria in Soil Solutions and other Natural Water Systems. Kearney Foundation of Soil Science, University of California, Riverside, CA.

Parker, D.R., Chaney, R.L., Norvell, W.A., 1995. Chemical equilibrium models: applications to plant nutrition. In: Loeppert, R.H., Schwab, A.P., Goldberg, S. (Eds.), Soil Chemical Equilibrium and Reaction Models, SSSA Special Publication 42. Am. Soc. Agron. and Soil Sci. Soc. Am., Madison, WI, p. 163–200.

Parker, D.R., Zelazny, L.W., Kinraide, T.B., 1987. Improvements to the program GEOCHEM. Soil Sci. Soc. Am. J. 51, 488–491.

Peterson, P.J., Alloway, B.J., 1979. Cadmium in soils and vegetation. In: Webb, M. (Ed.), The Chemistry, Biochemistry and Biology of Cadmium. Elsevier/North-Holland Biomedical Press, Amsterdam, pp. 46–92.

Pickering, W.F., 1981. Selective chemical extraction of soil components and bound metal species. CRC Crit. Rev. Anal. Chem. 12, 233–266.

Pickering, W.F., 1986. Metal ion speciation – Soils and sediments (a review). Ore Geol. Rev. 1, 83–146.

Pike, E.R., Graham, L.C., Foyden, M.W., 1975. An appraisal of toxic metal residue in the soils of a disused sewage farm II. J. Assoc. Public Anal. 13, 48–63.

Plant, J.A., Baldock, J.W., Smith, B., 1996. The role of geochemistry in environmental and epedemiological studies in developing countries: a review. In: Appleton, J.D., Fuge, R., McCall, G.J.H. (Eds.), Environmental Geochemistry and Health. Geological Society Spl. Pub. No. 113. The Geological Society, London, pp. 2–22.

Potash & Phosphate Institute/Potash and Phosphate Institute of Canada/Foundation for Agronomic Research, 1998. Heavy Metals in Soils and Phosphatic Fertilizers. PPI/PPIC/FAR Technical Bulletin 1998–2, Norcoss, Georgia.

Prasad, A.S., 1978. Toxic elements. In: Hemphill, D.D. (Ed.), Trace Elements and Iron in Human Metabolism. Plenum Press, New York, pp. 349–362.

Rappaport, B.D., Martens, D.C., Reneau, R.B., Simpson, T.W., 1988. Metal availability in sludge amended soils with elevated metal levels. J. Environ. Qual. 17, 42–47.

Rauser, W.E., 1990. Phytochelatins. Ann. Rev. Biochem. 59, 61–86.

Reuss, J.O., Dooley, H.L., Griffis, W., 1978. Uptake of cadmium from phosphate fertilizers by peas, radishes and lettuce. J. Environ. Qual. 7, 128–133.

Ritchie, G.S.P., Sposito, G., 1995. Speciation in soils. In: Ure, A.M., Davidson, C.M. (Eds.), Chemical Speciation in the Environment. Blackie Academic & Professional, London, pp. 201–233.

Robbins, L.M., Lyle, M., Heath, G.R., 1984. A sequential extraction procedure for partitioning elements among coexisting phases in marine sediments. College of Oceanography Report, 84–3. Oregon State University Corvallis, Ore., p. 55.

Roberts, A.H.C., Longhurst, R.D., Brown, M.W. 1994. Cadmium status of soils, plants and grazing animals in New Zealand. NZ J. Agri. Res. 37, 119–129.

Rose, A.W., Hawkes, H.E., Webb, J.S., 1979. Geochemistry in Mineral Exploration, second ed. Academic Press, New York.

Ross, S.M., Ed., 1994. Toxic Metals in Soil–Plant Systems. Wiley, New York.

Rothbaum, H.P., Gogirel, R.L., Johnstone, A.E., Mattingley, G.E.G., 1986. Cadmium accumulation in soils from long continued applications of super phosphate. J. Soil Sci. 37, 99–107.

Rule, J.H., Alden, R.W., 1992. Partitioning of cadmium in geochemical fractions of anaerobic estuarine sediments. Estuar. Coast. Shelf Sci. 34, 487–499.

Salomons, W., Foerstner, U., 1980. Trace metal analysis of polluted sediments. II. Evaluation of environmental impact. Environ. Technol. Lett. 1, 506–517.

Salomons, W., Foerstner, U., 1984. Metals in the Hydrocycle. Springer, New York.

Salt, D.E., Thurman, D.A., Tomsett, A.B., Sewell, A.K., 1989. Copper phytochelatins of mimulus guttatus. Proc. Roy. Soc. B 236, 79–89.

Sauerbeck, D.R., 1991. Plant, element and soil properties governing uptake and availability of heavy metals derived from sewage sludge. Water Air Soil Pollut. 57–58, 227–237.

Sauerbeck, D.R., Rietz, E., 1983. Soil chemical evaluation of different extractants for heavy metals in soils. Comm. Europe Communities Report EUR 8022. Environ. Eff, Org. Inorg. Contam. Sewage Sludge (CA 99: 193726), pp. 147–160.

Sauerbeck, D.R., Styperek, P., 1984. Predicting the cadmium availability from different soils by $CaCl_2$ extraction. In: L'Hermite, P., Ott, H. (Eds.), Processing and Use of Sewage Sludge. Reidel, Dordrecht, The Netherlands, pp. 431–435.

Sauve, S., Norvell, W.A., McBride, M.B., Hendershot, W., 2000. Speciation and complexation of cadmium in extracted soil solutions. Environ. Sci. Technol. 34, 291–296.

Schalscha, E.B., Morales, M., Ahumada, I., Schirado, T., Pratt, P.F., 1980. Fractionation of Zn, Cu, Cr and Ni in waste water solids and in soil. Agrochimica 24, 361–368.

Schalscha, E.G., Marlaes, M., Vergara, I., Chang, A.C., 1982. Chemical fractionation of heavy metals in waste-water effected soils. J. Water Pollut. Control Fed. 54, 175–180.

Scheller, H.V., Huang, B., Hatch, E., Goldsbrough, P.B., 1987. Phytochelatin synthesis and glutathione levels in response to heavy metals in tomato cells. Plant Physiol. 85, 1031–1414.

Schindler, P.W., Furst, R.D., Wolf, P.U., 1976. Ligand properties of surface silanol groups. 1. Surface complex formation with Fe^{3+}, Cu^{2+}, Cd^{2+} and Pb^{2+}. J. Colloid Interface Sci. 55, 469–475.

Schnitzer, M., Hansen, E.H., 1970. Organo-metallic interactions in soils: 8. An evaluation of methods for the determination of stability constants of metalfulvic acid complexes. Soil Sci. 109, 333–340.

Schoer, J., Eggersgluess, D., 1982. Chemical forms of heavy metals in sediments and suspended matter of Weser, Elbe and Ems rivers. Mitt. Geol. Paleontol. Inst., Univ. Hamburg 52, 667–685.

Schroeder, H.A. 1965a. The biological trace elements or peripatetics through the periodic table. J. Chronic Dis. 18, 217–228.

Schroeder, H.A., 1965b. Cadmium as a factor in hypertension. J. Chronic Dis. 18, 647–656.

Schulze, D.G., Bertsch, P.M., 1995. Synchrotron X-ray techniques in soil, plant and environmental research. Adv. Agron. 55, 1–66.

Schwertmann, U., 1964. Differenzierung der Eisenoxide des Bodens durch Extraktion mit Ammoniumoxalat-losung. Z. Pflanzenernahr. Dung Bodenk. 105, 194–202.

Selinas, O., Frank, A., Galgan, V., 1996. Biogeochemistry and metal biology. In: Appleton, J.D., Fuge, R., McCall, G.J.H. (Eds.), Environmental Geochemistry and Health. Geological Society Spl. Pub. No. 113. The Geological Society, London. pp. 81–89.

Sharma, R.P., Shupe, J.L., 1977. Lead, cadmium and arsenic residues in animal tissues in relation to their surrounding habitat. Sci. Total Environ. 7, 53–62.

Sheppard, M.I., Stephenson, M., 1997. Critical evaluation of selective extraction methods for soils and sediments. In: Prost, R. (Ed.), Contaminated Soils. Proceedings of the 3rd International Conference Biogeochem. Trace Elements. Paris, May 15–19, 1995. INRA, Paris, France, pp. 69–97.

Shuman, L.M., 1979. Zinc, manganese and copper in soil fractions. Soil Sci. 127, 10–17.

Shuman, L.M., 1982. Separating soil iron- and manganese oxide fractions for microelement analysis. Soil Sci. Soc. Am. J. 46, 1099–1102.

Shuman, L.M., 1983. Sodium hypochlorite methods for extracting microelements associated with soil organic matter. Soil Sci. Soc. Am. J. 47, 656–660.

Shuman, L.M., 1985. Fractionation method for soil microelements. Soil Sci. 140: 11–22.

Sillanpaa, M., Jansson, H., 1992. Status of Cadmium, Lead, Cobalt and Selenium in Soils and Plants of Thirty Countries. FAO Soils Bull. 65. FAO, Rome, Italy.

Sillen, L.G., Martell, A.E., 1971. Stability Constants of Metal–ion Complexes. Chem. Soc., London.

Silviera, D.J., Sommers, L.E., 1977. Extractability of Cu, Zn, Cd and Pb in soils incubated with sewage sludge. J. Environ. Qual. 6, 47–52.

Singh, B.R., 1991. Unwanted components of commercial fertilizers and their agricultural impact. Proc. Fertilizer Soc., London, Dec. 1991, pp. 2–28.

Singh, B.R., Narwal, R.P., Jeng, A.S., Alamas, A., 1995. Crop uptake and extractability of cadmium in soils naturally high in metals at different pH levels. Commun. Soil Sci. Plant Anal. 26, 2123–2142.

Smilde, K.W., van Luit, B., 1983. The effect of phosphate fertilizer cadmium on cadmium in soils and crops. Inst. Bodemvruchtbaarheid, Rapp. No. 6–83.

Smith, R.M., Martell, A.E., 1976. Critical Stability Constants, vol. 4. Inorganic Complexes. Plenum Press, New York.

Smith, R.M., Martell, A.E., 1977. Critical Stability Constants, vol. 3. Other Organic Ligands. Plenum Press, New York.

Snedecor, G.W., Cochran, W.G., 1980. Statistical Methods. The Iowa State University Press, Ames, IA.

Soltanpour, P.N., Schwab, A.P., 1977. A new soil test for the simultaneous extraction of macro- and micro-nutrients in alkaline soils. Commun. Soil Sci. Plant Anal. 8, 195–207.

Sommers, L., Van Volk, V., Giardono, P., Sopper, W.E., Bastian, R., 1987. Effects of soil properties on accumulation of trace elements by crops. In: Page, A.L., Logan, T.J.,Ryan, J.A., (Eds.), Land Application of Sludge. Food Chain Implications. Lewis Publishers, Chelsea, MI, pp. 5–24.

Sparks, D.L., 2000. Soil chemistry: past achievements and new frontiers. In: Kheoruenromme, I., Theerawong, S. (Eds.), Soil Science: Accomplishments and Changing Paradigm Towards the 21st Century. Proceedings of International Symposium on Soil Science, Bangkok, Thailand, April 17–18, 2000, pp. 15–28.

Sposito, G., 1981. The Thermodynamics of Soil Solutions. Clarendon Press, Oxford.

Sposito, G., 1983. The chemical form of trace metals in soils. In: Thornton, I., (Ed.), Applied Environmental Geochemistry. Academic Press, Geology Series, London, pp. 123–170.

Sposito, G., 1986. Distribution of potentially hazardous trace metals. In: Siegel, H. (Ed.), Metal Ions in Biological Systems, vol. 20. Marcel Dekker, New York.

Sposito, G., Bingham, F.T., 1981. Computer modeling of trace metal speciation in soil solutions: correlation, with trace metal uptake by higher plants. J. Plant Nutr. 3, 81–108.

Sposito, G., Coves, J., 1988. SOILCHEM: A Computer Program for the Calculation of Chemical Speciation in Soils. Kearney Foundation of Soil Science, University of California, Berkeley, CA.

Sposito, G., Lund, L.J., Chang, A.C., 1982. Trace metal chemistry in arid-zone field soils amended with sewage sludge. I. Fractionation of Ni, Cu, Zn, Cd and Pb in solid phases. Soil Sci. Soc. Am. J. 46, 26–264.

Sposito, G., Mattigod, S.V., 1980. GEOCHEM: A Computer Program for the Calculation of Chemical Equilibria in Soil Solutions and other Natural Water Systems. Kearney Foundation of Soil Science, University of California, Berkeley, CA.

Sposito, G., Page, A.L., Frink, M.E., 1980. Effects of Acid Precipitation on Soil Leachate Quality – Computer Calculations. Environ. Res. Lab., US EPA, Corvallis, ORE.

Sposito, G., Page, A.L., Mattigod, S.V., 1979. Trace Metal Speciation in Saline Waters Affected by Geothermal Brines. Kearney Foundation of Soil Science, University of California, Berkeley, CA.

Sposito, G., Reginato, R.J., 1992. Opportunities in Basic Soil Science Research. Soil Sci. Soc. Am., Madison, WI.

Steffens, J.C., 1990. The heavy metal binding peptides in plants. Ann. Rev. Plant Physiol. Plant Mol. Biol. 41, 553–575.

Stenstrom, T., Vahter, M., 1974. Cadmium and lead in Swedish commercial fertilizers. Ambio 3, 91–92.

Stevenson, F.J., Fitch, A., 1986. Chemistry of complexation of metal ions with soil solution organica. In: Huang, P.M., Schnitzer, M. (Eds.), Interaction of Soil Minerals with Natural Organics and Microbes. SSSA Spec. Pub., vol. 17. SSSA, Madison, WI, pp. 29–58.

Stout, M.G. Rawson, R.W., 1981. Minerals, trace elements and cancer. Proc. Cancer Res. Ther. 17, 243–272.

Stover, R.C., Sommers, L.E., Silviera, D.J., 1976. Evaluations of metals in waste water sludge. J. Water Poll. Control Fed. 48, 2165–2175.

Street, J.J., Lindsay, W.L., Sabey, B.R., 1977. Solubility and plant uptake of cadmium in soils amended with cadmium and sewage sludge. Environ, J., Qual. 6, 72–77.

Stumm, W., 1992. Chemistry of the solid-water interface. Processes at the Mineral-Water and Particle-Water Interface in Natural Systems. Wiley, New York.

Stumm, W., Hohl, H., Dalang, F., 1976. Interaction of metal ions with hydrous oxide surfaces. Croatica Chimica Acta 48, 491–499.

Stumm, W., Huang,C.P., Jenkins, S.R., 1970. Specific chemical interactions affecting the stability of dispersed systems. Croatica Chimica Acta 42, 223–231.

Stumm, W., Kummertt, R., Sigg, L., 1980. A ligand exchange model for adsorption of inorganic and organic ligands at hydrous oxide interfaces. Croatica Chimica Acta 53, 291–312.

Syers, J.K., Mackay, A.D., Brown, M.W., Currie, L.D., 1986. Chemical and physical characteristics of phosphate rock materials of varying reactivity. J. Sci. Food Agri. 37, 1057–1064.

Symeonides, C., McRae, S.G., 1977. The assessment of plant available cadmium in soils. J. Environ. Qual. 6, 120–123.

Szmigielska, A.M., Van Rees, K.C.J., Cieslinski, G., Huang, P.M., 1997. Comparison of liquid and gas chromatography for analysis of low molecular weight organic acids in rhizosphere soil. Comm. in Soil Sci. and Plant Analysis 28, 99–111.

Szmigielska, A.M., Van Rees, K.C.J., Cieslinski, G., Huang, P.M., Knott, D.R., 1995. Determination of low molecular weight dicarboxylic acids in root exudates by gas chromatography. J. Agri. Food Chem. 43, 956–959.

Temminghoff, E.J.M., Van Der Zee, S.E.A.T.M., De Haan, F.A.M., 1995. Speciation and calcium competition effects on cadmium sorption by sandy soil at various pHs. Eur. J. Soil Sci. 46, 649–655.

Tessier, A., Campbell, P.G.C., 1988. Partitioning of trace metals in sediments. In. Kramer, J.R., Allen, H.E. (Eds.), Metal Speciation: Theory, Analysis and Application. Lewis Publishers, Chelsea, MI, pp. 545–569.

Tessier, A., Campbell, P.G.C., Bissom, M., 1979. Sequential extraction procedure for the speciation of particulate trace metals. Anal. Chem. 51, 844–850.

Tessier, A., Turner, D.R., 1995. Metal Speciation and Bioavailability in Aquatic System. IUPAC series on Analytical and Physical Chemistry of Environmental Systems, vol. 3. Wiley, Chichester.

Thatcher, R.W., Lester, M.L., McAlaster, R., Horst, R., 1982. Effects of low levels of cadmium and lead on cognitive functioning in children. Arch. Environ. Health. 37, 159–166.

Thornton, I. 1992. Sources and pathways of cadmium in the environment. In: Nordberg, G.F., Alessio, L., Herber, R.F.M., (Eds.), Cadmium in the Human Environment. International Agency for Research on Cancer (IARC), Lyon.

Tiller, K.G., 1983. Micronutrients. In: Soils: an Australian Viewpoint. CSIRO/Academic Press, New York, pp. 365–387.

Tiller, K.G., Honeysett, J.L., de Bries, M.P.C., 1972. Soil zinc and its uptake by plants. II. Soil chemistry in relation to prediction of availability. Austr. J. Soil Res. 10, 165–182.

Tiller, K.G., McLaughlin, M.J., Roberts, A.H.C., 2000. Environmental impacts of heavy metals in agroecosystems and amelioration strategies in Oceania. In: Huang, P.M., Iskander, I.K. (Eds.) Soils and Groundwater Pollution and Remediation. CRC Press, Boca Raton, FL, pp. 1–41.

Tourtelot, H.A., Huffman, C., Rader, L.F., 1964. Cadmium in samples of the Pierre shale and some equivalent stratigrahic units, Great Plains region. US Geol. Survey Prof. Paper 475-D, 73–78.

Turner, D.R., 1995.Problems in trace metal speciation modeling. In: Tessier, A., Turner, D.R., (Eds.), Metal Speciation and Bioavailability in Aquatic Systems. Wiley, New York, pp. 149–203.

Tyler, L.D., McBride, M.B., 1982. Influence of Ca, pH and humic acid on cadmium uptake. Plant Soil 64, 259–262.

Ure, A.M., 1996. Single extraction schemes for soil analysis and related applications. Sci. Total Environ. 178, 3–10.

Van Driel, W., Smilde, K.W., van Luit, B., 1987. Comparison of the Heavy Metal Uptake of Cyperus esulentus and of Agronomic Plants Grown on Contaminated Dutch Sediments. Miscellaneous Paper D-83-1, Institute for Soil Fertility, US Army Engineer Waterways Experiment Station, Vicksburg, MS.

Van Riemsdijk, W.H., Bolt, G.H., Koopal, L.K., Blackmeer, J., 1986. Electrolyte adsorption on heteregeneous surfaces: adsorption models. J. Colloid Interface Sci. 109, 219–228.

Van Riemsdijk, W.H., De Witt, J.C.M., Koopal, L.K., Bolt, G.H., 1987. Metal ion adsorption on heteregeneous surfaces: adsorption models. J. Colloid Interface Sci. 116, 511–522.

Verner, J., Ramsey, M., Helios Rybicka, E., Jedrzejczyk, B., 1994. Heavy metal contamination of soils in Bukowno, Poland. Abstracts, 3rd Intern. Symp. Environ. Geochem. 12–15 Sep. 1994. Cracow, Poland, pp. 426–427.

Vogeli-Lange, R., Wagner, G.J., 1990. Subcellular localization of cadmium and cadmium-binding peptides in tobacco leaves. Plant Physiol. 92, 1086–1093.

VSBo (Verordnung uber Schadstoffgehalt im Boden),1986. Swiss Ordinance on Pollutants in Soils. Nr 814.12, Publ. Eidg. Drucksachen und Materialzentrale, Bern, Switzerland.

Waketa, H., Schmitt, R.A., 1970. Cadmium. In: K.H. Wadepohl (Ed.), Handbook of Geochemistry, vol. 11–2. Verlag, Berlin, p. 48 B.

Waldbodt, G.L., 1978. Systemic Poisons. In: Health Effects of Environmental Pollutants. second ed. C.V. Mosby Co., St. Louis, MO, p. 160–178.

Wang, M.K., Chiu, C.Y., Chen, C.T., Chen, B.T., 2003. Effects of soil heavy metal contaminants on Millet (*Setaria italica* L. Bean) root exudates of low molecular weight organic acids in rhizosphere soils. Proc. Seventh Intern. Conf. Biogeochem. Trace Elements. (Uppsala, Sweden). Extended Abstracts, vol. 2. pp. 196–197.

Weast, C. (Ed.). 1979. Handbook of Chemistry and Physics. 60th ed., CRC Press, Boca Raton, FL.

Webb, M. (Ed.). 1979. The Chemistry, Biochemistry and Biology of Cadmium. Elsevier/North Holland Biomedical Press, Amsterdam.

Westall, J.C., Zachary, J.L., Morel, F.M.M., 1976. MINEQL: A Computer Program for the Calculation of Chemical Equilibrium Composition of Aqueous Systems. Tech. Note 18, Ralph M. Parsons Laboratory, Massachussets Institute of Technology, Cambridge, MA.

Whitten, M.G., Ritchie, G.S.P., 1991. Calcium chloride extractable cadmium as an estimate of cadmium uptake by subterraean clover. Austr. J. Soil Res. 29, 215–221.

Williams, C.H., David, D.J., 1973. The effect of superphosphate on the cadmium content of soils and plants. Austr. J. Soil Res. 11, 43–56.

Williams, C.H., David, D.J., 1976. The accumulation in soil of cadmium residues from phosphate fertilizers and their effect on the cadmium content of plants. Soil Sci. 121, 86–93.

Woolhouse, H.W., 1983. Toxicity and tolerance in the response of plants to metals. In: Lange, L., Nobel, P.S., Osmond, C.B., Ziegler, H., (Eds.), Encyclopedia of Plant Physiology. (New Series), vol. 12c, Verlag, Berlin, pp. 245–300.

Xian, X., Shokohifard, G., 1989. Effect of pH on chemical forms and plant availability of cadmium, zinc and lead in polluted soils. Water Air Soil Pollut. 45, 265–273.

Zachara, J.M., Smith, S.C., Resch, C.T., Cowan, C.E., 1992. Cadmium sorption to soil separates containing layer silicates and iron and aluminum oxides. Soil Sci. Soc. Am. J. 56, 1074–1084.

Part II: SPECIATION, BIOAVAILABILITY, AND PHYTOTOXICITY OF TRACE ELEMENTS

Biogeochemistry of Trace Elements in the Rhizosphere
P.M. Huang and G.R. Gobran (Editors)

Chapter 8

Speciation and bioavailability of trace metals (Cd, Cu, Ni, Pb, Zn) in the rhizosphere of contaminated soils

P. Legrand[a], M.-C. Turmel[a], S. Sauvé[b], and F. Courchesne[a]

[a]Département de géographie, Université de Montréal, C.P. 6128, succursale Centre-Ville, Montréal, Québec, Canada H3C 3J7
E-mail: francois.courchesne@umontreal.ca

[b]Département de Chimie, Université de Montréal, C.P. 6128, succursale Centre-Ville, Montréal, Québec, Canada H3C 3J7

ABSTRACT

The unique characteristics of the rhizosphere, combined with the relevance of its function in the soil:plant system, greatly contribut to the marked interest in this microenvironment. Specific changes occur in the rhizosphere as a consequence of root activity, which in turn impact on the speciation and bioavailability of nutrients and trace metals. In order to better understand the processes involved in the fractionation of trace metals in the rhizosphere, the objectives of this study are (i) to contrast the solid phase fractionation of trace metals (Cd, Cu, Ni, Pb and Zn) between the inner rhizosphere, the outer rhizosphere and the bulk components of forest mineral soils along a soil contamination gradient and, (ii) to determine Cu^{2+} activity in the liquid phase and establish the relationships with pH, dissolved organic carbon (DOC) and total dissolved Cu. Three sites located at distances of 2.5, 15 and 43 km from a Cu–Ni smelter in the Sudbury area of Ontario, Canada, were selected on the basis of their level of soil contamination. The fine roots from a number of white birch trees (*Betula papyrifera* Marsh.) were sampled in the B horizon along with the bulk soil (n = 6, 4 and 3 at sites 1, 2 and 3, respectively). The partition of the rhizosphere material from the roots was performed in the laboratory where the rhizosphere was separated into two components: the outer and the inner rhizosphere. The five trace metals were extracted using H_2O, 0.1 M $BaCl_2$, 0.1 M $Na_4P_2O_7$ and an HNO_3–HCl digestion. The Cu^{2+} activity in the water extract was measured with an ion-selective electrode (Cu-ISE). The pH and DOC concentrations were also measured on all samples. Results indicate that the potentially available pools of Cd, Cu and Ni, measured in the H_2O and $BaCl_2$ extracts, reflect the level of soil contamination and follow the gradient site 1 > site 2 > site 3. The extent of the difference between the bulk soil, the outer and the inner rhizosphere with respect to metal concentrations varies according to the trace metal, the metal fraction and the sampling site. In most cases, either the outer or the inner rhizosphere contains significantly higher metal concentrations than the bulk soil. In addition, the rhizosphere enrichment ratio (inner rhizosphere/bulk soil metal concentrations) is greater for the water-soluble and $BaCl_2$-exchangeable fractions. This potentially available metal pool represents a small fraction of the total recoverable (HNO_3–HCl digestion) metal content both in the bulk soil (3%) and the inner rhizosphere (7%), and is systematically higher in the rhizosphere. The Cu^{2+} activities in the water extract decrease with distance from the smelter, but are very similar between the three soil components for a given site. The inner rhizosphere contains the highest DOC concentrations and the most acidic pH values. The water-soluble Cu/DOC molar ratios are similar for the bulk soil, the outer and the inner rhizosphere of any given site, suggesting that the complexation power of organic matter is of a relatively similar magnitude in the three soil components. The pCu^{2+} activities are best predicted using DOC in the bulk soil (r^2=0.707, p = 0.0003), with pH and water-soluble Cu in the outer rhizosphere (r^2 = 0.731, p = 0.001), and with pH and DOC in the inner rhizosphere (r^2 = 0.903, p = 0.00001).

1. INTRODUCTION

The rhizosphere has been recognized as a distinct microenvironment in which the properties and the intensity of soil processes differ from those of the bulk soil. Indeed, unique physical, chemical and biological characteristics emerge in the rhizosphere due to its proximity to roots and to the sites of nutrient uptake by plants (Marschner, 1995). The most widely reported changes occurring in the rhizosphere as compared with the bulk soil include acidification, organic matter (OM) enrichment in both the solid and liquid phase, intensified mineral weathering, as well as a higher microbial biomass and activity (Gobran *et al.*, 1998). Interestingly, the pH and the OM content are two of the most influential variables in the determination of metal speciation (Lindsay, 1979; McBride, 1994), which is defined as the distribution of an element into its various chemical forms, such as free ions, soluble and exchangeable forms, organic complexes, inorganic fractions adsorbed to or co-precipitated with neoformed solid phases, as well as species included in primary mineral structures. Herein, the subdivision of an element into its various chemical forms will be referred to, on the one hand, as fractionation, for all the analyses in the solid phase and, on the other hand, as speciation, for the analyses in the liquid phase.

The specific physicochemical and biological characteristics of the rhizosphere, such as the pH, the OM content and the exudation of organic substances by roots, mycorrhizae and microorganisms, all contribute to the establishment of distinct trace metal species and metal pools covering a range of bioavailability levels (McLaughlin *et al.*, 1998; Assadian and Fenn, 2001). Metal bioavailability varies with the chemical form and can be defined as the capacity of a metal to be transferred from a soil compartment to living organisms, such as plants (Kabata-Pendias and Pendias, 1992). Because the total soil metal content does not necessarily reflect the bioavailability of metals and their potential phytotoxicity to plants and soil organisms, the fractionation of trace metals has to be established in order to understand the bioavailability of metals (Vaughan *et al.*, 1993; McBride, 1994). The dissolved free metal ion in the soil solution is generally considered to be the most readily available, and therefore, the most toxic chemical form (Wang and Benoit, 1996; Weng *et al.*, 2001). Yet, total soluble and exchangeable metal forms in soils as well as some organo-metallic complexes are also considered to be mobile and potentially bioavailable chemical forms because of their transfer into plant roots on a time scale corresponding to the rate of plant assimilation (Løbersli *et al.*, 1991; Kabata-Pendias and Pendias, 1992). The determination of trace metal bioavailability is also strongly related to the specific characteristics of the plant studied and to the various soil conditions under which it grows (Kabata-Pendias and Pendias, 1992).

Many studies have shown that the free metal fraction in the soil solution correlates best with metal concentrations found in plant roots and shoots. For example, Sauvé *et al.* (1996) showed that Cu concentrations in plant tissues of radish (*Raphanus sativus* cv. Cherry Belle), lettuce (*Lactuca sativa* cv. Buttercrunch) and ryegrass (*Lolium perenne* cv. Barmultra) correlated best with Cu^{2+} activity in the solution, which was considered a better indicator of bioavailability than total soil or total dissolved Cu concentrations. Minnich *et al.* (1987) also found that Cu accumulation in roots and shoots of young snapbeans (*Phaseolus vulgaris* L.) correlated well with Cu^{2+} activity in soils, and also with DTPA-extractable Cu. In addition, Shenker *et al.* (2001) showed that uptake of Cd by wheat (*Triticum aestivum* L.) and barley (*Hordeum vulgare* L.) was most influenced by Cd^{2+} activity in solution, while Knight *et al.* (1997) found a strong correlation ($r^2 = 0.80$; $\alpha = 0.05$) between the shoot Zn concentrations of *Thlaspi caerulescens* (J&C Presl.) and Zn^{2+} activity in solution. The importance of the uptake by plants of metals present as metal–organic complexes must also be stressed, as several studies have shown evidence of the assimilation of this metal fraction by higher plants. For instance, Laurie *et al.* (1991) showed that the chelating agent EDTA increased the Cu, Zn and Mn concentrations in plant tissues of barley (*H. vulgare* L.). Cieśliński *et al.* (1998) showed phytoaccumulation of Cd complexed with low-molecular-weight organic acids (LMWOAs) in durum wheat (*T. turgidum* var. *durum*), while Huang *et al.* (1997) found increased Pb concentrations in shoots of corn (*Zea mays* L. cv. Fiesta) and pea (*Pisum sativum* L. cv. Sparkle) with the addition of organic chelates in Pb-contaminated soils. Actually, the chemical forms in which preferential uptake of micronutrients by plants occurs vary with plant type and are still a matter of debate (Laurie and Manthey, 1994).

There is currently no consensus on whether rhizospheric processes augment or impoverish the bioavailable metal pool. The conceptual model developed by Gobran and Clegg (1996) to assess nutrient availability in the mineral soil:root system suggests that roots and their associated organisms maintain a higher level of macro- and micronutrient availability in the rhizosphere as compared with the bulk soil of forested mineral soils mainly through the release, transport and accumulation of reactive OM. Accordingly, the interaction of soil, roots and microorganisms creates a distinct soil environment characterized by enhanced moisture content, mineralization rates and exchange sites, all of which contribute to increase the size of the bioavailable nutrient pool in the rhizosphere. Other authors have also shown that the bioavailable trace metal fraction is usually higher in the rhizosphere than in the bulk soil (Marschner and Römheld, 1996; Hinsinger, 1998; Marschner, 1998). For instance, Courchesne *et al.* (2001) found a higher content of $BaCl_2$-exchangeable metals (Al, Fe, Mn, Cu and Zn) in the rhizosphere of balsam fir (*Abies balsamea* L.) and black spruce (*Picea mariana* Mill.) than in the bulk soil, suggesting the existence of a higher bioavailable metal pool close to roots and to the sites of elemental uptake by plants. From different experimental contexts, such as a rhizobox study (Chino *et al.*, 1999) and a field study on salt cedar (*Tamarix gallica*) and bermudagrass (*Cynodon dactylon*) (Fenn and Assadian, 1999), a higher soluble Cu content has been reported in the rhizosphere as compared with the bulk soil. This enhanced elemental bioavailability in the rhizosphere was also observed for many nutrients, such as Ca, K, Mg and Mn, and is attributed to the presence of root exudates, organic substances, higher microorganism activity and to the proximity of microsites enriched in OM (Wang and Zabowski, 1998; Schöttelndreier and Falkengren-Grerup, 1999).

Contradictory findings have been reported in other studies, where the rhizosphere is depicted as an environment depleted of bioavailable metal pools. Cherrey *et al.* (1999) found the rhizosphere of ryegrass (*L. perenne* cv. Aubisque) to be impoverished of extractable Cu (EDTA, DTPA or $CaCl_2$). Furthermore, Lorenz *et al.* (1997) reported lower concentrations of soluble Cd and Zn in the rhizosphere than in the bulk soil. In a comparison of the soil solution concentrations of Cd and Zn before and after the growth of the hyperaccumulator *T. caerulescens*, Knight *et al.* (1997) observed a decrease in the total Cd and Zn concentrations in the soil solution of the rhizosphere, as well as in the free Cd^{2+} and Zn^{2+} concentrations.

The need to assess the speciation of trace metals and to estimate their bioavailability in the plant root environment is crucial in order to establish the extent of trace metal contamination in soils and to address their potential adverse effects. Although the bioavailability of metals is best predicted by their liquid phase speciation, the assessment of the trace metal fractionation in the solid phase is still essential because the replenishment rate of metals in the soil solution is controlled by solid phase metals (Minnich *et al.*, 1987; Zhang *et al.*, 2001; Krishnamurti *et al.*, 2002). For instance, Zhang *et al.* (2001) showed that Cu concentrations in plant parts correlated best with the "effective Cu concentration," which was defined as the Cu concentration in the soil solution combined with a calculated concentration representing the amount of Cu transferred from the solid phase to the soil solution. McGrath *et al.* (1997) and Knight *et al.* (1997) also showed that 90–99% of the total Zn taken up by *T. caerulescens* originated from the non-mobile fractions of the solid phase (the mobile fractions being described here as the soluble and exchangeable fractions).

Most studies focusing on the differences in elemental speciation between the rhizosphere and the bulk soil have been performed under controlled conditions, either in the laboratory or in greenhouses. Moreover, the majority of these studies have focused on the speciation of macronutrients, such as Ca, K and Mg, and have looked at the rhizosphere and bulk soil of cultivated agricultural species. Very few studies have looked at the contrast in trace metal speciation between the rhizosphere and the associated bulk soil in forested ecosystems. This gap in knowledge may be due to the practical difficulties associated with the separation of soil components and with the collection of rhizosphere materials in the field. We believe it is essential to integrate the rhizosphere as a distinct biogeochemical compartment in field studies of forested ecosystems. Indeed, its crucial influence on the biochemical cycles of nutrients and trace metals is large compared to the small volume it occupies in soils. In this context, the specific objectives of this study are (i) to contrast the solid phase fractionation of trace metals (Cd, Cu, Ni, Pb and Zn) between the inner rhizosphere, the outer rhizosphere and the bulk components of forest mineral soils along a soil contamination gradient; and (ii) to determine Cu^{2+} activity in the liquid phase and establish the relationships with pH, DOC and total dissolved Cu. The emphasis is put on easily extractable metal forms and on liquid phase speciation because readily dissolved metal complexes and free metal ions are assumed to represent most of the potentially bioavailable metal fractions in soils. In addition, the use of a soil contamination gradient offers a working range of metal concentrations in soils to test whether the rhizosphere effect varies with the metal content of the soil.

2. MATERIALS AND METHODS

2.1. Study area

The study area is located in the Sudbury region, Ontario, Canada. This area is host to some of the largest Ni and Cu mine sites in the world. The exploitation of these metals began at the end of the 19th century and is still very active today. Thus, forests and soils of the Sudbury area have been affected for over a century by smelting activities and their environmental consequences, such as logging, fire and atmospheric emissions of gaseous and particulate pollutants (Freedman, 1978). Since the building of a 380 m smokestack in the1970s, levels of pollutants in the area surrounding Sudbury have reduced greatly. This has progressively led to the recolonization of plant species such as birch trees, where conditions permitted their growth (Freedman, 1978). This "superstack" has also caused the dispersion of emitted pollutants over a much larger region. For instance, less than 2% of the emitted sulphur and only 40% of Cu and Ni were deposited within a 60 km radius of the smelter, causing this area and the area beyond it to suffer from a regional acidity problem (Freedman and Hutchinson, 1980).

2.2. Experimental design, site description and sample collection

In order to have a gradient of soil contamination, we selected three sites along the prevailing wind direction, i.e. from west to east. The sites are respectively located at 2.5, 15 and 43 km, from the Falconbridge Cu–Ni smelter. Site 1, at 2.5 km, is located on a fluvial deposit. The vegetation is composed of planted white pines (*Pinus strobus* L.) with small and naturally regenerating white birch (*Betula papyrifera* Marsh.). Approximately 20% of the canopy is occupied. Site 2, at 15 km, is located on a glacio-fluvial deposit. The mixed vegetation is composed of white pine, paper birch, trembling aspen (*Populus tremuloides* Michx.), red oak (*Quercus rubra* L.), red maple (*Acer rubrum* L.), balsam fir (*Abies balsamea* L.) and white spruce (*Picea glauca* Moench.), and the canopy coverage is 100%. Site 3, at 43 km, is located on a fluvial deposit, and the vegetation is dominated by mixed balsam fir, paper birch, trembling aspen and white spruce, with a canopy coverage of 100%. All three sites have acidic, mild, mesic soils designated as orthic humo-ferric podzols according to the Canadian System of Soil Classification (Soil Classification Working Group, 1998).

Rhizosphere and bulk soil samples were collected at each site. To assess field variability as well as to ensure the collection of sufficient quantities of rhizospheric material, the B horizon of 6, 4 and 3 white birch trees was sampled at sites 1, 2 and 3, respectively. The average age of the trees was 12, 30 and 15 years at sites 1, 2 and 3, respectively. Mineral horizons were selected because the majority of the living fine roots in forested ecosystems are found in the subsurface horizons (Persson *et al.*, 1995). The rhizosphere and bulk soil acquisition is adapted from the method presented in Gobran and Clegg (1996). The separation of the rhizosphere from the bulk soil was performed on-site, where the collected roots were gently shaken. The detached soil material was defined as the bulk soil. The roots were subsequently stored in sterile plastic bags and brought back to the laboratory, where the rhizosphere soil was further separated from the roots. All bulk soil samples were air-dried and sieved within 1 week upon arrival in the laboratory, whereas the root samples were stored at 4°C until further treatment. In order to further refine the gradient between the bulk soil and the rhizosphere, we divided the rhizosphere into two components: the outer and the inner rhizosphere. The outer rhizosphere corresponds to the soil component defined as the rhizosphere in Gobran and Clegg (1996), which is obtained by gently shaking the roots to recover all fallen rhizospheric soil particles. The inner rhizosphere corresponds to the component defined in Gobran and Clegg (1996) as the soil–root interface, which represents the soil still adhering to the shaken roots. This material is obtained by gently scrubbing off the remaining soil particles. Any remaining root material in the soil samples was retrieved using plastic tweezers. All bulk and rhizosphere soil samples were sieved to 0.5 mm with a stainless steel screen in order to homogenize the particle-size distribution of the different soil components.

2.3. Laboratory analyses

2.3.1. Chemical and physical soil properties

To characterize the mineral soil, selected chemical and physical analyses were performed on the bulk soil of one field replicate for each site. The pH and electrical conductivity (EC) of the bulk soil were measured using a Fisher Scientific Accumet® pH meter-10 with a glass body electrode and a Radiometer CDM 83 conductivity meter, respectively. The pH was measured using 3 g of soil with 30 mL H_2O, while EC was measured using 5 g of soil with 10 mL H_2O. Cation exchange capacity (CEC) was determined after measuring the concentrations of Ca, Na, Mg, K, Fe, Al and Mn using 3 g of soil with 30 mL 0.1 M $BaCl_2$ (Hendershot and Duquette, 1986). Cation concentrations were measured with an atomic absorption spectrophotometer (AAS; Varian AA-1475). Organic carbon in the solid phase was measured by a $K_2Cr_2O_7$ titration with $FeSO_4$, according to the modified Walkley–Black method (Carter, 1993). The Fe, Al and Mn contents in Na-pyrophosphate ($Na_4P_2O_7$), acid ammonium oxalate (AAO) and dithionite-citrate (DC) extractions were measured in order to evaluate the accumulation of weathering products (Sheldrick, 1984). Analyses were performed by AAS. For the $Na_4P_2O_7$ extraction, 0.3 g of soil was used with 30 mL of 0.1 M $Na_4P_2O_7$. The AAO extraction was performed using 0.25 g of soil and 10 mL of 0.2 M $(NH_4)_2C_2O_4H_2O$. As for the DC extraction, 0.5 g of soil with 25 mL of 0.68 M $Na_3C_6H_5O_72H_2O$ and 0.4 g of $Na_2S_2O_4$ was used. The mineralogy of the clay-sized particles was determined by X-ray diffraction (XRD) after pretreatment with dithionite–citrate–bicarbonate (DCB) to remove Fe and Al sesquioxides. Samples were saturated with K and Mg, mounted in preferential orientation and analyzed following different treatments (K at room temperature, K heated to 300 and 550°C, Mg at room temperature and Mg saturated with ethylene-glycol) (Brindley and Brown, 1980; Dixon and Weed, 1989). The mineral abundance of each sample was determined using an I/I_{QZ} ratio, where the integrated intensity of each mineral (I) is normalized to the intensity of the quartz peak (I_{QZ}) at $d = 0.426$ nm (Courchesne and Gobran, 1997). Proportions of sand, silt and clay in the bulk soil were estimated using the hydrometer technique after the addition of 5 g L^{-1} Na-metaphosphate. The samples were pretreated with DCB in order to remove all Fe oxides.

2.3.2. Solid phase trace metal fractionation

Three distinct extractions combined with an acid digestion were performed to operationally establish the fractionation of Cd, Cu, Ni, Pb and Zn in the solid phase of the bulk soil, the outer rhizosphere and the inner rhizosphere. The extractants used were H_2O, 0.1 M $BaCl_2$ and 0.1 M $Na_4P_2O_7$, while a HNO_3–HCl mixture was used for the acid digestion. These extractions yield a reliable estimation of the potential mobility and bioavailability of trace metals in the solid phase (Morgan and Stumm, 1991; Assadian and Fenn, 2001; Lasat, 2002). The H_2O extraction provides an estimation of the water-soluble metal species, while the $BaCl_2$ extraction provides information on the exchangeable metal fraction of the solid phase. These two fractions are considered to include the most mobile chemical forms (Kabata-Pendias and Pendias, 1992) and will hereafter be referred to as the potentially available metal pool. The $Na_4P_2O_7$ extraction estimates the metal fraction associated with OM. This fraction is not considered as readily available, as the OM usually forms stable complexes with metals (Løbersli *et al.*, 1991). The acid digestion dissolves most metal forms, which, in this case, include the metals from the three previous fractions as well as residual chemical forms, such as metals associated with amorphous and crystalline oxides. However, this acid digestion does not completely dissolve metals from the structures of primary minerals. Nevertheless, in this study, the HNO_3–HCl digested metals will be considered to reflect the total-recoverable metal content of any soil component. Hence, the power of the various extractants used follows the order: HNO_3–HCl digestion $>$ $Na_4P_2O_7$ $>$ $BaCl_2$ $>$ H_2O.

Based on the method used by Séguin *et al.* (2003), the H_2O extraction was performed using 3 g of soil with 30 mL of ultra-pure (UP) water. The samples were placed on an end-over-end automatic shaker for 2 h, and centrifuged for 10 min at 1500 ***g***. The solution was filtered using 0.45 μm polyethersulfone (PES) Millipore Durapore® filters, and acidified with 2% Trace Metal Grade (TMG) HNO_3. Non-acidified aliquots of all H_2O extracts were set aside for the subsequent measurement of DOC, pH and pCu^{2+}. All samples were analyzed on an inductively coupled plasma–mass spectrometer (ICP–MS) to ensure optimal detection limits. Quality control was enforced by using both inner and outer laboratory controls, as well as spiked samples whose recovery rates varied between 90 and 110%. All soil samples were analyzed in triplicates. The $BaCl_2$ extraction was performed as indicated previously. The triplicate samples were shaken for 2 h, centrifuged at 1500 ***g*** for 10 min, then filtered and acidified in the same way as the water extract. All trace metals were analyzed by ICP–MS, except for Zn, which was measured on an AAS. For the $Na_4P_2O_7$ extraction, the samples were shaken for 16 h. At the end of this period, 0.5 mL of a 0.1% Superfloc solution (Rotterdam, Netherlands) was added to maximize the centrifugation procedure (10 min at 1500 ***g***). Samples were subsequently filtered and acidified as in the previous extractions, and analyzed on an inductively coupled plasma–atomic emission spectrometer (ICP–AES). Triplicates were extracted for all bulk soil samples, whereas the outer and the inner rhizosphere samples were analyzed in duplicates. For the acid digestion, the amount of soil used varied between 0.2 and

0.5 g, depending on the OM content. As such, 0.2 g of soil was used for the inner rhizosphere soil samples, while 0.5 g was needed for the bulk soil samples. These quantities were combined with 9 mL of HNO_3 and 2 mL of HCl. The acid digestion method was modified from the US-EPA SW3051-A method (1998). The digestion was performed in a CEM MDS-2000 microwave. Triplicates of all samples were digested at 690 kPa, with an average temperature of 155°C for 20 min. Afterward, the digestion volume was adjusted to 40 mL with UP water, and the samples were centrifuged for 10 min at 900 ***g***. Solutions were slowly decanted to retain only the supernatant. Trace metal concentrations of all triplicates were measured on an AAS.

2.3.3. Liquid phase speciation of total dissolved Cu

In order to fulfill the second objective of this study, DOC, pH and the activity of free Cu^{2+} ions (pCu^{2+}) were measured in the liquid phase of all soil components. As previously mentioned, aliquots were taken from the non-acidified H_2O extracts to measure DOC concentrations, pH and pCu^{2+}, and were subsequently analyzed for Cu, here termed total water-soluble Cu. To optimize pCu^{2+} measurements, potassium nitrate (KNO_3) was added to the water extracts in order to homogenize the ionic strength of the solutions and to obtain a final concentration of 0.01 M KNO_3. The DOC concentrations were measured with a Shimadzu TOC analyzer (Kyoto, Japan), whereas pH values were obtained with a Radiometer PHM 82. The pCu^{2+} measurements were conducted with an Orion 9629 *ionplus* Series™ cupric ion selective electrode (Cu-ISE) and a Fisher Scientific Accumet® 910 pH meter. The determination of pCu^{2+} values as well as the electrode calibration are based on a protocol developed by Sauvé *et al.* (1995). Modifications to this method include the use of a magnetic stirrer that was placed in the water extract in order to speed up electrode equilibration time. Moreover, the calibration equations that were calculated for each day of analysis ($n = 7$) in order to convert voltage readings (mV) into pCu^{2+} values were all integrated into a single mean equation. The purpose of this integration was to reduce the effect of the daily fluctuations of the standards on the calibration equations. The external and the internal laboratory quality controls gave similar results, irrespective of whether we used the average or the daily calibration equations. At the end of each day, two standards used for electrode calibration were reanalyzed in order to evaluate the accuracy of readings and to identify any potential instrumental drift.

2.3.4. Statistical analyses

The non-parametric Friedman test was used to evaluate the significance of differences between the three soil components for a given site, metal and extractant, as well as between sites for a given soil component. Non-parametric statistics were chosen because of the small number of samples available for the statistical analyses. Indeed, the sample numbers were not high enough to meet the normality assumption of parametric tests. The non-parametric Wilcoxon-signed rank test was used to evaluate the significance of differences between the bulk soil and the inner rhizosphere only when the Friedman test failed to yield any significant differences between the three soil components. The significance level used for the Friedman and the Wilcoxon tests was $\alpha \leq 0.10$. Stepwise multiple linear regressions were performed with pCu^{2+}, pH, DOC and total water-soluble Cu. For this purpose, the data for DOC and total water-soluble Cu were log-transformed. The significance level used was $\alpha < 0.05$. All statistical analyses were performed using the SPSS 10.0 software for Windows.

3. RESULTS AND DISCUSSION

3.1. Soil properties

The characteristics of the bulk soil, measured on one field replicate at each site, are presented in Tables 1 and 2. The pH values of the bulk soil are more acidic at site 1, close to the smelter, and gradually increase toward site 3. The deposition of atmospheric pollutants, such as sulfur compounds, is most probably responsible for the soil acidification observed close to the smelter. The EC and CEC values of the three sampling locations do not follow any specific trend, although site 1 exhibits both the highest EC value and the lowest CEC. Soil organic C content increases with distance from the smelter, reflecting the increase in canopy density. The amounts of Fe and Al extracted by AAO and DC also increase gradually from site 1 to 3. The proportion of sand in the bulk soil decreases from site 1 to 3 as the silt and the clay contents increase.

Table 2 presents the mineralogy of the clay-sized particles of the bulk soil at each sampling site. The ordination of the various minerals is based on their relative abundance with respect to quartz (I/I_{QZ}). The relative abundance of each mineral follows the order: quartz $\approx$ plagioclase $>$ chlorite $\approx$ mica $\approx$ K-feldspar $>$ amphibole $>$ mica-vermiculite. The mineralogical data underline the homogeneity of the clay-sized particles in the bulk soil component of the three sites. The similarity in trace metal concentrations in the acid extraction further emphasizes the homogeneity of the B horizons along the gradient as shown for total-recoverable Cd and Pb in Fig. 1.

3.2. Solid phase trace metal fractionation

3.2.1. Soil contamination gradient at Sudbury

Tables 3 and 4 show that metal concentrations decrease from site 1 to 3 for a number of metal-extractant combinations. This observation points toward the existence of a soil contamination gradient, with metal concentrations in the B horizon decreasing with distance from the smelter. Fig. 2 illustrates this gradient for water-soluble Cd, Cu and Ni concentrations. In these cases, this soil contamination

Table 1
Chemical and physical properties of the bulk soil in the B horizon at the three sampling sites

Site	pH	EC (μS cm^{-1})	CEC (cmol (+)kg^{-1})	Org C (%)	Fe			Al			Sand	Silt	Clay
					Na-pyro	Oxal.	DC	Na-pyro	Oxal.	DC		(%)	
					(%)								
1	4.84	65.1	0.66	0.97	0.13	0.42	0.58	0.21	0.55	0.47	64.6	30.0	5.4
(2.5 km)			(0.08)		(0.01)	(0.01)	(0.03)	(0.00)	(0.02)	(0.04)			
2	4.88	43.4	1.25	1.61	0.31	0.57	0.73	0.37	0.69	0.77	47.5	43.5	9.0
(15 km)			(0.05)		(0.02)	(0.01)	(0.01)	(0.02)	(0.02)	(0.04)			
3	5.26	52.4	1.24	2.12	0.27	0.62	1.01	0.42	1.05	1.05	42.9	44.1	13.0
(43 km)			(0.10)		(0.00)	(0.03)	(0.01)	(0.01)	(0.05)	(0.03)			

Notes: Values in parentheses are standard deviations calculated from laboratory triplicates.
(1) EC = Electrical conductivity.
CEC = cation exchange capacity.
Org C = organic carbon in the solid phase,
(2) Fe and Al contents (%) were determined using 0.1 M $Na_4P_2O_7$ (Na-pyro), 0.2 M $(NH_4)_2C_2O_4 \cdot H_2O$ (Oxal.) and 0.68 M $Na_3C_6H_5O_72H_2O$ with $Na_2S_2O_4$ (DC).

Table 2
Mineral abundance in the clay-sized particles of the B horizons at the three sites

Sites	Quartz	Plagioclase	Chlorite	Mica	K-feldspar	Amphibole	Mica-vermiculite
1 (2.5 km)	+++[a]	+++	++	++	++	+	tr
2 (15 km)	+++	+++	++	++	++	+	tr
3 (43 km)	+++	++	+++	++	+	+	tr

[a] +++: dominant; ++: major; +: minor; tr: trace.

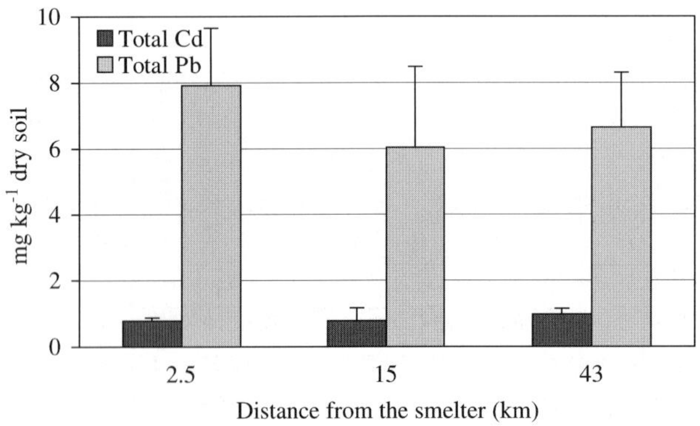

Fig. 1. Concentrations of total-recoverable Cd and Pb (total) in the bulk soil determined by HNO_3–HCl digestion at each sampling site. Mean values and standard deviations are calculated from the field and the laboratory replicates. The three sites did not show any significant differences in either total-recoverable Cd or Pb.

gradient is significant for the three soil components, with the level of significance varying from $p \leq 0.10$ to 0.05 (Table 4).

When all extractants are considered, the soil contamination gradient is most significant for Cu and Ni, the two dominant metals in the smelter emissions. In fact, almost all Cu fractions show a significant soil contamination gradient in each of the three components. As for Ni, the water-soluble and $BaCl_2$-exchangeable fractions from the three soil components exhibit a significant soil contamination gradient (Table 4). The total-recoverable Cu and Ni concentrations reflect a clear soil contamination gradient between sites 1 and 2 (Table 3). However, at site 3, their concentrations are higher than at sites 1 or 2. The relatively higher chlorite content at site 3 (Table 2) may explain the high total-recoverable concentrations of Cu and Ni, as these elements can be integrated in the chlorite mineralogical structure, and thus dissolved with the acid digestion (Dixon and Weed, 1989).

Although a decrease in metal concentrations with distance from the smelter is observed in a number of cases, some exceptions occur where a reversed concentration gradient exists. For example, the $BaCl_2$-exchangeable Zn concentrations in the outer and the inner rhizosphere are higher at site 3, whereas the lowest concentrations are found at site 1, close to the smelter. Other significant inverted concentration gradients occur in the outer rhizosphere for the organo-complexed Cd fraction and for the total-recoverable Pb fraction (Table 4).

The decrease in soil metal concentrations observed with distance from the smelter is in agreement with results from other studies performed in the Sudbury area. For instance, Freedman (1978) measured Cu, Ni and Zn concentrations in organic horizons (L, F and Ah) at various distances from the Coppercliff smelter and found higher total Cu and Ni concentrations at sites closer to the smelter. Moreover, a strong positive spatial correlation existed between Cu and Ni ($r = 0.91$, $p \leq 0.001$), indicating the fact that these two metals shared the same origin. Other studies also show that Cu and Ni concentrations are higher in the vicinity of the Sudbury smelters (Dudka *et al.*, 1995; Ontario Ministry of the Environment, 2001). Near Rouyn–Noranda, another smelting area in Eastern Canada, Dumontet *et al.* (1992) observed that Cd, Cu, Ni, Pb and Zn concentrations in mineral soils decreased with an increase in distance from a Cu–Zn smelter. Although the surface litter was the most contaminated soil horizon, the 0–15 cm mineral soil also appeared contaminated by these trace metals. Séguin *et al.* (2004) measured water-soluble trace metal concentrations along a soil contamination gradient near the same Cu–Zn smelter in Rouyn–Noranda. The spatial distribution of the metals analyzed (Cd, Cu, Ni, Pb and Zn) did not follow a specific trend with respect to metal contamination in soils. Séguin *et al.* (2004) explain the absence of a trend by the fact that their samples were collected in the mineral soil, a region of the soil profile that is often too deep to be affected by metal contamination, notably in clayey soils (Kabala and Singh, 2001; Ontario Ministry of the Environment, 2001). The absence of a contamination gradient observed for metals like Cd and Pb in the mineral horizons of our study might be explained similarly, as these metals are not known to be emitted in substantial amounts by the smelters of the Sudbury area (Freedman, 1978; Ontario Ministry of the Environment, 2001). Freedman (1978) also found that the total Zn concentrations in organic horizons slightly increased toward forest sites at greater distances from the smelter, which is concordant with our results. He suggested that the depleted total Zn concentrations at sites close to the smelter could be due to, among other factors, the competitive replacement of Zn by Cu or Ni at cation-exchange sites

Table 3
Mean trace metal concentrations of field and laboratory replicates for the three soil components and at each site

Extraction	Cd (mg kg^{-1})				Cu (mg kg^{-1})				Ni (mg kg^{-1})				Pb (mg kg^{-1})				Zn (mg kg^{-1})			
	Bulk	Outer rz[a]	Inner rz		Bulk	Outer rz	Inner rz		Bulk	Outer rz	Inner rz		Bulk	Outer rz	Inner rz		Bulk	Outer rz	Inner Rz	
[Site 1 (2.5 km)]																				
H_2O	0.011	0.024	0.028	**[b]	0.156	0.386	0.490	***	1.34	1.47	2.01	**	0.003	0.008	0.003		0.180	0.219	0.251	●●
(water-soluble)	(0.004)	(0.007)	(0.017)		(0.102)	(0.132)	(0.174)		(1.27)	(0.77)	(1.13)		(0.004)	(0.003)	(0.003)		(0.015)	(0.090)	(0.067)	
$BaCl_2$	0.058	0.136	0.187	**	0.750	2.59	2.49	***	3.20	4.75	6.06		0.092	0.054	0.054	●	0.409	0.618	0.735	***
(exchangeable)	(0.013)	(0.054)	(0.098)		(0.522)	(0.94)	(0.88)		(3.66)	(2.79)	(4.88)		(0.049)	(0.018)	(0.024)		(0.189)	(0.229)	(0.317)	
Na-pyro	0.861	1.41	1.09	**	11.39	25.73	25.71	***	5.32	5.36	7.02		7.45	6.93	6.83	●	3.21	1.48	1.34	●●
(organo-compl.)	(0.282)	(1.05)	(0.37)		(5.05)	(7.19)	(7.07)		(3.81)	(1.87)	(4.07)		(2.11)	(1.90)	(2.03)		(2.02)	(0.78)	(0.63)	
Acid	0.783	1.07	1.33	**	21.90	41.59	40.41	**	39.11	31.65	39.23		7.92	7.59	20.83	**	22.96	23.02	23.67	
(total-recoverable)	(0.086)	(0.23)	(0.39)		(7.51)	(9.26)	(10.54)		(14.85)	(5.65)	(11.91)		(1.74)	(2.81)	(3.59)		(6.48)	(9.56)	(7.82)	
[Site 2 (15 km)]																				
H_2O	0.004	0.005	0.007	**	0.024	0.061	0.075	●[c]	0.585	0.929	1.24	**	0.008	0.007	0.004		0.112	0.142	0.181	**
(water-soluble)	(0.003)	(0.003)	(0.004)		(0.013)	(0.025)	(0.035)		(0.332)	(0.483)	(0.63)		(0.003)	(0.002)	(0.001)		(0.043)	(0.054)	(0.073)	
$BaCl_2$	0.046	0.059	0.075		0.056	0.115	0.160	●	2.14	5.25	6.09	**	0.049	0.064	0.085	●	0.330	1.00	1.16	**
(exchangeable)	—	—	(0.018)		(0.033)	(0.090)	(0.099)		(0.574)	(1.15)	(1.45)		(0.005)	(0.013)	(0.023)		(0.084)	(0.18)	(0.06)	

Na-pyro	1.12	1.66	1.56	**	1.79	3.26	3.55	**	3.37	5.02	4.75	**	9.09	10.42	9.97		2.65	1.39	0.627	
(organo-compl.)	(0.15)	(0.25)	(0.31)		(0.74)	(1.58)	(1.76)		(0.80)	(2.01)	(1.74)		(1.56)	(0.86)	(1.19)		(2.15)	(0.78)	(0.504)	
Acid	0.769	1.29	0.888		5.47	9.52	8.70		24.30	28.46	25.45		6.04	18.19	20.05	**	18.25	21.61	20.36	**
(total-recoverable)	(0.403)	(0.209)	(0.220)		(1.06)	—	(2.87)		(3.89)	(5.81)	(5.32)		(2.44)	(6.02)	(2.31)		(2.61)	(3.03)	(3.49)	
[Site 3 (43 km)]																				
H_2O	0.001	0.004	0.003		0.013	0.058	0.062	*	0.073	0.201	0.244	**	0.009	0.009	0.007	**	0.062	0.227	0.205	*
(water-soluble)	(0.0001)	(0.001)	(0.002)		(0.010)	(0.024)	(0.036)		(0.020)	(0.030)	(0.031)		(0.003)	(0.003)	(0.002)		(0.029)	(0.049)	(0.050)	
$BaCl_2$	0.050	0.091	0.101		0.008	0.061	0.075	*	0.583	2.32	2.26	*	0.022	0.066	0.066		1.01	3.25	3.19	*
(exchangeable)	(0.009)	(0.023)	(0.009)		(0.005)	(0.006)	(0.057)		(0.102)	(0.47)	(0.61)		(0.012)	(0.013)	(0.007)		(0.51)	(1.34)	(1.98)	
Na-pyro	1.01	1.71	1.54	*	1.42	3.24	2.90	*	1.88	3.22	2.28		8.66	9.74	8.62	*	3.12	5.73	4.16	*
(organo-compl.)	(0.35)	(0.61)	(0.70)		(0.04)	(0.86)	(0.47)		(0.17)	(0.25)	(0.52)		(1.64)	(2.48)	(2.50)		(2.17)	(3.71)	(3.02)	
Acid	0.973	0.780	1.03		13.79	19.59	10.94		33.08	38.93	33.29		6.64	33.54	33.29	*	34.84	33.28	32.43	
(total-recoverable)	(0.171)	(0.344)	(0.04)		(5.29)	(1.74)	(4.86)		(8.49)	(6.04)	(5.04)		(1.67)	(5.17)	(3.02)		(7.17)	(10.73)	(7.51)	

Notes: Values in parentheses are standard deviations.
[a] rz = rhizosphere. [b] Results of the Friedman test, where $^{*}\ p \leq 0.10$; $^{**}\ p \leq 0.05$; $^{***}\ p \leq 0.01$. The results of the test illustrate the significant differences found between the metal concentrations of the bulk soil, the outer rhizosphere and the inner rhizosphere for each extraction and sampling site. [c] Results of the Wilcoxon signed rank test, performed only following a negative Friedman test, where • $p \leq 0.10$ and •• $p \leq 0.05$. The results illustrate the significant differences found between the metal concentrations of the bulk soil and the inner rhizosphere for each extraction and sampling site.

Table 4
Differences[a] in metal concentrations[b] between the three sampling sites for a given soil component and extraction

Extraction	Cd			Cu			Ni			Pb			Zn			Cu^{2+}		
	Bulk	Outer rz	Inner rz	Bulk	Outer rz	Inner rz	Bulk	Outer rz	Inner rz	Bulk	Outer rz	Inner rz	Bulk	Outer rz	Inner rz	Bulk	Outer rz	Inner rz
H_2O		*[c]	**	*	*	*	*	**	**		*		*			*	**	**
$BaCl_2$				**	*	*	*	**	*				*	**	**	—	—	—
Na-pyro		*[d]		*	*	*	*		*						*	—	—	—
Acid			*	*		*	**				**	*				—	—	—

[a] Results of the Friedman test, where * $p \leq 0.10$; ** $p \leq 0.05$. [b] The values used in the Friedman test are means of the field and laboratory replicates. [c] *, ** metal concentrations follow the gradient: site 1 > site 2 > site 3. [d] *, ** metal concentrations either follow the gradient: site 3 > site 2 > site 1, or do not show an apparent gradient.

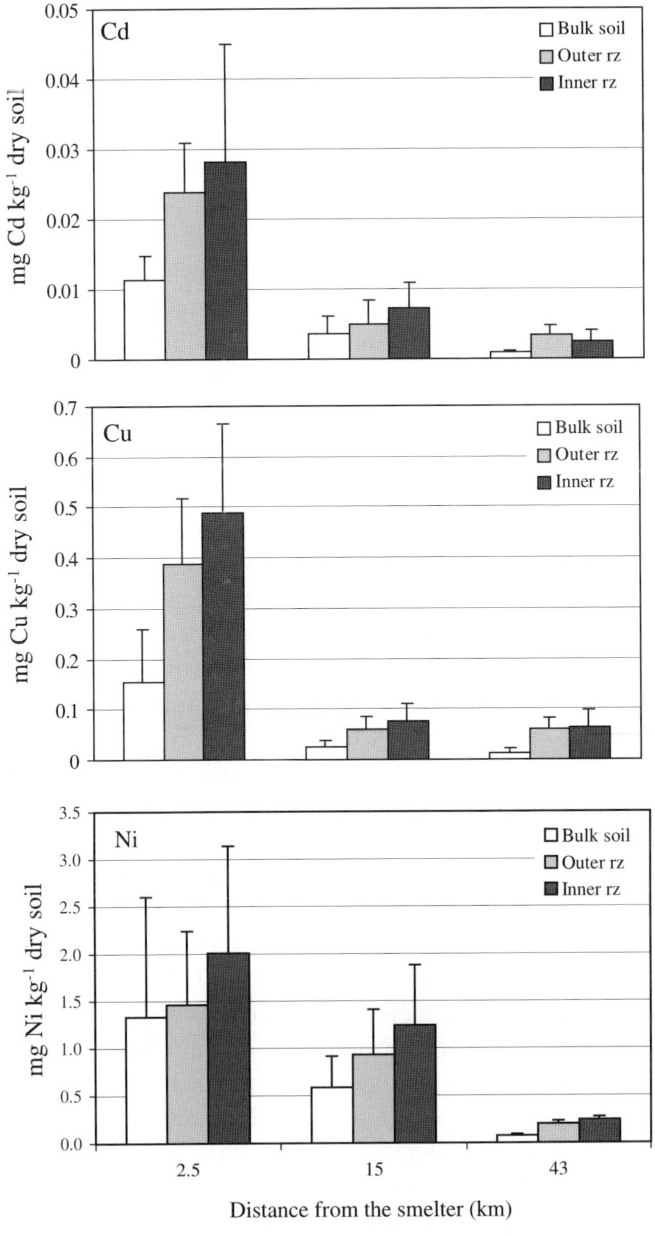

Fig. 2. Concentrations of water-soluble Cd, Cu and Ni in the three soil components at each site. Values and standard deviations are means of the field and laboratory replicates.

in the litter, and to the higher leaching rates measured in sites receiving higher levels of acidic precipitations.

3.2.2. Soil component gradient

Regardless of the statistical significance of the differences, the outer rhizosphere (25 cases) and the inner rhizosphere materials (29 cases) contain metal concentrations that are, in 90% of all cases (five metals × four extracts × three sites), systematically higher than the levels found in the bulk soil (Table 3). Moreover, the six cases out of 60 that have higher metal concentrations in the bulk soil than in either the outer or the inner rhizosphere concern Pb and Zn. These exceptions are mostly associated with the $BaCl_2$-exchangeable and the organo-complexed fractions for Pb and Zn. However, among these six exceptions, the only cases showing a significantly higher metal content in the bulk soil than in the inner rhizosphere originate from site 1.

3.2.2.1. Absolute concentrations. The Friedman test revealed a majority of situations in which the three soil components have significantly different metal concentrations, regardless of the extraction and the sampling location. In fact, 40 cases out of 60 show a statistically significant soil component differentiation with respect to metal concentrations, whether these concentrations are higher in the bulk soil, in the outer rhizosphere or in the inner rhizosphere. The trace metals associated with the largest number of significant soil component differentiation follow the order Cu > Zn > Pb > Cd > Ni.

The four Cd extracts in the three soil components are significantly different at site 1, close to the smelter where the Cd concentrations are the highest (Table 3). The only significantly different Cd concentrations at site 2 are for the water-soluble and organo-complexed fractions, whereas only the organo-complexed Cd differs at site 3. In all cases, Cd concentrations are higher in the rhizosphere.

On the other hand, Cu concentrations are statistically significant for almost every Cu extract and sampling location (Table 3). The acid digestions performed at sites 2 and 3 are the only extracts that do not yield a significant difference. The outer or the inner rhizosphere, nonetheless, contain higher Cu concentrations than the bulk soil, in every extract and sampling location.

With the exception of the water-soluble Ni fraction, the soil components at site 1 are not significantly different based on Ni concentrations (Table 3). This absence of significant differences for most Ni extracts at site 1 could be explained by the high heterogeneity of Ni concentrations between the field replicates. At site 2, the water-soluble, $BaCl_2$-exchangeable and organo-complexed Ni fractions are statistically different between soil components ($p \leq 0.05$), whereas the differentiation of the soil components is only significant for the water-soluble and $BaCl_2$-exchangeable Ni fractions at site 3. The total-recoverable Ni concentrations in the three soil components are very homogenous and do not follow a specific soil component trend.

Despite the few situations where their concentrations are higher in the bulk soil, Pb and Zn generally follow the soil component differentiation observed for the other trace metals (Table 3). Lead is the only trace metal for which significantly different total-recoverable concentrations are observed at all sites, where the total-recoverable Pb concentrations in the rhizosphere are largely superior to the levels found in the bulk soil. In the case of Zn, its water-soluble and $BaCl_2$-exchangeable fractions at all sites are statistically different in the three soil components. One of the two rhizosphere components systematically contains the highest Zn concentrations. However, the significance level of this differentiation gradually decreases along the sampling transect (from $p \leq 0.01$ at site 1 to $p \leq 0.10$ at site 3).

3.2.2.2. Enrichment ratios. Table 5 presents the average inner rhizosphere/bulk soil metal concentrations (mg kg^{-1}) ratios for each trace metal, chemical fraction and sampling location. For all sites and most metals (with the exception of Pb),

Table 5
Average Inner rhizosphere/bulk soil ratios[a] for trace metal concentrations

Extraction	Cd	Cu	Ni	Pb	Zn
[Site 1 (2.5 km)]					
H_2O	2.5 (1.5)	5.2 (4.3)	2.2 (1.1)	2.3 (2.7)	1.5 (0.4)
$BaCl_2$	3.9 (1.8)	6.5 (6.3)	3.4 (2.5)	0.8 (0.2)	1.9 (0.8)
Na-pyro	1.3 (0.1)	3.1 (2.2)	1.5 (0.8)	0.9 (0.1)	0.5 (0.4)
Acid	1.7 (0.4)	2.2 (1.2)	1.0 (0.2)	2.8 (0.9)	1.0 (0.1)
[Site 2 (15 km)]					
H_2O	2.2 (0.5)	3.6 (1.6)	2.2 (0.7)	0.7 (0.4)	1.6 (0.2)
$BaCl_2$	2.2	4.7 (4.6)	2.9 (0.3)	1.7 (0.3)	3.7 (0.7)
Na-pyro	1.4 (0.2)	2.1 (0.7)	1.4 (0.3)	1.1 (0.1)	0.7 (1.2)
Acid	1.6 (1.2)	1.6 (0.6)	1.1 (0.3)	3.8 (1.6)	1.1 (0.1)
[Site 3 (43 km)]					
H_2O	3.0 (1.5)	5.9 (2.7)	3.6 (1.2)	0.7 (0.2)	3.6 (1.5)
$BaCl_2$	2.0 (0.2)	14.2 (14.9)	3.9 (1.1)	2.3	3.1 (0.7)
Na-pyro	1.5 (0.3)	2.0 (0.3)	1.2 (0.2)	1.0 (0.1)	1.3 (0.7)
Acid	1.1 (0.2)	0.8 (0.3)	1.1 (0.5)	5.2 (1.3)	0.9 (0.1)

[a] Inner rhizosphere metal concentration (mg kg^{-1})/bulk soil metal concentration (mg kg^{-1}). Values in parentheses are standard deviations of the field replicates.

the water-soluble and $BaCl_2$-exchangeable metal fractions show the greatest rhizosphere enrichment factors. Indeed, although higher absolute organo-complexed and total-recoverable metal concentrations are measured, the rhizosphere enrichment factor for these metal fractions is not as high as the enrichment factor of the more potentially available pools (water-soluble and $BaCl_2$-exchangeable fractions). In addition, Cu shows the highest rhizosphere enrichment ratios in water-soluble, $BaCl_2$-exchangeable and organo-complexed fractions, at all sites (Table 5). However, all of these high Cu enrichment ratios have very high standard deviation values (from 50 to 100%).

Because they help evaluate the effect of soil contamination on the magnitude of the soil component differentiation (the rhizosphere effect), these enrichment ratios could constitute useful indicators of the soil response to environmental stresses. As such, the enrichment ratio values of the two main pollutants in the Sudbury area, Cu and Ni, evolve in a different way along the contamination gradient (Table 5). For instance, no clear increase or decrease in the Ni enrichment ratios is discernable along the sampling transect, regardless of the chemical fraction. Indeed, the rhizosphere enrichment factor is higher for the water-soluble and $BaCl_2$-exchangeable Ni fractions as compared to the organo-complexed and total-recoverable fractions, but these ratios are similar at all sites; moreover, the field variability of this rhizosphere enrichment is small. The rhizosphere enrichment ratios of all Cd fractions are also rather similar at all sites. On the other hand, the Cu, Pb and Zn rhizosphere enrichment ratios show larger fluctuations along the transect. For instance, the enrichment ratios of the $BaCl_2$-exchangeable Cu, Pb and Zn fractions increase along the contamination gradient. The Cu ratios, however, show great field variability, which could exaggerate this trend. On the other hand, the enrichment ratios of total-recoverable Cu show a decreasing trend along the soil contamination gradient, while the total-recoverable Pb ratios increase from sites 1 to 3.

The existence of higher levels of metal pools in the rhizosphere has been reported in other studies of the rhizospheric environment. For instance, in a study on soil component differentiation under trembling aspen (*Populus tremuloides* Michx.) along a soil contamination gradient, Séguin *et al.* (2004) showed that water-soluble Cd, Cu, Ni, Pb and Zn concentrations were higher in the inner rhizosphere and decreased toward the outer rhizosphere and the bulk soil. In addition, the rhizosphere enrichment in soluble metal species was confirmed at every sampling site, irrespective of the level of soil contamination. In a study on uncontaminated forest soils, Courchesne *et al.* (2001) found higher $BaCl_2$-exchangeable contents of Al, Fe, Mn, Cu and Zn in the rhizosphere than in the bulk soil. In a greenhouse experiment, using a rhizobox with planted wheat (*T. aestivum* L.), Wang *et al.* (2002) showed that the Cd, Cr, Cu, Ni, Pb and Zn bioavailable pools (combining soluble, exchangeable and carbonate-bound metals) followed the order near rhizosphere > near bulk soil > rhizosphere > bulk soil.

Other studies focused on the rhizosphere enrichment in nutrients. As such, Gobran and Clegg (1996) showed that in soils of a Norway spruce stand (*Picea abies* (L.) Karst.), soluble and exchangeable cations Ca, Mg, Na and K increased following the order bulk < rhizosphere < soil–root interface. In a field study comparing soil metal concentrations under five different herb species, Schöttelndreier and Falkengren-Grerup (1999) found higher concentrations of $BaCl_2$-exchangeable K, Mg, Mn and Ca in the rhizosphere than in the bulk soil. In agricultural soils, Chung and Zasoski (1994) showed significant enhancement of K and Al extracted with AAO in the rhizosphere than in the bulk soil. A number of factors contribute to the enrichment of bioavailable metal pools in the rhizosphere. As such, root exudates, mainly comprising organic and amino acids, have been shown to increase the solubility of metals and nutrients in the rhizosphere (Mench and Martin, 1991; Jones and Darrah, 1994; Wenzel *et al.*, 2003). The preferential accumulation of OM close to roots generates a higher number of cation-exchange sites, and can thus enhance the exchangeable metal pool of the rhizosphere (Arocena *et al.*, 1999; Schöttelndreier and Falkengren-Grerup, 1999). The lower pH of the rhizosphere can also increase the solubility of metals and nutrients. For example, the increased extractability (1 M NH_4NO_3) of mobile Zn in the rhizosphere has been attributed to the acidification of this environment (McGrath *et al.*, 1997).

However, some studies found opposite results, where the rhizosphere was found to be more depleted of metals and nutrients than the bulk soil (Dieffenbach *et al.*, 1997; Lorenz *et al.*, 1997; Göttlein *et al.*, 1999; Wang *et al.*, 2001). In a pot experiment, Lorenz *et al.* (1997) showed that the growth of radish (*Raphanus sativus* cv. "Crystal Ball") depleted the rhizosphere of water-soluble K, Ca, Mg, P, Mn, Cd and Zn, suggesting that elemental uptake by plants was an important factor contributing to the depletion of elements near the roots. Dieffenbach *et al.* (1997) also showed a decrease in K and Mg concentrations in the soil solution near root tips and elongation zones of Norway spruce. The concentrations of metals and nutrients in the rhizosphere depend on both soil and plant properties. The soil-transfer and root-uptake mechanisms are affected by multiple factors such as mass flow and diffusion, solid phase and soil solution concentrations of metals and nutrients, plant type and developmental stage, root size and morphology, root exudates soil pH and the presence of mycorrhizae (Jungk, 1996).

3.2.3. Metal fractionation

The chemical fractionation of each metal for the inner rhizosphere and the bulk soil is presented in Fig. 3. As the extraction values in this figure are determined by differencing, the total-recoverable metal fraction will hereafter be referred to as the residual fraction. In most cases, the organo-complexed and the residual fractions are the two dominant chemical fractions for the bulk soil and the inner rhizosphere. The water-soluble and $BaCl_2$-exchangeable metal fractions

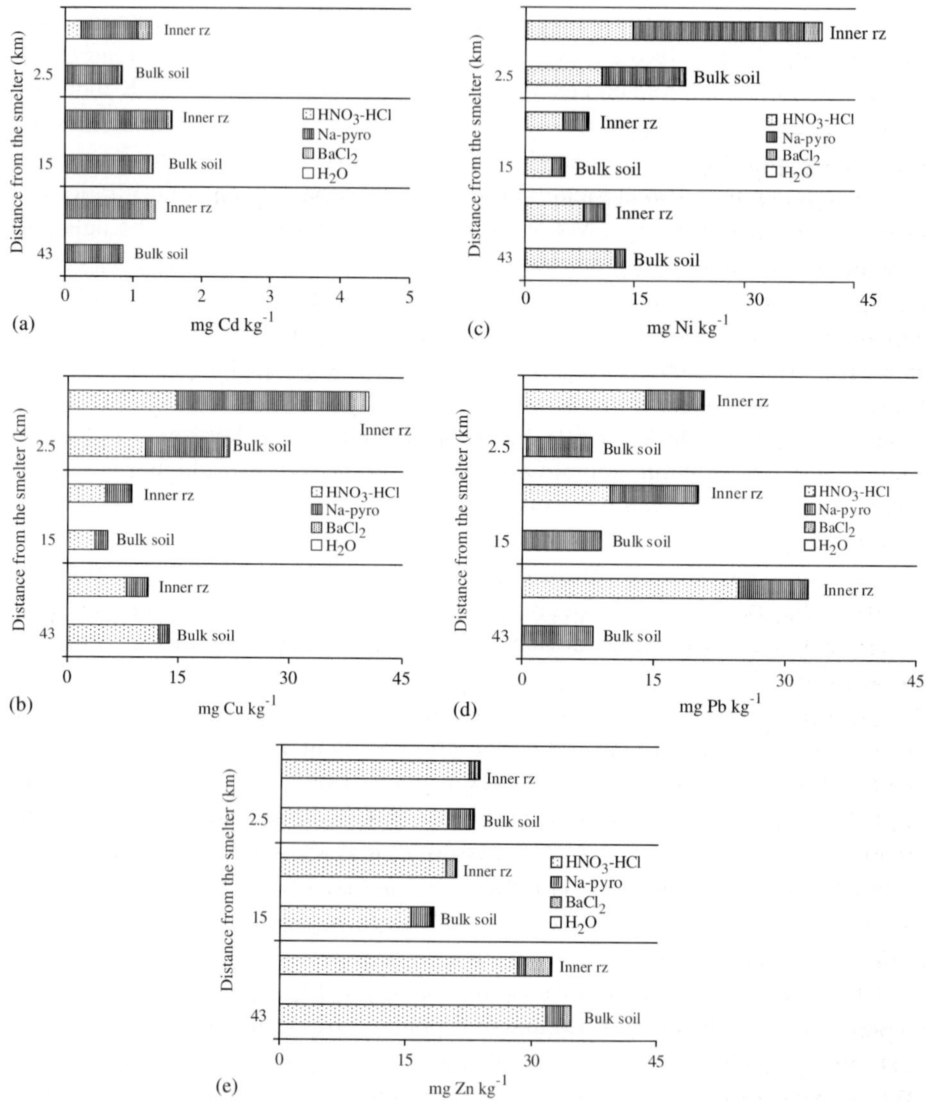

Fig. 3. Trace metal fractionation for (a) Cd, (b) Cu, (c) Ni, (d) Pb and (e) Zn in the bulk soil and the inner rhizosphere at each sampling site. The values for each fraction were determined by differencing, where the $BaCl_2$-exchangeable fraction ($BaCl_2$) = $BaCl_2 - H_2O$; the organo-complexed fraction (Na-pyro) = Na-pyro $- BaCl_2 - H_2O$; the residual fraction (HNO_3–HCl) = HNO_3–HCl $-$ Na-pyro $- BaCl_2 - H_2O$.

represent only a very small percentage of the total-recoverable metal fraction, a fraction that is, however, almost systematically superior in the inner rhizosphere than in the bulk soil. In the bulk soil, the average proportion of water-soluble and $BaCl_2$-exchangeable metal forms is around 3%, with the highest proportion of this potentially available metal pool being 8.8% for Ni at site 2, and the lowest ones being less than 1% for Pb and Cu at most sites. In the inner rhizosphere, the proportion of total-recoverable metal present as water-soluble or $BaCl_2$-exchangeable forms averages 7% and fluctuates wildly from 23.9% for Ni at site 2 and 0.2% for Pb at site 3 (Fig. 3).

These results are consistent with other studies showing that the soluble and exchangeable metal fractions represent only a small fraction of the total metal content in soils. In the bulk soil of contaminated soils, Tao *et al.* (2003) showed that the $MgCl_2$-exchangeable Cu fraction represented only 0.5% of the total Cu content, while Cu associated with OM and sulfides were the dominant fractions at 47% of total Cu. Moreover, in a soil profile study on contaminated sites, Kabala and Singh (2001) showed that the subsurface horizons of contaminated soils contained less than 10% of Cu, Pb and Zn in the mobile fraction, which comprised the water and exchangeable metal fractions. However, the contribution of the mobile fraction was higher in the surface horizons of the sites that were located closer to a smelter.

The cumulative proportion of water-soluble and $BaCl_2$-exchangeable metals also varies along the contamination gradient. Indeed, higher proportions of water-soluble and $BaCl_2$-exchangeable forms are found for most metals at site 1, closer to the smelter, and these percentages decrease toward site 3. For instance, this potentially available Cu pool at site 1 represents 3.4 and 6.2% in the bulk soil and the inner rhizosphere, respectively (Fig. 3b). It gradually decreases to 0.06 and 0.7%, respectively, at site 3. This could be related to the more acidic conditions found at site 1 (Table 1), which would favor the dissolution of Cu and its retention on the exchange complex. The only exception to this decreasing trend is Zn (Fig. 3e). This observation tallies with previous results showing a significant increase in the water-soluble and $BaCl_2$-exchangeable absolute Zn concentrations from site 1 to 3.

Lead is mostly present in the organo-complexed and residual fractions, and its potentially available metal pool represents less than 1.2% in both soil components and at all sites (Fig. 3d). This reflects the well-recognized affinity of Pb for OM (Kabata-Pendias and Pendias, 1992; Wang and Benoit, 1996). The potentially soluble and exchangeable Pb forms would therefore be complexed rapidly by OM, either in the dissolved or solid states. Cadmium is mainly present in the organo-complexed fraction, and its potentially available pool is less than 7% in the bulk soil and 14% in the inner rhizosphere, at all sites (Fig. 3a). Copper is also largely bound into organic complexes, especially at site 1 where it comprises over 50% of all Cu fractions in both soil components (Fig. 3b). Nickel, however, shows a greater proportion of $BaCl_2$-exchangeable forms than of organo-complexes in the inner rhizosphere, as its affinity with OM is less than that of Cu or Pb (Vaughan *et al.*, 1993) (Fig. 3c). In the case of Zn, the fractionation in the bulk soil as well as in the inner rhizosphere is dominated by the residual chemical form (Fig. 3e).

3.3. Liquid phase Cu speciation

3.3.1. Cu^{2+} activity

The data from the analyses conducted on the liquid phase are summarized in Table 6. The results of the Friedman test, performed to evaluate the influence of soil contamination level on Cu^{2+} activities, are presented in Table 4. In all soil components, the Cu^{2+} activities are significantly influenced by the level of soil contamination, with the absolute Cu^{2+} activities decreasing with distance from the smelter (Table 6 and Fig. 4a). As shown in Table 6 and Fig. 4a, a significant soil component difference ($p \leq 0.10$) is found only at site 3, where the bulk soil hosts relatively higher mean free Cu^{2+} activities, while the lowest amounts are measured in the outer rhizosphere. Thus, unlike the results presented for Cu in the solid phase, the inner rhizosphere is not the soil component where the highest Cu^{2+} activities are found. In fact, no general trend emerges for free Cu^{2+} activities between the three soil components (Fig. 4a).

Average inner rhizosphere/bulk soil ratios of Cu^{2+} concentrations (mg kg^{-1}) were calculated for each site. The field replicates at site 1 give an average ratio of 1.3. This mean ratio decreases to 0.8 at site 2 and to 0.4 at site 3. Therefore, the relative Cu^{2+} content of the inner rhizosphere varies with distance from the smelter, an observation that is probably associated with an increase in OM content in a rhizosphere and a decrease in Cu concentration.

The fraction of water-soluble Cu that is present in the free Cu^{2+} ion form (% Cu) is shown in Fig. 4b. At all sites, this fraction is systematically higher in the bulk soil than in the outer or the inner rhizosphere, with 28.6 , 29.2 and 16.9% of the water-soluble Cu in the bulk soil being present in the ionic form at sites 1, 2 and 3, respectively. At site 1, the percentage of water-soluble Cu occurring as Cu^{2+} is 3–4 times higher in the bulk soil than in the associated outer and inner rhizosphere, respectively. At sites 2 and 3, this percentage of the bulk soil fraction is 2–10 times higher than in the inner or the outer rhizosphere. These values well illustrate the complexity of assessing metal bioavailability in soils. In the water extracts, the inner rhizosphere tends to hold the highest Cu concentrations, which may be related, among other factors, to its slightly more acidic pH. However, Cu^{2+} activities do not follow this trend, as the inner rhizosphere values are never the highest measured. Moreover, the free Cu^{2+} activities are not well contrasted between the three soil components (Fig. 4a). In the inner rhizosphere, the lower Cu^{2+} activities could be explained by the large amounts of dissolved organic substances near the roots. Indeed, these dissolved organic substances can bind with dissolved Cu to form organic-metal complexes. Very few studies have contrasted the free metal activities in the rhizosphere and the bulk soil, whether it is in a forested, agricultural or laboratory environment. A study comparing the ionic activities of Cd and Zn in the rhizosphere of radish and bulk soil of contaminated soils showed that the bulk soil proportionally contained more free

Cd^{2+} and Zn^{2+} ions (Lorenz *et al.*, 1997). The lower activities of Cd^{2+} and Zn^{2+} in the rhizosphere were attributed to higher DOC contents in this environment, which induced the complexation of the free ions by organic substances.

3.3.2. DOC concentrations

The DOC concentrations in the water extracts are presented for all sampling sites in Table 6 and Fig. 5a. For a given soil component, the DOC concentration increases from site 1 to 3 (Fig. 5a), but this trend is significant only for the bulk soil ($p \leq 0.10$). This apparent DOC enrichment along the sampling transect relates well to the enhancement observed for soil organic carbon content (Table 1) as shown by the positive correlations between these two variables ($r^2 = 0.67$, $p = 0.001$; $r^2 = 0.61$, $p = 0.002$; $r^2 = 0.75$, $p = 0.0001$ in the bulk soil, the outer rhizosphere and the inner rhizosphere, respectively). Again, the increased vegetation density might explain the enhancement in both solid and dissolved organic C along the transect.

The DOC concentrations are systematically higher in the inner rhizosphere, while the lowest values are found in the bulk soil at all sites (Fig. 5a). This differentiation of soil components based on DOC concentrations is statistically significant at all sampling sites (Table 6). The average inner rhizosphere/bulk soil ratios for DOC concentrations are rather similar at the three sites, with values of 3.7 at site 1, 3.2 at site 2 and 3.1 at site 3.

The enhancement of DOC levels in the rhizosphere has long been recognized in the literature (Bowen and Rovira, 1991; Marschner and Romheld, 1996; Hinsinger, 1998), and our results are in agreement with other studies on forested ecosystems that have shown the existence of higher DOC concentrations in the soil solution of the rhizosphere than of the bulk soil (Gobran and Clegg, 1996; Wang and Zabowski, 1998; Courchesne *et al.*, 2001; Séguin *et al.*, 2003). This increase in DOC can be attributed to the proximity of roots and to exudates, as well as to the decomposition products of roots and their associated fungi and microorganisms (Vaughan *et al.*, 1993; Gobran *et al.*, 1998; Arocena *et al.*, 1999).

The affinity of Cu for solid and dissolved OM is well known (Kabata–Pendias and Pendias, 1992). Acidic forest soils provide an environment in which Cu can readily be associated with OM in a complex or chelate form. This affinity induces a strong Cu retention in the presence of a high soil organic carbon content, which, in turn, can diminish the bioavailable Cu pool by leaving a lower dissolved Cu content and a small percentage of free Cu^{2+} (Løbersli *et al.*, 1991; Sauvé *et al.*, 1995). As such, the Cu-organic chelates often represent the dominant Cu form in solution (Sanders, 1982; Kabata–Pendias and Pendias, 1992). For example, studies focusing on the speciation of Cu in the soil solution showed that over 80% of total dissolved Cu is complexed with organic substances (Dudley *et al.*, 1987; Sauvé *et al.*, 1997; Al-Wabel *et al.*, 2002; Krishnamurti and Naidu, 2002).

The fact that Cu^{2+} activity is similar in all soil components (Fig. 4a) suggests that the ability of dissolved organic matter to complex Cu is of similar magnitude

Table 6
The pCu^{2+} value, Cu^{2+} and total H_2O-soluble Cu concentrations, pH and DOC values measured in the three soil components at various distances from the smelter

Site	pCu^{2+}			Cu^{2+} (μg kg^{-1})			Cu (H_2O) (mg kg^{-1})				pH				DOC (mg L^{-1})			
	Bulk	Outer rz[a]	Inner rz	Bulk	Outer rz	Inner rz	Bulk	Outer rz	Inner rz		Bulk	Outer rz	Inner rz		Bulk	Outer rz	Inner rz	
[Site 1 (2.5 km)]																		
1-A	7.26	7.20	7.47	35.1	40.4	21.3	0.05	0.33	0.52	***	5.05	4.85	4.68	***	7.07	26.63	29.00	***
	(0.35)			(42.3)			(0.01)		(0.13)		(0.1)				(0.19)	(1.00)	(1.17)	
1-B	7.59	7.33	7.23	16.2	29.4	37.2	0.12	0.37	0.62		4.82	4.71	4.62		9.31	34.08	51.53	
							(0.03)	(0.02)							(0.30)			
1-C	7.25	7.45	7.40	35.9	22.4	25.0	0.25	0.32	0.41		4.81	5.15	4.78		5.52	20.07	24.22	
							(0.03)	(0.01)	(0.01)						(0.29)	(0.44)	(0.58)	
1-D	6.91	6.99	7.09	77.5	65.3	51.5	0.29	0.60	0.71		4.71	4.75	4.65		6.58	17.38	20.80	
	(0.26)			(41.2)			(0.02)	(0.02)			(0.02)				(0.43)	(0.33)		
1-E	7.47	7.66	7.33	21.4	13.8	29.3	0.18	0.22	0.21		4.94	4.92	4.82		6.99	14.47	17.42	
							(0.05)	(0.03)	(0.03)						(0.12)	(0.11)	(0.32)	
1-F	7.61	7.22	7.31	15.6	37.8	30.8	0.04	0.48	0.46		4.79	4.79	4.68		8.52	18.56	20.92	
	(0.28)			(8.8)			(0.01)	(0.03)	(0.01)						(0.08)	(0.11)	(0.38)	

[Site 2 (15 km)]																		
2-A	8.39	7.59	8.26	2.6	16.4	3.5		0.01	0.05	0.05	●[c]	4.95	4.99	4.88		13.31	35.12	45.47	**
								(0.01)	(0.01)	(0.01)						(0.11)	(0.54)	(0.54)	
2-B	7.80	7.87	7.93	10.1	8.6	7.4		0.03	0.09	0.12		4.73	4.81	4.68		12.79	30.04	39.40	
		(0.17)			(3.3)			(0.002)	(0.01)	(0.03)						(0.21)	(0.29)	(0.69)	
2-C	7.87	7.85	8.05	8.6	9.0	5.7		0.04	0.04	0.05		4.90	4.85	4.72		21.11	35.52	47.81	
		(0.32)			(5.7)				(0.003)	(0.01)						(0.34)	(0.79)	(0.33)	
2-D	8.01	7.73	8.27	6.2	11.7	3.4		0.02	0.06	0.08		4.96	5.05	5.00		12.30	47.89	51.40	
								(0.004)		(0.01)						(0.33)	(0.47)	(1.37)	
[Site 3 (43 km)]																			
3-A	8.06	8.82	8.48	5.6	1.0	2.1	*[b]	0.02	0.06	0.07	*	5.27	5.25	5.08	**	17.62	31.95	38.:)2	**
								(0.004)		(0.03)						(0.64)	(0.73)		
3-B	7.89	8.78	8.08	8.2	1.1	5.3		0.004	0.03	0.03		5.14	5.03	5.01		13.33	34.40	40.84	
								(0.004)	(0.01)	(0.003)						(0.17)	(0.60)	(1.04)	
3-C	8.74	9.21	9.40	1.2	0.4	0.3		0.01	0.08	0.10		5.42	5.31	5.26		20.52	57.85	84 89	
								(0.001)		(0.02)						(0.48)	(1.40)	(1.43)	

Values in parentheses are standard deviations of laboratory triplicates.
[a] rz = rhizosphere. [b] Results of the Friedman test, where * $p \leq 0.10$; ** $p \leq 0.05$; *** $p \leq 0.01$. The results illustrate the significant differences found between the three soil components at each sampling site. [c] Result of the Wilcoxon signed rank test, where ● $p \leq 0.10$. This result illustrates the significant difference found between the $Cu(H_2O)$ concentrations of the bulk soil and of the inner rhizosphere at site 2.

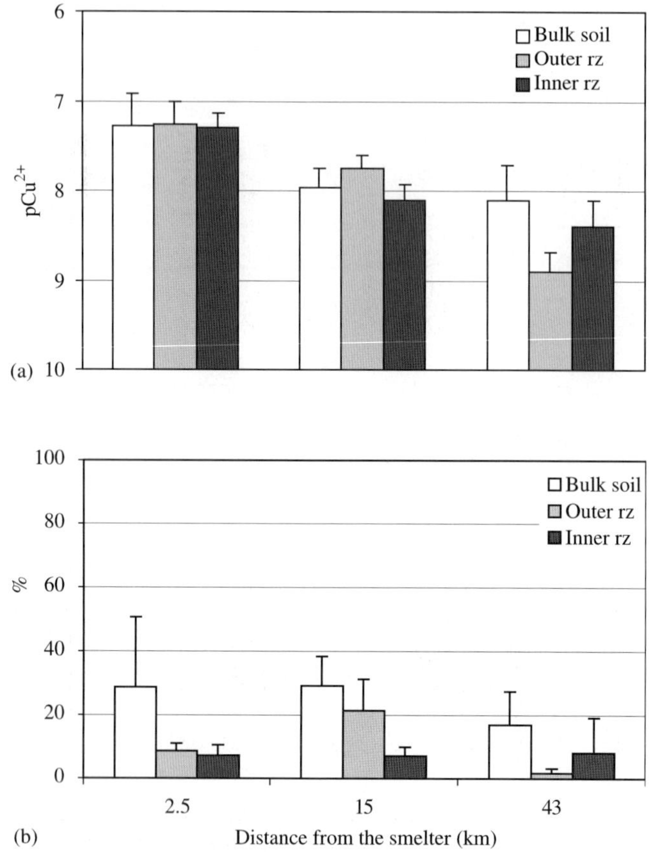

Fig. 4 (a) Cu^{2+} activities and (b) percentage of water-soluble Cu ($Cu(H_2O)$) present as Cu^{2+} in the three soil components along a soil contamination gradient. Values and standard deviations are means of the field and laboratory replicates.

in the rhizosphere and the bulk soil. To verify this hypothesis, the water-soluble Cu/DOC molar ratio ($Cu(H_2O)$ (μmol kg^{-1})/DOC (μmol kg^{-1}) $* 10^6$) was calculated. The purpose of this molar ratio is to test whether the amounts of water-soluble Cu normalized with respect to DOC concentrations differ among the three soil components. The ratios measured at site 1 are 450 and 370 for the bulk soil and the inner rhizosphere, respectively, while they range from 30 to 32 at site 2 and from 14 to 22 at site 3. The decrease of the water-soluble Cu/DOC molar ratios along the sampling transect follows the decrease observed in water-soluble Cu concentrations combined with the concomitant DOC increase from site 1 to site 3. In turn, the similarity of the water-soluble Cu/DOC molar ratios between the soil components at a given distance suggests that the metal complexation

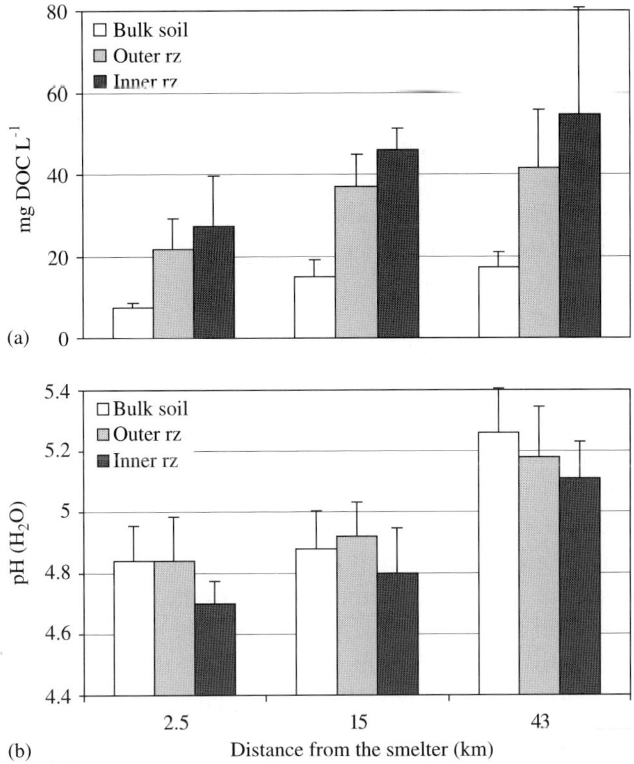

Fig. 5. Values of (a) DOC in the water extract and (b) pH(H_2O) in the three soil components along a soil contamination gradient. Values and standard deviations are means of the field and laboratory replicates.

capacity of the dissolved OM is of a relatively similar magnitude in the rhizosphere and in the bulk soil, and thus, that the free Cu^{2+} form should be similar if pH is equivalent.

It has been mentioned that the free ionic metal form is often the best predictor of bioavailability and toxicity to plants. However, some studies have shown that dissolved metal–DOC complexes can also effectively correlate with bioavailability to plants, as they can enhance metal concentrations in plant shoots and roots. The free metal ion would, therefore, not be the only chemical fraction readily adsorbed and translocated into plants. For instance, in a pot experiment, Antoniadis and Alloway (2002) found that DOC amendments in soils increased the amount of $CaCl_2$-extractable Cd, Ni and Zn, as well as enhancing the assimilation of these trace metals by ryegrass (*L. perenne* L.). Nigam *et al.* (2001) also found increased Cd assimilation by plants when combined with higher organic acid concentrations. These authors also mention that Cd^{2+} translocation in plants is probably limited because of the rapid adsorption of this cation onto roots cell walls. Cd–organic ligand complexes would thus present greater bioavailability and a more efficient transport into plant shoots. Other authors conclude that organic acids play an important role in preventing trace metal assimilation by complexation and chelation mechanisms (Manthey *et al.*, 1994). Some plant mechanisms such as metal sequestration by phytochelation or adsorption onto cell parts such as vacuoles could also prevent metal uptake (Ross and Kaye, 1994). In a culture experiment, Shenker *et al.* (2001) found that Cd bound to organic complexes was only slightly assimilated by plants, as Cd^{2+} assimilation seemed to be favored.

3.3.3. pH values

The solution pH is a key variable for the speciation of Cu. Table 6 and Fig. 5b present pH values measured in the three soil components. For each soil component, the pH values are lower at site 1 closer to the smelter, and gradually increase toward site 3. This pH gradient along the sampling transect is statistically significant for all three soil components ($p \leq 0.10$). At most sampling sites, the inner rhizosphere is the soil component where the average pH is most acidic, whereas the bulk soil has the highest pH values. The pH values of the three soil components are statistically different at sites 1 ($p \leq 0.01$) and 3 ($p \leq 0.10$).

The acidification of the rhizosphere has been reported in a range of studies from pot experiments to field investigations (Marschner, 1986; Smith and Pooley, 1989; Courchesne *et al.*, 2001; Wang *et al.*, 2001). The acidification is considered to be mainly induced by the response of roots to ionic charge imbalances in the soil solution (Nye, 1986; Haynes, 1990). This imbalance is caused by the preferential uptake of cations or anions, as selected by plant roots. The acidification of the soil solution therefore results from a release of H^+ by roots in response to an ionic charge imbalance caused by the preferential uptake of cations such as NH_4^+. Other factors, such as the exudation of organic substances by roots and CO_2 enhancement by microorganism respiration, are also known to contribute to the acidification of the rhizosphere.

The solution pH plays a key role in controlling the solubility of OM, the solubilization or precipitation of metals and the liquid phase speciation of metals. In general, metal solubility and desorption increase at lower soil pH values (Lindsay, 1979). Accordingly, Wang and Benoit (1996) showed that the desorption and solubilisation of Pb^{2+} occurred in soils with low pH values (pH < 4). Vulkan *et al.* (2000) showed that the Cu^{2+} fraction of dissolved Cu varied between 0.02 and 96%, and that the percentage was strongly dependent on pH values. The highest percentages of Cu^{2+} were found in acidic soils, where the pH was below 5.5 and where dissolved Cu content was high. Sauvé *et al.* (1995) further illustrated the determinant influence of pH on the free Cu^{2+} activity in low-Cu soils that nonetheless had high Cu^{2+} concentrations because of their severe acidity.

3.3.4. Multiple regression analyses

Simple linear regressions were first performed for every soil component in order to assess the individual associations between pH, DOC, water-soluble Cu and total-recoverable Cu on the one hand, and Cu^{2+} activity in soils on the other. Table 7 presents the results of this analysis. Note that DOC, water-soluble Cu and total-recoverable Cu concentrations were log-transformed in order to obtain a linear relation with pCu^{2+} values.

The four variables tested individually show strong relationships with pCu^{2+} in every soil component. Indeed, the regression coefficients of the four variables are all statistically significant, with significance levels varying from $p = 0.06$ to 0.0001 (Table 7). In the rhizosphere, pH appears as the key variable in the prediction of pCu^{2+} as it alone can account for up to 59 and 76% of the pCu^{2+} variation in the outer and the inner rhizosphere, respectively. In the bulk soil, where DOC is the main variable (71%), pH appears less important (46%) but is nonetheless significant ($p = 0.011$). Whether in the bulk soil, the outer rhizosphere or the inner rhizosphere, water-soluble Cu systematically yields higher r^2 values than total-recoverable Cu. Since water-soluble and total-recoverable Cu are strongly correlated, both cannot be integrated into the same multiple regression analysis. The higher r^2 of water-soluble Cu in the three soil components justifies its use together with pH and DOC.

The data from the three sampling sites are integrated in Fig. 6 to illustrate the relationships between pCu^{2+} and pH and DOC concentrations. Regardless of the soil component, the combination of low pH and DOC concentration yields the highest Cu^{2+} activity values. Conversely, high pH and DOC concentrations are associated with lower Cu^{2+} activities. Fig. 6 also shows the differentiation of the rhizosphere and bulk components with respect to DOC concentrations.

Table 7
Regression coefficients (r^2) and level of significance (p) of pCu^{2+} values against pH in H_2O, $\log_{10}$ (DOC) in H_2O, $\log_{10}$ Cu dissolved in H_2O (Cu(H_2O)) and $\log_{10}$ total-recoverable soil Cu (total Cu)

pCu^{2+} against	Bulk soil (n = 13)		Outer rz (n = 13)		Inner rz (n = 13)	
	r^2	p	r^2	p	r^2	p
pH (H_2O)	0.460	(0.011)	0.585	(0.002)	0.761	(0.0001)
$\log_{10}$ DOC[a]	0.707	(0.0003)	0.374	(0.026)	0.634	(0.001)
$\log_{10}$ Cu (H_2O)[a]	0.605	(0.002)	0.493	(0.007)	0.520	(0.005)
$\log_{10}$ Total Cu[a]	0.350	(0.033)	0.415	(0.061)	0.506	(0.006)

[a] $\log_{10}$ DOC values are in mg L^{-1}, and $\log_{10}$ Cu(H_2O) and $\log_{10}$ total Cu are in mg kg^{-1}.

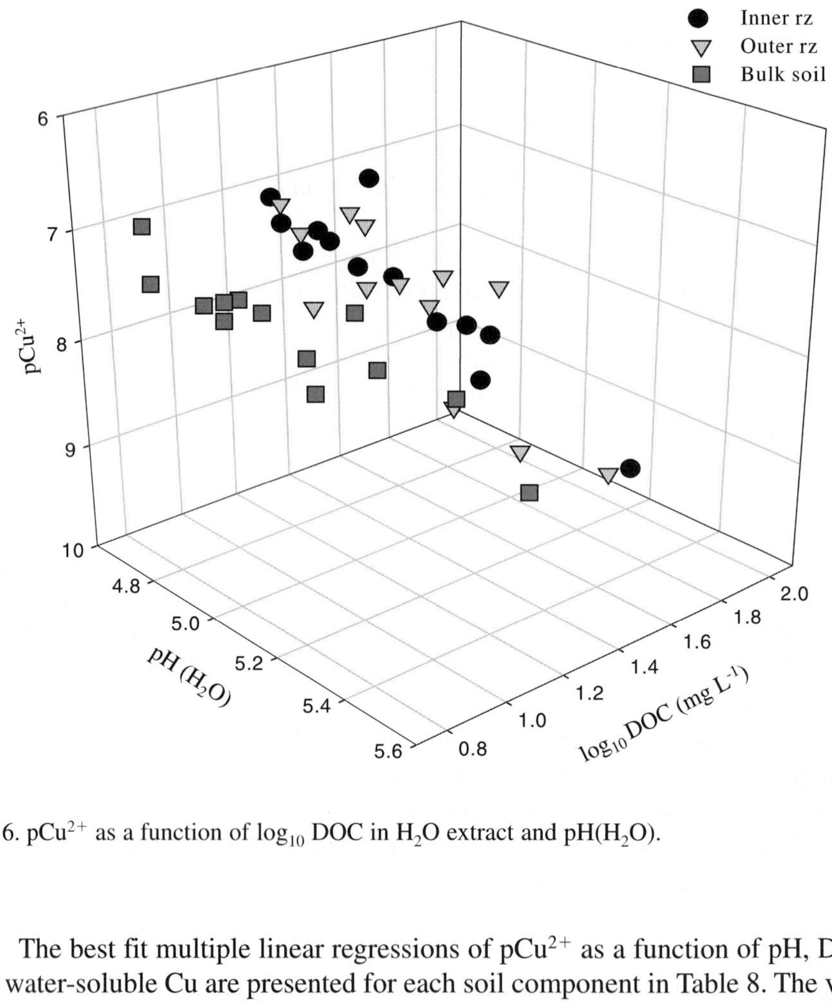

Fig. 6. pCu^{2+} as a function of $\log_{10}$ DOC in H_2O extract and $pH(H_2O)$.

The best fit multiple linear regressions of pCu^{2+} as a function of pH, DOC and water-soluble Cu are presented for each soil component in Table 8. The variables used in the analysis vary with the soil component. The regression equation yielding the highest significance level in the bulk soil only incorporates DOC, as this variable alone explains up to 71% of the pCu^{2+} variation. Although integrating water-soluble Cu and pH in this regression model can increase r^2 to 0.793, these variables did not contribute to the prediction of pCu^{2+} since their respective regression coefficients were not significant. In the outer rhizosphere, however, pH and water-soluble Cu together account for over 73% of the pCu^{2+} variation. The integration of DOC does not improve the predictive capacity in a significant way ($r^2 = 0.74$, $p = 0.586$ for DOC). The combination of pH and DOC is used in the regression equation for the inner rhizosphere. These two variables significantly explain as much as 90% of the pCu^{2+} variation. The addition of water-soluble Cu to this equation does not significantly increase the model's

Table 8
Best-fit multiple linear regressions of pCu^{2+} values against pH in H_2O, $\log_{10}$ dissolved organic carbon in H_2O (DOC) and $\log_{10}$ Cu dissolved in H_2O ($Cu(H_2O)$)

	Constant	Variables		r^2
Bulk soil				$n = 13$
pCu^{2+}	$= 5.537 \pm 0.438^a$	$+2.141 \pm 0.416 \log_{10} DOC^c$		0.707
	$(0.0000001)^b$	(0.0003)		(0.0003)
Outer rz				$n = 13$
pCu^{2+}	$= -2.744 \pm 3.239$	$+2.012 \pm 0.675$ pH	$- 0.671 \pm 0.287 \log_{10} Cu(H_2O)^c$	0.731
	(ns)	(0.014)	(0.042)	(0.001)
Inner rz				$n = 13$
pCu^{2+}	$= -4.377 \pm 1.629$	$+2.053 \pm 0.389$ pH	$+ 1.491 \pm 0.389 \log_{10} DOC$	0.903
	(0.023)	(0.0004)	(0.003)	(0.00001)

[a] ± Values represent standard errors. [b] Values in parentheses are the *p* values (level of significance) for each variable, and ns = non significant. [c] $\log_{10}$ DOC are in mg L^{-1} and $\log_{10} Cu(H_2O)$ are in mg kg^{-1}.

prediction capacity ($r^2 = 0.913$, $p = 0.355$ for water-soluble Cu). The relatively small contribution of water-soluble Cu in the pCu^{2+} regression equations is not surprising considering that this variable is itself largely controlled by pH and DOC. The strong influence of these two variables overrules any impact of water-soluble Cu on Cu^{2+} concentrations.

Studies in which multiple regression analyses were performed for pCu^{2+} predictions show contrasting results (McBride *et al.*, 1997; Sauvé *et al.*, 1997; Vulkan *et al.*, 2000). In most cases, the same variables (pH, DOC and soluble or total soil metal content) are used, although their regression coefficients and significance levels vary. For instance, Sauvé *et al.* (1997) found that 85% of the pCu^{2+} variation in a wide variety of soils could be explained by incorporating total Cu and pH in a regression model. McBride *et al.* (1997) showed that pCu^{2+} was best predicted using a regression equation that incorporated pH, OM and total Cu in soil samples from long-contaminated sites. In a series of soils with varying degrees of Cu contamination, Vulkan *et al.* (2000) found that Cu^{2+} activity was best controlled by pH and total Cu ($r^2 = 0.89$, $p < 0.001$) and, to a lesser extent, by DOC. In diverse unamended soils, Yin *et al.* (2002) calculated distribution coefficients for free Cu^{2+} between soil organic matter and solution ($K_{d\text{-}Cu^{2+}}$), and showed that free Cu^{2+} could be predicted based on three variables: total metal concentration, SOM content and pH.

4. CONCLUSIONS AND FUTURE PROSPECTS

The results of this study allow for an assessment of trace metal bioavailability with respect to the level of soil contamination as well as to soil component differentiation in forested ecosystems. The potentially available pools (water-soluble and $BaCl_2$-exchangeable fractions) of Cd, Cu and Ni, measured along the sampling transect reflect, in most cases, the level of soil contamination induced by atmospheric emissions. In general, the potentially available Cd, Cu and Ni pools, irrespective of the soil component, followed the gradient: site 1 > site 2 > site 3. Because of local factors, the water-soluble and $BaCl_2$-exchangeable Pb fractions measured at the three sites did not reflect a clear soil contamination trend, and the $BaCl_2$-exchangeable Zn concentrations showed an inverted trend, where the highest concentrations were found at site 3.

The degree of differentiation between the metal concentrations of the bulk soil, the outer rhizosphere and the inner rhizosphere varied depending on the trace metal and the chemical fraction. In most cases, either the outer rhizosphere or the inner rhizosphere contained the highest metal concentrations and were significantly different from the concentrations found in the bulk soil. Only a minority of cases occurred where the Pb and Zn concentrations were higher in the bulk soil than in the outer or inner rhizosphere. Irrespective of the trace metal, the water-soluble and $BaCl_2$-exchangeable chemical fractions were more significantly differentiated among the three soil components as compared to the organo-complexed or total-recoverable metal fractions. In addition, the level of soil contamination seemed to influence the extent of the soil component differentiation, as the concentrations of metals like Cd and Cu were more contrasted among the bulk soil, the outer and the inner rhizosphere at site 1, closer to the smelter. The water-soluble and $BaCl_2$-exchangeable fractions of most metals (except Pb) were the metal pools in which the greatest rhizosphere enrichment ratios (inner rhizosphere/bulk soil metal concentrations) were found. In addition, the rhizosphere enrichment ratio of this potentially available fraction was greater for Cu than any other metal, at all sites.

The metal concentrations of the four extractions varied with the trace metal, the soil component and the distance from the smelter. Nonetheless, for all metals and sampling sites, the potentially available metal pool always represented a very small fraction of the total-recoverable metal content in the bulk soil (3%) and the inner rhizosphere (7%), a fraction that was systematically higher in the rhizosphere. Accordingly, the organo-complexed and the residual pools were the dominant metal fractions for Cd, Cu, Ni, Pb and Zn in the bulk soil and the rhizosphere.

The speciation of Cu in the liquid phase and the measurement of associated variables (pH and DOC) also highlighted the effect of soil contamination and the uniqueness of each soil component in some cases. The soil contamination gradient was reflected in the decreasing activities of pCu^{2+} from site 1 to site 3 in the bulk soil, the outer and the inner rhizosphere. Except for a small but significant difference in the pCu^{2+} values of the three soil components at site 3, no other differentiation of the soil components with respect to free Cu^{2+} activity was found. However, the inner rhizosphere/bulk soil ratio of Cu^{2+} concentrations, although very small, was found to decrease from 1.3 at site 1 to 0.4 at site 3. The bulk soil, the outer and the inner rhizosphere contained similar amounts of water-soluble Cu when these values were normalized with respect to DOC concentrations (water-soluble Cu/DOC molar ratio), which could explain the similar Cu^{2+} activities of the three soil components. This further indicates that the apparent complexation power of OM probably has the same magnitude in the bulk soil and in the rhizosphere. DOC and pH were significantly different in each soil component at most sites, with the inner rhizosphere being systematically more acidic and having a higher DOC content than the bulk soil. The pH, DOC and water-soluble Cu all showed a very strong relationship with pCu^{2+} individually, but they were not all significant when integrated into a multiple regression analysis ($p \leq 0.05$). As such, the pCu^{2+} was best predicted by DOC in the bulk soil ($r^2 = 0.707$), by pH and water-soluble Cu in the outer rhizosphere ($r^2 = 0.731$) and by pH and DOC in the inner rhizosphere ($r^2 = 0.903$).

Although the separation of these three soil components was operationally defined, the results show that the bulk soil and the inner rhizosphere clearly differ with respect to metal enrichment and bioavailability. Considering the rhizosphere as a distinct environment is crucial for studies focusing on metal bioavailability in soils. Indeed, the results of this study establish the fact that the metal pools of the bulk soil do not correspond to the actual levels of potential metal availability found at the soil:root interface, and thus, do not accurately reflect the environment in which plant roots evolve and assimilate nutrients and contaminants. An approach that integrates the rhizosphere needs to be developed to estimate the exposure to and the uptake of trace metals by the soil biota and by higher plants. Accordingly, the soil properties and mechanisms underlying the contrasting results found between the rhizosphere and the bulk components of forested soils now need to be validated using field or greenhouse experiments, and analytical tools that are specifically designed to operate at the spatial scale of the rhizospheric environment.

ACKNOWLEDGMENTS

We thank Bruno de Passillé, Hélène Lalande, Véronique Séguin and Julie Turgeon for their help with the field and laboratory work. Funding for this research was provided by Metals in the Environment–Research Network (MITE–RN), by the Natural Science and Engineering Research Council of Canada (NSERC) and by the Fonds Québécois de la Recherche sur la Nature et les Technologies (FQRNT).

REFERENCES

Al-Wabel, M.A., Heil, D.M., Westfall, D.G., Barbarick, K.A., 2002. Solution chemistry influence on metal mobility in biosolids-amended soils. J. Environ. Qual. 31, 1157–1165.

Antoniadis, V., Alloway, B.J., 2002. The role of dissolved organic carbon in the mobility of Cd, Ni and Zn in sewage sludge-amended soils. Environ. Poll. 117, 515–521.

Arocena, J.M., Glowa, K.R., Massicotte, H.B., Lavkulich, L., 1999. Chemical and mineral composition of ectomycorrhizosphere soils of subalpine fir (*Abies lasiocarpa* (Hook.) Nutt.) in the Ae horizon of a Luvisol. Can. J. Soil Sci. 79, 25–35.

Assadian, N., Fenn, L.B., 2001. Rhizosphere chemical changes enhance heavy metal absorption by plants growing in calcareous soils. In: Gobran, G.R., Wenzel, W.W., Lombi, E. (Eds.), Trace Elements in the Rhizosphere. CRC Press, Boca Raton, FL, pp. 43–60.

Bowen, G.D., Rovira, A.D. 1991. The rhizosphere. The hidden half of the hidden half. In: Waisel, Y., Eshel, A., Kafkafi, U. (Eds.), Plant Roots: The Hidden Half. Marcel Dekker, New York, pp. 641–669.

Brindley, G.W., Brown, G., 1980. Crystal Structures of Clay Minerals and their X-ray Identification. Mineralogical Society, London, UK.

Carter, M.R., 1993. Soil Sampling and Methods of Analysis. Lewis Publishers, Boca Raton, FL.

Cherrey, A., Chaignon, V., Hinsinger, P., 1999. Bioavailability of copper in the rhizosphere of rape and ryegrass cropped in vineyard soils. In: Wenzel, W.W., Adriano, D.C., Alloway, B., Doner, H.E., Keller, C., Lepp, N.W., Mench, M., Naidu, R., Pierzynski G.M. (Eds.), Proceedings of Extended Abstracts, Fifth International Conference on the Biogeochemistry of Trace Elements, Vienna, pp. 196–197.

Chino, M., Goto, S., Youssef, R., Miah, Y., De Long, C., 1999. Behavior of micronutrients in the rhizosphere. In: Wenzel, W.W., Adriano, D.C., Alloway, B., Doner, H.E., Keller, C., Lepp, N.W, Mench, M., Naidu, R., Pierzynski, G.M. (Eds.), Proceedings of Extended Abstracts, Fifth International Conference on the Biogeochemistry of Trace Elements, Vienna, pp. 180–181.

Chung, J.-B., Zasoski, R.J., 1994. Ammonium–potassium and ammonium–calcium exchange equilibria in bulk and rhizosphere soil. Soil Sci. Soc. Am. J. 58, 1368–1375.

Cieśliński, G., Van Rees, K.C.J., Szmigielska, A.M., Krishnamurti, G.S.R., Huang, P.M., 1998. Low-molecular-weight organic acids in rhizosphere soils of durum wheat and their effect on cadmium bioaccumulation. Plant Soil 203, 109–117.

Courchesne, F., Gobran, G.R., 1997. Mineralogical variations of bulk and rhizosphere soils from a Norway spruce stand. Soil Sci. Soc. Am. J. 61, 1245–1249.

Courchesne, F., Séguin, V., Dufresne, A., 2001. Solid phase fractionation of metals in the rhizosphere of forest soils. In: Gobran, G.R., Wenzel, W.W., Lombi, E. (Eds.), Trace Elements in the Rhizosphere. CRC Press, Boca Raton, FL, pp. 189–206.

Dieffenbach, A., Göttlein, A., Matzner., E., 1997. *In situ* soil solution chemistry in an acid forest soil as influenced by growing roots of Norway spruce (*Picea abies* (L.) Karst.). Plant Soil 192, 57–61.

Dixon, J.B., Weed, S.B., 1989. Minerals in Soil Environments, second ed. Book Series # 1. Soil Science Society of America, Madison, WI.

Dudka, S., Ponce-Hernandez, R., Hutchinson, T.C., 1995. Current level of total element concentrations in the surface layer of Sudbury's soils. Sci. Total Environ. 162, 161–171.

Dudley, L.M., McNeal, B.L., Baham, J.E., 1987. Characterization of soluble organic compounds and complexation of copper, nickel, and zinc in extracts of sludge-amended soils. J. Environ. Qual. 16, 341–348.

Dumontet, S., Dinel, H., Lévesque, P.E.N., 1992. The distribution of pollutant heavy metals and their effect on soil respiration and acid phosphatase activity in mineral soils of the Rouyn–Noranda region. Québec. Sci. Total Environ. 121, 231–245.

Fenn, L.B., Assadian., N., 1999. Can rhizosphere chemical changes enhance heavy metal absorption by plants growing in calcareous soil? In: Wenzel, W.W., Adriano, D.C., Alloway, B., Doner, H.E., Keller, C., Lepp, N.W., Mench, M., Naidu, R., Pierzynski, G.M. (Eds.), Proceedings of Extended Abstracts, Fifth International Conference on the Biogeochemistry of Trace Elements, Vienna, pp. 154–155.

Freedman, W., 1978. Effects of smelter pollution near Sudbury, Ontario, Canada, on surrounding forested ecosystems. Ph.D. Thesis, University of Toronto, Canada.

Freedman, W., Hutchinson, T.C., 1980. Pollutant inputs from the atmosphere and accumulation in soils and vegetation near a nickel–copper smelter at Sudbury, Ontario, Canada. Can. J. Bot. 58, 108–132.

Gobran, G.R., Clegg, S., 1996. A conceptual model for nutrient availability in the mineral soil–root system. Can. J. Soil Sci. 76, 125–131.

Gobran, G.R., Clegg, S., Courchesne., F., 1998. Rhizospheric processes influencing the biogeochemistry of forest ecosystems. Biogeochem. 42, 107–120.

Göttlein, A., Heim, A., Matzner, E., 1999. Mobilization of aluminium in the rhizosphere soil solution of growing tree roots in an acidic soil. Plant Soil 211, 41–49.

Haynes, R.J., 1990. Active ion uptake and maintenance of cation–anion balance: a critical examination of their role in regulating rhizosphere pH. Plant Soil 126, 247–264.

Hendershot, W.H., Duquette, M., 1986. A simple barium chloride method for determining cation exchange capacity and exchangeable cations. Soil Sci. Soc. Am. J. 50, 605–608.

Hinsinger, P., 1998. How do plant roots acquire mineral nutrients? Chemical processes involved in the rhizosphere. Adv. Agron. 64, 225–265.

Huang, J.W., Chen, J., Berti, W.R., Cunningham, S.D., 1997. Phytoremediation of lead-contaminated soils: role of synthetic chelates in lead phytoextraction. Environ. Sci. Technol. 31, 800–805.

Jones, D.L., Darrah, P.R., 1994. Role of root derived organic acids in the mobilization of nutrients from the rhizosphere. Plant Soil 166, 247–257.

Jungk, A.O., 1996. Dynamics of nutrient movement at the soil–root interface. In: Waisel, Y., Eshel, A., Kafkafi, U., (Eds.), Plant Roots : The Hidden Half, second ed. Marcel Dekker, New York, pp. 529–556.

Kabala, C., Singh, B.R., 2001. Fractionation and mobility of copper, lead, and zinc in soil profiles in the vicinity of a copper smelter. J. Environ. Qual. 30, 485–492.

Kabata-Pendias, A., Pendias, H., 1992. Trace Elements in Soils and Plants, second ed. CRC Press, Boca Raton, FL.

Knight, B., Zhao, F.J., McGrath, S.P., Shen, Z.G., 1997. Zinc and cadmium uptake by the hyperaccumulator *Thlaspi caerulescens* in contaminated soils and its effects on the concentrations and chemical speciation of metals in soil solution. Plant Soil 197, 71–78.

Krishnamurti, G.S.R., Naidu, R., 2002. Solid-solution speciation and phytoavailability of copper and zinc in soils. Environ. Sci. Technol. 36(12), 2645–2651.

Lasat, M.M., 2002. Phytoextraction of toxic metals: a review of biological mechanisms. J. Environ. Qual. 31 109–120.

Laurie, S.H., Manthey, J.A., 1994. The chemistry and role of metal ion chelation in plant uptake processes. In: Manthey, J.A., Crowley, D.E., Luster, D.G. (Eds.), Biochemistry of Metal Micronutrients in the Rhizosphere. CRC Press, Boca Raton, FL, pp. 165–182.

Laurie, S.H., Tancock, N.P., McGrath, S.P., Sanders, J.R., 1991. Influence of complexation on the uptake by plants of iron, manganese, copper and zinc. I. Effect of EDTA in a multi-metal and computer simulation study. J. Exp. Bot. 42, 509–514.

Lindsay, W.L., 1979. Chemical Equilibria in Soils. Wiley, New York.

Løbersli, E., Gjengedal, E., Steinnes, E., 1991. Impact of soil acidification on the mobility of metals in the soil–plant system. In: J.-P. Vernet (Ed.), Heavy Metals in the Environment. Elsevier, New York, pp. 37–53.

Lorenz, S.E., Hamon, R.E., Holm, P.E., Domingues, H.C., Sequeira, E.M., Christensen, T.H., McGrath, S.P., 1997. Cadmium and zinc in plants and soil solutions from contaminated soils. Plant Soil 189, 21–31.

Manthey, J.A., Luster, D.G., Crowley, D.E., 1994. Biochemistry of metal micronutrients in the rhizosphere: an introduction. In: Manthey, J.A., Crowley, D.E., Luster, D.G. (Eds.), Biochemistry of Metal Micronutrients in the Rhizosphere. CRC Press, Boca Raton, FL, pp. 1–9.

Marschner, H., 1986. Mineral Nutrition of Higher Plants. Academic Press, London.

Marschner, H., 1995. Mineral Nutrition of Higher Plants, second ed. Academic Press, London.

Marschner, H., 1998. Soil–root interface: biological and biochemical processes. In: Huang, P.M., Adriano, D.C., Logan, T.J., Checkai, R.T. (Eds.), Soil Chemistry and Ecosystem Health. SSSA Special Publication 52, Soil Science Society of America, Madison, WI, pp. 191–231.

Marschner, H., Römheld, V., 1996. Root-induced changes in the availability of micronutrients in the rhizosphere. In: Waisel, Y., Eshel, A., Kafkafi, U. (Eds.), Plant Roots: The Hidden Half, second ed. Marcel Dekker, New York, pp. 557–579.

McBride, M.B., 1994. Environmental Chemistry of Soils. Oxford University Press, New York.

McBride, M.B., Sauvé, S., Hendershot, W.H., 1997. Solubility control of Cu, Zn, Cd and Pb in contaminated soils. Eur. J. Soil Sci. 48, 337–346.

McGrath, S.P., Shen, Z.G., Zhao, F.J., 1997. Heavy metal uptake and chemical changes in the rhizosphere of *Thlaspi caerulescens* and *Thlaspi ochroleucum* grown in contaminated soils. Plant Soil 188, 153–159.

McLaughlin, M.J., Smolders, E., Merckx, R., 1998. Soil-root interface: physicochemical processes. In: Huang, P.M., Adriano, D.C., Logan, T.J., Checkai, R.T. (Eds.), Soil Chemistry and Ecosystem Health. SSSA Special Publication 52, Soil Science Society of America, Madison, WI, pp. 233–277.

Mench, M., Martin, E., 1991. Mobilization of cadmium and other metals from two soils by root exudates by *Zea mays* L., *Nicotina tabacum* L., and *Nicotina rustica* L. Plant Soil 132, 187–196.

Minnich, M.M., McBride, M.B., Chaney, R.L., 1987. Copper activity in soil solution: II. Relation to copper accumulation in young snapbeans. Soil Sci. Soc. Am. J. 51, 573–578.

Morgan, J.J., Stumm, W., 1991. Chemical processes in the environment, relevance of chemical speciation. In: Merian, E. (Ed.), Metals and their Compounds in the Environment: Occurrence, Analysis, and Biological Relevance. VCH Publisher, New York, pp. 67–103.

Nigam, R., Srivastava, S., Prakash, S., Srivastava, M.M., 2001. Cadmium mobilisation and plant availability – The impact of organic acids commonly exuded from roots. Plant Soil 230, 107–113.

Nye, P.H., 1986. Acid–base changes in the rhizosphere. In: Tinker, B., Lauchli, A. (Eds.), Advances in Plant Nutrition, vol. 2. Praeger, New York, pp.129–153.

Ontario Ministry of the Environment, 2001. Metals in soil and vegetation in the Sudbury area (Survey 2000 and additional historic data). Prepared by: Morra, L., Jones, R., Crawford, G., and McLaughlin, D., Ontario Ministry of the Environment, Toronto, Canada.

Persson, H., Von Fricks, Y., Majdi, H., 1995. Root distribution in a Norway spruce (*Picea abies* (L.) Karst.) stand subjected to drought and ammonium sulfate application. Plant Soil 168–169, 161–165.

Ross, S.M., Kaye, K.J., 1994. The meaning of metal toxicity in soil–plant systems. In: Ross, S.M. (Ed.), Toxic Metals in Soil–Plant Systems. Wiley Chichester/New York, pp. 27–61.

Sanders, J.R., 1982. The effects of pH upon the copper and cupric ion concentrations in soil solutions. J. Soil Sci. 33, 679–689.

Sauvé, S., McBride, M.B., Hendershot, W.H., 1995. Ion-selective electrode measurements of copper (II) activity in contaminated soils. Arch. Environ. Contam. Toxicol. 29, 373–379.

Sauvé, S., Cook, N., Hendershot, W.H., McBride, M.B., 1996. Linking plant tissue concentrations and soil copper pools in urban contaminated soils. Environ. Poll. 94(2), 153–157.

Sauvé, S., McBride, M.B., Norvell, W.A., Hendershot, W.H., 1997. Copper solubility and speciation of *in situ* contaminated soils: effects of copper level, pH and organic matter. Water Air Soil Poll. 100, 133–149.

Schöttelndreier, M., Falkengren-Grerup, U., 1999. Plant induced alteration in the rhizosphere and the utilisation of soil heterogeneity. Plant Soil 209, 297–309.

Séguin, V., Gagnon, C., Courchesne, F., 2004. Changes in water extractable metals, pH and organic carbon concentrations at the soil–root interface of forested soils. Plant Soil, 260, 1–17.

Sheldrick, B.H., 1984. Analytical Methods Manual 1984. Land Resource Research Institute, Contribution 84-30, Agriculture Canada, Ottawa.

Shenker, M., Fan, T.W.-M., Crowley, D.E., 2001. Phytosiderophores influence on cadmium mobilization and uptake by wheat and barley plants. J. Environ. Qual. 30, 2091–2098.

Smith, W.H., Pooley, A.S., 1989. Red spruce rhizosphere dynamics: spatial distribution of aluminum and zinc in the near-root soil zone. For. Sci. 35(4), 1114–1124.

Soil Classification Working Group, 1998. The Canadian System of Soil Classification. Agriculture and Agri-food Canada, Ottawa, Canada.

Tao, S., Chen, Y.J., Xu, F.L., Cao, J., Li, B.G., 2003. Changes of copper speciation in maize rhizosphere soil. Environ. Poll. 122, 447–454.

US-Environmental Protection Agency (EPA), 1998. Microwave Assisted Acid Digestion for Sediments and Soil. SW 3051-A, Washington, DC.

Vaughan, D., Lumsdon, D.G., Linehan, D.J., 1993. Influence of dissolved organic matter on the bio-availability and toxicity of metals in soils and aquatic systems. Chem. Ecol. 8, 185–201.

Vulkan, R., Zhao, F.-J., Barbosa-Jefferson, V., Preston, S., Paton, G.I., Tipping, E., McGrath, S.P., 2000. Copper speciation and impacts on bacterial biosensors in the pore water of copper-contaminated soils. Environ. Sci. Technol. 34(24), 5115–5121.

Wang, E.X., Benoit, G., 1996. Mechanisms controlling the mobility of lead in the spodosols of a northern hardwood forest ecosystem. Environ. Sci. Technol. 30, 2211–2219.

Wang, X., Zabowski, D., 1998. Nutrient composition of Douglas-fir rhizosphere and bulk soil solutions. Plant Soil 200, 13–20.

Wang, Z., Göttlein, A., Bartonek, G., 2001. Effects of growing roots of Norway spruce (*Picea abies* (L.) Karst.) and European beech (*Fagus sylvatica* L.) on rhizosphere soil solution chemistry. J. Plant Nutr. Soil Sci. 164, 35–41.

Wang, Z., Shan, X., Zhang, S., 2002. Comparison between fractionation and bioavailability of trace elements in rhizosphere and bulk soils. Chemosphere 46, 1163–1171.

Weng, L., Temminghoff, E.J.M., Van Riemsdijk, W.H., 2001. Determination of the free ion concentration of trace metals in soil solution using a soil column Donnan membrane technique. Eur. J. Soil Sci. 52, 629–637.

Wenzel, W.W., Bunkowski, M., Puschenreiter, M., Horak, O., 2003. Rhizosphere characteristics of indigenously growing nickel hyperaccumulator and excluder plants on serpentine soil. Environ. Poll. 123, 131–138.

Yin, Y., Impellitteri, C.A., You, S.-J., Allen, H.E., 2002. The importance of organic matter distribution and extract soil:solution ratio on the desorption of heavy metals from soils. Sci. Total Environ. 287, 107–119.

Zhang, H., Zhao, F.-J., Sun, B., Davison, W., McGrath, S.P., 2001. A new method to measure effective soil solution concentration predicts copper availability to plants. Environ. Sci. Technol. 35(12), 2602–2607.

Biogeochemistry of Trace Elements in the Rhizosphere
P.M. Huang and G.R. Gobran (Editors)

Chapter 9

Influence of willow (*Salix viminalis* L.) roots on soil metal chemistry: Effects of clones with varying metal uptake potential

M. Greger

Department of Botany, Stockholm University, SE-106 91 Stockholm, Sweden
E-mail: gregerm@botan.su.se

ABSTRACT

Various Salix clones show differing efficiencies of metal uptake. Some clones accumulate certain metals to a significant extent, while excluding others. The aim of this study was to find out if the tendency to have low or high uptake of Zn, Cd or Cu was due to the varying composition of root exudates. A study of pH, Cd, Cu or Zn, organic acids and peptides in root exudates from different Salix clones was performed using a rhizobox-like method containing control soil or soil amended with Zn, Cd or Cu. The study shows that the release of metals from soil colloids differs between metals, as well as between willow clones with different properties of metal accumulation. Most of these mechanisms are found in low accumulators. Mechanisms involving pH seem to be important for both Cd and Cu accumulation. High pH decreases Cd release in soil with low Cd content and an increase in pH decreases the release of Cu from soil with high Cu levels by low-Cu-accumulating clones. In the case of Zn, organic acids and/or peptides seem to be important in reducing Zn uptake in soil with a low Zn content, while increasing pH functions as a mechanism to decrease Zn uptake in soil with a high Zn content. Thus, rhizosphere processes may partly account for the differences in the ability of willow clones to accumulate Cd, Zn and Cu.

1. INTRODUCTION

Heavy metal uptake by plants in relation to soil metal concentration has been studied in two ways: from the point of view of the soil or from the point of view of the plant. Soil scientists have tried to find an extraction medium that mimics the plant-available fraction of metals in the soil. Various extractants have been tested and compared with plant uptake of a specific metals, mostly without success (e.g. Ross, 1994). It may never be possible to accurately characterize the plant-available fraction of a metal, since many plants appear to regulate metal uptake by influencing soil chemistry (Greger and Landberg, 1995). Even though soil pH, organic matter content, cation-exchange capacity (CEC), bioavailability and total metal contents affect the uptake of metals by plants, the relationships of these factors to metal uptake are not always evident (Ross, 1994; Greger and Landberg, 1995). Often, they are overshadowed by the plants own regulation of the release and uptake of metal from the soil in the rhizosphere.

Plant physiologists have, on the other hand, disregarded the role played by soil and have studied metal uptake in nutrient solutions, which should imitate soil solutions. This seems to be an opportune strategy, since there is often a positive relationship between metal content of the soil solution and plant uptake (Denaix *et al.*, 2002). However, none of the methods presented above illustrate how plant roots may influence the release of metals from soil, and what proportion of these released metals are actually taken up by plants (Stoltz and Greger, 2002). How do plants regulate metal uptake, and how do external factors, such as soil metal levels, influence this regulation? In addition, are there differences in efficiencies of metal uptake between various metals as well as between various plant genotypes?

2. WILLOW CLONES VARY IN THEIR PROPERTIES TO TAKE UP METALS

Different plant species and various genotypes of a species can show differing efficiencies of metal uptake. This has been shown for various *Salix* clones (Landberg and Greger, 1994; Greger and Landberg, 1996; Greger *et al.*, 2001). Some clones accumulate certain metals to a significant extent, at the same time excluding others (Landberg and Greger, 1994). Translocation of a metal to the shoot of a plant varies between clones, and even if one element is translocated to a significant extent in one clone, this is not necessarily the case for another metal (Landberg and Greger, 1994). Moreover, a high accumulating clone is not necessarily tolerant to that metal (Greger and Landberg, 1999). Properties of metal accumulation appear to be independent of high biomass production (Greger *et al.*, 2001). Thus, there is a significant difference between clones with respect to accumulation, translocation and tolerance to metals (Greger and Landberg, 1999). Owing to the variation in these properties, one can find willow clones with a combination of properties suited to specific situations, for example to work as phytoextractors of Cd (Greger and Landberg, 1999).

Plants take up metals into the root free space and further transport them apoplastically into the xylem and, to a lesser extent, into the cell cytoplast. The cell wall fraction of metals is very high (Beauford *et al.*, 1977; Zornova *et al.*, 2002) due to high cell wall CEC. In the root, the suberized endodermis forms a barrier to the free, inward diffusion of dissolved ions and molecules to the vascular tissues within the stele (Mauseth, 1988). However, this barrier is absent in the root tip and the further away from the tip this barrier is formed, the greater is the potential for inward ion diffusion to the xylem. Differences in the distance between the barrier and the root tip in willow clones can explain variations in the accumulation and translocation of metals (Lux *et al.*, 2004).

Clonal differences in Cd uptake are independent of whether the plant is cultivated in nutrient solution or soil; however, the relation between clones in accumulation of other metals, such as Cu and Zn, may differ depending on these two media (Greger *et al.*, 2001). In the case of Cu and Zn, the presence of soil may trigger roots to initiate rhizosphere processes (Stoltz and Greger, 2002), which is absent in solution culture. It is therefore important to investigate differences in root rhizosphere mechanisms between *Salix* clones.

3. ROOT EXUDATES

The effects of roots on the soil are manifested in the rhizosphere, seldom in the bulk soil (Gobran *et al.*, 1999). For example, Murányi *et al.* (1994) showed that there was a pH decrease of approximately 1 unit in the first mm of soil from the root surface. Proton exudation from roots will release cations from soil colloids, and thus increase cation concentration in soil water and cation uptake into the root-free space (Fig. 1). Some plants also increase rhizosphere pH by taking up NO_3^{2-} instead of NH_4^+ (Taylor and Foy, 1985), thus decreasing metal release from soil colloids and free space metal uptake.

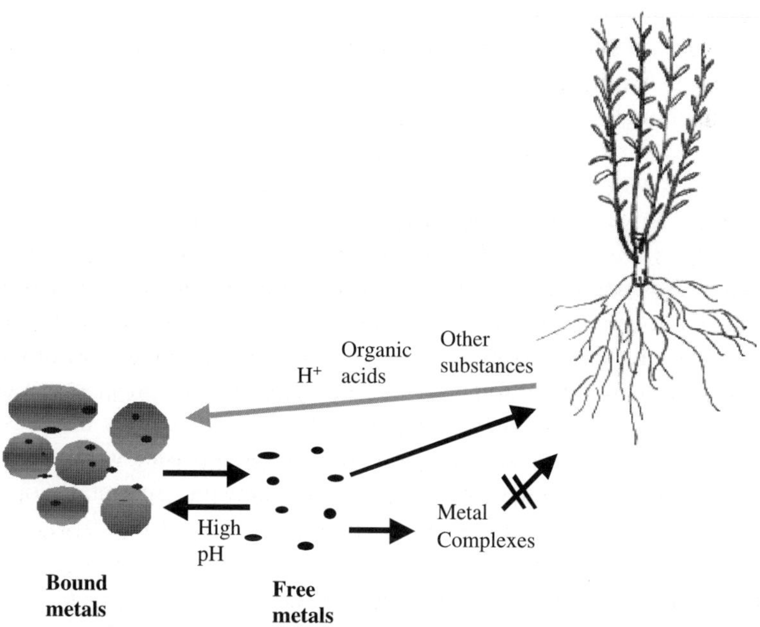

Fig. 1. Release of metals from soil colloids by various compounds in root exudate.

Plants release various substances from the root tips; about 5–21% of photosynthetically fixed carbon is transferred to the rhizosphere as root exudate (Marschner, 1995). Root exudates include both high- and low-molecular-weight compounds. The first mentioned include polysaccharides and proteins, while the latter include, among others, amino acids, organic acids, sugars and phenolics (Marschner, 1995). Root exudates mobilize or bind metals to an extent that depends on the ligand, the solid phase and the metal involved (Fig. 1), e.g. *Zea mays* exudates are able to bind Cd (Mench *et al.*, 1987). There is increasing evidence that soluble root exudates increase the solubility of metals in the rhizosphere, and that these phenomena may depend on plant cultivars (Bromfield, 1958; Römheld and Marschner, 1990). Increased solubilization of phosporus and micronutrients in the presence of root exudates may increase the uptake of these elements by nutrient-deficient plants (Marschner, 1995). On the other hand, increased solubility of metals by complex formation with substances in root exudates may decrease metal uptake in plants characterized by low uptake of and low tolerance to particular metal. For example, in Al-tolerant plants, organic acids may restrict Al uptake as a result of the formation of Al–organic acid complexes (Ma *et al.*, 2001). Also, the uptake of other metals may be inhibited by the release of organic acids from the roots (Jackson *et al.*, 1991). The free organic acid complex might enter the root-free space, but if the size of the molecule is too large further translocation in the plant body, such as translocation to the shoot, is probably prevented (Marschner 1995). The release of organic acids from roots into the rhizosphere has been implicated in many mechanisms for either increasing or decreasing the availability of nutrients to the plants (Uren and Reisenauer, 1988; Marschner, 1995).

4. EXUDATES FROM ROOTS OF WILLOW CLONES WITH DIFFERENT METAL UPTAKE PROPERTIES

Variations in the composition of root exudates may be one cause of difference in metal uptake between plant species and genotypes (Mench and Martin, 1991). It is important to investigate if the differences in accumulation capacities for metals by willow clones can be related to differences in rhizosphere processes. It is necessary to ask, therefore, how soil chemistry is influenced by the presence of willow roots, if soil chemistry varies between willow clones with properties of low and high metal accumulation and if the mechanisms are metal – or concentration – specific.

4.1. Materials and methods

Metal-tolerant clones of *Salix viminalis* with properties for accumulation of either low or high concentrations of Cd, Zn or Cu (Table 1) were placed in

Table 1
Clones used in this investigation and their properties to accumulate Cd, Zn or Cu shown as concentration of the metal in shoot and roots after metal treatment. Data are from a screening where different *Salix* clones were treated with 112 mg Cd, 18.9 mg Cu or 441 mg Zn m^{-3}

Clone	Metal concentration ($mg\ kg^{-1}$)		Root fresh weight (g)	
	Root	Shoot	Untreated soil	Treated soil
Cadmium				
CdL	16	0.96	6.24^{a}	7.52^{ab}
CdH	156	8.53	5.97^{a}	4.38^{a}
Zinc				
ZnL	143	18	4.39^{a}	6.21^{a}
ZnH	1320	1326	8.70^{ab}	6.45^{a}
Copper				
CuL	14	0.10	8.77^{ab}	13.10^{b}
CuH	197	2.40	5.65^{ab}	7.61^{a}

Notes: The root fresh weights from the present investigation are also presented. Significant differences ($p \leq 0.05$) are denoted by different letters (Student's *t*-test).

a rhizobox-like system (Fig. 2; modified from Youssef and Chino, 1988) for 48 h. These contained either untreated control soil or control soil spiked with Cd, Cu or Zn, depending on the clone being tested. Prior to transfer, willows were grown from cuttings for 3 months in nutrient medium, and the root mass of the plants were very similar (Table 1). The non-sterile soils (80% peat and 20% clay) were spiked with 19, 600 or 400 mg kg^{-1} $CdCl_2$, $ZnCl_2$ or $CuCl_2$, respectively, 2 weeks in advance. Pretests showed that all clones tolerated these conditions with no toxic effects. The metal concentrations in the untreated soil were 0.42, 2.47 and 2.72 mg Cd, Cu and Zn kg^{-1} dry weight, respectively. After 48 h, roots were dipped and gently shaken in water for 30 s in order to collect the exudate from the root surface. Root exudates were then filtered through a 0.45 μm Millipore filter and analyzed for pH, metal, organic acids and peptides. Three replicates were made. Organic acids and peptides were not analyzed in the soil samples. The control soil had an initial pH of 4.8.

Succinic acid showed very little variation in all clone exudates in all treatments, but the concentration of H^{+} (pH) and other organic acids and level of hydrophilic peptides were shown to vary within the rhizosphere at different soil metal levels and between *Salix* clones (Tables 2–4). Differences were also found when comparing metals.

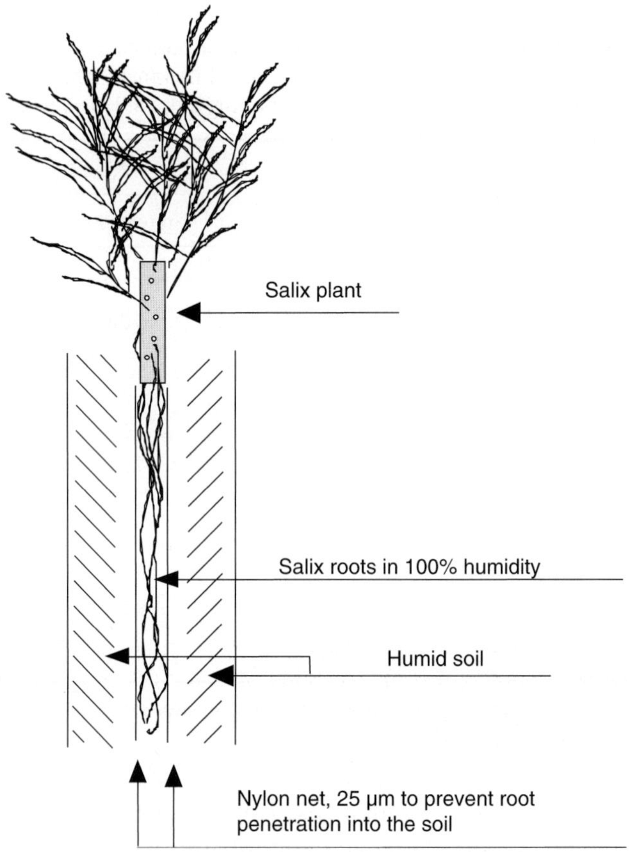

Fig. 2. Experimental set-up.

4.2. Zinc

Differences in Zn accumulation between high- and low-Zn-accumulating clones were not due to pH, since neither in the presence of untreated nor Zn-spiked soil were pH differences evident between clones (Table 2).

Organic acids and peptide concentrations seem to play an important role in Zn accumulation in the low-Zn-accumulating clone grown in the presence of untreated soil. These compounds release Zn into the rhizosphere by forming soluble complexes keeping Zn in solution. It is possible that the size of the complex prevents Zn uptake into the plant tissue, and at least decreases the translocation in root and shoot tissues. This is a proposition to why the level of Zn in the rhizosphere of the low Zn accumulator tends to be higher than in the case of the high-Zn accumulator and why the low-Zn accumulator accumulates less Zn than the high-Zn accumulator. Organic acids have earlier been found to play a role in Zn accumulation (e.g. Fan *et al.*, 2001), and has been found to increase the Zn

Table 2
pH, Zn concentration (ng/g fresh weight root), concentrations of organic acids (mg/kg fresh weight root) and level of hydrophilic peptides of rhizosphere solution, i.e. root exudate after 48 h in the presence of soil untreated or treated with 600 mg Zn kg^{-1}

Clone	pH	Metal	Citric Acid	Malic Acid	Succinic Acid	Peptides
Untreated soil						
ZnL	3.79 (0.22)[a]	259 (55)[a]	58 (32)[a]	151 (81)[a]	247 (7)[a]	Medium
ZnH	3.95 (0.18)[a]	59 (17)[a]	ND[b]	9 (7)[b]	110 (110)[a]	Low
Zn-treated soil						
ZnL	5.19 (0.26)[b]	5705(629)[c]	ND[b]	ND[b]	168 (14)[a]	Low
ZnH	5.34 (0.09)[b]	4722(593)[c]	9 (7)[a]	ND[b]	110 (9)[a]	Low

Notes: L is a low accumulating clone and H is a high accumulating clone of the indicated metal. ND, not detected. $n=3$, (SE). Significant differences ($p \leq 0.05$) in each column are denoted by different letters (Student's *t*-test).

uptake in Zn-deficient plants (Marschner, 1995). In the present case, however, there was probably no mechanism initiated by Zn deficiency. The reason for the low uptake in the low-Zn-accumulating clone may be due to complexes with organic acids and/or peptides preventing the uptake, in a similar manner as in Al-tolerant plants, where organic acids prevents Al uptake (Ma *et al.*, 2001).

At high soil Zn levels, another mechanism is used; the pH in the rhizosphere increases, which in turn decreases the release of Zn from the soil colloids. However, there is a slightly lower Zn concentration in the rhizosphere solution of the high-Zn accumulator than that of the low-Zn accumulator. The similarity in mechanisms between the two clone types in the presence of Zn-spiked soil points to the fact that the differences between the clones may be due to a higher Zn uptake from soil solution by the high-Zn accumulator. The proposed rhizosphere mechanisms are shown in Fig. 3.

4.3. Cadmium

The difference in Cd accumulation between high- and low-Cd-accumulating clones in untreated soil seems to be dependent on pH regulation of the rhizosphere in the presence of low-Cd-accumulating clones. The latter has a significantly higher pH in the rhizosphere than the high accumulator, which decreases the release of Cd from the soil showing a lower Cd concentration in the rhizosphere of the low-Cd accumulator than of the high-Cd-accumulator (Table 3). In root exudate, however, the organic acids appear not to be involved in the Cd release differences.

When the soil was spiked with Cd, the low-Cd-accumulating clone had a higher rhizospheric Cd concentration than the high-Cd-accumulating clone.

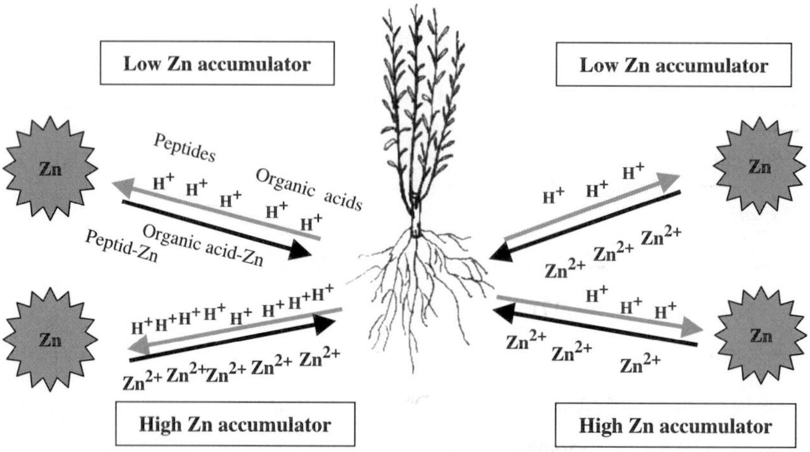

Fig. 3. Schematic diagram showing proposed Zn-releasing processes in rhizosphere when roots of high- and low-Zn-accumulating clones are grown in soil with low (left) or high (right) Zn levels.

Table 3
pH, Cd concentration (ng/g fresh weight root), concentrations of organic acids (mg/kg fresh weight root) and level of hydrophilic peptides of the rhizosphere solution, i.e. root exudate after 48 h in the presence of soil untreated or treated with 19 mg Cd kg^{-1}

Clone	pH	Metal	Citric Acid	Malic Acid	Succinic Acid	Peptides
Untreated soil						
CdL	4.95 (0.22)[a]	ND[a]	ND	ND	128 (3)[a]	Low
CdH	3.62 (0.08)[b]	5 (2)[a]	ND	ND	142 (10)[a]	Low
Cd-treated soil						
CdL	4.23 (0.22)[b]	468 (4)[b]	ND	ND	256 (144)[a]	Medium
CdH	3.82 (0.11)[b]	151 (52)[c]	ND	ND	380 (120)[a]	Low

Notes: L is a low accumulating clone and H is a high accumulating clone of the indicated metal. ND, not detected. $n=3$, (SE). Significant differences ($p \leq 0.05$) in each column are denoted by different letters (Student's t-test)

The reason may be binding of Cd in the soil solution to macromolecules such as peptides, which were found at an elevated level in the rhizosphere solution of low-Cd-accumulating clones. The proposed rhizosphere processes are illustrated in Fig. 4.

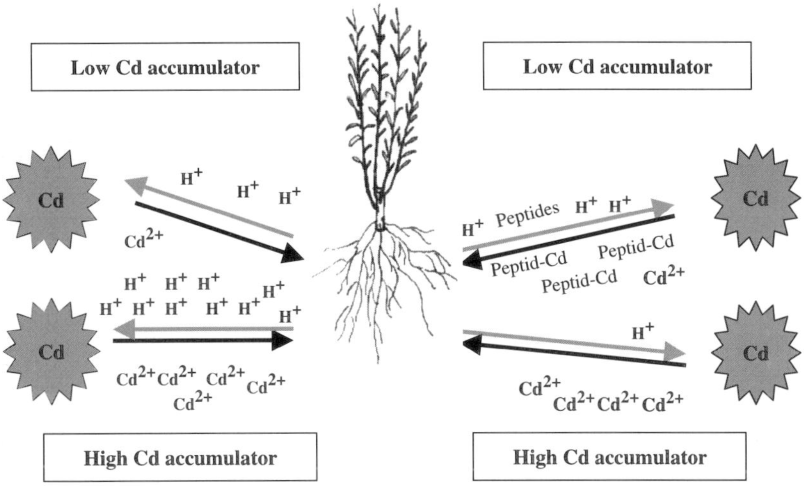

Fig. 4. Schematic diagram of proposed Cd-releasing processes in rhizosphere when roots of high- and low-Cd-accumulating clones are grown in soil with low (left) or high (right) Cd levels.

4.4. Copper

The pH of the rhizosphere could not explain the difference in Cu accumulation between the clone types in the presence of untreated soil, since Cu level and rhizosphere pH were similar for both low- and high-Cu-accumulating clones (Table 4). However, at higher external Cu levels (Cu-spiked soil), a pH-regulating mechanism was probably present in the rhizosphere of the low Cu-accumulating clone, since the rhizosphere pH of this clone increased from 3.84 to 5.22, when compared with the control soil. Such a pH increase was not found in the rhizosphere of the high-Cu-accumulating clone resulting in a lower pH and a higher Cu level in the rhizosphere of the high-Cu-accumulating clone compared with low-Cu accumulator. Thus, high Cu-accumulating clones could accumulate more Cu than low-Cu accumulating ones. None of the measured organic acids appear to be involved in the difference in release of Cu from soil between clones. The proposed rhizosphere processes are illustrated in Fig. 5.

5. CONCLUSIONS

The conclusion one can draw from this work is that rhizosphere mechanisms involved in the release of metals from soil colloids may differ between metals as well as between willow clones with different properties of metal accumulation. Most of these mechanisms are found in low accumulators. Mechanisms involving pH seem to be important for both Cd and Cu accumulation. High pH decreases Cd release in soil with low Cd content and an increase in pH decreases the release of Cu from soil with high Cu level by low-Cu-accumulating clones.

Table 4
pH, Cu concentration (ng/g fresh weight root), concentrations of organic acids (mg/kg fresh weight root) and level of hydrophilic peptides of the rhizosphere solution, i.e. root exudate after 48 h in the presence of soil untreated or treated with 400 mg Cu kg^{-1}

Clone	pH	Metal	Citric Acid	Malic Acid	Succinic Acid	Peptides
Untreated soil						
CuL	3.84 (0.12)[a]	9 (7)[a]	ND	ND	117 (15)[a]	Low
CuH	4.44 (0.38)[a]	9 (2)[a]	ND	ND	118 (33)[a]	Low
Cu-treated soil						
CuL	5.22 (0.40)[b]	64 (6)[a]	ND	ND	148 (20)[a]	Low
CuH	4.09 (0.22)[a]	163(40)[b]	ND	ND	324 (65)[a]	Low

Notes: L is a low accumulating and H is a high accumulating clone of the indicated metal. ND, not detected. $n=3$, (SE). Significant differences ($p \leq 0.05$) in each column are denoted by different letters (Student's *t*-test).

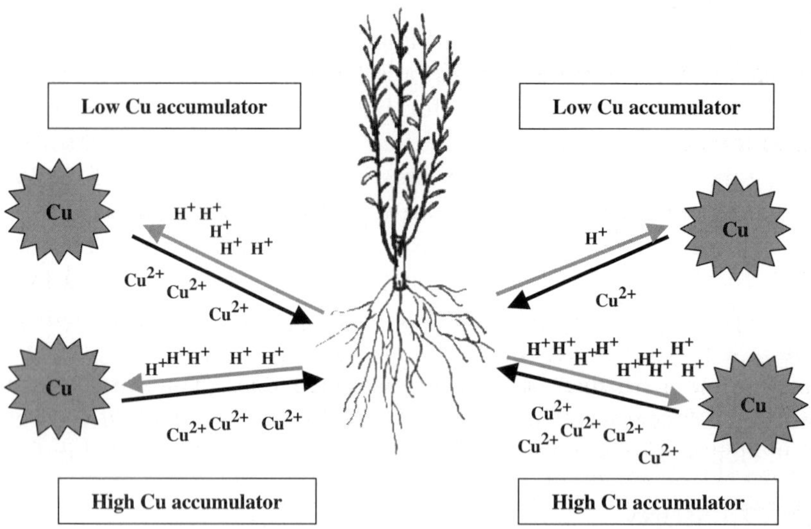

Fig. 5. Schematic diagram of proposed Cu-releasing processes in rhizosphere when roots of high- and low-Cu-accumulating clones are grown in soil with low (left) or high (right) Cu levels.

Changes in soil pH due to living organisms are normally the result of proton extrusion during active uptake across cell membranes. Reduced active uptake as a result of root toxicity could, however, not explain the changes noted with low Cd and Cu clones in the spiked soil since there was no effect on the root growth.

In the case of Zn, organic acids and/or peptides seem to be important in reducing Zn uptake in soil with a low Zn content, while increasing pH function as a mechanism to decrease Zn uptake in soil with a high Zn content. Thus, rhizosphere processes may partly account for the differences in the ability of willow clones to accumulate Cd, Zn and Cu.

The differences between clones may depend on a combined effect of plant exudate and microbial effects on the exudate (Marschner, 1995). In studies under nonsterile conditions, rhizosphere microbes may alter the chemical composition of root exudates. Therefore, the differences between high and low metal soil condition as well as different metals in spiked soil can be due to toxic metal effects or effects resulting from an excess of chloride on microbes. When comparing various clones the differences in exudate composition could have been due to various microbe–clone relationships. One should, however, keep in mind that a microbe–plant relationship is present in real environment where we also find these metal-accumulation differences between clones. Whether the differences in rhizosphere processes are due to plants alone or a combination with microbial interactions has to be further investigated.

ACKNOWLEDGMENTS

This study was funded by the Geological Survey of Sweden.

REFERENCES

Beauford, W., Barber, J., Barringer, A.R., 1977. Uptake and distribution of mercury within higher plants. Physiol. Plant 39, 261–265.

Bromfield, S.M., 1958. The solution of gamma-MnO_2 by substances released from soil and from root or oats and vetch in relation to Mn availability. Plant Soil 10, 147–160.

Denaix, L., Lamy, I., Masson, P., Mench, M., 2002. Composition or speciation in the soil solution. In: Mench, M. (Ed.), COST 837 Meeting on Risk Assessment and Sustainable Land Management Using Plants in Trace Element Contaminated Soil. INRA, Bordeaux, p. 18.

Fan, T.W.-M., Lane, A.N., Shenker, M., Bartley, J.P., Crowley, D., Higashi, R.M., 2001. Comprehensive chemical profiling of gramineous plant root exudates using high-resolution NMR and MS. Phytochemistry 57, 209–221.

Gobran, G.R., Clegg, S., Courchesne, F., 1999. The rhizosphere and trace element acquisition. In: Selim, H.M., Iskander, A. (Eds.), Fate and Transport of Heavy Metals in the Vadouse Zone, CRC Press, Boca Raton, FL, pp. 225–250.

Greger, M., Landberg, T., 1995. Kadmiumhalten i Salix relaterat till kadmiumhalten i jorden. Report, Vattenfall utveckling AB, 1995/9.

Greger, M., Landberg, T., 1996. Use of willow clones with high accumulating properties in phytoremediation of agricultural soils with elevated Cd levels. In: Prost, R. (Ed.), Proceedings of the third International Congress on Biogeochemistry, INRA, Paris, pp. 505–511.

Greger, M., Landberg, T., 1999. Use of willow in phytoremediation. Int. J. Phytorem. 1, 115–123.

Greger, M., Landberg, T., Berg, B., 2001. Salix Clones with Different Properties to Accumulate Heavy Metals for Production of Biomass. Akademitryck AB, Edsbruk.

Jackson, P.J., Unkefer, P.J., Delhaize, E., Robinson, N.J., 1991. Mechanisms of trace metal tolerance in plants. In: Katterman, F. (Ed.), Environmental Injury to Plants, Academic Press, San Diego, pp. 231–258.

Landberg, T., Greger, M., 1994. Can heavy metal tolerant clones of *Salix* be used as vegetation filters on heavy metal contaminated land? In: Perttu, K., Aronsson, P. (Eds.), "Willow Vegetation Filters for Municipal Wastewaters and Sludges, Rapport 50, Swedish University of Agricultural Sciences, Uppsala, pp. 133–144.

Lux, A., Sottnikova, A., Opatrná, J., Greger, M., 2004. Differences in structure of adventitious roots in Salix clones with contrasting characteristics of cadmium accumulation and sensitivity. Phytiol. Plant 120, 537–545.

Ma, J.F., Ryan, P.R., Delhaize, E., 2001. Aluminium tolerance in plants and the complexing role of organic acids. Trends in Plant Sci. 6, 273–278.

Marschner, H., 1995. Mineral Nutrition of Higher Plants. Second ed. Academic press, London.

Mauseth, J.D., 1988. Plant Anatomy. The Benjamin/Cummings, Menlo Park, CA, pp. 269–293.

Mench, M., Martin, E., 1991. Mobilization of cadmium and other metals from two soils by root exudates of *Zea mays* L., *Nicotiana tabacum* L. and *Nicotiana rustica* L. Plant Soil 132, 187–196.

Mench, M., Morel, J.L., Guckert, A., 1987. Metal binding properties of high molecular weight soluble exudates from maize (*Zea mays* L.) roots. Biol. Fertil. Soils 3, 165–169.

Murányi, A., Seeling, B., Ladewig, E., Jungk, A., 1994. Acidification in the rhizosphere of rape seedlings and in bulk soil by nitrification and ammonium uptake. Z. Pflanzenernähr. Bodenk. 157, 61–65.

Ross, S., 1994. Toxic Metals in Soil-Plant Systems. Wiley, Chichester.

Römheld, V., Marschner, H., 1990. Genotypical differences among graminaceous species in release of phytosiderophores and uptake of iron phytosiderophores. Plant Soil 123, 147–153.

Stoltz, E., Greger, M., 2002. Accumulation properties of As, Cd, Cu, Pb and Zn by four wetland plant species growing on submerged mine tailings. Environ. Exp. Bot. 47, 271–280.

Taylor, G.J., Foy, C.D., 1985. Mechanisms of aluminum tolerance in *Triticum aestivum* (wheat). IV, the role of ammonium and nitrate nutrition. Can. J. Bot. 63, 2181–2186.

Uren, N.C., Reisenauer, H.M., 1988. The role of root exudates in nutrient acquisition. Adv. Plant Nutr. 71, 469–477.

Youssef, R.A., Chino, M., 1988. Development of a new rhizobox system to study the nutrient status in the rhizosphere. Soil Sci. Plant Nutr. 34, 361–465.

Zornova, P., Vásquez, S., Esteban, E., Fernández-Pascual, M., Carpena, R., 2002. Cadmium-stress in nodulated white lupin: strategies to avoid toxicity. Plant Physiol. Biochem. 40,1003–1009.

Biogeochemistry of Trace Elements in the Rhizosphere
P.M. Huang and G.R. Gobran (Editors)

Chapter 10

Fractionation and bioavailability of copper, cadmium and lead in rhizosphere soil

S. Tao, W.X. Liu, Y.J. Chen, J. Cao, B.G. Li, and F.L. Xu

College of Environmental Sciences, Peking University, Beijing 100871, China
E-mail: taos@urban.pku.edu.cn

ABSTRACT

Chemical forms of copper, cadmium and lead in the rhizosphere of several plant species were investigated using rhizobox cultivation, sequential extraction and a set of specially designed experiments to examine dynamic changes in metal fractionation and bioavailability in the rhizosphere, and to experimentally evaluate the factors governing the variation in metal fractionation. The results from the dynamic change experiments demonstrated that there were continuous changes in metal fractionation within the maize rhizosphere. Initially, the amount of exchangeable copper and cadmium increased before dropping toward or below the initial levels after 40–70 days. Carbonate-associated copper followed a similar trend, but at a slower pace than the exchangeable fraction, while carbonate-bound cadmium and lead leveled off after an initial increase. The accumulation of the metals in the maize plant showed biomass-dependent characteristics. The amount of metals accumulated inside the plant material exceeded the initial quantity of the exchangeable metals in the soil, indicating a transformation from less bioavailable to more bioavailable forms. During cultivation, a decrease in redox potential, increases in pH and microbial activity, and an increase followed by a decrease in dissolved organic carbon in the maize rhizosphere were observed. The increase pattern of exchangeable copper and cadmium was affected by the relative magnitude of mobilization and bioaccumulation, which occured in opposite directions. The effects of acidification, alkalization and generation of root exudates were studied by the addition of acid, alkali or root exudates from the solution cultures of several plants to soil prior to incubation, and metal fractionation measurements. Changes in metal fractionation in the rhizospheres of unsterilized and sterilized soils were studied using rhizoboxes cultivated with seedlings of maize, wheat, pea and soybean. The comparison was also conducted for the soil receiving root exudate from solution cultures of the four plant species. For the calcareous soil and the plant species studied, pH was not the main cause of the corresponding alterations. The resulting significant influence of root exudates on metal fractions appears to have been brought about through complexation rather than acidification. Microbial activity also had significant effects on the metal fractionation changes.

1. INTRODUCTION

Heavy metal contamination of agricultural land is an increasingly significant environmental concern in China, especially in the northern region. Among the various anthropogenic sources, wastewater irrigation and sludge application, common practices in the suburbs of northern Chinese cities, provide a significant input of heavy metals to the farmland (Zhang *et al.*, 1988; Zhang and Gong, 1996). The uptake of metals from agricultural soil by crops and vegetables is an important pathway through which metals in contaminated soils impose health threats to organisms. On the other hand, the capacity of plant roots to remove heavy metals from contaminated soils is an emerging environmental cleanup and remedial biotechnology. In order to evaluate the risks of metal contamination in the area, it is essential to understand metal bioavailability, which depends on a metal's chemical form in the soil, rather than on the total amount accumulated (Allen, 1997; Zemberyova *et al.*, 1998).

The conditions of rhizosphere soil are considerably different from those of bulk soil (Nye, 1981). Many researchers have focused on this zone while addressing issues concerning metal fractionation and bioavailability in relation to various kinds of cultivation practice (Krishnamurti *et al.*, 1996; McGrath *et al.*, 1997; Shuman and Wang, 1997; Zoysa *et al.*, 1997). To date, the dependence of metal bioavailability on its chemical fractionation is well documented (Krishnamurti *et al.*, 1995; Krishnamurti and Naidu, 2000, 2002; Adriano, 2001). Nevertheless, the complexity of the soil–plant relationship may induce some changes in the properties of the soil rhizosphere, thereby affecting metal fractionation (Jefferey and Uren, 1983; Levesque and Mathur, 1986; Hamon *et al.*, 1995). Metals associated with soil minerals, such as carbonates or oxides, can be desorbed under acidic or reducing conditions, and root-induced changes in pH and Eh can thereby influence the bioavailability of trace metals in the rhizosphere (Marschner and Römheld, 1996). In addition, metal cations are also released via the formation of metal-organic chelates with phytosiderophores and organic acids excreted by plant roots or microbes (Krishnamurti *et al.*, 1996, 1997). Krishnamurti *et al.* (1996) investigated speciation and bioavailability of cadmium in wheat rhizosphere soil and found that the level of NH_4Cl-extractable cadmium was 2–9 times higher in the rhizosphere soil than in the bulk soil at the 2-week growth stage. At the 7-week plant growth stage, however, such differences were not observed.

To understand the changes in metal fractionation within the rhizosphere, we need more insight into the key factors influencing metal fractionation. It has been assumed that the factors affecting metal fractionation and bioavailability in soil include root-induced pH changes, metal binding by root exudates, root-induced microbial activities and root depletion as a consequence of plant uptake (Ernst, 1996). The processes responsible for root-induced rhizosphere pH changes include the evolution of CO_2 by root respiration, the release of root exudates, the excretion or reabsorption of H^+ or HCO_3^- and the microbial production of organic acids (Marschner and Römheld, 1996). In the soil complex surroundings, many trace metals are associated with minerals that can either be solubilized under acidic conditions or precipitated out under alkaline conditions (Jefferey and Uren, 1983). Plant roots are known to release considerable amounts of organic carbon into the rhizosphere. Some of the root exudates have strong complexation capacity for metal binding (Krishnamurti *et al.*, 1997). Metal fractionation in the rhizosphere is, therefore, subject to the influence of complexation. Given the high supply of organic carbon generated by the roots, the population density of microorganisms, especially with regard to bacteria, is much higher in the rhizosphere than in the bulk soil (Lynch and Whipps, 1990). Microorganisms may exert their influence on the bioavailability of trace metals by affecting (1) the growth and morphology of roots; (2) the physiology and development of plants; (3) the fractionation of metals; and (4) the root uptake process (Bowen and Rovira, 1991). Therefore, to evaluate metal dynamics in the soil–plant system, it is essential to characterize the relative importance of these factors individually.

Certain factors controlling metal fractionation in the soil rhizosphere are time-dependent. This is one of the reasons explaining some different results of root-induced rhizospheric changes in metal fractionation. For instance, using a rhizobox system, Chino *et al.* (1999) found that soluble copper increased near the root, while its total content showed little or no change. Cherrey *et al.* (1999), however, reported a depletion of copper extracted with EDTA, DTPA or $CaCl_2$ in the rhizosphere soil.

The primary objective of this study was to investigate the fractionation of copper, cadmium and lead in the rhizosphere soil in an attempt to obtain a better understanding of their availability and subsequent uptake by plants. In this regard, the present chapter focuses on (1) root-induced changes in various fractions of copper, cadmium and lead in the maize rhizosphere using rhizobox and sequential extraction techniques; and (2) influences of pH, root exudates and microbial activity on metal fractionation in the rhizosphere of maize, wheat, pea and soybean, based on the results of specially designed experiments.

2. EXPERIMENTAL DESIGN

2.1. Soil and plant materials

The calcareous soil sample (0–20 cm) was collected from a vegetable plot in Tianjin, China. The field had been irrigated with wastewater for more than 30 years, with occasional municipal and industrial sludge applications. The sample was air-dried and crushed to pass through a 2-mm plastic sieve. The general properties of the soil were pH 8.12, cation exchange capacity (CEC) 32.2 cmol kg^{-1}, organic carbon 37.4 g kg^{-1}, $CaCO_3$ 14.2 cmol kg^{-1}, and a soil microbial biomass (MBM) of 82.5 mg kg^{-1}. The total concentrations of copper, cadmium and lead were 126, 2.24 and 139.7 mg kg^{-1}, respectively.

Seeds from maize (*Zea mays* L.), wheat (*Triticum aestivum* L.), pea (*Pisum sativum* L.) and soybean (*Glycine max* L. Merr.) were obtained from Chinese Agricultural University and surface-sterilized by soaking in a mixture of H_2O_2 (3%) and $CaSO_4$ (saturated) for 30 min.

2.2. Dynamic change of metal fractionation in maize rhizosphere

Dynamic change in metal fractionation was observed in the rhizoboxes. A rhizobox consisted of two parts, each packed with 150 g of soil. The bottom of the upper compartment was sealed with a nylon cloth. For each rhizobox, eight pregerminated maize seedlings were transferred to the upper compartment, and 25 mL of deionized water was added to each rhizobox. The box was cultivated and watered twice a day with deionized water. The maize was grown under a cycle of 14 h by day at 30°C and 10 h by night at 26°C in a growth-chamber (HPG-280B, 9000 LX). Following the formation of a root pad, a layer (20.0 g, *ca.* 2 mm thick) of soil was added in between the upper and bottom parts. This layer, separated by nylon cloth from other soils, served as the rhizosphere soil. The two parts were stacked together. The controlling boxes were set up in an identical fashion without maize seedlings. The schematic representation of the rhizobox is shown in Fig. 1.

After the addition of 2 mm soil to the lower box, the box devices were cultivated for 0, 15, 24, 30, 36, 42, 53, 57 and 100 days, respectively. There were 6, 6, 4, 6, 2, 2, 6, 2 and 2 replicates for each of the 9 treatments, and 2 duplicates for each of the 8 corresponding controls (except day 0), respectively. At the termination of the cultivation period, the maize seedlings were harvested and the

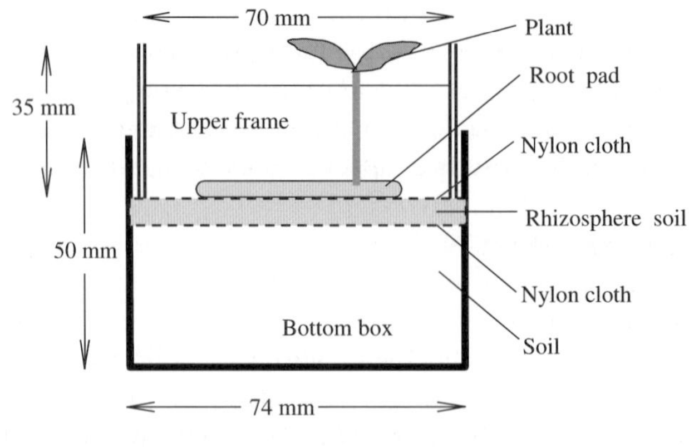

Fig. 1. Schematic representation of the rhizobox as the culture device.

soil samples (2 mm layer between the upper and the bottom parts) were collected (two from each rhizobox for duplicate analysis). The plant samples were washed and rinsed thoroughly with deionized water, then dried at 60°C for 24 h. The dry mass was recorded before determination.

2.3. The influence of root exudates on metal fractionation

A set of experiments was designed to evaluate the effect of root exudates on metal fractionation in soil without plant roots. Root exudates were collected from solution cultures of maize, wheat, pea and soybean. The seedlings were cultivated in the culture dishes under a cycle of 12 h by day at 30°C and 12 h by night at 26°C in a growth-chamber (HPG-280B, 9000 LX). The solution in the culture dishes (30 mL) was replaced twice a day, i.e. 30 mL distilled water (sterilized) in the morning and 30 mL fresh nutrient solution in the evening. The compositions of the nutrient solution can be found in the literature (Tao *et al.*, 2004). Five soil samples (20 g each) were held in separate culture dishes without plants. Beginning at the 8th day after germination, the day-culture-solution was collected, brought to a volume of 30 mL and added to dishes holding soil sample once a day for 4 days. The four dishes received the solution from different plant cultures and distilled water was added to the fifth dish as control. One day after the fourth addition, pH and metal (copper, cadmium and lead) fractionation in the soil were measured in duplicate.

To distinguish the effects of pH change and complexation, both induced by root exudates, on metal fractionation, the direct effect of pH change was evaluated by adding acid or alkaline solutions to the soil sample prior to incubation. Aqueous solutions with various pH values were prepared using HNO_3 or NaOH and were added to 11 beakers each holding an aliquot of 10 g of the soil sample. After thorough mixing, the soils in the beakers were incubated at room temperature for 12 days and measured for pH and metal fractionation in duplicate using two samplings from each breaker. The pH values of the treated soil ranged from 6.17 to 10.12.

2.4. The influence of microorganisms on metal fractionation

To study the additive effect of microorganisms on the influence of root exudates, the exudate-addition experiments mentioned above were repeated using sterilized soil. To examine the direct effect of microorganisms on fractionation of copper, cadmium and lead in rhizosphere soil, two sets of rhizoboxes were prepared with either unsterilized or sterilized soil samples. For each rhizobox, seedlings (16 each except 60 for wheat) were transferred to the upper compartment. Chloroform fumigation was employed for soil sterilization. The fumigation procedure lasted for 40 h, and any remaining chloroform was removed by means of a vacuum pump. At the termination of the cultivation period of 28 days, two soil samples were collected from each rhizobox for duplicate determinations of metal fractionation, pH and Eh.

2.5. Analysis

All metal determinations were implemented with a flame atomic absorption spectrophotometer (Hitachi 180-80). Plant samples were digested with 10 mL of 70% HNO_3 in a microwave oven (CEM-MDS 2000) at 50% energy (630 W) and 120 psi for 60 min. Extracts for the determination of total metal concentrations were obtained by treating 0.2 g of plant samples in the microwave oven (CEM-MDS 2000) with 5 mL of concentrated HNO_3, 2 mL of concentrated $HClO_4$ and 3 mL of HF for 60 min at 50% energy (630 W) and 150 psi.

To measure metal fractionation, the soil samples were consecutively extracted using a modified Tessier's procedure (Tessier *et al.*, 1979). Four fractions were gained in the following sequence by using a 1.0 g of soil sample: (1) exchangeable (a 5-mL volume of 1.0 M $MgCl_2$ for 2 h at room temperature); (2) carbonate-associated (a 5-mL aliquot of NaOAc for 6 h at room temperature); (3) amorphous iron–manganese oxides (5 mL of 0.04 M $NH_2OH{\cdot}HCl$ in 25% HOAc for 6 h at 96°C); and (4) organic bound (3 mL of 30% H_2O_2 and 2 mL of 0.02 M HNO_3 adjusted to pH 2 at 85°C for 1 and 2 h in sequence, followed by 5 mL of a mixture of 0.8 M NH_4Ac and 20% HNO_3 for 30 min). The soil sample was simultaneously measured for the total and five fractions, including the residue after the sequential extractions of copper, cadmium and lead in triplicate. The recoveries of the sequential extraction as the sum of five fractions were between 92.1 and 113.8% for the three metals.

Soil pH was measured using 10.0 g of air-dried soil suspended in 25 mL deionized water. The redox potential was measured with a platinum electrode and normalized to pH 7. Total organic carbon (TOC) and dissolved organic carbon (DOC) were determined using a TOC analyzer (Shimadzu TOC 5000A). Soil MBM was measured using a fumigation–extraction procedure (Jone and Mollison, 1948). CEC was determined as the NH_4^+-equivalents found in a NaCl leachate following saturation with NH_4^+ (NH_4OAc, pH 7). All chemicals were of analytical grade. All plastics and glassware were washed before use, soaked in 10% HNO_3 and fully rinsed with deionized water.

3. RESULTS AND DISCUSSION

3.1. Changes in metal fractionation in the rhizosphere soil

The total concentrations of copper, cadmium and lead in the soil sample used in the study were measured as 125.8, 2.24 and 139.7 mg kg^{-1}, respectively (the mean of three measurements with corresponding standard deviations of 6.15, 0.18 and 10.02 mg kg^{-1}, respectively). The background values of the three metals in uncontaminated soil (same series) in the area were only 29.0 ± 9.0, 0.090 ± 0.023 and 21.0 ± 5.33 mg kg^{-1} on average, respectively (China Environmental Monitoring Center, 1990). Accordingly, the sample was considered to represent a typical heavy-metal-contaminated soil in the wastewater-irrigated area.

Results of the five-step sequential extraction are presented in Fig. 2, which illustrates the distribution of exchangeable (Exc.), carbonate-associated (Car.), Fe Mn oxides-bound (Oxi.), organic-bound (Org.) and residual (Res.) metals found in the soil sample.

Although the distribution of the metals among the various fractions determined by the sequential extraction scheme is not chemically clear-cut, the results of the fractionation do provide an understanding of the metals' relative potential mobility and bioavailability. It has been reported that the exchangeable fraction is probably the most significant available fraction for plant uptake (Sparks, 1983). Krishnamurti and Naidu (2002) studied soil solution speciation and solid-phase fractionation of copper and zinc in 11 uncontaminated soils of South Australia, and revealed significant correlations between metal phytoavailability and fulvic complex copper and exchangeable zinc. They also emphasized the role of solid-phase fractions in heavy metal phytoavailability.

According to Fig. 2, copper bound to organics was the dominant fraction in the soil sample, accounting for nearly half of the total amount. The concentration of exchangeable copper in the soil was very low in comparison with the amount of copper in the other forms, and constituted only 0.5% of the total fractionated copper. This small but active fraction, however, may provide some indication of the form of copper that is most accessible for plant uptake (Sparks, 1983; Grzebisz *et al.*, 1997). The general distribution pattern of various fractions of lead was similar to that of copper, except that no exchangeable lead was detectable. On the other hand, the fractionation of cadmium was quite different. The exchangeable cadmium in the soil was much higher than that of copper and lead, contributing 16.1% of the total, while organic-bound cadmium measured only around 5%. To some extent, this situation indicates the difference in geochemical behavior of cadmium from copper and lead.

The residual fractions of copper, cadmium and lead varied from 21.9% to 30.8% of the total concentrations. It is expected that the residual fraction in contaminated soils is relatively lower than that in uncontaminated soils, since the introduced metals tend to go into the more reactive forms (Nyamangara, 1998).

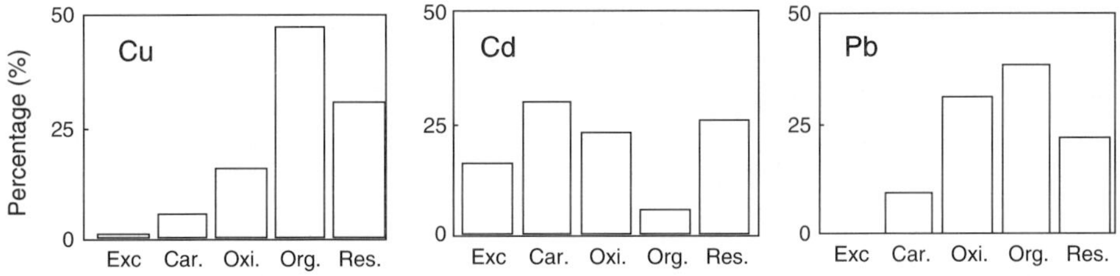

Fig. 2. Distribution of the five fractions of copper, cadmium and lead in the soil from a wastewater irrigated area in Tianjin.

For example, Schramel *et al.* (2000) found that the copper bound in the residual phase in an uncontaminated soil was around 65–85%, but that the percentages dropped to 40–50% because of the contamination. Kabala and Singh (2001) also reported low percentages of residual copper in highly contaminated soils near a smelter in Poland.

The root-induced changes in copper, cadmium and lead fractionation in the soil were examined based on the results of the sequential extraction over the cultivation time. Fig. 3 depicts the changes in metal fractionation in the maize rhizosphere by plotting the concentrations of individual metal fractions (means plus standard deviations after control correction) against cultivation time ranged from 0 to 100 days. The results of one-way ANOVA indicated significant differences (α=0.05) for all cases. Tukey multiple comparison was also conducted; the homogenous subsets are illustrated in Fig. 3 as horizontal bars. Each bar covers a set of data with no significant difference among them at 0.05 significant level.

The content of exchangeable copper clearly shows a tendency to increase during the first 3 weeks, attaining a maximum after 20 days and decreasing thereafter. After 40 days or so, the amount of exchangeable copper in the rhizosphere dropped below its initial level. A similar trend can be observed for exchangeable cadmium, which first increased and then decreased after 40 days as well.

As shown in Fig. 3, the content of copper associated with carbonate increased and then decreased, following a similar but less significant trend of change to that of the exchangeable form. The increase did not become evident until the day 42. The net loss of the fraction as compared to the initial level occurred between 60 and 100 days. For both cadmium and lead, the carbonate fraction gradually increased

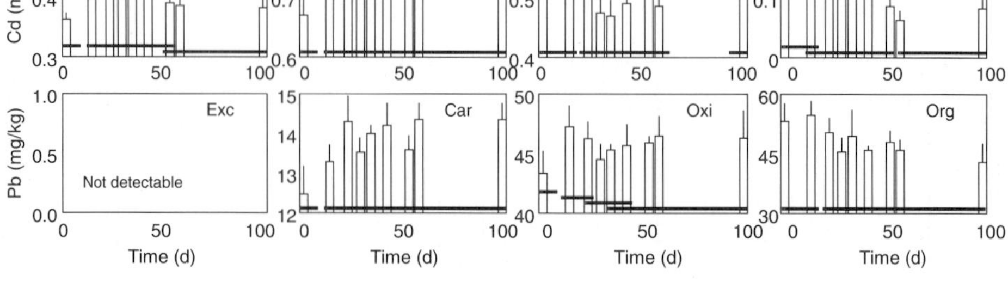

Fig. 3. Change in metal fractions in the maize rhizosphere during the cultivation period ranged from 0 to 100 days. The means and standard deviations are presented as vertical bars while homogeneous subsets based on Tukey multiple comparison are illustrated using horizontal lines.

at the beginning and leveled off afterwards. Owing to their relatively large quantities when compared to the quantities of metal transformed among the fractions, the variation in the oxides and organic-associated fractions was not as clear as that in the exchangeable and carbonate-bound ones. Still, an initial increase in the oxide-associated copper and lead, and an initial decrease in the organic-bound copper can be seen in Fig. 3. The organic-bound cadmium and lead, however, demonstrated a general decreasing trend over the experimental period.

The exchangeable fraction of copper and cadmium in the rhizosphere soil did not further increase after 20–40 days. Since it is believed that the mobilization processes (exudate release, microbial activity, root respiration, etc.) should last during the entire 100-day period of the experiment, there must be other processes that reduced the amount of exchangeable metals in the soil. Plant uptake is presumably one of them. Reisenauer (1988) studied the uptake of manganese by maize and found a manganese influx of 1.7×10^{-7} nmol cm^{-1} s^{-1} to maize. At the same time, the manganese concentration in the soil solution at the root surface decreased from an initial 0.20 to 0.13 μmol L^{-1} due to plant uptake (Reisenauer, 1988). Metals in soil solution were not determined separately during the present research, and were considered to be included in the exchangeable fractions. A reasonable explanation for the decrease in exchangeable copper and cadmium is, therefore, plant uptake. Fig. 4 displays the accumulation pattern for copper, cadmium and lead in maize in terms of the average quantity of metals (μg) accumulated in the plants from individual culture devices.

According to the results shown in Fig. 4, a considerable amount of metal accumulation occurred after 20 days of cultivation. This was roughly equal to the time when the level of exchangeable copper in the soil began to drop (see Fig. 3), which seems to further support the postulation that plant uptake acts as a process leading to the consumption of bioavailable copper within the rhizosphere. However, taking copper as an example, a simple calculation indicated that even if all of the soils in the rhizosphere disk (20 g) and upper frame (150 g) were taken into account, the amount of copper accumulated in the maize (around 300 μg) still exceeded the total amount of the exchangeable fraction that initially occurred in the soil (92–143 μg). Compared with copper, the exchangeable cadmium

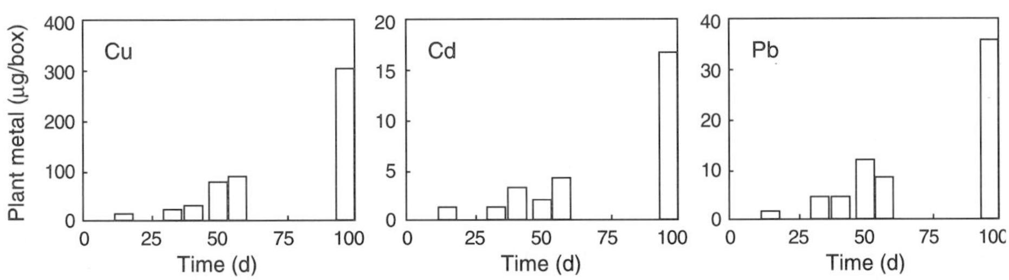

Fig. 4. Accumulation of copper, cadmium and lead in maize during 100 days of cultivation.

accounted for a much larger fraction (16.1%) at the beginning, whereas accumulation in the plant did not cause a liable decrease of the exchangeable fraction in the rhizosphere (Fig. 3). Although it is generally accepted, over the short term at least, that metal uptake by plants from the soil is associated with the free ion activities (Sparks, 1983) or metal complex in soil solution (Krishnamurti and Naidu, 2002), there is little doubt that the transformation of copper from less bioavailable pools (most likely the carbonate-associated copper) to the more bioavailable fraction took place in the maize rhizosphere.

The major soil properties that contribute to the changes in copper availability are pH, redox potential, DOC (including root exudates as chelates) and microbial activity (Horak, 1982; Hamon *et al.*, 1995; Marschner and Römheld, 1996). To monitor the change in these properties during cultivation, the differences in levels of pH, Eh, DOC and MBM in rhizosphere and bulk soils were measured, and are depicted in Fig. 5.

pH change in the rhizosphere has been demonstrated in many papers (Nye, 1981; Marschner and Römheld, 1983). Lorenz *et al.* (1994) observed a pattern of pH change similar to ours in a potted radish rhizosphere during an experiment lasting 45 days. The root-induced change in pH occurs chiefly as a consequence of differential rates in the uptake of cations and anions by plants. As shown in

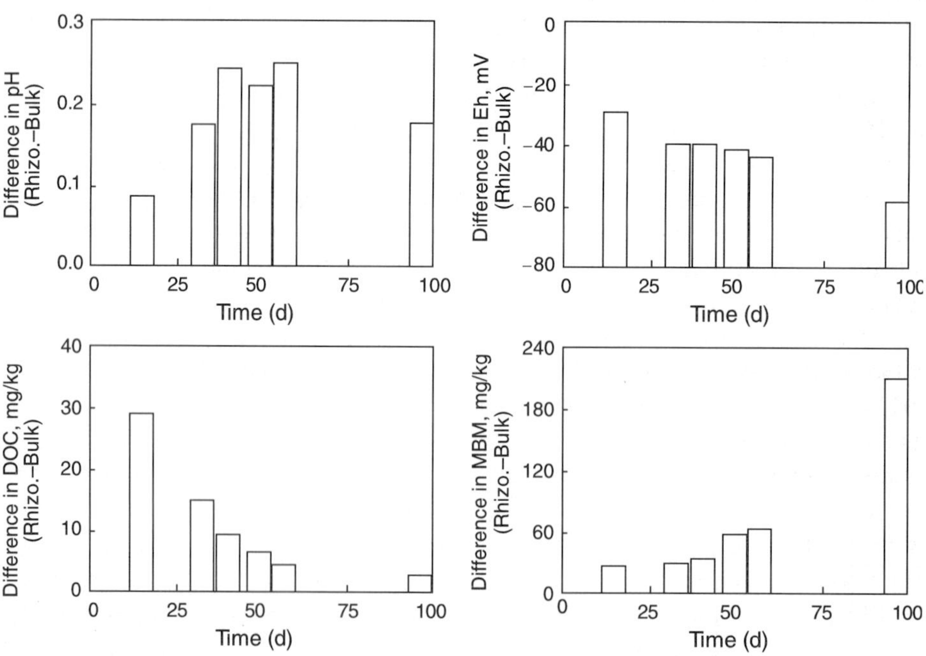

Fig. 5. Differences of pH, Eh, DOC and MBM between the rhizosphere (Rhizo.) and the bulk soils (Bulk) during the 100-day cultivation. For all parameters presented in this figure, the values are zero at the beginning.

Fig. 5, pH in the maize rhizosphere increased up to 0.2 unit compared with the bulk soil. Although it has been observed in some cases that roots can make the neighboring soil more acid (Rappart *et al.*, 1987), Nye (1981) concluded that, in general, roots must release HCO_3^- into the soil to maintain electrical neutrality across the root–soil interface, making the adjacent soil more alkaline. RoÈmkens *et al.* (1999) also observed an increase in soil pH in planted pots and suggested that it was probably related to changes in the soil's calcium concentration. Generally, metals are more mobile in the soil solution under acidic conditions (Rappart *et al.*, 1987). However, pH change as small as 0.2 unit is not likely to be the reason for metal mobilization within the maize rhizosphere in this study. The corresponding evidence and a discussion will be presented later in this chapter.

Because it is difficult to make a precise measurement of redox potential in soil, especially in well aerated soil, reliable data on redox potential in the rhizosphere in this study are lacking. The results presented in Fig. 5 may be inaccurate; however, a trend of change toward negative Eh in the rhizosphere soil is inferred. Rappart *et al.* (1987) indicated that greater amounts of exchangeable copper have been detected in soils at low rather than high redox potential. It is assumed that in the maize rhizosphere in this study, the reduced redox potential was favorable to metal mobilization.

The changes in DOC and MBM within the rhizosphere during cultivation are also illustrated in Fig. 5. There are various components involved in the release of organic carbon into the rhizosphere (Rovira *et al.*, 1979). Some components of DOC are active substances exhibiting strong affinities to trace elements (Krishnamurti and Naidu, 2002). The increasing content of DOC may facilitate change in other stable copper fractions. As a matter of fact, DOC change in the rhizosphere over the cultivation period is quite similar to that of the exchangeable or bioavailable fraction of copper and cadmium, which increased sharply at first and reached a maximum at day 20 or so, followed by a continuous decrease.

The overall effect of plant–microbe interaction shows an increase in MBM in the rhizosphere, owing to the high supply of organic carbon by roots (Lynch and Whipps, 1990). As shown in Fig. 5, enhanced microbial activity was manifested by a steady increase in MBM. Horak (1982) demonstrated that mobilization of copper in the rhizosphere of peas should be a direct outcome of low-molecular-weight root exudates and of an indirect effect via microbial activity in the rhizosphere. The results of the present study show that changes in the major soil properties, including redox potential, DOC and microbial activities, are all in favor of a transformation of metals from less available to more available fractions, leading to variations in various metal fractions in the maize rhizosphere.

In summary, there are two main categories of processes occurring during cultivation, mobilization and bioaccumulation. The mobilization induced by plant roots and microbial activities releases metals from the relatively inert fractions to the exchangeable fraction, while bioaccumulation reduces the exchangeable pool by depleting metals from the soil solution. Bioaccumulation could be ignored during the early stage (around 20 days) of the cultivation when the seedlings were little. During this stage, a general increase in the exchangeable fraction was observed. Bioaccumulation became more and more substantial as the biomass of the plant progressively increased. The depletion process finally overtook the root-induced mobilization, resulting in a decrease in the exchangeable copper and cadmium in the rhizosphere. The procedures involved in metal uptake, therefore, could be: (1) delivery of copper from less bioavailable pools to more bioavailable fractions; and (2) uptake of the exchangeable fraction by the maize plant. In the case of exchangeable copper and cadmium, the trend of increase followed by decrease (Fig. 3) was a result of the two processes acting in opposite directions. The rise or fall in the concentrations depended on the relative intensity between them (Fig. 6).

Evidence for the dependence of changes in the exchangeable fraction on time can be found in the literature. For instance, in a similar study that lasted for 105 days, McGrath *et al.* (1997) discovered that levels of mobile cadmium and zinc (extracted using 1 M NH_4NO_3) in the rhizosphere of *Thlaspi caerulescens* and *T. ochroleucum* were lower than in the bulk soil. Krishnamurti *et al.* (1996) discovered that NH_4Cl-extractable cadmium was higher in wheat rhizosphere at the 2-week growth stage than at the 7-week growth stage.

3.2. Influence of root exudates on metal fractionation in rhizosphere soil

At least three processes triggered by root exudates can induce certain changes in metal fractionation in soil. The introduction of root exudates to a soil can cause pH change, provide ligands for metal complexation and enhance microbial activity. Each of these processes may mobilize or immobilize metals in the soil. In this study, these processes were investigated separately. To examine

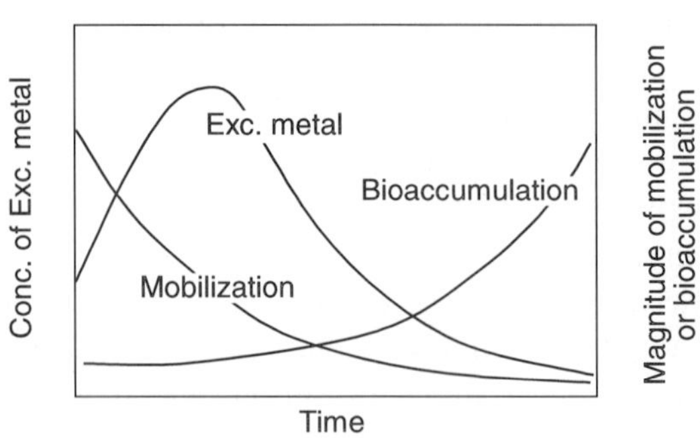

Fig. 6. Combined effects of mobilization and bioaccumulation on the levels of the exchangeable (Exc.) fraction of metals as a function of time.

the possible effects of root exudates without other influences of roots, exudates from aqueous cultures of maize, wheat, pea and soybean were collected and applicd to soil without growing the plant. It should be pointed out that the use of the solution to mimic the selected root actions has certain drawbacks. Biomolecules from pure aqueous cultures of the plants may be quite different from those in the presence of the soil. Moreover, the exudates were poorly buffered and diluted in the distilled water. Therefore, only the direction and the relative extent, rather than the exact degree of the fractionation changes, are meaningful. Fig. 7 reveals the pH and organic carbon contents of the root exudates collected from the aqueous cultures of these plants. The fresh weights of maize, wheat, pea and soybean at the end of the cultivation were 20.3, 11.9, 36.0 and 20.8 g, respectively.

It is well documented in the literature that the low molecular weight free exudates from root secretion include organic acids (Marschner and Römheld, 1996), which could alter the soil pH. For the plant species studied, the pH values of the root exudates originating from various species were lower than that of soil (8.12), ranging from 4.6 for maize to 7.4 for soybean. The pH values shown in Fig. 7 (left) are negatively correlated with organic carbon contents of the exudates, which are presented on the right side of Fig. 7. The coefficient of correlation between them is 0.882, which is significant at a level of 0.05. The concentrations of the exudates in terms of organic carbon collected from the solution cultures of various plants varied with different plant species. The largest amount was detected in the sample from maize, even though the biomass of maize was not the highest.

The fractionation changes in copper, cadmium and lead induced by the addition of root exudates from solution cultures of maize, wheat, pea and soybean were studied. Fig. 8 presents the changes in copper, cadmium and lead fractionation observed in the sterilized (top) and unsterilized (bottom) soils receiving the root exudates from the solution cultures. The results are presented as the differences of various fractions relative to the controls to which distilled water,

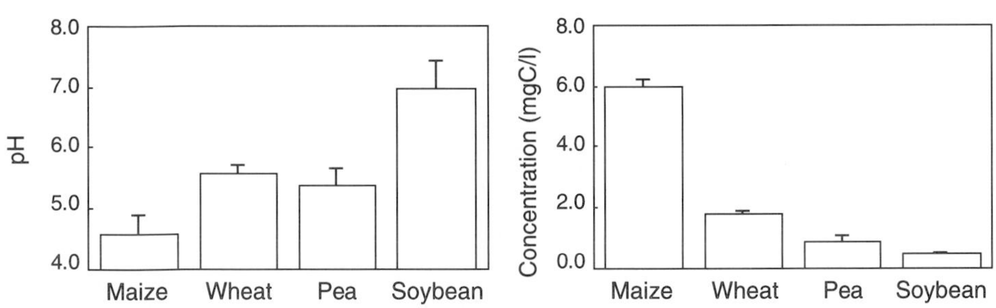

Fig. 7. The pH values and organic carbon contents of the root exudates collected from the aqueous cultures of maize, wheat, pea and soybean. Means and standard deviations are shown.

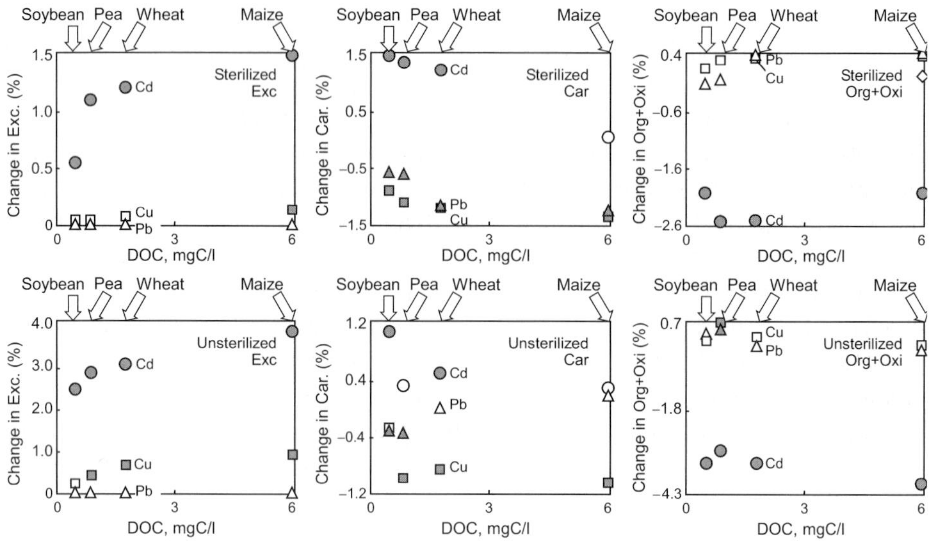

Fig. 8. Effects of the root exudates from solution cultures of various plants on the metal fractionation in the unsterilized (bottom) and the sterilized (top) soils. Four DOC levels from low to high (from left to right) correspond to four plant species: soybean, pea, wheat and maize.

instead of root exudates, was added. The relatively inert fractions (organic and Fe–Mn oxides) are combined. Since the contents of exudates from various plant species were considerably different from one another (Fig. 7, right), the horizontal axis of Fig. 8 is scaled based on the DOC concentrations of the exudates from various plants, which correspond to the four plant species. The plant species are also labeled in Fig. 8.

One sample *t*-test with Ho: $\mu = 0$ (no influence) was conducted to evaluate the influence of exudate addition on metal fraction. The data points, which are significantly different from zero (significant influence), were marked with solid symbols in Fig. 8. It appears that there were exudate-induced changes in fractionation of copper, cadmium and lead with (top) or without (bottom) sterilization. The directions of changes are similar to the results presented in the previous section, in which the dynamic changes in various metal fractions in the rhizosphere of maize were recorded. For the exchangeable fraction, cadmium in all samples and copper in most of the unsterilized sample and one sterilized sample increased. The change in cadmium was much greater than that of copper. The changes in the exchangeable fractions of copper and cadmium increased with an increase in the DOC content of the exudates, and the most profound changes were observed for the soil receiving exudates from maize. Exudate-induced changes in carbonate fractions of the metals studied showed different directions with an increase in cadmium and decreases in copper and lead. The change in carbonate-bound cadmium, but not copper and lead, is similar to that observed is in the rhizobox (see Fig. 3), in which an increase in carbonate-bound cadmium and lead was demonstrated around day 28 of the experimental period.

In the experiment with exudate addition, it was expected that the variations in metal fractionation would be produced by either pH change or ligand complexation. To distinguish the two possibilities, another set of experiments using the same soils, with inorganic acid or alkali added, was conducted. After a 12-day incubation period, the equilibrium pH values of 11 soil samples mixed with various aqueous solutions of acid or alkali ranged from 6.17 to 10.12. The changes in pH from the original value of 8.12 were −1.95, −1.53, −1.20, −0.91, −0.62, 0.00, 0.92, 1.27, 1.60, 1.84 and 2.00, respectively. The measured changes in the four fractions of copper, cadmium and lead, in the form of percentages, are plotted against pH change in Fig. 9. It should be noted that the variations of the duplicate measurements, both for pH and metal fractionation, are too small to be shown in the figure.

According to the results shown in Fig. 9, within the range of pH changes studied (±2.0 units), the maximum change in a single copper fraction could reach 2% (amorphous Fe–Mn oxides-bound copper at pH 10.12). The maximum change in carbonate-bound lead was 3.5%, and in exchangeable cadmium, 23%. However, when a small pH change occurred, in either an acidic or alkaline direction, there was a very moderate shift in the fractionation of copper, cadmium and lead. For instance, when the pH of the soil (8.12) decreased to 7.50 or increased to 9.04, individual copper fractions changed less than 0.25%. Based on these results, when pH change is less than half a unit by either acidification or alkalization, it is unlikely to induce significant fractionation changes of copper, cadmium and lead in the soil.

To quantify pH change as a result of root exudate addition, the root exudates from the solution cultures of the four plant species were applied to either unsterilized or sterilized soils. The pH values for the root exudates and the receiving soils (after 4 days of incubation with addition of root exudates once a day) were measured and are presented in Fig. 10 (left). The soil pH values were

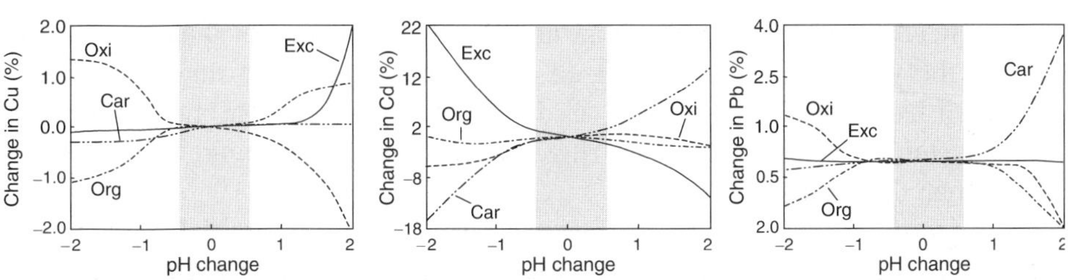

Fig. 9. Direct effects of acidification or alkalization on metal fractionation in the soils incubated with either acid or alkaline solutions. The shaded areas in the middle indicate a range of ±0.5 unit of pH change.

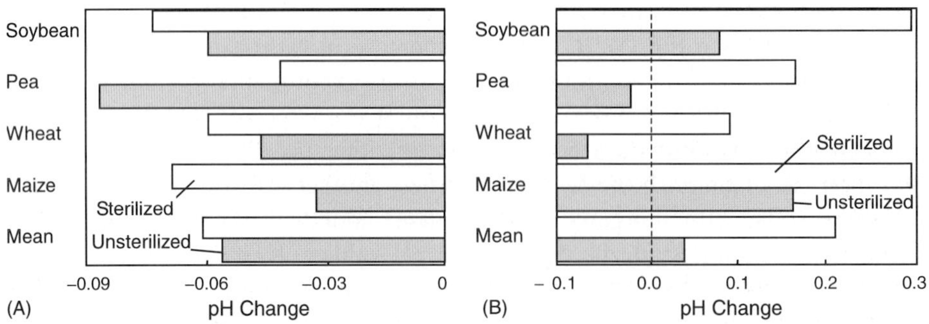

Fig. 10. The pH changes of unsterilized and sterilized soils induced by A(left): addition of root exudates from the solution cultures of maize, wheat, pea and soybean, and by B (right): root pad in the rhizoboxes of maize, wheat, pea and soybean. Mean values of the four plant species are also provided.

compared with a no-exudate-addition control to which distilled water instead of root exudates was applied. The pH changes monitored during the plant cultivation using rhizobox are also shown in Fig. 10 (right) for comparison.

Despite the low pH of the added solution (Fig. 7), only very slight changes in the pH of the receiving soils could be found, indicating the strong buffering capacity of the calcareous soil. For either the unsterilized or the sterilized soils, pH change generally occurred in the acidification direction. This was confirmed by a non-parametric randomization test at a significant level of 0.05 with all sterilized and unsterilized data pooled together. The difference between the sterilized and the unsterilized soils was not significant at the level of 0.05, according to the results of a paired randomization test. Furthermore, the results of a Spearman rank correlation test revealed no significant correlation between the pH of the root exudates (Fig. 7, left) and the resulting soil pH (Fig. 10, left).

The root-induced pH changes, shown on the right side of Fig. 10, were very different from those caused by addition of root exudates alone (Fig. 10, left). In addition to acidification effect of root exudates, root-induced changes in pH occur as a consequence of root respiration and the excretion or reabsorption of H^+ or HCO_3^- (Nye, 1981; Marschner and Römheld, 1996). As shown in Fig. 10 (right), the pH values in all sterilized and some unsterilized rhizosphere soils increased less than 0.3 units compared with the bulk soil. The two exceptions were unsterilized wheat and pea rhizosphere soils, which were a little more acidic than the bulk soil.

According to the results of the rhizobox experiment and the root exudate addition experiment, although pH change did occur in the rhizosphere under the influence of plant roots, the change could hardly achieve half a pH unit. The magnitude of the change was too small to cause any significant change in metal fractionation. It seems that the high buffering capacity of the calcareous soil is an important reason for the slight pH change and subsequent effect. In the case of exudate addition, the pH changes in the soil were slight, with an average drop of 0.1~0.2 unit. As such, no significant change in metal fractionation was expected from the acidification caused by the exudates. On the basis of the results shown in Fig. 9, a ±0.2-unit change in soil pH corresponds to the following changes in copper fractionation in the studied soil: −0.02 ~ 0.01%, −0.06 ~ 0.00%, −0.06 ~ 0.00% and 0.05 ~ 0.07% for exchangeable, carbonate-bound, Fe–Mn oxides associated and organic-bound fractions, respectively. The magnitudes of change in cadmium and lead fractions were similar to those of copper within that pH range. Although the pH changes that occurred in the rhizoboxes were much higher than those induced by exudate addition alone, the maximum difference was no more than 0.3 unit. Again, such a pH change was too small to cause any significant variation in metal fractionation (see Fig. 9).

Based on the results of zinc and cadmium uptake by the *T. caerulescens* in contaminated soils, Knight *et al.* (1997) observed the mobilization of metals along with a significant increase in pH in the rhizosphere, which facilitated the dissolution of humic substances in the soil matrix. They also suggested that acidification of the soil solution was not the reason for the increase in zinc availability (Knight *et al.*, 1997). A similar conclusion was also reached in a study performed by McGrath *et al.* (1997).

Compared with the actual metal fractionation changes observed (Fig. 3), root–induced acidification is not important in terms of metal fractionation change for the calcareous soil studied. Since the influence of acidification on metal fractionation was slight according to the pH change (Fig. 10) and pH effect (Fig. 9) on individual metal fractions, the variation of metal fractionation shown in Fig. 3 could not be accounted for by root exudate-induced acidification. Therefore, other root exudate-induced processes must be seriously involved. Some root exudates have been known to exhibit a strong affinity for metal binding (Krishnamurti *et al.*, 1997). Many previous reports have demonstrated that trace metals in the rhizosphere exist mainly as complexed forms (Merckx *et al.*, 1986; Hamon *et al.*, 1995). Jefferey and Uren (1983) found that copper predominantly bound to organic matter in both solid phase and soil solution. Therefore, for the sterilized soils, the complexation of organic ligands from root exudates plays an essential role in metal fractionation change. It appears that the addition of root exudates from the solution cultures of various plants would facilitate the mobilization of copper in soil through a complexation reaction. Moreover, the extent of the influence on the exchangeable fraction generally decreased in the following order: maize > wheat > pea > soybean (Fig. 8). This situation also correlates well with the quantity of root exudates released from various plants studied during the same period of experiment. In addition to the difference in quantity of exudates, the quality of the exudates from the four plant species may be different in terms of binding capacity for metals. It was indicated that monocots could release exudates with strong efficiency in metal complexation (Marschner and Römheld, 1996). This is another reason causing larger changes in exchangeable copper and cadmium exemplified by wheat and maize in the present study (Fig. 8).

3.3. Influence of microbial activity on metal fractionation in rhizosphere soil

The notable changes in various metal fractions are not likely to be caused solely by non-microbial processes. Microorganisms may also play an important role in these changes. Like plant roots, soil microbes can release organic acids leading to the mobilization of metal through the formation of metal-organic chelates. Horak (1982) indicated that mobilization of micronutrients in the rhizosphere soil of peas could be a direct outcome of low molecular weight root exudates, an indirect effect via microorganisms in the rhizosphere, or a combination of both. As suggested by Marschner and Römheld (1996), the direction of influence of microorganisms on micronutrients in the rhizosphere depends on the circumstances. The microbial utilization of root exudates may reduce the quantity of chelators and have negative effects on the metal mobilization, while consumption of sugars in root exudates by microorganisms can produce more metal chelators as by-products.

It is well established that microorganisms may affect the bioavailability of trace metals (Lynch and Whipps, 1990; Bowen and Rovira, 1991). However, the major challenge that remains is to distinguish between the effects of microbial activity and the effects of plant root activities.

The influence of microorganisms on the metal fractionation was investigated in two ways in this study: (1) by comparing the sterilized and unsterilized soils receiving root exudates; and (2) by comparing the sterilized and unsterilized soils in the rhizoboxes under direct influence of plant roots.

If the results from the sterilized soil in Fig. 8 (top) indicate the influence of the complexation of exudates, then the difference between the sterilized and the unsterilized soils (Fig. 8, top vs. bottom) reveals the indirect effect of the root exudates on metal fractionation through microorganisms. This difference is explicitly illustrated in Fig. 11 as the difference of the metal fractionation changes induced by addition of root exudates between the sterilized and the unsterilized soils. The differences were calculated by subtracting the metal fractionation changes of the sterilized soil from those of the unsterilized soil. Since the metal fractionation changes were either positive or negative, a negative value of the difference indicates a stimulative effect of microbial activity on the positive fractionation change, or an inhibitory effect of microbial activity on the negative fractionation change. The relationship among the effects of microbial activity, the direction of the fractionation change ('+' or '−') and the sign of the difference value (positive or negative) are summarized in Table 1.

There are two types of information provided in Fig. 11. A '+' or '−' (refer to the first row in Table 1) indicates significant root exudate-induced changes in

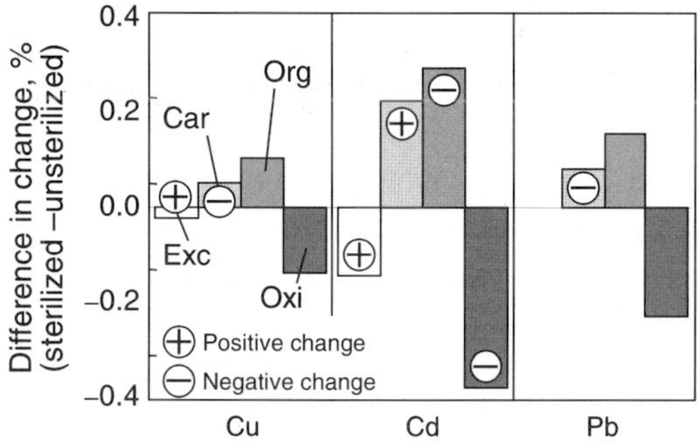

Fig. 11. Differences (sterilized – unsterilized) of the metal fractionation changes induced by addition of root exudates from the solution between the sterilized and unsterilized soils. The results are average values of the four plant species. (Positive or negative changes indicates increase or decrease of certain fraction).

Table 1
Effect of microbial activity on metal fractionation changes

Directions of the root exudate-induced fractionation change in both sterilized and unsterilized soils	Increase (+)	Decrease (−)
Negative difference of the fractionation changes between the sterilized and the unsterilized soils (sterilized − unsterilized)	Stimulative effect on positive fractionation change	Inhibitory effect on negative fractionation change
Positive difference of the fractionation changes between the sterilized and the unsterilized soils (sterilized − unsterilized)	Inhibitory effect on positive fractionation change	Stimulative effect on negative fractionation change

metal fractionation in both sterilized and unsterilized soils based on the results of the t-test (absence of '+' or '−' means that no significant difference, either positive or negative, was found). The differences of the fractionation changes between the sterilized and unsterilized soils (refers to the first column of Table 1) are illustrated by the bars. The influences of microbial activity are revealed by studying the two together.

Although the extents of the differences were quite different among the three metals studied, the differences of individual fractions, in terms of plant species average, were in the same direction (similar bar patterns for the three metals in

Fig. 11). As shown in Fig. 8, exchangeable copper and cadmium (the exchangeable lead was below the direction limit) increased in both sterilized and unsterilized soils ('+' in Fig. 11). Negative differences shown in Fig. 11 indicate that the changes were more pronounced when bacteria were present. For the changes of carbonate-bound metals, positive differences were derived for all three metals, indicating the inhibitory effect of microbial activity on the positive change of cadmium (Fig. 8) and a stimulative effect on the negative changes of copper and lead (Fig. 8).

When rhizobox experiments were used for the comparison, pH, Eh and copper fractionation in the rhizosphere soil with or without sterilization were determined. The influence of sterilization on pH change in the rhizosphere can be seen in Fig. 10 (right). In comparison with the non-rhizosphere control, the pH of the rhizosphere soil increased (the unsterilized soils with maize and soybean and all sterilized soils) or decreased (the unsterilized soils with wheat and pea) slightly depending on plant species. Although sterilization brought the soil pH up in all the cases, the magnitude of the changes was not large enough to cause significant effects on metal fractionation (Fig. 9).

Fig. 12 displays the influence of sterilization on the metal fractionation changes in rhizoboxes. The results are presented in a similar way to that of Fig. 11 (root exudate-induced changes). Differences of the fractionation changes between the sterilized and unsterilized soils are shown as bars for the three metals, the four

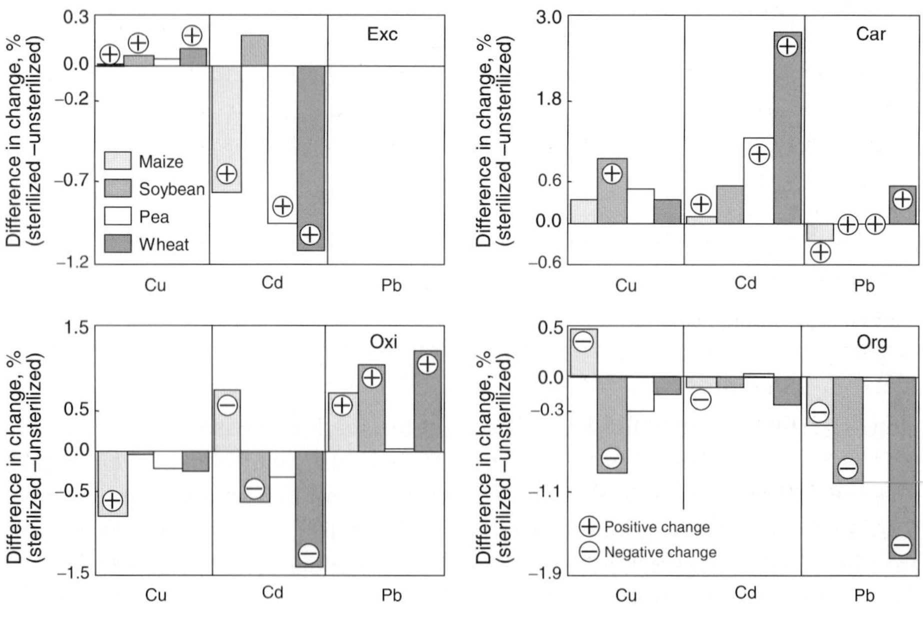

Fig. 12. Differences (sterilized–unsterilized) of the metal fractionation changes in the rhizoboxes of the four plant species between the sterilized and unsterilized rhizosphere soils (positive or negative changes indicate increase or decrease of certain fraction).

plant species and the four fractions. Again, a negative value of the difference indicates stimulative effect of microbial activity on the positive metal fractionation change or inhibitory effect of microbial activity on the negative metal fractionation change (Table 1). Significant changes of metal fractionation of specific metals and fractions in both sterilized and unsterilized soils are also marked with either '+' or '−', indicating the directions of the root exudate-induced changes.

As for the exchangeable fractions of cadmium, the influence of the microbial effect was same for the soils under the influence of root exudates (Fig. 11) and root pad (Fig. 12), for example, microbial activity facilitated the mobilization of cadmium in the soil rhizosphere ('+' and '−' bars). Even though both root pad and addition of root exudate caused increases in the exchangeable copper fractions (as shown by '+' in Figs. 11 and 12), the effects of microbial activity were in opposite directions in the two cases. It appears that microorganisms inhibited the copper mobilization in the rhizobox but accelerated the change in the soil receiving root exudate.

The inhibitory effect of the presence of microorganisms on the positive change of the carbonate cadmium in the rhizobox was again similar to that of the root-exudate addition experiment. However, inhibitory microbial effect on the increase of carbonate copper in the rhizobox soil was different from that of the root-exudate-added soil, in which negative fractionation change was stimulated by microbial activity.

The most significant effects of microorganisms on lead fractionation changes in the rhizobox experiment occurred for the amorphous Fe–Mn oxides and organic-bound fractions. The increase in the former and the decrease in the lather were both inhibited by the presence of microorganisms (refer to Table 1). In the soil receiving root exudate, however, the fractionation changes were not significant statistically, although the differences (Fig. 11) were in the same directions as those in the rhizobox experiment.

There seems to be a number of possible mechanisms responsible for microbial influence on metal fractionation, either indirectly (through pH and Eh change or plant uptake) or directly. Microbial activity related to pH change leads to changes in copper fractionation. However, the pH differences between the unsterilized and the sterilized soils were all less than 0.3 unit in this study (see Fig. 10), which is not expected to have a significant influence on metal fractionation according to the results presented in Fig. 9. An increase in redox potential by sterilization could contribute to the slight decrease in exchangeable copper. Rappart *et al.* (1987) detected greater amounts of exchangeable copper in soils with low rather than high redox potentials. Since the soil may not maintain sterile status throughout the experiment, the difference between the sterilized and the unsterilized results could be even larger than what was observed. More studies that elaborate the corresponding mechanisms of microbial influences on metal fractionation are necessary.

4. CONCLUSIONS

Dynamic changes in fractionation of copper cadmium, and lead in maize rhizosphere soil were observed during a 100-day cultivation period. The initial increases in exchangeable copper and cadmium indicate the effect of root-induced metal mobilization. Decreases in the metal-exchangeable fractions occurred after plant root absorption dominated the change. In the studied soil, pH did not have a significant influence on the metal fractionation. Complexation of root exudates and microbial activity in the rhizosphere were the major causes of the corresponding alterations of metal fractions in the rhizosphere soils.

REFERENCES

Adriano, D.C., 2001. Trace Elements in Terrestrial Environments: Biogeochemistry, Bioavailability, and Risks of Metals. Spriger, New York, p. 866.

Allen, H.E., 1997. Importance of fractionation of metals in natural waters and soils to risk assessment. In: Report of International Workshop on Risk Assessment of Metals and their Inorganic Compounds. International Council on Metals and the Environment, Ottawa, pp. 141–157.

Bowen, G.D., Rovira, A.D., 1991. The rhizosphere, the hidden half of the hidden half. In: Waisel, Y., Eshel, A., Kafkafi, Y. (Eds.), Plant Roots, the Hidden Half. Marcel Dekker Inc., New York, pp. 641–669.

Cherrey, A., Chaignon, V., Hinsinger, P., 1999. Bioavailability of copper in the rhizosphere of rape and ryegrass cropped in vineyard soils. In: Wenzel, W.W., Adriano, D.C., Alloway, B., Doner, H.E., Keller, C., Lepp, N.W., Mench, M., Naidu, R., Pierzynski, G.M. (Eds.), Proceedings of Extended Abstracts of Fifth International Conference on the Biogeochemistry of Trace Elements, Vienna, Austria, pp. 196–197.

China Environmental Monitoring Center, 1990. Soil background values of elements in China. Chinese Press Environ. Sci. p. 346.

Chino, M., Goto, S., Youssef, R., Miah, Y., 1999. Behaviour of micronutrients in the rhizosphere. In: Wenzel, W.W., Adriano, D.C., Alloway, B., Doner, H.E., Keller, C., Lepp, N.W., Mench, M., Naidu, R., Pierzynski, G.M. (Eds.), Proceedings of Extended Abstracts of Fifth International Conference on the Biogeochemistry of Trace Elements, Vienna, Austria, pp. 180–181.

Ernst, W.H.O., 1996. Bioavailability of heavy metals and decontamination of soils by plants. Appl. Geochem. 11, 163–167.

Grzebisz, W., Kocialkowski, W.Z., Chudzinski, B., 1997. Copper geochemistry and availability in cultivated soils contaminated by a copper smelter. J. Geochem. Explor. 58, 301–307.

Hamon, R.E., Lorenz, S.E., Holm, P.E., Christensen, T.H., McGrath, S.P., 1995. Changes in trace metal species and other components of the rhizosphere during growth of radish. Cell Environ. 18, 749–756.

Horak, O., 1982. Die geziehung zwischen der pflanzlichen Aufnahme einiger Mikroelemente und deren wasserlöslichem Anteil im Boden, Landwirtsch, Forsch. Sonderh 39, 404–414.

Jefferey, J.J., Uren, N.C., 1983. Copper and zinc species in the soil solution and the effects of soil pH. Aust. J. Soil Res. 21, 479–488.

Jone, P.C.T., Mollison, J.E.A., 1948. Technique for the quantitative estimation of soil microorganisms. J. Gen. Microbiol. 2, 54–69.

Kabala, C., Singh, R.R., 2001. Fractionation and mobility of copper, lead, and zinc in soil profiles in the vicinity of a copper smelter. J. Environ. Qual. 30, 485–492.

Knight, B., Zhao, F.J., McGrath, S.P., Shen, Z.G., 1997. Zinc and cadmium uptake by the hyperaccumulator *Thlaspi caerulescens* in contaminated soils and its effects on the concentration and chemical fractionation of metals in soil solution. Plant Soil 197, 71–78.

Krishnamurti, G.S.R, Cieslinski, G., Huang, P.M., VanRees, K.C.J., 1997. Kinetics of cadmium release from soils as influenced by organic acids: implication in cadmium availability. J. Environ. Qual. 26, 271–277.

Krishnamurti, G.S.R., Huang, P.M., VanRees, K.C.J., 1996. Studies on soil rhizosphere: speciation and availability of Cd. Chem. Spec. Bioavail. 8, 23–28.

Krishnamurti, G.S.R., Huang, P.M., VanRees, K.C.J., Kozak, L.M., Rostad, H.P.W., 1995. Speciation of particulate-bound cadmium of soils and its bioavailability. Analyst 120, 659–665.

Krishnamurti, G.S.R., Naidu, R., 2000. Speciation and phytoavailability of cadmium in selected surface soils of South Australia. Aust. J. Soil Res. 38, 991–1004.

Krishnamurti, G.S.R., Naidu, R., 2002. Solid-solution speciation and phytoavailability of copper and zinc in soils. Environ. Sci. Technol. 36, 2645–2651.

Levesque, M.P., Mathur, S.P., 1986. Soil tests for copper, iron and zinc in histosols. 1. The influence of soil properties, iron, manganese, and zinc on the level and distribution of copper. Soil Sci. 142, 153–163.

Lorenz, S.E., Hamon, R.E., McGrath, S.P., 1994. Differences between soil solutions obtained from rhizosphere and non-rhizosphere soils by water displacement and soil centrifugation. J. Soil Sci. 45, 431–438.

Lynch, J.M., Whipps, J.M., 1990. Substrate flow in the rhizosphere. Plant Soil 129, 1–10.

Marschner, H., Römheld, V., 1983. In-vivo measurement of root-induced pH changes at the soil–root interface: effect of plant species and nitrogen source. Z. Pflanzenphysiol 111, 241–251.

Marschner, H., Römheld, V., 1996. Root-induced changes in the availability of micronutrients in the rhizosphere. In: Waisel, Y., Eshel, A., Kafkafi, U. (Eds.), Plant Roots, The Hidden Half, second Ed. Marcel Dekker, NY.

McGrath, S.P., Shen, Z.G., Zhao, F.J., 1997. Heavy metal uptake and chemical changes in the rhizosphere of *Thlaspi caerulescens* and *Thlaspi ochroleucum* grown in contaminated soils. Plant Soil 188, 153–159.

Merckx, R., Cremers, A., Ginkel, J.H., Sinnaeve, J., 1986. Plant-induced changes in the rhizosphere of maize and wheat. 2. Complexation of cobalt, zinc and manganese in the rhizosphere of maize and wheat. Plant Soil 96, 95–107.

Nyamangara, J., 1998. Use of sequential extraction to evaluate zinc and copper in a soil amended with sewage sludge and inorganic metal salts. Agr. Ecosyst. Environ. 69, 135–141.

Nye, P.H., 1981. Changes of pH across the rhizosphere induced by roots. Plant Soil 61, 7–26.

Rappart, B.D., Martens, D.C., Reneau, R.B., Simpson, T.W., 1987. Metal accumulation in corn and barley grown on a sludge-amended typic ochraqualf. J. Environ. Qual. 16, 29–33.

Reisenauer, H.M., 1988. Determination of plant-available soil manganese. In: Graham, R.D., Hannam, R.J., Uren, N.C. (Eds.), Manganese in Soils and Plants. Kluwer Academic Publishers, The Netherlands, pp. 87–95.

RoÈmkens, P.F.A.M., Bouwman, L.A., Boon, G.T., 1999. Effect of plant growth on copper solubility and fractionation in soil solution samples. Environ. Pollut. 106, 315–321.

Rovira, A.D., Foster, R.C., Martin, J.K., 1979. Origin, nature and nomenclature of the organic materials in the rhizosphere. In: Harley, J.L., Russell, R.S. (Eds.), The Soil–Root Interface. Academic Press, London, pp. 1–4.

Schramel, O., Michalke, B., Kettrup, A., 2000, Study of the copper distribution in contaminated soils of hop fields by single and sequential extraction procedures. Sci. Total Environ. 263, 11–22.

Shuman, L.M., Wang, J., 1997. Effect of rice variety on zinc, cadmium, iron, and manganese content in rhizosphere and non-rhizosphere soil fractions. Comm. Soil Sci. Plant Anal. 28,23–36.

Sparks, D.L., 1983. Ion activities: an historical and theoretical overview. Soil Sci. Soc. Am. J. 48, 514–518.

Tao, S., Liu, W.X., Chen, Y.J., Xu, F.L., Dawson, R., Li, B.G., Cao, J., Wang, X.J., Hu, J.Y., Fang, J.Y., 2004. Evaluation of factors influencing root-induced changes of copper fractionation in rhizosphere of a calcareous soil. Environ. Pollut. 129, 15–12.

Tessier, A., Campbell, P.G.C., Bisson, M., 1979. Sequential extraction procedure for the fractionation of particulate trace-metals. Anal. Chem. 51, 844–851.

Zemberyova, M., Zwaik, A.A.H., Farkasovska, I., 1998. Sequential extraction for the fractionation of some heavy metals in soils. J. Radioanal. Nucl. Chem. 229, 56–71.

Zhang, M., Gong, Z.T., 1996. Contents and distribution of some heavy metal elements in the vegetable cultivated soil in China. Acta Pedologica Sin. 33, 85–93.

Zhang, X.X., Wang, L.P., Song, S.H., 1988. Study on heavy metal contamination of soil and crops from wastewater irrigated farmLand of Tianjin. China Environ. Sci. 8, 20–26.

Zoysa, A.K.N., Loganathan, P., Hedley, M.J., 1997. A technique for studying rhizosphere processes in tree crops: soil phosphorus depletion around camellia (*Camellia japonica L.*) roots. Plant Soil 190, 253–265.

Biogeochemistry of Trace Elements in the Rhizosphere
P.M. Huang and G.R. Gobran (Editors)

Chapter 11

Bioavailability and extractability of copper and zinc in a soil amended with pig slurry: Effect of iron deficiency in the rhizosphere of two grasses

S. Thomas[a], D. Mahammedi[a], M. Clairotte[a], M.F. Benedetti[b], M. Castrec-Rouelle[b], F. Persin[c], P. Peu[d], J. Martinez[d], and P. Hinsinger[a]

[a]INRA-ENSA.M, UMR 1222 Rhizosphère & Symbiose, Place Viala, 34060 Montpellier, Cedex 1, France
E-mail: philippe.hinsinger@ensam.inra.fr

[b]CNRS-UPRESA, 7047 – UMPC Lab., Géochimie & Métallogénie, Case 124, 75252 Paris, Cedex 05, France

[c]Université Montpelier II, Lab. GPSA – Equipe Génie des procédés CC024, Place E. Bataillon, 34095, Montpellier, Cedex 5, France

[d]CEMAGREF-UR Gestion des Effluents d'Elevage et des Déchets Municipaux 17, Avenue de Cucillé 35044, Rennes Cedes, France

ABSTRACT

The application of pig slurry in agricultural land is a potential source of Cu and Zn contamination of soils. This study was conducted at the Solepur experimental site in Brittany, northwestern France, which received massive, controlled applications of pig slurry over 5 consecutive years. A first objective was to evaluate the effect of pig slurry application on the bioavailability and chemical extractability of soil Cu and Zn. The bioavailability was assessed (i) either *in situ* via the analysis of ryegrass shoots (ii) or *ex situ* via a biotest with two grasses, ryegrass and wheat, which enabled easy access to plant roots and to the rhizosphere. A second objective was to examine precisely how rhizosphere processes could affect the bioavailability and chemical extractability of soil Cu and Zn, with a particular emphasis on the exudation of phytosiderophores as related to Fe deficiency. It was found that amounts of extractable Cu and Zn significantly increased as a consequence of heavy applications of pig slurry in the topsoil of the field site, regardless of the chemical extractant. However, the bioavailability of soil Cu and Zn, as assessed by plant analysis, was not always affected by pig slurry application. When assessed *in situ*, Cu and Zn concentrations in ryegrass shoots did not consistently increase with increasing applications of pig slurry. For Cu, however, the concentrations were occasionally far above the threshold concentration in forage for sensitive animals such as sheep, suggesting that the direct contamination of plant shoots by particles of pig slurry renders such an agricultural practice questionable for grazing or forage production. Plant biotests that accounted for both shoot and root metal contents showed a significant increase in the bioavailability of Cu in the soil treated with pig slurry. In contrast, no significant increase in Zn bioavailability was found. No evidence of phytotoxicity was recorded over the short term of the biotest. In the two grasses being studied, we found, as expected, that Fe deficiency increased the exudation of phytosiderophores from roots. Such root exudates dramatically affected Cu and Zn speciation and complexation in aqueous solution, as evidenced by differential pulse anodic stripping voltametry. This may explain the enhanced Cu and Zn acquisition that occurred under Fe deficiency in both ryegrass and wheat. Root-induced changes in Cu and Zn extractability in the rhizosphere revealed a depletion of two major fractions of soil Cu and Zn, i.e. metals bound either to metal oxides or to organic matter, as determined by a sodium dithionite– acetate–citrate extraction or by a sodium pyrophosphate extraction, respectively. The amount of bioavailable Cu as deduced from plant uptake was in good agreement with the amount of Cu depleted from the metal oxide-bound fraction of soil Cu (determined by a sodium dithionite–acetate–citrate extraction) in the rhizosphere of ryegrass and wheat. Additional research is, however, needed to ascertain the precise mechanisms responsible for Cu and Zn acquisition by these two grasses, and the involvement of root exudates such as phytosiderophores.

1. INTRODUCTION

The availability of plant nutrients and trace elements is dependent on chemical processes and factors that determine the actual concentration and speciation of nutrients and trace elements in the soil solution and the ability of the soil solid phase to replenish the soil solution (soil buffer power). Plant roots contribute to these chemical processes via changes of rhizosphere pH and redox potential, and exudation of metal-complexing molecules (Mench, 1990; Hinsinger, 1998; McLaughlin *et al.*, 1998; Hinsinger, 2001, 2004). Many of these rhizosphere processes may be stimulated by Fe deficiency in higher plants, as further explained below (Marschner and Römheld, 1994; Marschner, 1995; Hinsinger, 2001). These Fe deficiency-induced processes are likely to operate ubiquitously in soils because of limited bioavailability of Fe, especially in neutral to alkaline soils. Iron deficiency in crops may thus affect metal speciation in soil, and ultimately enhance the uptake of metals by plants. In the present study, we examined this hypothesis in the case of copper (Cu) and zinc (Zn), which are micronutrients and also potential metal contaminants.

Among trace elements, Cu and Zn have been classified by Chaney *et al.* (1998) as part of Group 3: elements that are essentially phytotoxic. This classification, which is based upon the "soil plant barrier model," ranks the trace elements according to the risk they pose to the food chain, and points to the propensity of plants to actively exclude some trace elements from uptake, or mitigate their transfer to organs consumed by higher trophic levels. Trace elements of Group 3 cause plants to die of phytotoxicity prior to produce concentrations that would be toxic to humans. When plants are grown in soils that are contaminated by metals such as Cu and Zn, phytoxicity is thus a more relevant issue than risks of food chain contamination. Copper is more phytotoxic than Zn and is therefore of greater concern for the environment.

The uptake activity of the root *per se* can result in considerable increases or decreases in nutrient and trace element concentrations in the rhizosphere. This may have a dramatic effect on the reaction equilibria governing the dynamics of cationic nutrients in soils, e.g. by shifting cation exchange equilibria as shown for major metal cations such as potassium (Hinsinger, 1998) or for trace metals (Hinsinger, 2001). However, to what extent metal uptake by plant roots can shift reaction equilibria involving Cu and Zn remains unknown, although it may well affect the exchangeable fraction of soil Cu and Zn. Many studies have shown the rather poor selectivity of metal cation transporters in the so-called ZIP family, which includes zinc (ZRT) and iron (IRT) transporters (Kochian, 1999; Clemens, 2001); Fe transporters (IRT) have been implied in the uptake of other divalent metal cations, such as Cd, Mn and Zn. As a result of this poor selectivity, the uptake of these metals can be enhanced in Fe-deficient plants owing to the stimulation of IRT, compared with Fe-sufficient plants. In contrast, Cu ions do not seem to be transported by IRT. It is hypothesized that specific Cu transporters that do not belong to the ZIP family exist (Clemens, 2001). However, Cohen *et al.* (1998) reported an increased uptake of Cd, Mn, Zn and also Cu in response to Fe deficiency in pea.

Root-induced changes in pH are more likely to play a role in the fate of Cu and Zn in the rhizosphere, as already evidenced by a few authors. Soil pH is known to strongly affect the dynamics of metals in soils. Roots and rhizosphere microbes contribute to considerable changes in pH by releasing (i) protons (i.e. rhizosphere acidification) or hydroxyls (i.e. rhizosphere alkalinisation) to counterbalance an excess of cation over anion uptake or an excess of anion over cation uptake, respectively; (ii) organic anions that are accompanied by an equivalent release of protons (i.e. rhizosphere acidification); and (iii) CO_2 via respiration leading to a build-up of rhizosphere concentration of H_2CO_3, which dissociates and thereby contributes to a decrease in rhizosphere pH in all but acid soils (Marschner, 1995; Hinsinger, 1998, 2001; Hinsinger *et al.*, 2003). As a consequence, the rhizosphere pH can differ by up to 1–2 pH units from bulk soil pH. This can have dramatic effects on the speciation of Cu and Zn in the soil solution, on the solid phase and ultimately on the bioavailability of Cu and Zn (e.g. Chaignon *et al.*, 2002a; Loosemore *et al.*, 2004; Seguin *et al.*, 2004).

Under Fe-deficient conditions, Strategy I plant species (i.e. all plant species but grasses) show a typical enhanced release of protons (and thus rhizosphere acidification) behind the apical root zone (Marschner *et al.*, 1982; Römheld,

1987; Marschner and Römheld, 1994; Marschner, 1995; Vansuyt *et al.*, 2003). This process helps Strategy I plants acquire more Fe under Fe-deficient conditions. The impact of this Fe-acquisition strategy on the acquisition of other metals by agricultural plants has not been addressed, although it is likely to play a role. A further complication of the response of Strategy I plant species to Fe deficiency is that it combines enhanced proton release with enhanced reductase activity (Brown, 1978; Römheld, 1987; Marschner and Römheld, 1994; Marschner, 1995). Although this response is well documented, its consequences for the redox potential in the rhizosphere remain questionable, as the reduction occurs at the root cell-membrane level (membrane-bound reductases). The potential effect of this phenomenon on the dynamics of Fe oxides and of metals bound to Fe oxides thus remains to be demonstrated. Changes in redox potential in the rhizosphere can also arise from root and microbial respiration and subsequent depletion of soil O_2. Root-induced reduction has been reported to occur in the vicinity of the apical portion of a growing root of a soil-grown legume (Fischer *et al.*, 1989). In contrast, the oxidation of the rhizosphere of lowland rice is well documented (Flessa and Fischer, 1992; Begg *et al.*, 1994; Zhang *et al.*, 1998). It occurs as a consequence of aerenchyma formation and O_2 leakage from the roots. Its effect on the dynamics of Zn has been shown by several authors (Kirk and Bajita, 1995; Zhang *et al.*, 1998).

Other major processes occurring in the rhizosphere and likely to affect metal dynamics are the complexation/chelation processes (Mench, 1990; Mench and Martin, 1991; Marschner, 1995; Hinsinger, 1998; McLaughlin *et al.*, 1998). Both roots and microbes release a whole range of exudates and metabolites that can complex or chelate metals in the rhizosphere. In particular, complexing compounds released by roots have been shown to increase the bioavailability of metals (e.g. Mench *et al.*, 1988; Mench and Martin, 1991; Chaignon *et al.*, 2002b). These complexing compounds contain carboxylic anions such as citrate or malate, and sometimes oxalate. However, stronger ligands, such as siderophores in microbes and phytosiderophores exuded by Strategy II plant species, i.e. grasses (graminaceous species), are also produced. This strategy is based on an enhanced release of phytosiderophores as a response to Fe deficiency (Römheld, 1987; Marschner and Römheld, 1994; Marschner, 1995) and on the existence of specific transporters of the ferric phytosiderophore chelate (von Wirén *et al.*, 1994; Curie *et al.*, 2001). Although phytosiderophores have been identified for their peculiar capability to chelate Fe, they have been shown to strongly chelate other metals, including micronutrients such as Zn and Cu (Murakami *et al.*, 1989; Hinsinger, 2001). Their ability to form stable chelates with Cu is even greater than for any other metal cations including Fe^{III}. In addition, it has been established that Zn can be taken up by roots of grasses as a Zn-phytosiderophore chelate, i.e. in a similar manner as Fe (von Wirén *et al.*, 1996). In a Cu-contaminated calcareous soil, it has been recently shown that Fe-deficient wheat acquired three- to fourfold larger amounts of Cu than Fe-sufficient wheat as a consequence of a two- to threefold larger exudation of phytosiderophores (Chaignon *et al.*, 2002b). In addition, previous work also reported an increased concentration of micronutrients such as Mn, Cu and Zn in shoots of Fe-deficient barley (Treeby *et al.*, 1989). The present work aims at studying the effect of Fe deficiency and thereafter of phytosiderophore release by roots on the speciation and bioavailability of Cu and Zn in the rhizosphere of grasses grown in a soil that had received massive applications of pig slurry.

2. STUDY SITE, EXPERIMENTAL APPROACH AND ANALYTICAL TECHNIQUES

2.1. Solepur, a unique experimental study site

The study site is located in Plouvorn (northwestern Brittany, France) where the CEMAGREF designed the "Solepur" experimental site in 1990. Initially, Solepur was a treatment plant for pig slurry based on a soil epuration lysimeter designed to decrease the nitrogen content of pig slurry via nitrification–denitrification processes (Martinez, 1997; Martinez and Hao, 1996). For this purpose, a hydrologically isolated lysimeter was constructed (3280 m^2): the soil was excavated over a depth of 1 m, a polyane film was installed at this depth, a drainage network was installed and the soil was put back into the plot, horizon by horizon. Pig slurry was spread at elevated loads for 5 consecutive years. After nitrification occurred in the soil, the leachates collected from the drains were connected to large vessels where denitrification was expected to occur, and the final effluent could be safely used for watering adjacent fields. Besides the treated plot, a smaller plot was similarly equipped that did not receive any pig slurry and was used as a control plot. On the treated plot, 30 applications were realised between 1991 and 1995, amounting to a total of 4931 $m^3\ ha^{-1}$ of pig slurry, which represented about 30–100 years of application at an annual rate of 50–150 $m^3\ ha^{-1}$, as achieved by farmers in the region (Coppenet *et al.*, 1993). The treated and control plots were maintained with a permanent grass cover by regular sowing of ryegrass. The major interest of this site is that it provided a rather unique tool for assessing the cumulated effect of repeated applications of pig slurry on soils, and for monitoring the fate of Cu and Zn in soil.

2.2. Solepur, a unique collection of pig slurry-contaminated soil samples

During the 5 consecutive years of heavy application of pig slurry (1991–1996), soil was regularly sampled at 0–20, 20–40 and 40–60 cm soil depths; the amounts and concentrations of the applied pig slurries were precisely determined; and an accurate budget of major or trace elements was quantified (L'Herroux *et al.*, 1997; Martinez and Peu, 2000; Ponthieu, 2003). Total Cu and Zn concentrations significantly increased between 1991 and 1996 in the topsoil (0–20 cm horizon) and, to a lesser extent, at a depth of 20–40 cm (Fig. 1). After

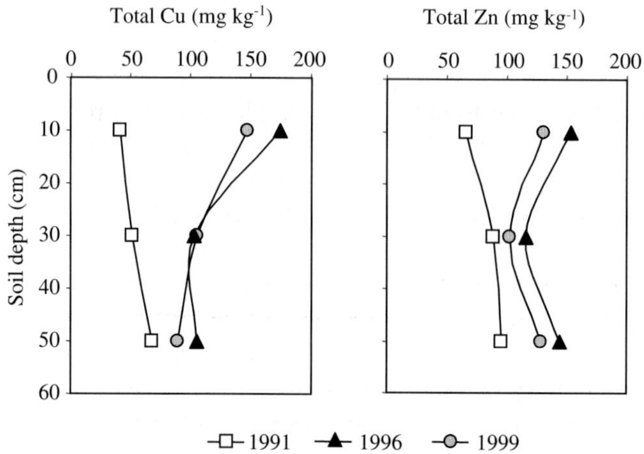

Fig. 1. Temporal evolution of total Cu and Zn concentrations in the soil sampled at three soil depths (0–20, 20–40 and 40–60 cm) in the treated Solepur plot. Massive application of pig slurry occurred from 1991 to 1996 and no further application occurred afterwards.

the last slurry application (1996), total Cu and Zn concentrations decreased significantly in the topsoil (0–20 cm layer).

Chemical extractions were performed in parallel (non-sequential extraction) on soil subsamples to characterize the speciation of trace metals: (1) the exchangeable fraction was determined by a $CaCl_2$ extract; (2) the fraction supposedly bound to iron and manganese oxides was determined by a Na-dithionite – Na-citrate extraction; (3) the fraction supposedly bound to organic matter was determined by a Na-pyrophosphate extraction; and (4) the acid-soluble fraction was determined by a 0.43 M HNO_3 extract.

Prior to slurry application (1991), <1% of total Cu and Zn have been found to be associated with the exchangeable fraction, and Cu and Zn were mainly found in the metal oxide fraction. After pig slurry applications, a large fraction of total Cu and Zn was found in the acid-soluble fraction, which decreased with soil depth. A part of this pool was transferred from the topsoil (0–20 cm) to the intermediate layer (20–40 cm) after 1996, especially for Cu.

In the present work, four samples from the topsoil (0–20 cm) were compared to evaluate the bioavailability of Cu and Zn: three were sampled in the treated plot in 1991 (i.e. prior to pig slurry application), 1996 (after the last heavy application of pig slurry) and 1999. A topsoil sample was also collected in 1999 from the control plot (without any application of pig slurry).

2.3. Experimental device and conditions of plant growth for the biotest

A biotest was conducted with an experimental device designed by Chaignon *et al.* (2002b) and Chaignon and Hinsinger (2003) in order to measure the bioavailability of Cu and Zn in grasses, to study the rhizosphere effect, and the chemical extractability of Cu and Zn. This was a two-step procedure. The first step was conducted in hydroponics for 21 days. Plants were grown on the surface of a mesh in order to obtain a dense, planar mat of roots. In order to test the hypothesis of an effect of Fe deficiency and phytosiderophore release, two treatments were compared: (i) a control treatment in which plants are supplied with an adequate level of Fe as 100 μM FeNaEDTA (thereafter called +Fe) and (ii) an Fe-deficient treatment in which no Fe was supplied to the nutrient solution over the last 5 days of the hydroponic step (thereafter called −Fe).

In the second step, plants were grown for 8 days on top of about 2-mm-thick disks of soil, with a supply of nutrient solution devoid of any Cu, Zn and Fe, so that these metal nutrients were only derived from the soil. Some additional samples of soils were incubated in a similar manner for 8 days to provide a control without plants. A major advantage of this device is that it provides easy access to rhizosphere chemical changes by comparing the soil disks obtained from both the control without plants and the planted soils (Guivarch *et al.*, 1999; Chaignon and Hinsinger, 2003). The whole soil disk is at a distance of less than 2 mm from the roots for the 8-day contact period and can thus be considered as rhizosphere soil (Chaignon and Hinsinger, 2003). Another major advantage of this biotest is that it enables easy access to the root compartment, in which metals such as Cu tend to accumulate rather than being transferred to plant shoots. This biotest was applied to perennial ryegrass (*Lolium perenne* L. cv Aubisque) and bread wheat (*Triticum aestivum* L. cv Aroona). Ryegrass was planted at the Solepur plots. However, little data in the literature reported on the ability of this species to release phytosiderophores. Therefore, bread wheat, which had been studied by several authors (Rengel and Römheld, 2000; Chaignon *et al.*, 2002b), was included for some of the biotests: two seeds of wheat were sown per pot. To obtain a comparable biomass of roots and shoots, 35 seeds of ryegrass were sown per pot. Five replicates (pots) were prepared for each treatment.

2.4. Methods for the collection and determination of the complexing properties of root exudates (phytosiderophores)

Root exudates were collected from intact plants prior to harvest at the end of each step of the bioassay. Roots were repeatedly rinsed with MilliQ deionized water prior to being transferred to Petri dishes containing 15 cm^3 of aerated MilliQ deionized water. A bacteriostatic, 0.01 g cm^{-3} of Micropur (Roth GmbH, Karlsruhe, Germany), was added to prevent microbial degradation of exudates (Gries *et al.*, 1998), a procedure followed by earlier researchers. The collection of exudates started 3 h after the onset of the light period in order to coincide with the peak period of exudation of phytosiderophores (Takagi *et al.*, 1984). After a 3-h collection period, the plants were removed and the solutions were filtered through ashless filter paper (Whatman 40). The amount of phytosiderophores in the root exudates was deduced from the amount of Cu mobilized from a Cu-loaded resin (Chelite P, Serva, Heidelberg, Germany) as previously described by several authors (Treeby *et al.*, 1989; Walter *et al.*, 1994; Cakmak *et al.*, 1994). The Cu-Chelite P was prepared by stirring 5 g of beads of Chelite P in 500 cm^3 of 50 mM $CuSO_4$ for 15 min. After filtration, the resin beads were washed with MilliQ deionized water until the water was free of Cu. Then, the Cu-loaded resin was equilibrated with 500 cm^3 of 10 mM MES buffer (pH 5). The assay was conducted by mixing 2 cm^3 of Cu-loaded resin suspended in MES buffer (pH 5), 2 cm^3 of exudate collection solution and 6 cm^3 of MilliQ deionized water for 45 min. After filtration through ashless filter paper (Whatman 40), the concentration of Cu in the filtrate was measured by flame atomic absorption spectrometry (Varian SpectrAA-600, Australia). The amount of phytosiderophores was expressed as Cu equivalents per pot per 3 h.

The effect of the collected exudates on the speciation and complexation of Cu and Zn was assessed by a voltametric approach with a Metrohm polarograph (663 VA Stand, equipped with the General Purpose Electrochemical Software) in DPASV (Differential Pulse Anodic Stripping Voltametry) mode. Three electrodes were combined: (i) a hanging-drop Hg electrode, which is the measurement electrode, the metal being reduced and then oxidized at a given potential that is metal species-dependent; (ii) an auxiliary Pt electrode; and (iii) a reference Ag/AgCl electrode (potential values that are given are actually obtained from the potential difference with this reference potential). Further details of the operating conditions are given in Table 1.

The method was applied to 11 cm^3 of analyte, i.e. milliQ deionized water in which roots had exuded for 3 h or not (control without exudates). This amount of analyte was obtained by bulking the five replicates. In order to increase the conductivity of the analyte and to work at a constant ionic strength, KNO_3 was added. The solutions were spiked with increasing amounts of Cu and Zn (as sulfate salts at a concentration of 10 mg dm^{-3}). Prior to the measurement, they were flushed with N_2 to be deoxygenated.

Table 1
Operating conditions for assessing the speciation of Cu and Zn by DPASV

Deposition potential/initial potential/final potential (V)	−1.6/−1.6/0.35
Flushing time duration (s)	300–360
Time duration/Equilibration time/Modulation time (s)	240/10/0.05
Superimposed potential (V)	50
Potential scanning rate (V s^{-1})	0.0102

2.5. Techniques of analysis of soil and plant samples

The shoots and roots were harvested and oven-dried separately at 105°C. The oven-dried roots and shoots were then digested separately in a 1:1 mixture of hot, concentrated HNO_3 and $HClO_4$. Concentrations of Cu and Zn in the digests were determined by flame atomic absorption spectrometry. At harvest, the soil was also collected and oven-dried at 105°C. Two 400 mg subsamples were then used for two parallel extractions by adding 4 cm^3 of the two following extractants. A preliminary study conducted with $CaCl_2$ showed that the exchangeable fraction of Cu and Zn was negligible in the soil sampled in the Solepur site. A first extraction was conducted with 0.1 M sodium pyrophosphate ($Na_4P_2O_7$) in order to extract metals bound to organic matter, although this extractant is not perfectly selective and likely to extract some metals bound to Mn oxides too (Chaignon *et al.*, 2003). After 24 h shaking on an end-over-end shaker, the suspensions were centrifuged at 17000 ***g*** for 10 min. The supernatants were then collected and centrifuged for another 5 min at 12800 ***g***. The other extraction was conducted with 0.1 M sodium dithionite ($Na_2S_2O_4$), 0.35 M sodium acetate and 0.2 M sodium citrate buffered at pH 4.8, in order to extract metals bound to reducible metal oxides (Mn and Fe oxides). The vials containing 400 mg of soil and 4 cm^3 of extractant were heated in a water bath at 70°C for 2 h, then shaken for 21 h on an end-over-end shaker, then heated for another hour in a water bath at 70°C. The suspensions were then centrifuged as described above. The supernatant was however filtered at 0.2 μm to eliminate any solid particles, prior to determining Cu and Zn concentrations by flame atomic absorption spectrometry.

3. PLANT BIOMASS AND COMPLEXING ROOT EXUDATES

3.1. Biomass of plants (biotest)

At the end of the hydroponic step of the biotest, no biomass differences between the two plant species were found for roots and shoots (Table 2), showing

Table 2
Biomass of shoots and roots at the end of the hydroponic step of the biotest (g DW per pot)

		Ryegrass	Wheat
Shoots	+Fe	0.31 ± 0.02^b	0.34 ± 0.05^b
	−Fe	0.21 ± 0.03^a	0.23 ± 0.06^a
Roots	+Fe	0.05 ± 0.01^a	0.06 ± 0.01^a
	−Fe	0.05 ± 0.04^a	0.06 ± 0.01^a

Note: Numbers within each subcolumn followed by different letters indicate a significant difference between treatments according to a Newman–Keuls test at $p<0.05$.

seed numbers (2 for wheat/35 for ryegrass) had adequately been selected for comparing the two plant species. For both species, Fe deficiency resulted in a significant decrease in shoot biomass, but not in root biomass. Visible symptoms of chlorosis were observed on the leaves of both grass species after 5 days of Fe deprivation.

After another 8 days of growth on soil disks (end of soil step of the biotest), few growth differences were found between the two species (Table 3). The depressive effect of Fe deficiency was significant for both shoot and root biomass. No significant difference was found for ryegrass when comparing biomass among various soil samples. In contrast, for Fe-sufficient wheat, a significantly smaller biomass in shoots (−16%) and roots (−23%) was found in the slurry-treated soil (1999) than in the control soil. This was not found however for Fe-deficient wheat.

3.2. Amounts of complexing exudates (Cu bioassay)

In both grass species, Fe deprivation for 5 days resulted in an enhanced release of phytosiderophores, as the amount of the recovered metal-complexing exudates was about fourfold greater in Fe-deficient than in Fe-sufficient plants (Fig. 2). This confirmed previous results obtained for wheat (e.g. Rengel and Römheld, 2000; Chaignon *et al.*, 2002b). No published data was available for ryegrass, which released half of the phytosiderophores than wheat did, regardless of Fe status, on a root biomass basis (Fig. 2 shows data on a per pot basis but wheat and ryegrass had the same root biomass, as shown in Table 2). On a per plant basis however, its exudation rate was far lower than wheat's, as each pot contained two wheat plants to compare with up to 35 ryegrass plants. Similar results were obtained with or without bacteriostatic addition, which validates

Table 3
Biomass of shoots and roots at the end of the soil step of the biotest (g DW per pot)

	Ryegrass				Wheat	
	1991	1996	1999	1999-control	1999	1999-control
Shoots						
+Fe	0.84 ± 0.09^{bcd}	0.94 ± 0.25^{cd}	1.05 ± 0.17^{d}	0.96 ± 0.17^{d}	0.79 ± 0.15^{bcd}	0.94 ± 0.08^{cd}
−Fe	0.71 ± 0.10^{ab}	0.61 ± 0.07^{ab}	0.64 ± 0.08^{ab}	0.65 ± 0.10^{ab}	0.58 ± 0.10^{ab}	0.51 ± 0.05^{a}
Roots						
+Fe	0.25 ± 0.02^{bcde}	0.25 ± 0.07^{cde}	0.26 ± 0.03^{de}	0.25 ± 0.03^{e}	0.27 ± 0.04^{bde}	0.35 ± 0.03^{f}
−Fe	0.19 ± 0.03^{abcd}	0.17 ± 0.02^{a}	0.16 ± 0.03^{ab}	0.18 ± 0.02^{abc}	0.19 ± 0.07^{abc}	0.17 ± 0.02^{a}

Note: For either shoots or roots, numbers followed by different letters indicate a significant difference between treatments according to a Newman–Keuls test at $p < 0.05$.

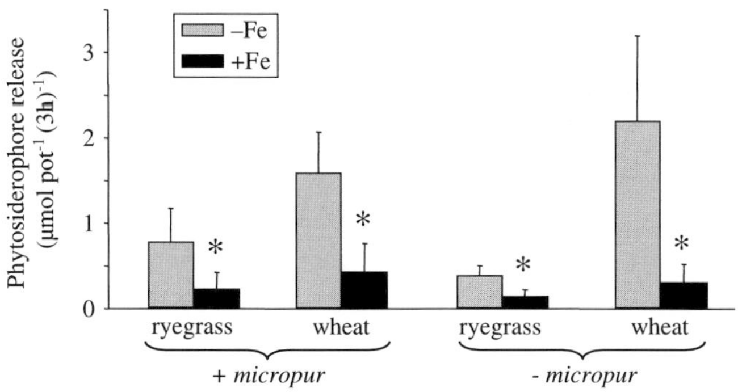

Fig. 2. Amounts of phytosiderophore released by plant roots over a 3-h collection period in milliQ deionized water with or without bacteriostatic (Micropur), as assessed by resin-Cu mobilization assay, as affected by the Fe status of the plants. The asterisk indicates a significant difference between the two Fe treatments according to a Student *t*-test at $p < 0.05$.

former research (e.g. Gries *et al.*, 1998; Chaignon *et al.*, 2002b), although Neumann and Römheld (2001) pointed out that some differences might arise as a consequence of Micropur addition, at least at high concentrations of Micropur.

The enhanced release of phytosiderophores as a response to Fe deficiency is characteristic of Strategy II plant species such as grasses (Marschner and Römheld, 1994). In the context of the present study, it provides a unique tool to test the hypothesis that phytosiderophores are implicated in the mobilization of metals other than Fe, as formerly suggested by a few studies on barley and wheat (Treeby *et al.*, 1989; Chaignon *et al.*, 2002b).

3.3. Effects of root exudates on the speciation of added Cu and Zn (spiking experiment)

The ability of root exudates, and especially phytosiderophores, to modify the speciation of Cu and Zn via complexation processes was assessed by voltametry (DPASV). The collecting solution, i.e. milliQ deionized water (with or without bacteriostatic), was placed in contact with plant roots for 3 h. After collection of the exudates, the collecting solution was spiked with known amounts of both Cu and Zn. A control (exudate-free) solution followed the same protocol, except the contact period with plant roots. In the control solution, DPASV polarograms showed that all the added Cu and Zn were recovered as labile species with an electrical potential (Ep) of, respectively, 0.011 and −1.011 V (Fig. 3). A slight peak was also observed, which was attributed to trace contamination of Pb (Ep = − 0.402 V). When Micropur was added, an extra peak was found at Ep = 0.278 V that could not be attributed to any known metal species. Apart from this, very similar results were observed with or without

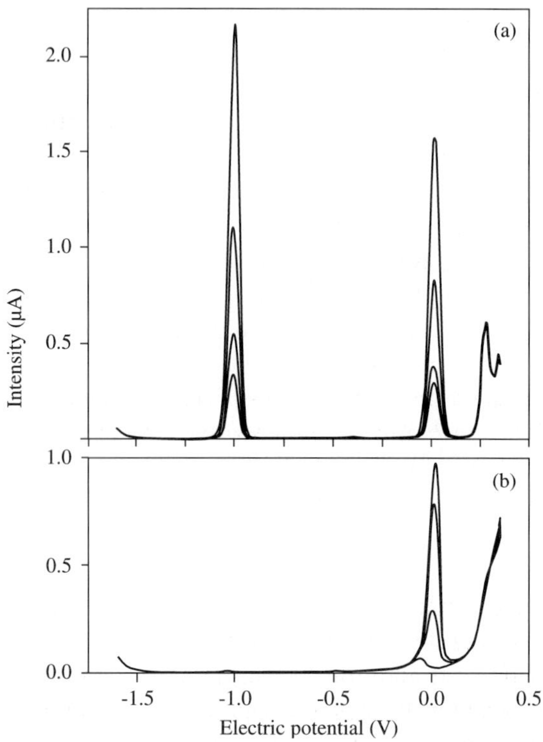

Fig. 3. DPASV polarograms obtained for (a) the control (exudate-free) and (b) the exudate solutions obtained from Fe-starved ryegrass, after additions of four known amounts of Cu and Zn. A bacteriostatic (Micropur) was added to the solutions prior to exudate collection (same for the control solution).

bacteriostatic addition (Thomas, 2002). Therefore, the results will focus on polarograms obtained with Micropur addition during the exudate collection.

Very different results were obtained for the exudate solutions in comparison with polarograms obtained in the exudate-free solutions (control). Regardless of plant species and Fe status of the plants, the observed polarograms systematically showed the following characteristics (as shown for Fe-starved ryegrass exudates in Fig. 3, for instance): (i) a considerable decrease in the peak height of labile Zn, with greater Zn additions needed to identify labile Zn relative to the control solution; (ii) the peak corresponding to Cu species was broader and shifted towards a lower electrical potential at low Cu additions while the peak shifted progressively to Ep values closer to that of labile Cu with increasing Cu additions. This clearly demonstrated that most of the added Cu and Zn were complexed by root exudates, presumably phytosiderophores. Labile Cu and Zn were found only for additions of > 100 and 450 mm^3 of the spiking solution

(Cu and Zn concentrations >1.43 and 6.26 μM, respectively). The broad peak at low and high Cu concentrations suggested a compound peak of several reducible Cu species (possibly CuL_1, CuL_2 and Cu^{2+}), L_1 and L_2 representing different organic ligands (phytosiderophores). Another major difference between the polarograms obtained with and without exudates is found at the highest electrical potentials. The broadening found in all exudate containing solutions is again the evidence for the presence of a complexing agent that would facilitate the oxidation of Hg (occurring at a lower Ep value (around 0.25 V).

Phytosiderophores were expected to be released from both Fe-deficient and Fe-sufficient plants, although in different amounts. The polarograms obtained for both cases looked fairly similar indeed, but a closer observation revealed some slight differences. The heights of Cu peaks were lower and electric potentials shifted towards lower values for solutions collected from Fe-starved ryegrass regardless of Cu additions (Fig. 4). At low Cu levels of 0.14 and 0.71 μM, peaks were flat for exudates of Fe-deficient ryegrass, while the peaks were closer to the Ep value of labile Cu (0.011 V) for exudates of Fe-sufficient ryegrass. This is another indication of a greater amount of Cu-complexing exudates, presumably phytosiderophores, as a response to Fe starvation, which is in agreement with the results of the resin-Cu mobilizing assay (Fig. 2).

When comparing the two plant species, once again the polarograms were fairly close to one another. For Zn species, especially, the Ep values were very similar for ryegrass and wheat exudates: in both species they were slightly more

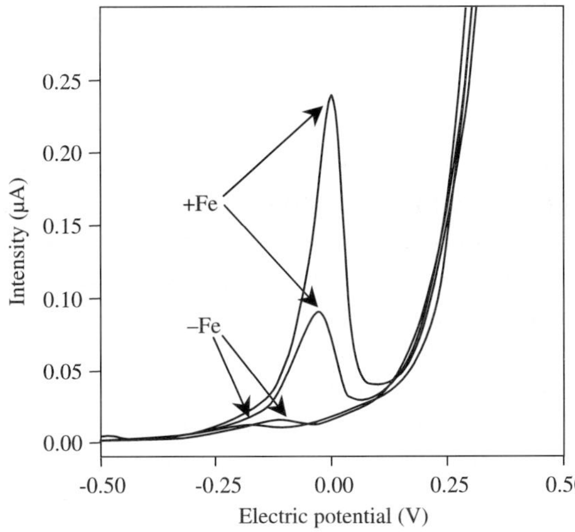

Fig. 4. DPASV polarograms obtained for the exudate solutions obtained from either Fe-deficient or Fe-sufficient ryegrass, at two levels of addition of Cu (0.14 and 0.71 μM). A bacteriostatic (Micropur) was added to the solutions prior to exudate collection.

Table 4
Electrical potential (Ep) values (V) of peaks corresponding to Cu and Zn species in DPASV polarograms of Cu- and Zn-spiked solutions of exudates of Fe-deficient ryegrass and wheat obtained at two levels of Cu and Zn addition

	Ryegrass		Wheat	
	Zn Ep	Cu Ep	Zn Ep	Cu Ep
#0.8 μM Cu and Zn addition	−1.081	−0.029	−1.081	−0.074
#2.8 μM Cu and Zn addition	−1.036	0.006	−1.031	0.001

negative than the Ep value of labile Zn, i.e. −1.011 V, more so at the low level of Zn addition. In contrast, slight differences in the positions of the peaks could be noticed for Cu species (Table 4). The Ep values found for Cu species were lower for wheat than ryegrass. This is consistent with the greater amount of phytosiderophores released by wheat than ryegrass (Fig. 2). This might also be due to differences in the exudate composition, as the nature of phytosiderophores might be different in wheat and ryegrass. However, we did not attempt to identify the types of phytosiderophores released by these two species, which would have required other methods of assessment. Additional measurements, based on HPLC analyses, would be needed to ascertain this point.

4. CONCENTRATION AND AMOUNT OF COPPER AND ZINC IN PLANTS AS INDICATORS OF BIOAVAILABILITY

4.1. Concentrations of Cu and Zn in plant roots and shoots (biotest)

At the end of the hydroponic step of the biotest, greater concentrations of Cu and Zn were found in Fe-deficient plants than Fe-sufficient plants (Table 5). For Cu, this was found only for roots, although they had a similar biomass in both Fe treatments (Table 2). This suggests that Fe deficiency contributes to enhanced acquisition of Cu in nutrient solution, possibly as a consequence of increased exudation of phytosiderophores by the two grass species (Fig. 2), as found earlier for wheat (Chaignon *et al.*, 2002b). For Zn, similar results were obtained for roots. Increased concentrations of Zn were also found in ryegrass and wheat shoots as a response to Fe deficiency. The approximate twofold increase in Zn concentrations in Fe-deficient plants could not be simply due to the observed decreases in shoot biomass, which amounted to about a third of that of Fe-sufficient plants (Table 2). In both ryegrass and wheat, Fe deficiency was thus responsible for an enhanced acquisition of Zn supplied in nutrient solution, possibly as a consequence of increased exudation of phytosiderophores (Fig. 2), as already shown for wheat (Chaignon *et al.*, 2002b).

After an additional 8 days of contact with the various soil samples (end of the biotest), similar results were still found (Table 6). There was a systematic increase in Cu and Zn concentrations in Fe-deficient relative to Fe-sufficient plants, both in roots and shoots. We also found that Cu concentrations in roots and shoots of both species were significantly affected by pig slurry application.

Table 5
Copper and Zn concentrations of shoots and roots at the end of the hydroponic step of the biotest (mg (kg DW) $^{-1}$)

		Ryegrass		Wheat	
		Cu	Zn	Cu	Zn
Shoots	+Fe	4.5 ± 1.7^{a}	47 ± 7^{a}	–	38 ± 9^{a}
	−Fe	4.0 ± 2.4^{a}	90 ± 14^{b}	5.5 ± 2.5^{a}	83 ± 20^{b}
Roots	+Fe	20 ± 11^{a}	43 ± 11^{a}	10 ± 6^{a}	84 ± 16^{a}
	−Fe	53 ± 28^{b}	221 ± 162^{ab}	32 ± 13^{ab}	269 ± 146^{b}

Note: For either shoots or roots, numbers followed by different letters indicate a significant difference of either Cu or Zn concentrations between treatments according to a Newman–Keuls test at $p<0.05$.

Table 6
Copper and Zn concentrations of shoots and roots at the end of the biotest (mg (kg DW) $^{-1}$)

			Ryegrass				Wheat	
			1991	1996	1999	1999-control	1999	1999-control
Cu	Shoots	+Fe	5.8 ± 1.0^{cd}	7.2 ± 0.9^{d}	6.2 ± 1.1^{cd}	3.2 ± 0.6^{ab}	4.1 ± 0.6^{abc}	2.6 ± 1.1^{a}
		−Fe	4.0 ± 1.0^{abc}	12.0 ± 1.7^{f}	9.4 ± 1.2^{e}	5.0 ± 2.0^{bcd}	11.5 ± 2.7^{f}	9.0 ± 0.9^{e}
	Roots	+Fe	8 ± 2^{a}	26 ± 4^{c}	23 ± 1^{bc}	5 ± 1^{a}	16 ± 4^{c}	3 ± 1^{a}
		−Fe	15 ± 3^{abc}	73 ± 13^{e}	48 ± 16^{d}	10 ± 5^{ab}	48 ± 17^{d}	10 ± 2^{ab}
Zn	Shoots	+Fe	19 ± 4^{a}	25 ± 8^{a}	21 ± 4^{a}	14 ± 2^{a}	25 ± 5^{a}	20 ± 3^{a}
		−Fe	41 ± 7^{bc}	56 ± 10^{c}	44 ± 11^{bc}	41 ± 13^{b}	73 ± 11^{d}	103 ± 16^{e}
	Roots	+Fe	27 ± 2^{abc}	31 ± 3^{abc}	27 ± 3^{abc}	21 ± 2^{a}	44 ± 12^{abc}	27 ± 6^{ab}
		−Fe	50 ± 13^{abc}	59 ± 14^{c}	60 ± 29^{bc}	49 ± 18^{abc}	113 ± 39^{d}	94 ± 11^{d}

Note: For either shoots or roots, numbers followed by different letters indicate a significant difference of either Cu or Zn concentrations between treatments according to a Newman–Keuls test at $p<0.05$.

The lowest Cu concentrations were found in soil samples without pig slurry addition. The highest Cu concentrations in roots and shoots were found from plants grown in soil sampled at the end of the 5-year period of heavy pig slurry application (1996). In contrast, the effect of pig slurry application on plant uptake of Zn was not clear.

Copper concentration in shoots did not exceed 12 mg (kg DW) $^{-1}$ in spite of the observed increase with pig slurry application and Fe deficiency, which is far below phytotoxic levels reported by Reuter and Robinson (1997) for ryegrass (21 mg (kg DW)$^{-1}$) and wheat (18–75 mg (kg DW)$^{-1}$). Copper concentrations were greater in roots than shoots as often reported in plants grown in Cu-contaminated soils (Brun *et al.*, 2001; Chaignon and Hinsinger, 2003; Chaignon *et al.*, 2002b), especially in Fe-deficient plants: up to 50 mg (kg DW)$^{-1}$ in 1999 and above 70 mg (kg DW)$^{-1}$ in 1996. However, we lack reference values for determining whether these levels of concentration are in the toxic range (Chaignon and Hinsinger, 2003). Concentrations of Zn in plant material varied much less and remained far below the reported phytotoxic levels compiled by Reuter and Robinson (1997) for ryegrass (221 mg (kg DW)$^{-1}$) and wheat (560 mg (kg DW)$^{-1}$), even in Fe-deficient plants.

4.2. Concentrations of Cu and Zn in shoots of plants sampled *in situ*

A ryegrass cover was maintained on the Solepur plots since 1991, with regular re-sowing every year or every second year. Cut ryegrass was either left in plots as a mulch or harvested and removed from the plots. Periodic sampling of ryegrass shoots was achieved over 1 m^2 surface areas in four random positions in the plot and analyzed for Cu and Zn. Table 7 shows no consistant increase of either Cu or Zn concentrations over the entire period of investigation. At some dates of sampling, increased concentrations of both Cu and Zn were observed, without visible symptoms of phytotoxicity: Cu concentrations exceeded the phytotoxic level of 21 mg (kg DW)$^{-1}$ reported by Reuter and Robinson (1997) and ranged, instead, from 50 to 100 mg Cu (kg DW)$^{-1}$. Elevated concentrations may be the consequence of a direct contamination of grass from sprayed pig slurry, suggesting that samples were inadequately washed. Cu concentrations were above the toxic level in forage for animals such as sheep, which are known to be especially sensitive to Cu toxicity: concentrations above 15 mg Cu (kg DW) $^{-1}$ should be avoided. The data demonstrate potential risks associated with heavy pig slurry application on grass land used either for pasture or herbage. When omitting these 4 dates at which elevated concentrations of Cu and Zn were found, no consistent trend toward an increase of either Cu or Zn concentration in grass shoots was found over time: they oscillated between 7 and 13 mg (kg DW)$^{-1}$ for Cu, and 25–45 mg (kg DW)$^{-1}$ for Zn, i.e. within adequate metal ranges. In control plots, forage concentrations ranged between 4 and 11 mg (kg DW)$^{-1}$ for Cu, and 16–46 mg (kg DW)$^{-1}$ for Zn. Therefore, it can be concluded that analyses of

Table 7
Copper and Zn concentrations in shoots of ryegrass harvested *in situ* in the treated plot of Solepur experiment, either during the period of heavy pig slurry application (1991–1995) or after the end of pig slurry application (1998)

Date of sampling (day/month/year)	Cu concentration (mg(kg DW)$^{-1}$)	Zn Concentration (mg(kg DW)$^{-1}$)
17/05/1991	7.8 ± 0.7	34 ± 3
01/08/1991	56.9 ± 21.2	146 ± 45
02/04/1992	10.8 ± 0.5	30 ± 5
19/05/1992	95.6 ± 39.1	109 ± 36
17/08/1992	45.0 ± 15.6	73 ± 23
29/06/1993	12.1 ± 0.4	38 ± 4
20/08/1993	12.8 ± 0.2	33 ± 3
30/04/1995	7.1 ± 0.7	25 ± 2
05/07/1995	41.5 ± 36.3	90 ± 60
22/09/1998	12.2 ± 1.6	44 ± 2

grass shoots sampled *in situ* showed no evidence of increased Cu and Zn bioavailability in the soil treated with massive pig slurry applications.

4.3. Bioavailable Cu and Zn as assessed by the biotest

On the basis of the biomass and Cu and Zn concentrations of roots and shoots, it is possible to compute the amounts of Cu and Zn in the plant material. When subtracting the amounts found in the plants at the end of the hydroponic step of the biotest, one can deduce the amount of Cu and Zn actually taken up by the plants over the 8-day contact period with soil. Expressing these values relative to the amount of soil supplied in each pot provided estimates of Cu and Zn bioavailability and a comparison to analytical soil testing procedures (Chaignon and Hinsinger, 2003). The heavy application of pig slurry resulted in a two- to four-fold increase in Cu bioavailability (Fig. 5), as shown when comparing the 1996 or 1999 samples with the samples corresponding to no pig slurry application (1991 and 1999 control). No significant difference was found between the two plant species. A significant increase in Cu bioavailability was found only in 1996 for ryegrass and 1999 for wheat. In contrast, pig slurry application did not result in any significant increase in Zn bioavailability. Iron deficiency significantly increased Zn bioavailability in wheat. The wheat genotype in the present experiment was selected for Zn acquisition as related to phytosiderophore exudation (Chaignon *et al.*, 2002b).

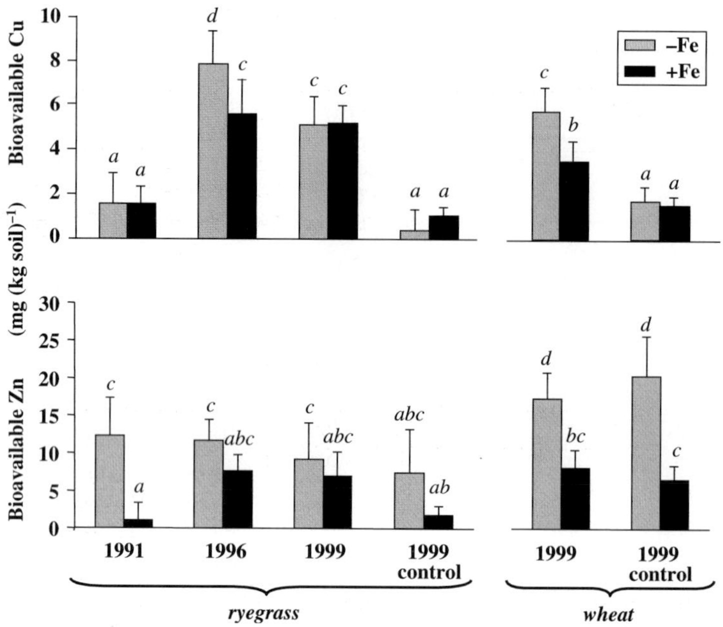

Fig. 5. Bioavailability of Cu and Zn as assessed by measuring the amounts of Cu and Zn taken up by plants over the 8-day period of contact with the soil samples (mg (kg soil)$^{-1}$). Different letters indicate a significant difference between treatments (according to a Newman–Keuls test at $p < 0.05$).

Iron deficiency was not a major factor affecting bioavailable Cu and Zn (Table 6, Fig. 5). Copper and Zn uptake by Fe-deficient plants was reduced because of the 25–50% decrease in biomass that occurred relative to Fe-sufficient plants (Table 3). Despite substantial decreases in biomass, Fe deficiency resulted in similar or greater amounts of Cu and Zn being absorbed by plants, relative to Fe-sufficient plants. This may be a consequence of enhanced exudation of phytosiderophores as a response to Fe deficiency (Fig. 2), as shown for wheat grown in a Cu-contaminated, calcareous soil (Chaignon *et al.*, 2002b).

Concentrations of bioavailable Cu and Zn computed from plant uptake measurements (Fig. 5) partially agreed with concentrations of Ethylene Diamine Tetra Acetate (EDTA) -extractable Cu and Zn (data not shown). Both soil EDTA-extractable Cu and Zn determinations steeply increased roughly 10- to 20-fold from 1991 to 1996 (Cu increasing from 3 to 35 mg kg^{-1} and Zn from 3 to 65 mg kg^{-1}), and tended to decrease between 1996 and 1999. This agrees to some extent with bioavailable Cu but contrasts with bioavailable Zn deduced from plant analyses. Therefore, EDTA appeared to be a poor soil extractant for predicting the bioavailability of Cu and Zn in grasses grown on pig slurry-amended soil. Other chemical extractants were thus compared with the estimates of metal bioavailability (see below).

5. CHEMICAL EXTRACTABILITY OF COPPER AND ZINC IN THE RHIZOSPHERE

5.1. Changes in chemical extractability of Cu and Zn in the rhizosphere

A major advantage of the biotest is that it enables a simple collection of the rhizosphere soil (Guivarch *et al.*, 1999; Chaignon and Hinsinger, 2003). After 8 days of contact between the root mat and the soil, it was assumed that the bulk of soil was influenced by root activity (Chaignon and Hinsinger, 2003). Chemical changes in the rhizosphere were determined by comparing soil samples collected in pots planted with ryegrass or wheat and those in unplanted pots. Two selective chemical extractions were performed in parallel: (i) with 0.1 M sodium pyrophosphate ($Na_4P_2O_7$) (the results refer to Cu_{PYR} or Zn_{PYR}); and (ii) with 0.1 M sodium dithionite ($Na_2S_2O_4$), 0.35 M sodium acetate and 0.2 M sodium citrate buffered at pH = 4.8 (the results refer to Cu_{DIT} or Zn_{DIT}).

A preliminary experiment showed that concentrations of exchangeable Cu and Zn extracted with 0.01 M $CaCl_2$ were almost negligible and varied between 0.1 and 0.9 mg (kg soil)$^{-1}$ for Cu and between 0.15 and 0.4 mg (kg soil)$^{-1}$ for Zn (Mahammedi, 2001). In comparison, greater concentrations of Cu and Zn were recovered with the two other selective extractions (Figs. 6 and 7) that consistently showed that concentrations of extractable Cu (Cu_{PYR} and Cu_{DIT}) and extractable

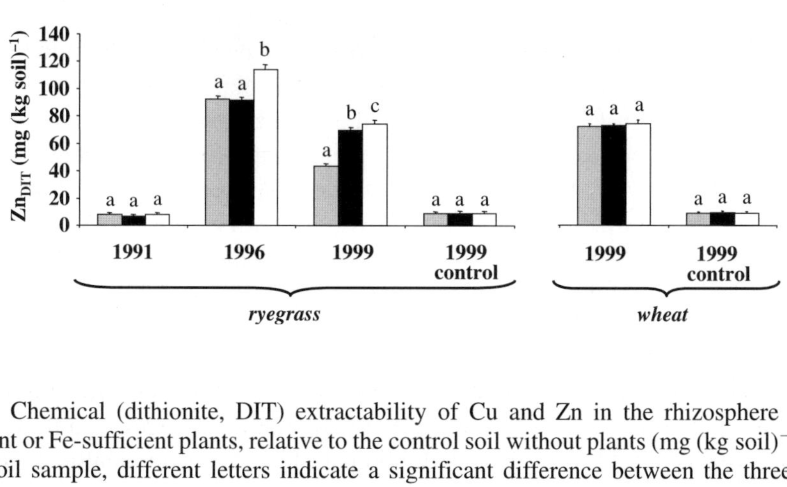

Fig. 6. Chemical (dithionite, DIT) extractability of Cu and Zn in the rhizosphere of Fe-deficient or Fe-sufficient plants, relative to the control soil without plants (mg (kg soil)$^{-1}$). For each soil sample, different letters indicate a significant difference between the three treatments (according to a Newman–Keuls test at $p < 0.05$).

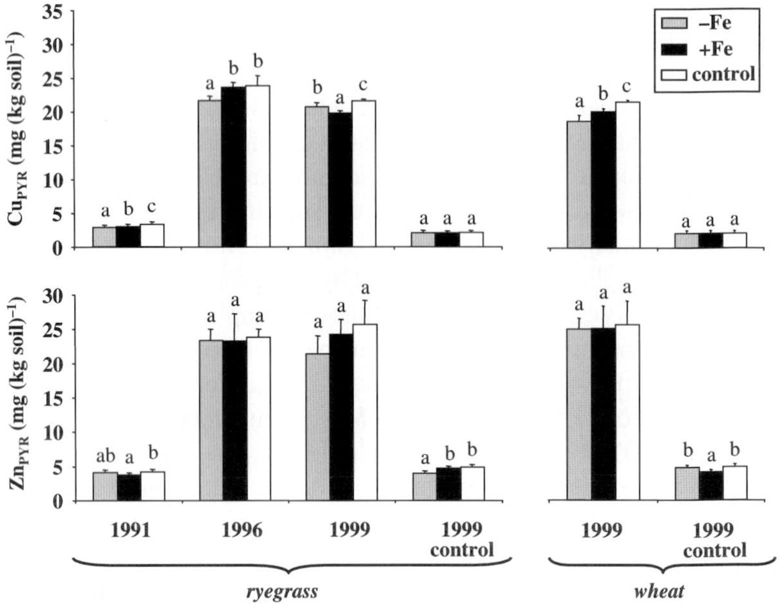

Fig. 7. Chemical (pyrophosphate, PYR) extractability of Cu and Zn in the rhizosphere of Fe-deficient or Fe-sufficient plants, relative to the control soil without plants (mg (kg soil)$^{-1}$). For each soil sample, different letters indicate a significant difference between the three treatments (according to a Newman–Keuls test at $p < 0.05$).

Zn (Zn_{PYR} and Zn_{DIT}) steeply increased after heavy application of pig slurry. These metal concentrations were in agreement with the results found in 1996 and 1999 soil samples and with those found in the soil that had not received any slurry (1991 and 1999 control). In the latter case, little or no significant change was found in the rhizosphere. In soil receiving heavy applications of pig slurry (1996 and 1999), a decrease in extractable Cu and Zn occurred systematically in the rhizosphere in comparison with the control soil without plants (Figs. 6 and 7). The observed decrease in Cu_{DIT} and Zn_{DIT} in the rhizosphere of both ryegrass and wheat was generally greater and more frequently significant than that of Cu_{PYR} and Zn_{PYR}, respectively. In general, this decrease was even greater in the rhizosphere of Fe-deficient plants than in the rhizosphere of Fe-sufficient plants. These results suggest that the two grasses were capable of depleting Cu and Zn bound to metal oxides (Cu_{DIT} and Zn_{DIT}) and possibly to organic matter (Cu_{PYR} and Zn_{PYR}).

5.2. Chemical extractability versus bioavailability of Cu and Zn

Depletion of extractable Cu and Zn in the rhizosphere was compared with the actual uptake of Cu and Zn by plants by comparing soil and plant data sets (Fig. 8). Negative values on the *x*-axis indicate that the extractability of metals

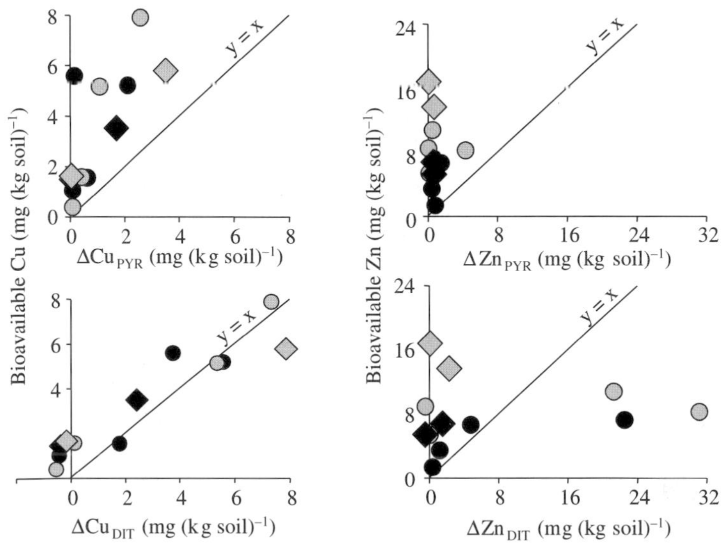

Fig. 8. Bioavailability of Cu and Zn as a function of the decrease in chemical extractability of Cu and Zn in the rhizosphere of Fe-deficient (grey symbols) or Fe-sufficient (black symbols) ryegrass (circles) and wheat (diamonds), relative to the control soil without plants.

increased in the rhizosphere rather than decreasing (depletion), suggesting redistribution among the various soil metal fractions. These always remained rather negligible. In this graph, a good match between the two data sets (experimental points close to the $y = x$ line) suggested that the depletion of extractable Cu or Zn in the rhizosphere accounted for the uptake of Cu or Zn. This typically occurred for Cu_{DIT} (Fig. 8), and indicated that both plant species depleted the fraction of soil Cu that is bound to metal oxides. In contrast, while bioavailable Cu closely matched with decreases in Cu_{DIT} in the rhizosphere, it was systematically greater than the observed decrease in Cu_{PYR} in the rhizosphere (Fig. 8). The depletion of Cu_{PYR} in the rhizosphere may have thus contributed a portion of the bioavailable Cu, although the depletion of Cu_{DIT} was on its own enough to explain the whole of the observed uptake of Cu by plants. This means that the uptake of Cu was less than the sum of Cu_{DIT} and Cu_{PYR} depleted from the rhizosphere. However, it should be remembered that these two fractions may overlap to some extent as the corresponding extractants are not perfectly selective (Chaignon *et al.*, 2003). Indeed, in the soil samples of the present experiment, the sum of Cu_{DIT} and Cu_{PYR} substantially exceeds total soil Cu. Therefore, these two fractions must overlap to some extent, which makes their sum rather meaningless.

For Zn, little depletion of Zn_{PYR} occurred in the rhizosphere, compared with the observed uptake of Zn by ryegrass and wheat (Fig. 7). A greater depletion of

Zn_{DIT} occurred, but did not account for bioavailable Zn. For wheat, the depletion of Zn_{DIT} was far less than the amount of Zn taken up, while for ryegrass it was sometimes greater than the actual uptake. There is no clear explanation of these results. It can simply be concluded that neither the depletion of Zn_{PYR} nor that of Zn_{DIT} explained the Zn bioavailability for the two studied grasses. This contrasts with the results obtained for Cu.

Analytical results were consistent with the observed effects of Fe deficiency on the bioavailability of soil Cu and the hypothesized involvement of the exudation of phytosiderophores in this process. Two possible pathways might be invoked: (i) the exudation of phytosiderophores would complex Fe, result in an increased dissolution of Fe-oxide and thereby increase the release of Cu bound to this fraction; or (ii) the exudation of phytosiderophores would complex Cu and thereby result in a greater release of Cu bound to any soil constituent. The present results favor the first pathway, but demonstration was inconclusive. It should be remembered that Cu_{DIT} and Zn_{DIT} not only represent metals bound to Fe-oxide, but also Mn-oxides, which are also expected to be dissolved by this extractant (Ponthieu, 2003). Only the second pathway can be advocated for in the case of metals bound to Mn-oxides.

Besides the exudation of complexing exudates such as phytosiderophores, plant roots are responsible for other chemical changes in the rhizosphere that might have contributed to the acquisition of metals such as Cu and Zn (Mench, 1990; Hinsinger, 2001). Beyond the depletion of metals due to metal uptake and adsorption in the root apoplasm, roots can alter the pH and the redox potential of the rhizosphere. One experiment attempted to evaluate the root-induced pH change that occurred in the rhizosphere of ryegrass grown on soil samples collected from the topsoil of the Solepur experimental plots (Mahammedi, 2001). Soil sampled in the treated plot in 1991–1996 and 1999, and in the control plot in 1999, generally showed little change in rhizosphere pH (as measured in the $CaCl_2$ extracts). A trend towards rhizosphere alkalization occurred whenever soil pH was below 6.7 (i.e. in soils sampled in years 1991, 1995 and 1996). In contrast, for soils having a pH above 6.7, a slight rhizosphere acidification occurred (which did not exceed 0.25 pH unit), which was significant in 1993 and 1994. A significant pH increase (exceeding 1 pH unit) was found in the rhizosphere of ryegrass grown in the 1999 control soil, which had not received any pig slurry. This soil had an acidic pH (close to 5.3). This suggested that the heavy application in pig slurry resulted in an increase in pH over the years, compared with the control plot. This type of behavior of plants leading to rhizosphere alkalization in acidic soils and rhizosphere acidification in neutral to alkaline soils has already been reported for several plant species, including grass (Youssef and Chino, 1989; Hinsinger *et al.*, 2003). Such changes in rhizosphere pH can however hardly explain the observed differences in Cu and Zn bioavailability between soil samples with and without heavy applications of pig slurry.

6. CONCLUSIONS

This work aimed first at evaluating whether the bioavailability of Cu and Zn in a soil was affected by pig slurry application. This was assessed at the Solepur experimental site in Brittany, northwestern France. At this site, the soil received massive, controlled applications of pig slurry over 5 consecutive years, which was equivalent to several decades of application at rates commonly used by farmers in the region. In this treated plot, as well as in the neighbouring control plot, the soil was sampled annually at three depths, and the grass cover was sampled regularly. The chemical extractability of Cu and Zn significantly increased as a consequence of heavy applications of pig slurry in the topsoil of the field site, regardless of the chemical extractant. However, the bioavailability of soil Cu and Zn as assessed by plant analysis was not always affected by pig slurry application. When assessed *in situ*, Cu and Zn concentrations in ryegrass shoots did not consistently increase with increasing application of pig slurry. In spite of the absence of a consistent trend, some samples showed elevated concentrations of Cu and Zn, which rather suggest a direct contamination of the forage by sprayed pig slurry or inadequate washing of plant shoots prior to analysis. For Cu in particular, the concentrations in forage for sensitive animals such as sheep were far above the threshold concentration. This suggests that the direct contamination of plant shoots by particles of pig slurry renders such an agricultural practice questionable for grazing or forage production in the case of sheeps, which are fairly sensitive to Cu toxicity.

Analysis of plant shoot contamination *in situ* was complemented by additional measurements in the laboratory. Indeed, according to the classification of Chaney *et al.* (1998), metals such as Cu and Zn belong to the group of phytotoxic metals, i.e. those more likely to be toxic to plants than to animals feeding on these plants. In addition, the transfer of Cu to shoots is often severely restricted, and rhizotoxicity can occur in the absence of any build-up of shoot Cu concentration. Therefore, in order to further evaluate whether the bioavailability of Cu and Zn was affected by pig slurry application in this soil, we conducted a biotest with two grasses, ryegrass and wheat. This biotest enabled easy access to plant roots in order to complement shoot analysis. It showed a significant increase in the bioavailability of Cu in the soil treated with pig slurry, relative to the control, untreated soil. In contrast, no significant increase in Zn bioavailability was found, in spite of the heavy rates of application of pig slurry at this site. No evidence of phytotoxicity was recorded over the short term of the biotest. However, only one soil was tested in this work and care should be taken when extrapolating these results to other situations. Our results suggest that long-term application of pig slurry poses the risk of an increase in the bioavailability and phytoxicity of Cu, while this risk is unlikely for Zn. Our findings call into question the sustainability of the heavy application of pig slurry as practised in several regions of intensive agriculture in Europe, such as Brittany in France, Catalonia in Spain, Denmark and The Netherlands.

The biotest was also designed for an easy sampling of rhizosphere soil, in order to evaluate rhizosphere processes implied in Cu and Zn acquisition by plant roots. The experimental design of the biotest more precisely aimed at assessing the impacts of Fe deficiency on the stimulation of grass root exudates such as phytosiderophores that might complex Cu and Zn. We found, as expected, that Fe deficiency increased root exudation of phytosiderophores in both ryegrass and wheat. Plant exudates dramatically affected Cu and Zn speciation and complexation in aqueous solution, as evidenced by DPASV. This may explain the enhanced Cu and Zn acquisition that occurred under Fe deficiency in both ryegrass and wheat. Root-induced changes in Cu and Zn extractability in the rhizosphere revealed a depletion of two major fractions of soil Cu and Zn, i.e. metals bound either to metal oxides or to organic matter, as determined by a sodium dithionite–acetate–citrate extraction or by a sodium pyrophosphate extraction, respectively. The amount of bioavailable Cu as deduced from plant uptake was in good agreement with the amount of Cu depleted from the metal oxide-bound fraction of soil Cu (determined by a sodium dithionite–acetate–citrate extraction) in the rhizosphere of ryegrass and wheat.

Additional research is, however, needed to ascertain the mechanisms responsible for Cu and Zn acquisition by these two grasses and the links between extractability and bioavailability of these two metals. Chemical interactions occurring in the rhizosphere, such as metal complexation by root exudates, seem to be of key importance for understanding the fate of Cu and Zn in contaminated soils and their potential effects on agricultural plants. In particular, the exudation of phytosiderophores by roots may be a potentially important process determining the fate of such metals in the rhizosphere of grasses. If confirmed, the use of grasses in metal-contaminated soils used for agricultural purposes may be questionable, especially in soils that are prone to induce Fe deficiency, such as calcareous soils. This is in line with results obtained in Cu-contaminated calcareous soils in vineyards of France or Switzerland (Coullery, 1997; Chaignon *et al.*, 2002b). Additional research is clearly needed to investigate further the mechanisms involved in changes in metal extractibility and speciation in the rhizosphere of pasture grasses and cereals, and the consequences for the bioavailability and phytotoxicity of Cu and Zn and thus, for the management of agricultural soils exposed to such metal contaminations.

ACKNOWLEDGMENTS

This work was conducted as part of a project granted by the CNRS-PEVS (Programme Environnement, Vie et Sociétés) research programme. The technical help of Nicole Balsera for conducting the biotests is gratefully acknowledged.

REFERENCES

Begg, C.B.M., Kirk, G.J.D., MacKenzie, A.F., Neue, H.U., 1994. Root-induced iron oxidation and pH changes in the lowland rice rhizosphere. New Phytol. 128, 469–477.

Brown, J.C., 1978. Mechanism of iron uptake by plants. Plant Cell Environ. 1, 249–257.

Brun, L., Maillet, J., Hinsinger, P., Pépin, M., 2001. Evaluation of copper availability to plants in copper contaminated vineyard soils. Environ. Pollut. 111, 293–302.

Cakmak, S., Gülüt, K.Y., Marschner, H., Graham, R.D., 1994. Effect of zinc and iron deficiency on phytosiderophore release in wheat genotypes differing in zinc efficiency. J. Plant Nutr. 17, 1–17.

Chaignon, V., Bedin, F., Hinsinger, P., 2002a. Copper bioavailability and rhizosphere pH changes as affected by nitrogen supply for tomato and oilseed rape cropped on an acidic and a calcareous soil. Plant Soil 243, 219–228.

Chaignon, V., Di Malta, D., Hinsinger, P., 2002b. Fe-deficiency increases Cu acquisition by wheat cropped in a Cu-contaminated, vineyard soil. New Phytol. 154, 121–130.

Chaignon, V., Hinsinger, P., 2003. A biotest for evaluating copper bioavailability to plants in a contaminated soil. J. Environ. Qual. 38, 824–833.

Chaignon, V., Sanchez-Neira, I., Herrmann, P., Jaillard, B., Hinsinger, P., 2003. Copper bioavailability and extractability as related to chemical properties of contaminated soils from a vine-growing area. Environ. Pollut. 123, 229–238.

Chaney, R.L., Brown, S.L., Angle, J.S., 1998. Soil-root interface: ecosystem health and human food-chain protection, In: Huang, P.M., Adriano, D.C., Logan, T.J., Checkai, R.T., (Eds.), Soil Chemistry and Ecosystem Health, Soil Science Society of America Inc., Madison, USA, pp. 279–311.

Clemens, S., 2001. Molecular mechanisms of plant metal tolerance and homeostasis. Planta 212, 475–486.

Cohen, C.K., Fox, T.C., Garvin, D.F., Kochian, L.V., 1998. The role of iron-deficiency stress responses in stimulating heavy-metal transport in plants. Plant Physiol. 116, 1063–1072.

Coppenet, M., Golven, J., Simon, J.C., Le Corre, L., Le Roy, M., 1993. The chemical evolution of soils in intensive breeding farms: the example of Finistere. Agronomie 13, 77–83.

Coullery, P., 1997. Gestion des sols faiblement pollués par des métaux lourds. Revue Suisse d'Agriculture 29, 299–305.

Curie, C., Panaviene, Z., Loulergue, C., Dellaporta, S. L., Briat, J.F., Walker, E.L., 2001. Maize yellow stripe1 encodes a membrane protein directly involved in Fe(III) uptake. Nature (London) 409, 346–349.

Fischer, W.R., Flessa, H., Schaller, G., 1989. pH values and redox potentials in microsites of the rhizosphere. Zeitschrift für Pflanzenernährung und Bodenkunde 152, 191–195.

Flessa, H., Fischer, W.R., 1992. Plant-induced changes in the redox potentials of rice rhizospheres. Plant Soil 143, 55–60.

Gries, D., Klatt, S., Runge, M., 1998. Copper-deficiency-induced phytosiderophore release in the calcicol grass *Hordelymus europaeus*. New Phytol. 140, 95–101.

Guivarch, A., Hinsinger, P., Staunton, S., 1999. Root uptake and distribution of radiocesium from contaminated soil and the enhancement of Cs adsorption in the rhizosphere. Plant Soil 211, 131–138.

Hinsinger, P., 1998. How do plant roots acquire mineral nutrients? Chemical processes involved in the rhizosphere. Adv. Agron. 64, 225–265.

Hinsinger, P., 2001. Bioavailability of trace elements as related to root-induced chemical changes in the rhizosphere. In: Gobran, G.R., Wenzel, W.W., Lombi, E. (Eds.), Trace Elements in the Rhizosphere, CRC Press, Boca Raton, USA, pp. 25–41.

Hinsinger, P., 2004. Nutrient availability and transport in the rhizosphere. In: Goodman, R.M. (Ed.), Encyclopedia of Plant and Crop Science, Marcel Dekker, Inc., New York, USA, pp. 1094–1097.

Hinsinger, P., Plassard, C., Tang, C., Jaillard, B., 2003. Origins of root-induced pH changes in the rhizosphere and their responses to environmental constraints: A review. Plant Soil 248, 43–59.

Kirk, G.J.D., Bajita, J.B., 1995. Root-induced iron oxidation, pH changes and zinc solubilization in the rhizosphere of lowland rice. New Phytol. 131, 129–137.

Kochian, L., 1999. The role of rhizosphere processes in the bioavailability of trace elements to plants. In: Wenzel, W.W., Adriano, D.C., Alloway, B., Doner, H.E., Keller, C., Lepp, N.W., Mench, M., Naidu, R., Pierzynski, G.M. (Eds.), Proceedings of the Fifth International Conference on the Biogeochemistry of Trace Elements, 11–15 July 1999, Vienna, Austria.

L'Herroux, L., Le Roux, S., Appriou, P., Martinez, J., 1997. Behaviour of metals following intensive pig slurry applications to a natural field treatment process in Brittany (France). Environ. Pollut. 97, 119–130.

Loosemore, N., Straczek, A., Hinsinger, P., Jaillard, B., 2004. Zinc mobilization from a contaminated soil by three genotypes of tobacco as affected by soil and rhizosphere pH. Plant Soil 260, 19–32.

Mahammedi, D., 2001. Spéciation et biodisponibilité du cuivre et du zinc dans un sol ayant subi des épandages massifs de lisiers de porc. Ph.D. Thesis, DEA Géosciences de l'Environnement, Universitat. Aix-Marseille.

Marschner, H., 1995. Mineral Nutrition of Higher Plants, 2nd ed. Academic Press, London.

Marschner, H., Römheld, V., 1994. Strategies of plants for acquisition of iron. Plant Soil 165, 261–274.

Marschner, H., Römheld, V., Ossenberg-Neuhaus, H., 1982. Rapid method for measuring changes in pH and reducing processes along roots of intact plants. Zeitschrift für Pflanzenphysiologie 105, 407–416.

Martinez, J., 1997. Solepur: a soil treatment process for pig slurry with subsequent denitrification of drainage water. J. Agri. Eng. Res. 66, 51–62.

Martinez, J., Hao, X.,. 1996. A field treatment plant for pig slurry. Water Sci. Technol. 34, 87–92.

Martinez, J., Peu, P., 2000. Nutrient fluxes from a soil treatment process for pig slurry. Soil Use Manage. 16, 100–107.

McLaughlin, M.J., Smolders, E., Merckx, R., 1998. Soil–root interface: physicochemical processes. In: Ituang, P.M., (Ed.), Soil Chemistry and Ecosystem Health, Special Publication no. 52. Soil Science Society of America, Madison, WI, pp. 233–277.

Mench, M., 1990. Transfert des oligo-éléments du sol à la racine et absorption. Compte Rendu de l'Académie d'Agriculture de France 76, 17–30.

Mench, M., Martin, E., 1991. Mobilization of cadmium and other metals from two soils by root exudates of *Zea mays* L., *Nicotiana tabacum* L. and *Nicotiana rustica* L. Plant Soil 132, 187–196.

Murakami, T., Ise, K., Hayakawa, M., Kamei, S., Takagi, S., 1989. Stabilities of metal complexes of mugineic acids and their specific affinities for iron(III). Chem. Lett. 12, 2137–2140.

Neumann, G., Römheld, V., 2001.The release of root exudates as effected by the plant's physiological status. In: Pinton, R., Varanini, Z., Nannipieri, P. (Eds.), The Rhizosphere, Marcel Dekker, Inc., New York, pp. 41–93.

Ponthieu, M., 2003. Spéciation des éléments traces métalliques dans des sols et des solutions du sol: du modèle au terrain. Ph.D Thesis, Universitat Aix-Marseille.

Rengel, Z., Römheld, V., 2000. Root exudation and Fe uptake and transport in wheat genotypes differing in tolerance to Zn deficiency. Plant Soil 222, 25–34.

Reuter, D.J., Robinson, J.B., 1997. Plant analysis: an interpretation manual. CSIRO Publishing, Australia.

Römheld, V., 1987. Different strategies for iron acquisition in higher plants. Physiologia Plantarum 70, 231–234.

Seguin, V., Gagnon, C., Courchesne, F., 2004. Changes in water extractable metals, pH and organic carbon concentrations at the soil–root interface of forested soils. Plant Soil 260, 1–17.

Takagi, S., Nomoto, K., Takemoto, T., 1984. Physiological aspect of mugineic acid, a possible phytosiderophore of graminaceous plants. J. Plant Nutri. 7, 469–477.

Thomas, S., 2002. Spéciation et biodisponibilité du Cu et du Zn dans la rhizosphère de graminées – Effet de phytosidérophores dans un sol soumis à des apports massifs de lisiers de porc (dispositif SOLEPUR, Plouvorn, Finistère). Ph.D Thesis, DEA Géosciences de l'Environnement, Universitat Aix-Marseille.

Treeby, M., Marschner, H., Römheld, V., 1989. Mobilization of iron and other micro-nutrient cations from a calcareous soil by plant-borne, microbial, and synthetic metal chelators. Plant Soil 114, 217–226.

Vansuyt, G., Souche, G., Straczek, A., Briat, J.F., Jaillard, B., 2003. Flux of protons released by wild type and ferritin over-expressor tobacco plants: effect of phosphorus and iron nutrition. Plant Physiol. Biochem. 41, 27–33.

von Wirén, N., Marschner, H., Römheld, V., 1996. Roots of iron-efficient maize also absorb phytosiderophore-chelated zinc. Plant Physiol. 111, 1119–1125.

von Wirén, N., Mori, S., Marschner, H., Römheld, V., 1994. Iron inefficiency in maize mutant ys1 (*Zea mays* L. cv. Yellow-Stripe) is caused by a defect in uptake of iron phytosiderophores. Plant Physiol. 106, 71–77.

Walter, A., Römheld, V., Marschner, H., Mori, S., 1994. Is the release of phytosiderophores in zinc-deficient plants a response to impaired iron utilization? Physiologia Plantarum 92, 493–500.

Youssef, R.A., Chino, M., 1989. Root-induced changes in the rhizosphere of plants. I. pH changes in relation to the bulk soil. Soil Sci. Plant Nutri. 35, 461–468.

Zhang, X.K., Zhang, F.S., Mao, D.R., 1998. Effect of iron plaque outside roots on nutrient uptake by rice (*Oryza sativa* L.). Zinc uptake by Fe-deficient rice. Plant Soil 202, 33–39.

Biogeochemistry of Trace Elements in the Rhizosphere
P.M. Huang and G.R. Gobran (Editors)

Chapter 12

Binding and electrostatic attraction of trace elements to plant root surfaces

U. Yermiyahu[a] and T.B. Kinraide[b]

[a]Agricultural Research Organization, Gilat Research Center, D.N. Negev 2 85280, Israel

[b]Appalachian Farming Systems Research Center, Agricultural Research Service, United States Department of Agriculture, Beaver, West Virginia 25813-9423
E-mail: tom.kinraide@ars.usda.gov

ABSTRACT

An important feature of the biogeochemistry of trace elements in the rhizosphere is the interaction between plant root surfaces and the ions in the soil solution. These ions may accumulate in the aqueous phases of cell surfaces external to the plasma membranes (PMs). In addition, ions may bind to cell wall (CW) components or to the PM surface with variable strength. In this chapter, we shall describe the distribution of ions among the extracellular phases using electrostatic models (i.e. Gouy–Chapman–Stern and Donnan–plus-binding models) for which parameters are now available. Many plant responses to ions correlate well with computed PM-surface activities, but only poorly with activities in the soil solution. These responses include ion uptake, ion-induced intoxication, and the alleviation of intoxication by other ions. We illustrate our technique for the quantitative resolution of multiple ion effects by inserting cell-surface activities into nonlinear equations.

1. INTRODUCTION

An important feature of the biogeochemistry of trace elements in the rhizosphere is the interaction between plant root surfaces and the ions present in the soil solution. Although the solid phases of the soil may be the reservoir from which many ions are derived, the soil solution is the phase with which the plant roots interact most directly. The ions, especially the cations, of the soil solution may accumulate in the aqueous phases of cell surfaces external to the plasma membranes

(PMs) – phases sometimes referred to as the "water free space" and the "Donnan free space" (Marschner, 1995). In addition, ions may bind to cell wall (CW) components or to the PM surface with variable binding strength. Accumulation may also entail intracellular uptake, as well as translocation away from the roots and accumulation in the shoots.

In this chapter, we shall discuss the distribution of ions among the extracellular phases. Extracellular accumulation is interesting and important in its own right, as it influences the value and safety of food crops, and induces or alleviates the intoxicating effect of ions whose site of action may lie in the apoplast (Horst, 1995). Extracellular accumulation of ions in the root also determines the intracellular accumulation and translocation of ions to the rest of the plant. The following diagram illustrates our conceptualization of ion interactions among phases of the rhizosphere and the plants.

CW phase ⇄ Physiological effects
↗↙
Soil solids ⇄ Soil solution ↑↓
↖↘
PM surface ⇄ Physiological effects
↑↓
Cell interior ⇄ Physiological effects

The diagram deliberately omits some possible interactions, such as a direct interaction between ions bound to soil solids and ions in the CW. The diagram is also deliberately ambiguous. That is, we are uncertain to what extent the CW influences the interaction between ions in the soil solution and the PM surface (Sattelmacher, 2001; Kinraide, 2004). Furthermore, the diagram is generous with regard to physiological effects. We envisage the possibility of physiological consequences of ion accumulation in each of the designated cell phases. For example, Al^{3+} may impair CW development by interference with cross-linking among CW polysaccharides independent of effects upon the PM, by inhibition of polysaccharide secretion through the PM independent of intracellular effects, or by intracellular inhibition of polysaccharide synthesis independent of external effects (Schmohl and Horst, 2002). In this chapter we shall discuss several aspects of ion binding to root surfaces, but will concentrate upon binding and electrostatic attraction of ions to the PM surface, which is our principal area of research.

The last 15 years have witnessed considerable progress in the interpretation of plant–ion interactions. We shall concentrate on that period and upon interpretations based upon electrostatic phenomena. Certainly the recent progress would not have been possible without earlier discoveries. Electrostatic theory, going back about 100 years, has been summarized by McLaughlin (1977), Barber (1980) and Tatulian (1999). Hille (1992) discusses both theory and the physiological effects of electrical phenomena in zoological tissues. Grignon and Sentenac (1991) reviewed their own contributions, and those of others, to present an understanding of "pH and ionic conditions in the apoplast" in an article of that title. A spate of ζ (zeta) potential measurements (measurements of electrical potential very near the membrane surface) occurred over a brief period (Nagata and Melchers, 1978; Gibrat *et al.*, 1985; Abe and Takeda, 1988; Obi *et al.*, 1989a, b, 1990), and the reported measurements of PM surface charge density (σ) have been surveyed by Brauer *et al.* (2000). Electrostatic theory was used to explain some physiological phenomena such as cell fusion, the activity of membrane-bound enzymes, function of thylakoid membranes in photosynthesis, membrane transport, and ion toxicity (Nagata and Melchers, 1978; Barber, 1980; Chang *et al.*, 1983; Gibrat *et al.*, 1985; Abe and Takeda, 1988; Wagatsuma and Akiba, 1989). Nevertheless, botanical research generally neglected membrane *surface* electrical effects prior to 15 years ago despite decades of interest in *trans*membrane electrical potentials. Fig. 1 illustrates electrical potentials at the PM.

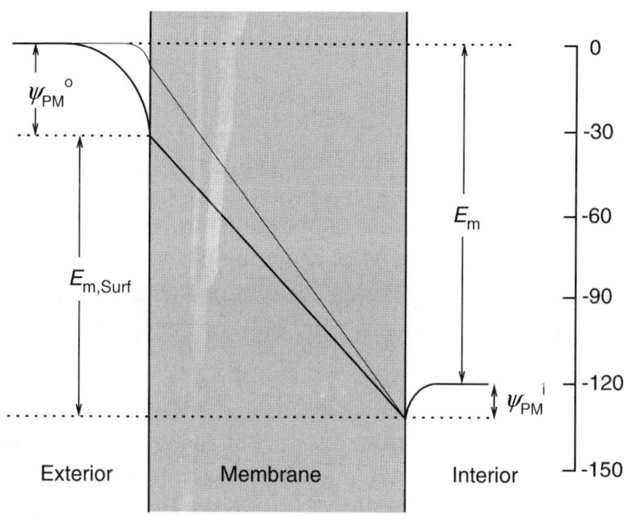

Fig. 1. Profile of the electrical potential across the PM. ψ_{PM} is the electrical potential at the PM surface. E_m is the transmembrane potential difference from bulk phase to bulk phase as measured by microelectrodes. $E_{m,Surf}$ is the transmembrane potential difference from surface to surface. The upper solid line represents the profile after the addition of solutes to the exterior medium that depolarize the outer face, thereby altering $E_{m,Surf}$, but not E_m. The figure is redrawn from Kinraide (2001).

2. ELECTROSTATIC MODELS FOR THE PLASMA MEMBRANE

2.1. Models to compute ion activities, binding, and electrical potentials at the PM surface

We have modeled the PM as though it were composed of negatively charged (R^-) and neutral (P^0) sites to which ions (I^Z, where Z is the charge on the ion) may bind. R^- and P^0 will be treated as single entities even though they include multiple components. Some positively charged components are also present, but their numbers are small, so R^- will refer to the number of negative sites minus the number of positive sites. The number of neutral sites is much larger than the number of negative sites, so neutral sites should be considered even though ion binding to them may be weak.

We consider that each of the PM binding sites may bind ions as expressed in the following equations, in which a 1:1 binding ratio is assumed. Other binding ratios are possible, but incorporating other ratios does not improve the performance of the model.

$$R^- + I^Z \rightleftarrows RI^{Z-1} \tag{1}$$

$$P^0 + I^Z \rightleftarrows PI^Z \tag{2}$$

Binding constants, specific for site and ion, may be expressed as

$$K_{R,I} = [RI^{Z-1}]/([R^-][I^Z]_{PM}) \tag{3}$$

$$K_{P,I} = [PI^Z]/([P^0][I^Z]_{PM}) \tag{4}$$

$[R^-]$, $[P^0]$, $[RI^{Z-1}]$, and $[PI^Z]$ denote membrane surface densities in mol m^{-2}. $[I^Z]_{PM}$ denotes the concentration in M of the unbound ion at the PM surface.

$[I^Z]_{PM}$ is computed by a Boltzmann equation, which describes the partitioning of an ion between two phases – the bulk-phase rooting medium and the PM surface.

$$[I^Z]_{PM} = [I^Z]\exp[-Z_i F\psi_{PM}/(RT)] \tag{5}$$

$[I^Z]$ denotes the concentration in M of ion I^Z in the bulk phase; ψ_{PM} denotes the electrical potential at the PM surface when the potential in the bulk phase is zero; alternatively, $\psi_{PM} = \psi_{PM\ surface} - \psi_{bulk\ phase}$; F, R, and T are the Faraday constant, the gas constant, and the temperature, respectively; $-Z_i F\psi_{PM}/(RT) = -Z_i\psi_{PM}/25.7$ at 25°C for ψ_{PM} expressed in mV.

$[R^-]$, $[P^0]$, $[RI^{Z-1}]$, and $[PI^Z]$ depend upon R_T and P_T, the total number of binding sites in each class. R_T and P_T would equal the surface densities of R^- and

P^0 if no solute ions were bound to those sites. If no solute ions were bound to the PM, then the intrinsic surface charge density (σ_0, in units C m^{-2}) would equal FR_T. The conditional σ reflects ion binding and equals the sums of all surface species times the charge of each species times F:

$$\sigma = \{-[R^-] + \Sigma_i(Z_i - 1)[RI^{Z-1}] + \Sigma_i Z_i[PI^Z]\}F \quad (6)$$

Eq. (6) represents the Stern portion of the Gouy–Chapman–Stern model. Stern reactions are strong interactions, represented as binding in Eqs. (1) and (2), between ions and the PM that alter σ. These associations are different from charge screening (described next), which does not eliminate surface charge but merely reduces the negativity of the potential.

The Gouy–Chapman portion of the model is expressed in the Müller equation, which is commonly but erroneously referred to as the Grahame equation (Tatulian, 1999),

$$\sigma^2 = 2\varepsilon_r\varepsilon_0 RT\Sigma_i[I^Z](\exp[-Z_i F\psi_{PM}/(RT)] - 1) \quad (7)$$

$2\varepsilon_r\varepsilon_0 RT = 0.00345$ at 25°C for concentrations expressed in M (ε_r is the dielectric constant for water, ε_0 is the permittivity of a vacuum).

To compute ψ_{PM}, trial values were assigned to it in Eqs. (5) and (7) until values for σ in Eqs. (6) and (7) converged. In order to do that, values for $K_{R,I}$, $K_{P,I}$, R_T, and P_T must be known in order to compute the variables in Eq. (6). Methods for the estimation of $K_{R,I}$, $K_{P,I}$, R_T, and P_T will be presented later. Once ψ_{PM} is known, PM-surface activities ($\{I^Z\}_{PM}$) may be computed from the Nernst variation of the Boltzmann equation. Fig. 2 illustrates the relationships.

$$\{I^Z\}_{PM} = \{I^Z\}\exp[-Z_i F\psi_{PM}/(RT)] \quad (8)$$

Table 1 presents parameter values for the Gouy–Chapman–Stern model. The experimental methods for their determination will be described later.

2.2. Experimental determination of ψ_{PM}

ψ_{PM} is usually measured as a ζ potential, which is the near-surface potential measured in electrophoresis experiments. For these measurements, protoplasts (whole viable cells enzymatically stripped of their CWs) or PM vesicles are prepared from plant tissue (Nagata and Melchers, 1978; Gibrat *et al.*, 1985; Abe and Takeda, 1988; Obi *et al.*, 1989a, b, 1990). The rate of migration of these particles are then measured in an electric field. The ζ potential is slightly smaller in magnitude than ψ_{PM} because the former is the potential a slight distance from the PM surface, that is, at the plane of shear. However, such measurements are few. In a compilation, Kinraide (1998) found only six studies of higher plants

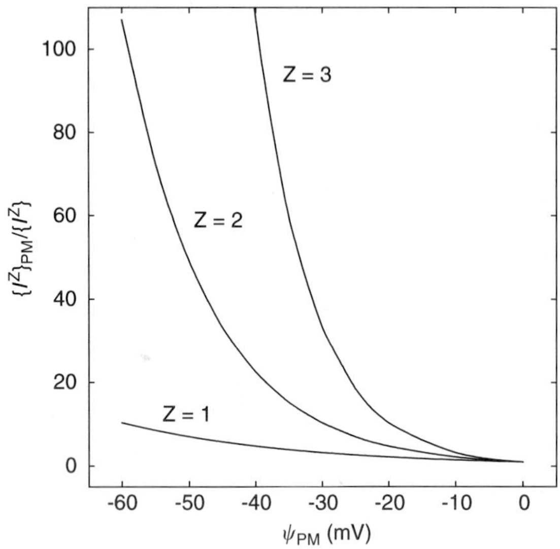

Fig. 2. The effect of ψ_{PM} and ion charge (Z) on ion activities at the PM surface relative to external activities. Values were computed from $\{I^Z\}_{PM} = \{I^Z\}\exp[-Z_i F\psi_{PM}/(RT)]$, Eq. (8).

Table 1

Parameter values for a Gouy–Chapman–Stern model for plant-root PMs. Parameters were determined by adsorption experiments using PM vesicles or by an evaluation of published ζ potential measurements using PM vesicles or protoplasts from several species and tissues

Model Parameters	Adsorption Experiments[a]	ζ potential Measurements[b]
R_T	0.3074 μmol m^{-2}	See Fig. 4(A)
P_T	nd[c]	2.4 μmol m^{-2}
$K_{R,Na}$, $K_{R,K}$	0.8–1 M^{-1}	0 M^{-1}
$K_{R,Zn}$	5	nd
$K_{R,Mg}$	9	24.8
$K_{R,Ca}$	30–50	29.3
$K_{R,Cu}$	400	nd
$K_{R,La}$	2200	2030
$K_{R,Al}$	20000	nd
$K_{R,H}$	21500	20200
$K_{P,I}$	nd	$K_{R,I}/1550$

[a] Yermiyahu *et al.* (1994, 1997b) and Vulkan *et al.* (2004). [b] Kinraide *et al.* (1998) and Kinraide (2001).
[c] not determined.

that met the criteria of measurements taken in at least four different solutions. This compilation yielded 35 measurements from several species and tissues. Later, 15 additional measurements from two unpublished studies were contributed by Dr. Qisen Zhang and Dr. Robert Reid and were incorporated into a new analysis of the compiled 50 measurements (Kinraide, 2001).

2.3. Experimental determination of binding constants: Adsorption experiments

The strength of binding between ions and the PM can be evaluated from adsorption experiments or from ζ-potential measurements. In the adsorption experiments, a known quantity of PM vesicles isolated from roots was added to isotonic sucrose solutions adjusted for initial ion concentration and pH. After equilibrium was achieved (30 min), the pH of the suspensions was measured, then PM vesicles were separated from the equilibrium solution using microfilter centrifuge tubes. The amount of ion adsorbed to the vesicles was measured directly (Yermiyahu *et al.*, 1994) or calculated from the reduction of ion concentration in the equilibrium solution (Yermiyahu *et al.*, 1997b). Adsorption is the sum of ion binding to the PM and ion attraction into the diffuse layer very near to the surface of the PM. (Eqs. (1)–(7) are sufficient for the computation of binding but not for adsorption [see Nir *et al.*, 1994].)

Results of an adsorption experiment are presented in Table 2, which lists measured and calculated values for Ca^{2+} adsorption to the PM. As expected, adsorbed Ca^{2+} was greater in suspending media with larger Ca^{2+} and smaller Na^+ concentrations, results that have implications for salinity toxicity. Fig. 3 illustrates the competition for adsorption between Al^{3+} and H^+, which can be varied independently in artificial media but cannot be so successfully varied in soils. The adsorbed Al^{3+} was greater in suspending media with larger Al^{3+} concentrations and greater pH values. In the same reactions, adsorbed H^+ was less in suspending media with larger Al^{3+} concentrations and greater pH values. These results help to explain the Al^{3+} enhancement of root elongation in acidic media and the H^+ enhancement of root elongation in Al^{3+}-containing media, as described below.

The results of adsorption experiments may be used to evaluate binding constants provided that R_T and, sometimes, P_T are known. We consider R_T to be the most uncertain of the parameters as it may vary with species, tissue, and preparation. Furthermore, direct measurements are rarely attempted. They would require an estimation of membrane area from a known quantity of vesicles together with the measured uptake of a solute assumed to bind quantitatively all the negative charges. Instead, R_T may be estimated indirectly using a fluorescent dye whose uptake is monitored by fluorescence quenching. Uptake is assumed to depend upon ψ_{PM}, which is assumed not to be changed by the dye. We shall not discuss the details of the methods here but refer readers to Brauer *et al.* (2000), who present the methods together with an assessment of their uncertainties. For

Table 2
Competition between Ca^{2+} and Na^+ at the PM surface of vesicles from Honey Dew melon. Adsorbed Ca^{2+} is expressed as ions per intrinsic charged site on the PM surface. Data are from Yermiyahu *et al.* (1999)

Treatment (mM)		Measured	Calculated[a]	
Ca^{2+}	Na^+	Adsorbed	Adsorbed[b]	Bound
0.05	1	0.095 ± 0.006[c]	0.108	0.102
0.05	20	0.048 ± 0.007	0.072	0.068
0.05	40	0.030 ± 0.007	0.044	0.042
0.05	100	0.009 ± 0.002	0.016	0.015
0.1	1	0.141 ± 0.023	0.215	0.198
0.1	20	0.092 ± 0.017	0.122	0.115
0.1	40	0.067 ± 0.007	0.076	0.072
0.1	100	0.035 ± 0.002	0.031	0.029
0.25	1	0.255 ± 0.027	0.414	0.328
0.25	20	0.135 ± 0.015	0.203	0.187
0.25	40	0.099 ± 0.013	0.138	0.129
0.25	100	0.040 ± 0.006	0.064	0.061
0.5	1	0.346 ± 0.018	0.452	0.342
0.5	20	0.205 ± 0.017	0.260	0.233
0.5	40	0.160 ± 0.009	0.192	0.177
0.5	100	0.104 ± 0.003	0.104	0.098
1.0	1	0.500 ± 0.029	0.465	0.348
1.0	20	0.311 ± 0.010	0.308	0.271
1.0	40	0.259 ± 0.028	0.245	0.223
1.0	100	0.203 ± 0.029	0.154	0.144

[a] The calculations used a surface area of 370 Å² per intrinsic charge. The binding constants $K_{R,Na}$ and $K_{R,Ca}$ were 0.8 and 50 M^{-1}, respectively.
[b] Adsorbed Ca^{2+} includes the bound Ca^{2+} plus the Ca^{2+} in the diffuse double layer.
[c] Standard deviation.

P_T we assigned a value of 2.4 μmol m^{-2} on the basis of the membrane surface area occupied by phosphatidic acids in the PM.

Finally, the binding constants may be computed. Readers may consult Yermiyahu *et al.* (1997b) for details, or we can direct readers to appropriate

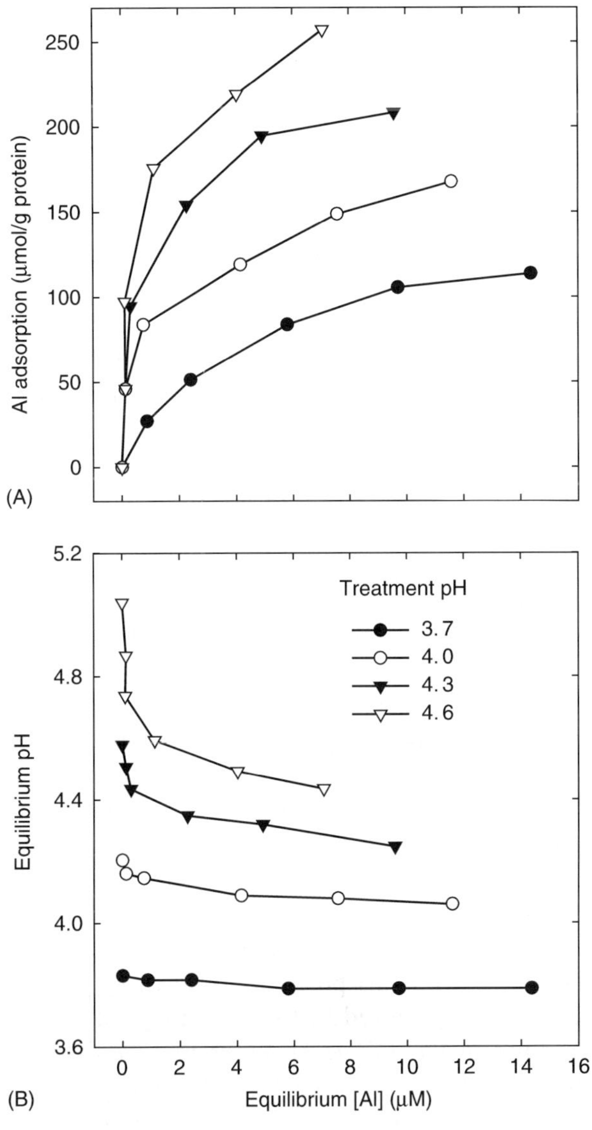

Fig. 3. Competition between Al^{3+} and H^+ for adsorption to PM. Vesicles were prepared from roots of Scout 66 wheat. For (A), adsorption was computed from Al^{3+} loss from the suspending medium. For (B), declines in pH from the initial values signify Al^{3+}-induced displacement of H^+. The figure is redrawn from Yermiyahu *et al.* (1997b).

computer programs. We shall, however, illustrate some results with a specific example (Yermiyahu *et al.*, 1997b). R_T was assigned a value of 0.3074 µmol m^{-2} (= 540 $Å^2$ per intrinsic charge) on the basis of the fluorescent dye method; P_T was ignored; and values for three binding constants ($K_{R,Cl} = 0$, $K_{R,K} = 1$, and $K_{R,Ca} = 30\ M^{-1}$)

were assumed on the basis of previous studies. Then, solutions of variable pH and $CaCl_2$ concentration in a background of 1 mM KCl were used for vesicle suspension. On the basis of Ca^{2+} and H^+ adsorption, the value of $K_{R,H}$ was determined. That was done by substituting values for $K_{R,H}$ into the Gouy–Chapman–Stern model to minimize the sum of differences squared between measured and calculated adsorption. With these four binding constants ($K_{R,Cl}$, $K_{R,K}$, $K_{R,Ca}$ and $K_{R,H}$) the values for other binding constants were determined (Table 1). Fig. 3 presents results from one of the adsorption experiments used in the computation of $K_{R,Al}$.

2.4. Experimental determination of binding constants: ζ-potential measurement

In addition to the adsorption method, we used an entirely different method to determine binding constants. ζ-potential measurements were compiled from several studies (see above) and provided enough information to compute binding constants for six cations (Table 1) according to the following procedure: 1. P_T was assigned a value of 2.4 μmol m^{-2} as noted above. 2. Initial values were assigned to R_T and to each of the $K_{R,I}$ constants. 3. Values for $K_{P,I}$ were assumed to be a constant fraction of $K_{R,I}$, so a value for Q ($= K_{P,I}/K_{R,I}$) was assigned. This assumption of proportionality seems reasonable given that the strength of ion binding to both negative and neutral phospholipids follows the order $M^{3+} > M^{2+} > M^{1+}$ (Tatulian, 1999). 4. A computer program computed ψ_{PM} using initial values for all parameter values but one. Values for that one parameter were raised incrementally so as to find its optimum value on the basis of the minimized sum of differences squared between computed ψ_{PM} and measured ζ potentials. 5. The program moved to the next parameter and found its optimum value. The program cycled through all parameters repeatedly, optimizing values at diminishing increment sizes until the sum of squares was no longer reduced.

This procedure, which used a single suite of parameter values for all eight studies, led to an R^2 of 0.706 for ψ_{PM} vs. ζ potential. However, inspection of the data showed that similar solutions produced different ζ potentials in different studies. Consequently, we optimized R_T for each study, but used a single suite of values for all other parameters (Table 1). When that was done, $R^2 = 0.930$ for ψ_{PM} vs. ζ potential (Fig. 4(A)). The high correspondence between the values for binding constants obtained independently from adsorption experiments and from ζ-potential measurements provides some reassurance that we can calculate values for ψ_{PM} that are at least proportional to the actual values. Given a value for ψ_{PM}, PM surface activities of ions can be computed by Eq. (8), and the amounts of bound ions can be computed as well.

The value of $K_{R,Al}$ is important for an assessment of toxic factors in acidic soils, but not a sufficient number of ζ-potential measurements are available for confirmation of the value 20,000 M^{-1} by the method above. Nevertheless, three studies lend credibility to that value (Akeson *et al.*, 1989; Wilkinson *et al.*, 1993; Jones and Kochian, 1997).

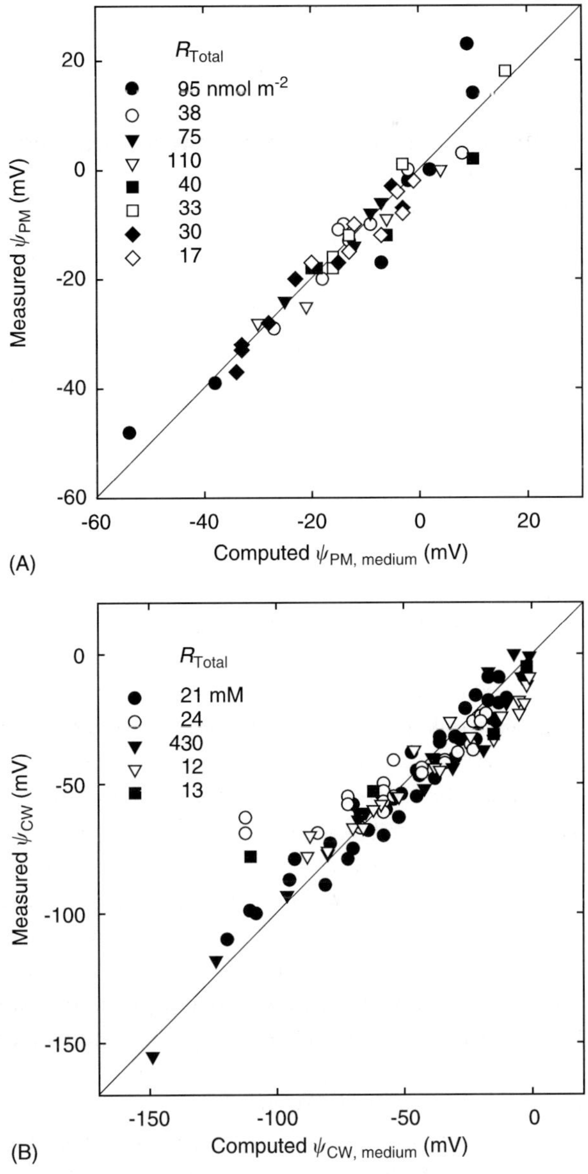

Fig. 4. A comparison of studies in which ψ_{PM} and ψ_{CW} were measured and computed. For (A), parameters for a Gouy–Chapman–Stern model were evaluated for eight studies; for (B), parameters for a Donnan-plus-binding model were evaluated for five studies. Optimized values for total negative sites (R_T) were computed for each study, but a single suite of binding constants was evaluated for the pooled PM data and for the pooled CW data. The figure is redrawn from Shomer *et al.* (2003).

3. ELECTROSTATIC MODELS FOR THE CW

3.1. Models to compute ion activities, binding, and electrical potentials in the CW

For the computation of ion activities, binding, and electrical potentials in the CW we use a Donnan-plus-binding model that has been presented in detail by Sentenac and Grignon (1981), Rufyikiri *et al.* (2003), Shomer *et al.* (2003), and others. The model incorporates ion binding and electrostatic attraction to CW components just as the Gouy–Chapman–Stern model incorporates ion binding and electrostatic attraction to PM components. One difference is that quantities in the Donnan-plus-binding model are expressed in volume units (i.e. concentration in units M) rather than in surface units (i.e. surface density in units mol m^{-2}). Consequently, some of the equations above are applicable except that R^- and P^0 refer to negatively charged and neutral components of the CW; $[R_T]$, $[P_T]$, $[R^-]$, $[P^0]$, $[RI^{Z-1}]$, $[PI^Z]$, and $[I^Z]_{CW}$ all refer to concentrations (M) in the Donnan phase; ψ_{PM} (now written ψ_{CW}) refers to the Donnan potential; and σ refers to charge concentration (M) in the Donnan phase. $\{I^Z\}_{CW}$ is computed by the equation

$$\{I^Z\}_{CW} = \{I^Z\}\exp[-Z_i F\psi_{CW}/(RT)] \quad (9)$$

3.2. Experimental determination of ψ_{CW}

The electrical potentials in the CW (ψ_{CW}) may be measured as ζ potentials. For these measurements CW fragments were prepared (O'Shea *et al.*, 1990), but ζ-potential measurements for CWs are even fewer than for PMs. More often, but still uncommonly, ψ_{CW} is measured by pushing microelectrodes (glass micropipette salt bridges connected to Ag/AgCl electrodes) against the CWs of whole cells either isolated or in whole tissues (Nagai and Kishimoto, 1964; Spanswick *et al.*, 1967; Saftner and Raschke, 1981). Shomer *et al.* (2003) presented two conflated but independent studies together with published data from three additional studies in which ψ_{CW} was estimated. These five studies provided over 100 individual measurements from several species and tissues.

3.3. Experimental determination of binding to CWs

CWs may be extracted from plant tissue and be subjected to adsorption experiments (Bush and McColl, 1987; Allan and Jarrell, 1989; Richter and Dainty, 1990; Grignon and Sentenac, 1991). On the basis of such experiments, Rufyikiri *et al.* (2003) found the following sequence of affinities of ions for CWs of banana root: $Al^{3+} > H^+ > Ca^{2+} > (Mg^{2+}$ or $K^+)$. Furthermore, CWs may be separated into fractions such as pectins, proteins, etc. prior to the experiments. For example, Franco *et al.* (2002) found the following sequence of affinities of ions for certain demethylated pectins: $Fe^{3+} > Al^{3+} > Cu^{2+} \approx Mn^{2+} > Zn^{2+} \approx Ca^{2+}$.

Shomer *et al.* (2003) observed that ions reduced ψ_{CW} negativity in the sequence $Al^{3+} > La^{3+} > H^+ > Cu^{2+} > Ni^{2+} > Ca^{2+} > Co^{2+} > Cd^{2+} > Mg^{2+} >$

Zn^{2+} > hexamethonium^{2+} > Rb^+ > K^+ > Cs^+ > Na^+. The order of depolarizing effectiveness corresponds substantially with the binding affinities mentioned above, with the values for $K_{R,I}$ (Table 1), and with the order in which ions are toxic to roots (see below). This order reflects both screening effectiveness and the strength of binding to the CWs, which appears to be a function of ion charge, ion size, and some other factors. For the ions just listed, the following equation applies:

$$Order = Z(2.91 + 1.15/Ionic\ radius) \qquad (10)$$

where *Order* ranges from 15 for Al^{3+} to 1 for Na^+; $R^2 = 0.830$; and each coefficient is highly significant. The relationship expressed in Eq. (10) is reliable within a charge class when size differences are great (e.g. H^+ vs. the other monovalent ions, Al^{3+} vs. La^{3+}, or hexamethonium^{2+} vs. the other divalent ions), but not when size differences are small.

Shomer *et al.* (2003) used ψ_{CW} measurements to estimate parameter values for a Donnan-plus-binding model. This was done in just the same way as the determination of parameter values from ζ-potential measurements described above. Measured ψ_{CW} provided no evidence for binding to neutral sites, a finding consistent with the failure of H^+ to convert ψ_{CW} from negative to positive values even at pH < 3 (binding of H^+ to R^- can only neutralize ψ_{CW}; binding to P^0 is required for conversion). A value of $K_{R,H} = 9{,}450$ was computed for H^+ binding. Otherwise, ψ_{CW} provided no basis for the computation of binding constants. That is, the ion-induced reductions in ψ_{CW} negativity could be accounted for almost entirely on the basis of screening. Despite these unexpected results, which the authors discussed at length, the model-computed values for ψ_{CW} corresponded well with measured ψ_{CW} (Fig. 4(B) and Table 3) and reasonably well with computed ψ_{PM} ($R^2 = 0.771$ for values in Table 3). The greater depolarizing effectiveness of La^{3+} and H^+ at the PM reflects the greater binding strength of these ions to the PM.

4. APPLICATION OF ELECTROSTATIC THEORY TO PLANT–ION INTERACTIONS

4.1. Ion transport

Ion transport through the PM is influenced by both the transmembrane electrical potential difference (E_m) and the surface potentials at the inner and outer surfaces of the membranes. Fig. 1 illustrates these potentials. The importance of E_m for ion transport has been recognized for many decades, but consideration of a role for ψ_{PM} in higher-plant studies is more recent (Zhang *et al.*, 2001; Kinraide, 2001, 2003a). In general, the influx of cations into cells is inhibited by treatments that reduce the negativity of ψ_{PM}, but the influx of anions is enhanced. The latter accounts for R^2 being greater in Fig. 5(B) than in 5(A).

Table 3
Measured and computed electrical potentials in some solutions. Data are taken from Shomer *et al.* (2003)

$LaCl_3$ (mM)	pH	$CaCl_2$ (mM)	NaCl (mM)	$\psi_{CW,measured}$ (mV)	$\psi^{a}_{CW,computed}$ (mV)	ψ^{b}_{PM} (mV)
0.001	5.6	0	0.1	−78.9	−72.0	−44.6
0.01	5.6	0	0.1	−55.7	−54.2	−25.4
0.1	5.6	0	0.1	−32.5	−35.5	−6.0
1	5.6	0	0.1	−9.3	−16.9	12.7
0	5.00	0	0.1	−87.4	−94.7	−71.0
0	4.00	0	0.1	−54.0	−59.1	−16.5
0	3.00	0	0.1	−9.4	−13.1	34.6
0	5.6	0.01	0.1	−88.6	−81.2	−85.1
0	5.6	0.1	0.1	−60.1	−56.6	−62.8
0	5.6	1	0.1	−31.6	−29.5	−36.1
0	5.6	0	0.1	−98.6	−110.6	−101.3
0	5.6	0	1	−57.5	−70.2	−85.5
0	5.6	0	10	−16.4	−22.5	−59.0

[a] Parameter values for the Donnan-plus-binding model: $R_{Total} = 0.0211$ M and $K_{R,H} = 9450$. [b] ψ_{PM} was computed by a Gouy–Chapman–Stern model in which the negative-site binding constants were $K_{R,Na} = 1$, $K_{R,Ca} = 30$, $K_{R,La} = 2200$, and $K_{R,H} = 21{,}500\ M^{-1}$.

Changes in surface electrical potential may affect ion transport by altering two components of the chemical potential difference ($\Delta\mu$) of an ion across the PM. First, the surface activity of the transported ion will change because of electrostatic attraction or repulsion as indicated in Eq. (8) and Fig. 2. Second, the surface-to-surface transmembrane potential difference will change. This latter potential ($E_{m,Surf}$ in Fig. 1) is different from the bulk-phase-to-bulk-phase transmembrane potential difference (E_m) measured with intracellular microelectrodes. These changes in the components of the chemical potential may change the flux of an ion through the membrane even if the surface-to-surface $\Delta\mu$ (equal to the bulk-phase-to-bulk-phase $\Delta\mu$) remains constant.

The Goldman–Hodgkin–Katz flux equation has long been in use and incorporates ion activities and electrical potentials (Nobel, 1991). In its usual form it considers potentials and activities in the bulk phases of the extracellular medium

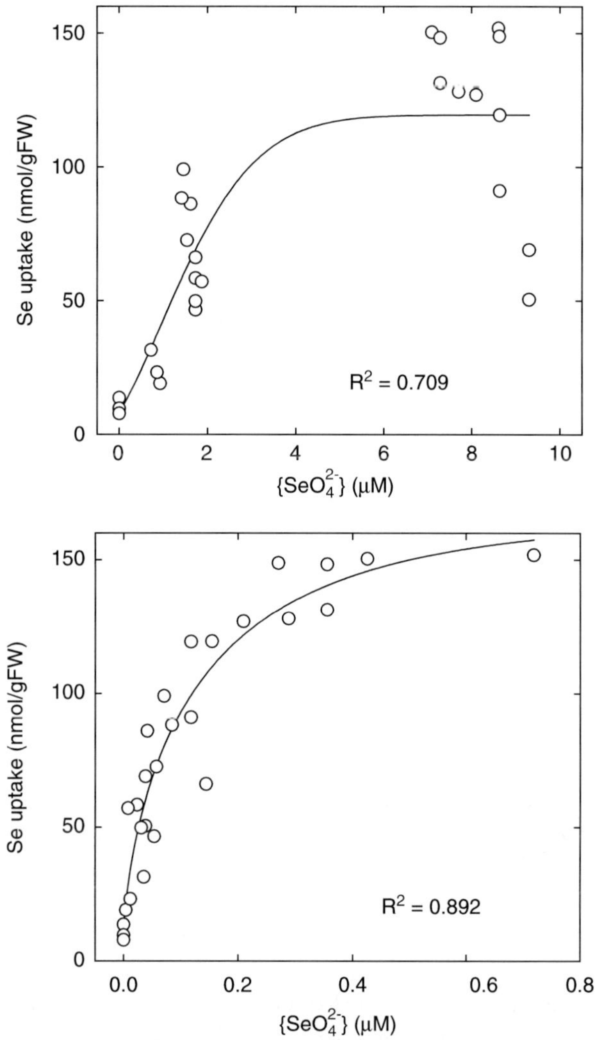

Fig. 5. Se uptake by roots in response to SeO_4^{2-} activities in the rooting medium (A) and at the PM surface (B). Atlas 66 wheat seedlings were cultured in media variously supplemented with $CaCl_2$ and $MgCl_2$, adjusted to several pH values. The figure is redrawn from Kinraide (2003a).

and the cell interior. It is written

$$J_i = J_i^{in} - J_i^{out} = P_i \upsilon_i(\{I^Z\}^{out} - \{I^Z\}^{in} \exp[\upsilon_i])/(\exp[\upsilon_i] - 1) \tag{11}$$

J_i, the net flux, is composed of the inward (J_i^{in}) and outward (J_i^{out}) fluxes of ion I^Z; $\upsilon_i = Z_i F E_m/(RT)$; $\{I^Z\}^{out}$ is the activity of ion I^Z in the tissue-bathing medium;

and $\{I^Z\}^{in}$ is the activity of ion I^Z in the cytoplasm. P_i is a permeability coefficient equal to $K_i u_i RT/\Delta x$ (a partition coefficient, an ionic mobility factor, RT, and the membrane thickness, respectively [Nobel, 1991]). J_i may be expressed in mol cm^{-2} s^{-1}, P_i in cm s^{-1}, activities in mol cm^{-3}, and υ_i without dimensions.

Eq. (11) does not take into account surface-potential effects, but it may be modified to incorporate surface potentials computed by a Gouy–Chapman–Stern model and surface ion activities computed by the Nernst equation. Thus $\{I^Z\}^{out}$ and $\{I^Z\}^{in}$ become $\{I^Z\}_{PM}{}^{out}$ and $\{I^Z\}_{PM}{}^{in}$, the outer and inner surface activities, and υ_i becomes $Z_i FE_{m,Surf}/(RT)$. The modified equation predicts successfully many transport phenomena not predicted by the standard Goldman–Hodgkin–Katz equation. Thus electrostatic effects at PM surfaces influence rate, saturation, *cis*- and *trans*inhibition, rectification, voltage gating, shifts in voltage optima, and other phenomena (Kinraide, 2001) that also respond to channel blockade, binding-site saturation, etc.

4.2. Ion toxicities

Many mineral cations and anions, including essential nutrient elements, may be phytotoxic at excessive concentrations. As in the case of membrane transport, toxicities are profoundly influenced by ψ_{PM}. The toxicity of cations is reduced by treatments that reduce the negativity of ψ_{PM} (Fig. 6), and the toxicity of anions is enhanced (Fig. 7). Both Figs. 6 and 7 show that root elongation is more directly influenced by surface activities than by activities in the medium.

The relative toxicities of different cations correlate with the strength with which the ions are attracted to and bind to the PM or CW. Compare the order of effectiveness with which ions reduce the negativity of ψ_{CW} reported by Shomer *et al.* (2003; see above) with the order with which the ions inhibit wheat root elongation (i.e. $AlO_4Al_{12}(OH)_{24}(H_2O)_{12}^{7+} > Sc^{3+}$ (may include some highly charged polynuclear species) $> Cu^{2+} >$ spermine$^{4+} > Al^{3+} > La^{3+} > H^+ > AlF^{2+} > Cs^+ > Rb^+ >$ tris(ethylenediamine)cobalt$^{3+} > Zn^{2+} > AlF_2^+ > Li^+ > Mg^{2+} > Ca^{2+} > K^+ > Na^+$ (citations in Shomer *et al.* [2003]). Here too, an influence of charge and size is seen (Eq. (10)).

Assessment of toxicities in terms of cell-surface activities holds many advantages over assessment of toxicities in terms of activities in the rooting medium. Fig. 8 illustrates the toxicities of five ions to wheat root elongation based upon toxicant activities at the PM surface. The curves are based on the equation

$$RRL = 100/\exp[(a_1\{\text{Toxicant}\}_0)^{b1}] \qquad (12)$$

According to the "strength" parameter, a_1, the intrinsic toxicity of Cu^{2+} is 2.4 times greater than the toxicity of Al^{3+} (here and elsewhere the term intrinsic means related to the PM surface). Relative intrinsic toxicity is less sensitive to changes in the composition of the rooting medium than is relative extrinsic toxicity. In contrast, comparisons on the basis of toxicant activities in external media

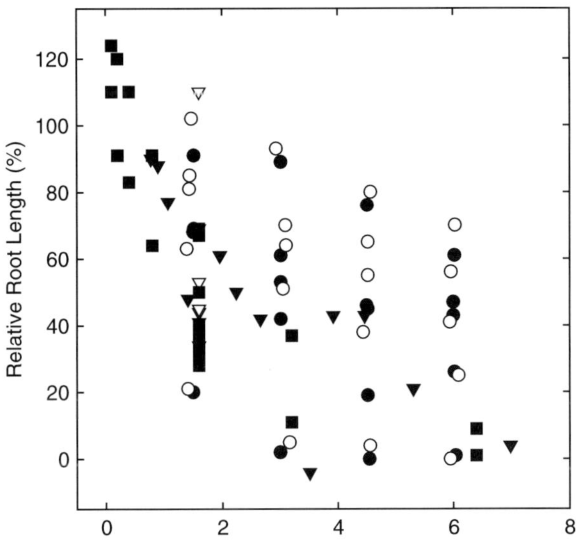

(A)

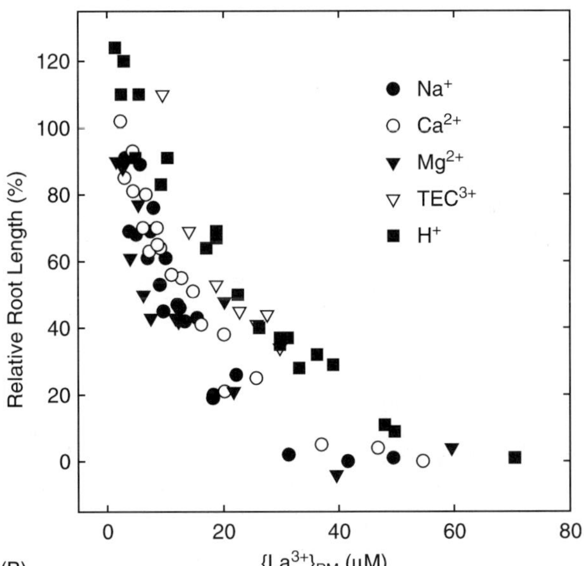

(B)

Fig. 6. Root elongation as a function of La^{3+} activity in the rooting medium (A) and at the PM surface (B). Seedlings of wheat, cv. Tyler, were cultured in media variously supplemented with chloride salts of the noted cations adjusted to several pH values. Relative root length = 100 (root elongation in the presence of toxicant)/(root elongation in toxicant-free medium). The figure is redrawn from Kinraide *et al.* (1992).

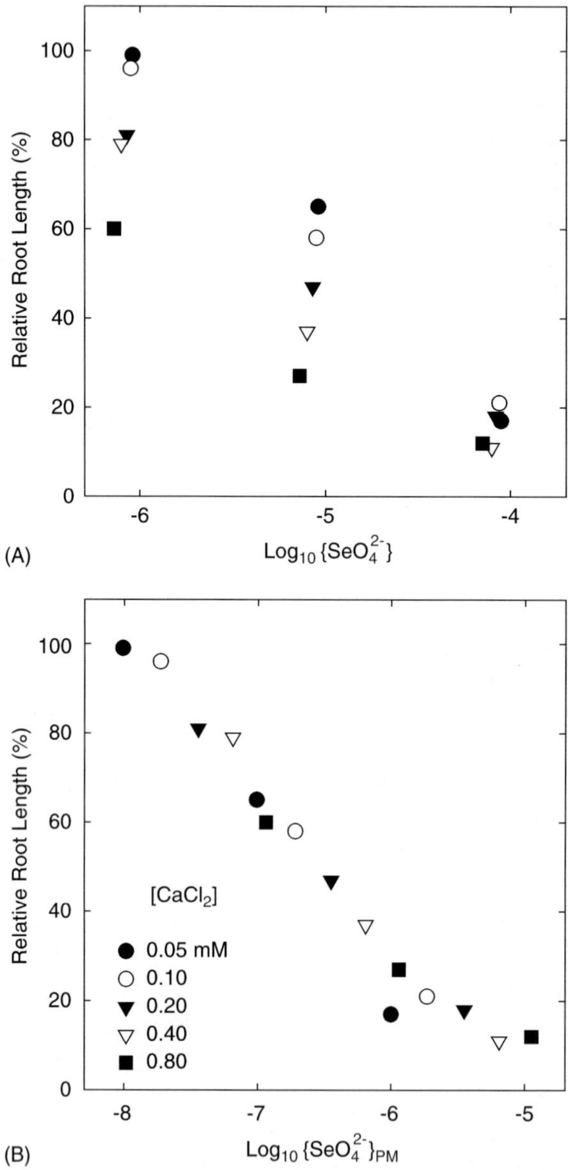

Fig. 7. Root elongation as a function of SeO_4^{2-} activity in the rooting medium (A) and at the PM surface (B). Atlas 66 wheat seedlings were cultured in media variously supplemented with $CaCl_2$. The figure is redrawn from Kinraide (1994).

yield different results in different media. In a medium composed of 1 mM Ca^{2+} at pH 4.3, Cu^{2+} would appear to be 5.6 times as toxic as Al^{3+}; but in a medium composed of 6 mM Ca^{2+} at pH 4.3, Cu^{2+} would appear to be 42.7 times as toxic as Al^{3+}. In the next section we shall see that the assessment of toxicities in terms

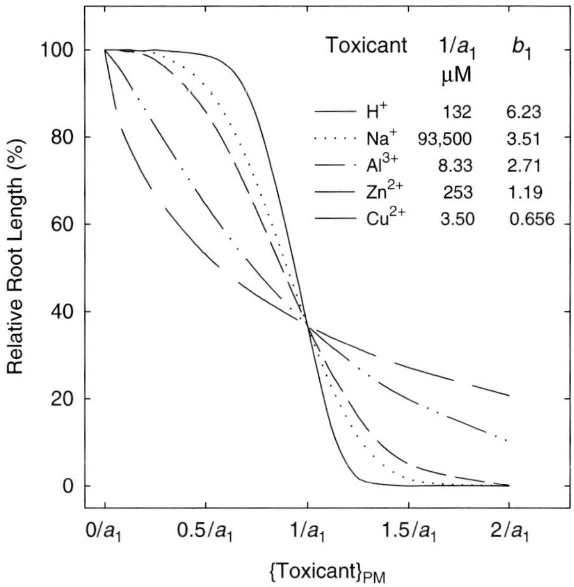

Fig. 8. Root elongation in wheat in response to toxic ions. The response curves conform to the equation $RRL = 100/\exp[(a_1\{\text{Toxicant}\}_{PM})^{b1}]$, Eq. (12), where a_1 is a strength coefficient and b_1 is a shape coefficient. The figure is redrawn from Kinraide *et al.* (2004).

of cell-surface activities is an aid in the resolution of separate and interacting toxicities caused by multiple toxicants.

As we have seen, greater charge and smaller size of cations generally increase their binding strength and intoxicating effectiveness. In fact, every polyvalent cation (charge $\geq +3$) is rhizotoxic, as far as we know. These facts have led to the hypothesis that depolarization itself is intoxicating. This tempting and potentially unifying hypothesis was considered by Ahn *et al.* (2001) who recorded Al binding, reductions in PM surface negativity, and reduced H^+-ATPase activity in the Al-sensitive region of squash roots. The hypothesis remains unresolved, but the cation alleviation of cation toxicity (see next section) militates against the hypothesis – if depolarization is the cause of intoxication, then further depolarization is unlikely to be a remedy.

4.3. Ion alleviation of ion toxicities

Ca^{2+} in the rooting medium is essential for root elongation, even in the absence of added toxicants. In the presence of rhizotoxic ions, supplementation of the medium with greater amounts of Ca^{2+} alleviates growth inhibition. Other ions may be ameliorative as well. For example, Al^{3+} toxicity may be alleviated by these ions in the following order: tris(ethylenediamine)cobalt^{3+} > Ca^{2+} = Mg^{2+} > Na^+ = K^+. Ca^{2+} and Mg^{2+} are sometimes equally ameliorative (as in the alleviation of Al^{3+} toxicity; but see Silva *et al.*, 2001); Ca^{2+} is sometimes more effective than Mg^{2+} (as in the alleviation of Na^+ and H^+ toxicities; Kinraide *et al.*, 2004); and Mg^{2+} is sometimes more effective than Ca^{2+} (as in the case of Zn^{2+} toxicity) (Pedler *et al.*, 2004). Thus, specific ion effects, i.e. effects other than general electrostatic effects, are sometimes important. Occasionally two ions may be simultaneously toxic and ameliorative. Al^{3+} sometimes stimulates root elongation in low-pH media, and H^+ sometimes stimulates root elongation in Al^{3+}-containing media. Consideration of ion effects in terms of PM surface activities of the toxicants has enabled the interpretation of most of these effects.

Consideration of electrostatic effects has enabled the identification of these three mechanisms for the Ca^{2+} alleviation of cationic toxicities (Kinraide, 1998). *Mechanism I* is the displacement of cell-surface toxicant by the Ca^{2+}-induced reduction of cell-surface negativity. *Mechanism II* is the restoration of Ca^{2+} at the cell surface if the surface Ca^{2+} has been reduced by the toxicant to growth-limiting levels. *Mechanism III* is the collective ameliorative effect of Ca^{2+} beyond Mechanisms I and II, such as the blockade of Na^+-permeable channels in the PM (Tyerman *et al.*, 1997). Other ions, such as Mg^{2+}, may have Mechanism I and III effects, but not Mechanism II effects. A corollary of Mechanism I is that Ca^{2+}, or other cations, may *enhance* the toxicity of anionic toxicants, an expectation that has been confirmed in the case of selenate (Fig. 7). Although Mechanism II rarely appears to be the principal mechanism of toxicity, it may sometimes be an important factor in salinity toxicity. Because Na^+ has low intrinsic toxicity, large concentrations are required for intoxication, and significant displacement of Ca^{2+} from the cell surface may occur. Table 2 illustrates this displacement of Ca^{2+} by Na^+.

Fig. 9 presents the results of growth experiments intended to assess the interactions among Al^{3+}, H^+, and Ca^{2+}. These ions as well as Mg^{2+} are critical determinants of soil acidity. In order to dissect the interactions among ions, we first compute surface activities, then write nonlinear equations describing possible interactions. One such model of interactions among ions assumes multiplicative effects. For Fig. 9 we assume that

$$RRL = 100RRL_{\mathrm{Al}} \cdot RRL_{\mathrm{H}} \cdot RRL_{\mathrm{Ca}} \quad (13)$$

(To interpret this relationship consider a case where Al^{3+} inhibits growth 30%, H^+ inhibits growth 20%, and Ca^{2+} insufficiency inhibits growth 40%. Then $RRL = 100 \times 0.7 \times 0.8 \times 0.6 = 33.6\%$.) Actually, Al^{3+} and H^+ are mainly, but not entirely, inhibitory, and Ca^{2+} is entirely ameliorative, but by more than one mechanism. The following equation, based upon Eq. (13), has successfully simulated some of the interactions:

$$RRL = 100/\exp[(a_1\{\mathrm{Al}^{3+}\}_{\mathrm{PM}})^{b1} + (a_2\{\mathrm{H}^+\}_{\mathrm{PM}})^{b2}](1 - 1/\exp[a_3\{\mathrm{Ca}^{2+}\}_{\mathrm{PM}})^{b3}]) \quad (14)$$

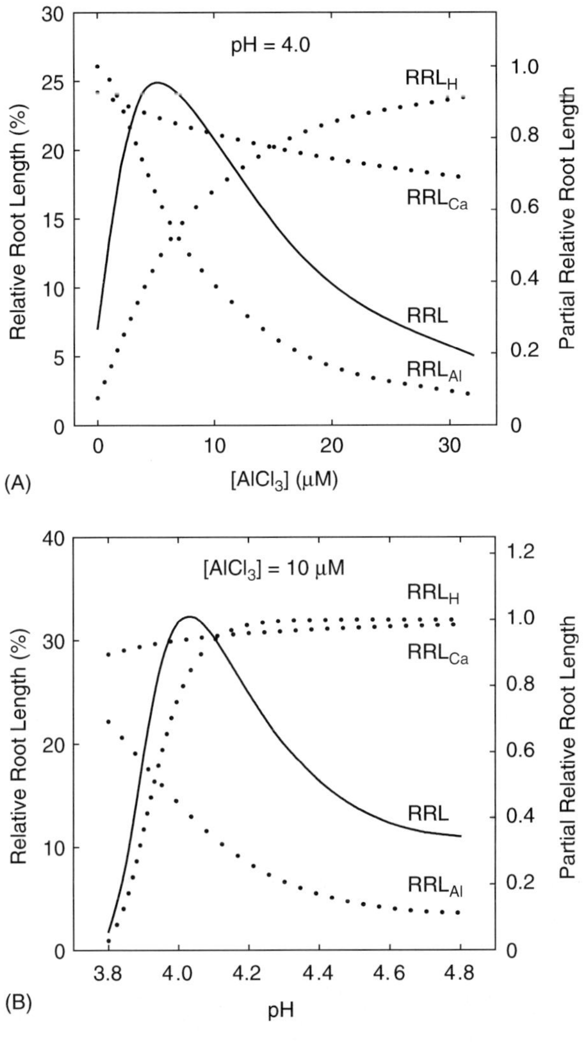

Fig. 9. Separating the components of intoxication and alleviation for root elongation of Scout wheat in solution culture. The solid lines refer to the measured relative root length (*RRL*), which is a product of the partial root responses ($RRL = 100RRL_{Al} \cdot RRL_H \cdot RRL_{Ca}$). The drawn curves conform to an expansion of Eq. (14) in which $a_1 = 245$, $b_1 = 2.13$, $a_{21} = 8.97$, $a_{22} = 0.170$, $b_2 = 7.23$, $a_3 = 5.52$, and $b_3 = 0.780$ (activities in mM). For (A), $CaCl_2$ = 0.5 mM and NaCl = 10 mM; for (B), $CaCl_2$ = 1 mM and NaCl = 10 mM. The figure is redrawn from Kinraide (2003b).

This equation is incomplete, however, because H^+ toxicity, but not Al^{3+} toxicity, is specifically alleviated by Ca^{2+}. That may be expressed by expanding the toxicant strength coefficient a_2 and writing it $a_{21}/(a_{22}\{Ca^{2+}\}_{PM} + 1)$ so that a_2 is reduced by Ca^{2+}. a_{22} is a strength coefficient for Mechanism III alleviation. Other studies have shown that a_{22} for Na^+ toxicity is greater than a_{22} for H^+ toxicity, and that $a_{22} = 0$ in the case of Al^{3+} toxicity (Kinraide 1998, 1999).

Listed here are the conclusions from an analysis of toxicity factors characteristic of acidic soils (Kinraide, 2003b; see Fig. 9) that employ Eq. (14).

1. Al^{3+} and H^+ are intrinsic toxicants.
2. Al^{3+} and H^+ are also extrinsic toxicants because of the electrostatic displacement of Ca^{2+} from the surface of the PM.
3. Al^{3+} and H^+ are also extrinsic ameliorants because each displaces the other from the surface of the PM. This accounts for the occasionally seen stimulation of growth by Al^{3+} (Fig. 9(A)) and low pH (Fig. 9(B)).
4. Ca^{2+} meets an intrinsic requirement that cannot be met by any other ion.
5. Ca^{2+} is also an extrinsic ameliorant because of the electrostatic displacement of Al^{3+} and H^+ from the surface of the PM.
6. Ca^{2+} is also an intrinsic ameliorant of intrinsic H^+ toxicity, but not intrinsic Al^{3+} toxicity.
7. Mg^{2+} resembles Ca^{2+} in item 5 but not items 4 and 6 in short-term cultures.

Similar analyses quantify the interactions among Ca^{2+}, Na^+, and K^+ in salinity toxicity (Yermiyahu *et al.*, 1997a; Kinraide, 1999).

5. CONCLUSIONS AND FUTURE PROSPECTS

We have attempted to introduce a quantitative approach for describing and interpreting the interactions presented as a diagram at the beginning of this chapter. Our main approach has been to compute ion activities at the cell surface (in the CW or at the PM surface) and then to relate these activities to physiological effects. Cell-surface ion activities inserted into nonlinear equations have successfully resolved multiple toxic and ameliorative effects observed in multi-ion systems, overcoming problems of often hidden intercorrelation among variables that have so long plagued studies of plant–ion interaction.

We are confident that a Gouy–Chapman–Stern model and a Donnan-plus-binding model may be used to compute values for ψ_{PM} and ψ_{CW} that are at least proportional to the actual values. With these electrical potentials, corresponding values for ion activities may be computed that are at least proportional to actual values also. Although the electrostatic theory is quite old, values for model parameters for plant cell surfaces have become available only recently. Computer programs for the electrostatic models may be requested from us.

Finally, many interesting practical problems remain to be studied for which an electrostatic approach may be helpful. Some theoretical problems remain unresolved, as mentioned in Barber (1980) and Kinraide (1994). One problem is the relationship of the CW to the PM. Is ψ_{PM} properly defined as the potential difference between the PM surface and the external rooting medium as though the PM were bathed directly in the medium with no influence from the CW? ψ_{PM} estimated on the basis of measurements from PM vesicles or protoplasts embody that assumption. In contrast, should the ψ_{PM} and $\{I^Z\}_{PM}$ be estimated as though the PM were bathed in the Donnan-phase solution of the CW? A recent analysis indicates that the CW has only small effects upon PM-to-medium surface-potential differences and PM-surface ion concentrations (Kinraide, 2004), but experimental verification of this conclusion is scant.

ACKNOWLEDGMENTS

This research was supported by the United States–Israel Binational Agricultural Research and Development Fund (BARD Grant No. IS-3120-99R).

REFERENCES

Abe, S., Takeda, J., 1988. Effects of La^{3+} on surface charges, dielectrophoresis, and electrofusion of barley protoplasts. Plant Physiol. 87, 389–394.

Ahn, S.J., Sivaguru, M., Osawa, H., Chung, G.C., Matsumoto, H., 2001. Aluminum inhibits the H^+-ATPase activity by permanently altering the plasma membrane surface potentials in squash roots. Plant Physiol.126, 1381–1390.

Akeson, M.A., Munns, D.N., Burau, R.G., 1989. Adsorption of Al^{3+} to phosphatidylcholine vesicles. Biochim. Biophys. Acta 986, 33–40.

Allan, D.L., Jarrell, W.M., 1989. Proton and copper adsorption to maize and soybean root cell walls. Plant Physiol. 89, 823–832.

Barber, J., 1980. Membrane surface charges and potentials in relation to photosynthesis. Biochim. Biophys. Acta 594, 253–308.

Brauer, D.K., Yermiyahu, U., Rytwo, G., Kinraide, T.B., 2000. Characteristics of the quenching of 9-aminoacridine fluorescence by liposomes made from plant lipids. J. Membrane Biol. 178, 43–48.

Bush, D.S., McColl, J.G., 1987. Mass-action expressions of ion exchange applied to Ca^{2+}, H^+, K^+, and Mg^{2+} sorption on isolated cell walls of leaves from *Brassica oleracea*. Plant Physiol. 85, 247–260.

Chang, T.Y., Senn, A., Pilet, P.E., 1983. Effect of abscisic acid on maize (*Zea maize* vs. LG-11) root protoplasts. Zeitschrift für Pflanzenphysiologie 110, 127–134.

Franco, C.R., Chagas, A.P., Jorge, R.A., 2002. Ion-exchange equilibria with aluminum pectinates. Colloids Surface A 204, 183–192.

Gibrat, R., Grouzis, J.-P., Rigaud, J., Grignon, C., 1985. Electrostatic characteristics of corn root plasmalemma: effect on the Mg^{2+}-ATPase activity. Biochim. Biophys. Acta 816, 349–357.

Grignon, C., Sentenac, H., 1991. pH and ionic conditions in the apoplast. Ann. Rev. Plant Physiol. Molecular Biol. 42, 103–128.

Hille, B., 1992. Ionic Channels of Excitable Membranes, second ed., Sinauer Associates, Sunderland, MA.

Horst, W.J., 1995. The role of the apoplast in aluminium toxicity and resistance of higher plants: a review. Zeitschrift für Pflanzenernährung und Bodenkunde 158, 419–428.

Jones, D.L., Kochian, L.V., 1997. Aluminum interaction with plasma membrane lipids and enzyme metal binding sites and its potential role in Al cytotoxicity. FEBS Lett. 400, 51–57.

Kinraide, T.B., 1994. Use of a Gouy-Chapman-Stern model for membrane-surface electrical potential to interpret some features of mineral rhizotoxicity. Plant Physiol. 106, 1583–1592.

Kinraide, T.B., 1998. Three mechanisms for the calcium alleviation of mineral toxicities. Plant Physiol. 118, 513–520.

Kinraide, T.B., 1999. Interactions among Ca^{2+}, Na^{+} and K^{+} in salinity toxicity: quantitative resolution of multiple toxic and ameliorative effects. J. Exp. Bot. 50, 1495–1505.

Kinraide, T.B., 2001. Ion fluxes considered in terms of membrane-surface electrical potentials. Aust. J. Plant Physiol. 28, 605–616.

Kinraide, T.B., 2003a. The controlling influence of cell-surface electrical potential on the uptake and toxicity of selenate (SeO_4^{2-}). Physiologia Plantarum 117, 64–71.

Kinraide, T.B., 2003b. Toxicity factors in acidic forest soils: attempts to evaluate separately the toxic effects of excessive Al^{3+} and H^{+} and insufficient Ca^{2+} and Mg^{2+} upon root elongation. Eur. J. Soil Sci. 54, 323–333.

Kinraide, T.B., 2004. Possible influence of cell walls upon ion concentrations at plasma membrane surfaces. Toward a comprehensive view of cell-surface electrical effects upon ion uptake, intoxication, and amelioration. Plant Physiol. 136, 3804–3813.

Kinraide, T.B., Pedler, J.F., Parker, D.K., 2004. Relative effectiveness of calcium and magnesium in the alleviation of rhizotoxicity in wheat induced by copper, zinc, aluminum, sodium, and low pH. Plant Soil 259, 201–208.

Kinraide, T.B., Ryan, P.R., Kochian, L.V., 1992. Interactive effects of Al^{3+}, H^{+}, and other cations on root elongation considered in terms of cell-surface electrical potential. Plant Physiol. 99, 1461–1468.

Kinraide, T.B., Yermiyahu, U., Rytwo, G., 1998. Computation of surface electrical potentials of plant cell membranes. Correspondence to published zeta potentials from diverse plant sources. Plant Physiol. 118, 505–512.

Marschner, H., 1995. Mineral Nutrition of Higher Plants, second ed., Academic Press, London.

McLaughlin, S., 1977. Electrostatic potentials at membrane – solution interfaces. In: Bronner, F., Kleinzeller, A., (Eds.), Current Topics in Membranes and Transport, Vol. 9. Academic Press, New York, pp. 71–144.

Nagai, R., Kishimoto, U., 1964. Cell wall potential in *Nitella*. Plant Cell Physiol. 5, 21–31.

Nagata, T., Melchers, G., 1978. Surface charge of protoplasts and their significance in cell-cell interaction. Planta 142, 235–238.

Nir, S., Rytwo, G., Yermiyahu, U., Margulies, L., 1994. A model for cation adsorption to clays and membranes. Colloid Poly. Sci. 272, 619–632.

Nobel, P., 1991. Physicochemical and Environmental Plant Physiology, Academic Press, San Diego.

Obi, I., Ichikawa, Y., Kakutani, T., Senda, M., 1989a. Electrophoresis, zeta potential and surface charges of barley mesophyll protoplasts. Plant Cell Physiol. 30, 129–135.

Obi, I., Ichikawa, Y., Kakutani, T., Senda, M., 1989b. Electrophoretic studies on plant protoplasts. I. pH dependence of zeta potentials of protoplasts from various sources. Plant Cell Physiol. 30, 439–444.

Obi, I., Kakutani, T., Imaizumi, N., Ichikawa, Y., Senda, M. 1990. Surface charge density of hetero-fused plant protoplasts: an electrophoretic study. Plant Cell Physiol. 31, 1031–1037.

O'Shea, P., Walters, J., Ridge, I., Wainright, M., Trinci, A.P.J., 1990. Zeta potential measurements of cell wall preparations from *Regnellidium diphyllum* and *Nymphoides peltata*. Plant Cell Environ. 13, 447–454.

Pedler, J.F., Kinraide, T.B., Parker, D.K., 2004. Zinc rhizotoxicity in wheat and radish is alleviated by micromolar levels of magnesium and potassium in solution culture. Plant and Soil 259, 191–199.

Richter, C., Dainty, J., 1990. Ion behavior in plant cell walls. IV. Selective cation binding by *Sphagnum russowii* cell walls. Can. J. Botany 68, 773–781.

Rufyikiri, G., Genon, J.G., Dufey, J.E., Delvaux, B., 2003. Competitive adsorption of hydrogen, calcium, potassium, magnesium, and aluminum on banana roots: experimental data and modeling. J. Plant Nutr. 26, 351–368.

Saftner, R.A., Raschke, K., 1981. Electrical potentials in stomatal complexes. Plant Physiol. 67, 1124–1132.

Sattelmacher, B., 2001. The apoplast and its significance for plant mineral nutrition. New Phytol. 149, 167–192.

Schmohl, N., Horst, W.J., 2002. Effect of aluminium on the activity of apoplastic acid phosphatase and the exudation of macromolecules by roots and suspension-culture cells of *Zea mays* L. J. Plant Physiol. 159, 1213–1218.

Sentenac, H., Grignon., C., 1981. A model for predicting ionic equilibrium concentrations in cell walls. Plant Physio. 68, 415–419.

Shomer, I., Novacky, A.J., Pike, S.M., Yermiyahu, U., Kinraide, T.B., 2003. Electrical potentials of plant cell walls in response to the ionic environment. Plant Physiol. 133, 411–422.

Silva, I.R., Smyth, T.J., Israel, D.W., Raper, C.D., Rufty, T.W., 2001. Magnesium is more efficient than calcium in alleviating aluminum rhizotoxicity in soybean and its ameliorative effect is not explained by the Gouy-Chapman-Stern model. Plant Cell Physiol. 42, 538–545.

Spanswick, R.M., Stolarek, J., Williams, E.J., 1967. The membrane potential of *Nitella translucens*. J. Exp. Bot. 18, 1–16.

Tatulian, S.A., 1999. Surface electrostatics of biological membranes and ion binding. In: Sørensen T.S., (Ed.), Surface Chemistry and Electrochemistry of Membranes, Marcel Dekker, New York, pp.871–922.

Tyerman, S.D., Skerrett, M., Garrill, A., Findlay, G.P., Leigh, R.A., 1997. Pathways for the permeation of Na^+ and Cl^- into protoplasts derived from the cortex of wheat roots. J. Exp. Bot. 48, 459–480.

Vulkan, R., Yermiyahu, U., Kinraide, T.B., Mingelgrin, U., Rytwo, G., 2004. Sorption of copper and zinc to plasma membrane of wheat root. J. Membrane Biol. 202, 97–104.

Wagatsuma, T., Akiba, R., 1989. Low surface negativity of root protoplasts from aluminum-tolerant plant species. Soil Sci. Plant Nutr. 35, 443–452.

Wilkinson, K.J., Bertsch, P. M., Jagoe, C. H., Campbell, G. C., 1993. Surface complexation of aluminum on isolated fish fill cells. Environ. Sci. Technol. 27, 1132–1138.

Yermiyahu, U., Nir, S., Ben-Hayyim, G., Kafkafi, U., 1994. Quantitative competition of calcium with sodium or magnesium for sorption sites on plasma membrane vesicles of melon (*Cucumis melo* L.) root cells. J. Membrane Biol. 138, 55–63.

Yermiyahu, U., Nir, S., Ben-Hayyim, G., Kafkafi, U., Kinraide, T.B., 1997a. Root elongation in saline solution related to calcium binding to root cell plasma membranes. Plant Soil 191, 67–76.

Yermiyahu, U., Nir, S., Ben-Hayyim, G., Kafkafi, U., Scherer, G.F.E., Kinraide, T.B., 1999. Surface properties of plasma membrane vesicles isolated from melon (*Cucumus melo* L.) root cells differing in salinity tolerance. Colloids Surfaces B 14, 237–249.

Yermiyahu, U., Rytwo, G., Brauer, D.K., Kinraide, T.B., 1997b. Binding and electrostatic attraction of lanthanum (La^{3+}) and aluminum (Al^{3+}) to wheat root plasma membranes. J. Membrane Biol. 159, 239–252.

Zhang, Q., Smith, F.A., Sekimoto, H., Reid, R.J., 2001. Effect of membrane surface charge on nickel uptake by purified mung bean root protoplasts. Planta 213, 788–793.

Biogeochemistry of Trace Elements in the Rhizosphere
P.M. Huang and G.R. Gobran (Editors)

Chapter 13

Model development for simulating the bioavailability of Ni to the hyperaccumulator *Thlaspi goesingense*

A. Schnepf[a], M.L. Himmelbauer[a], M. Puschenreiter[b], T. Schrefl[c], E. Lombi[d], W.J. Fitz[b], W. Loiskandl[a], and W.W. Wenzel[b]

[a]Department of Water, Atmosphere and Environment, Institute of Hydraulics and Rural Water Management, University of Natural Resources and Applied Life Sciences, Muthgasse 18, A-1190 Vienna, Austria
E-mail: andrea.schnepf@boku.ac.at

[b]Department of Forest and Soil Sciences, Institute of Soil Science, University of Natural Resources and Applied Life Sciences, Peter Jordan Strasse 82, A-1190 Vienna, Austria

[c]Institute of Solid State Physics, University of Technology, Wiedner Hauptstrasse 8-10, A-1040 Vienna, Austria

[d]CSIRO Land and Water, PMB2 Glen Osmond, SA 5064, Australia

ABSTRACT

Mathematical models on the single root scale were applied to simulate nickel (Ni) concentration gradients in the rhizosphere of *Thaspi goesingense*, which were previously measured in a rhizobox experiment. The model approaches included adaptations to the rhizobox conditions in terms of geometry and coverage of the membrane. Further model approaches included Ni uptake by root hairs, root exudation of chelating agents and chemical non-equilibrium. Only free Ni ions were considered to be taken up by *T. goesingense*. The required input parameters were partly measured for the experimental conditions of the rhizobox experiment and partly obtained from literature data. The curves that matched best measured labile Ni were those obtained by the most simple initial model, thc model including release from a fixed Ni phase and the two-stage sorption model. The simulation results for each of the three models were quite similar, indicating that the influence of the additionally included processes was relatively small under the assumptions of equilibrium between fixed phase and solution in the bulk soil. The models including both root hairs and exudation overestimated depletion of labile Ni close to the root. Future simulation work will benefit by including more detailed soil chemistry input parameters. Such approaches would involve multi-ion and multi-species models.

1. INTRODUCTION

Explicative mathematical models are an instrument for the purpose of identifying and understanding fundamental mechanisms underlying the behaviors of a system (Saltelli *et al.*, 2000). Mechanistic rhizosphere models have been applied to simulate the uptake of nutrients by plant roots (Tinker and Nye, 2000). In principle, they can also be used to simulate rhizosphere processes of hyperaccumulators, plant species that are capable of phytoextraction. It is known that the rhizosphere has a great influence on the bioavailability of trace elements, as roots affect the soil chemically, physically and microbiologically (Marschner, 1995; Wenzel *et al.*, 1999; Uren, 2001). Metal hyperaccumulators contain extremely high metal concentrations in their plant tissues (Brooks, 1998). Several *Thlaspi* species are known to be hyperaccumulators, e.g. *T. goesingense* has concentrations of > 1000 mg kg^{-1} Ni in plant dry mass (Reeves and Baker, 1984). The role of certain rhizosphere processes in hyperaccumulation is controversially discussed (McGrath *et al.*, 2001). Root exudates and pH effects have been considered as un important to metal mobilization in hyperaccumulator plants (Salt *et al.*, 2000; Zhao *et al.*, 2001). However, experiments on the role of root exudates released by hyperaccumulators have only been conducted in hydroponic experiments, where plant roots may show effects different from what they would show in soil. pH has been investigated only on the scale of the whole root system. However, it is well known that root tips behave differently from the basal roots. Whether root tips of hyperaccumulators show different activities in terms of acidification or the release of root exudates has never been investigated. It has been confirmed that *T. caerulescens* develops root hairs abundantly (Whiting *et al.*, 2003), which can be assumed to contribute significantly to metal uptake. A recent study has shown enhanced metal accumulation in *T. caerulescens* grown on soil inoculated with rhizosphere bacteria (Whiting *et al.*, 2001). Microbial activity may lead to increased availability of metals in plants. The effect of microbial activity on heavy metal bioavailability for *T. goesingense* is currently being studied. *T. goesingense* is usually known to be non-mycorrhizal (Regvar *et al.*, 2003).

Mathematical modeling of these processes can be used to test whether they are accurately understood by model validation, or to test different hypotheses at low costs and help design new experiments. This is a new and innovative approach to hyperaccumulation. Recently, modeling approaches on the whole plant and at field level for the estimation of the phytoextraction potential of hyperaccumulators have been presented (Gonnelli *et al.*, 2000; Robinson *et al.*, 2003). On the single root level, Whiting *et al.* (2003) used an analytical solution of an approximate steady-state model to simulate Zn uptake by *T. caerulescens*.

1.1. Objectives

Single root uptake models are generally based on solute transport theory, and calculate solute transport toward the root surface and uptake of solutes by the root. However, models on the single root scale are available that deal with root exudation and chemical non-equilibrium. Puschenreiter *et al.* (2005) presented a rhizobox experiment and model simulation on Ni uptake by *T. goesingense*. They tested whether a "classical" single root uptake model with or without root hairs (Barber, 1995; Tinker and Nye, 2000) could explain the concentration gradients of labile Ni measured with respect to distance from the root layer of the rhizobox. Their results suggest that root activities, such as the exudation of organic acids, are key processes to be addressed in future modeling and in the related experiments. In particular, it might induce replenishment of soluble Ni from sources other than the adsorbed fraction through interactions with the soil solid phase.

The specific objectives of this chapter are:

(1) To give an overview of mathematical models used to describe rhizosphere processes, including transport of ions in soil, uptake by plant root, root exudation and chemical equilibrium and non-equilibrium sorption.
(2) To present a modeling case study of the simulation of Ni in the rhizosphere of the hyperaccumulator *T. goesingense* based on data of Puschenreiter *et al.* (2005) using:
    (a) Models that assume only free Ni as phytoavailable with or without uptake by root hairs.
    (b) Models that consider root exudation and chemical non-equilibrium using input parameters from the literature, with a discussion of their effect on model output.
(3) To perform a sensitivity analysis for testing the influence of different model parameters on simulated concentration gradients.

A corroborated explicative mathematical model could be used as a future research tool, for example for the design of phytoremediation technologies.

2. MODEL DEVELOPMENT

Rhizosphere modeling remains difficult and complex, as it combines technical know-how from several fields such as plant physiology, soil physics, soil chemistry and mathematics. Mechanistic rhizosphere models do not always operate with adequate precision (Rengel, 1993; Darrah and Roose, 2001). Two main fields of application of mechanistic rhizosphere models are carbon flow in the rhizosphere and nutrient uptake by plants. While carbon flow models study the exudation of carbon compounds into the soil and its consequences on the microbial population, uptake models focus on the transport and uptake of ions by roots. In the following sections, we will concentrate on uptake models on the single root scale.

Generally, single root models rely on solute transport theory to determine the supply of nutrients to the root. Solutes in soil are assumed to be moved by the additive and simultaneous processes of diffusion and convection. The governing equation in radial coordinates is (Nye and Marriott 1969; Barber 1995)

$$\frac{\partial C_1}{\partial t} = \frac{1}{r}\frac{\partial}{\partial r}\left(r\mathrm{D}_\mathrm{e}\frac{\partial C_1}{\partial r} + \frac{r_0 q_0 C_1}{b}\right) \tag{1}$$

where C_1 is the solution concentration, D_e the effective diffusion coefficient, r_0 the root radius, q_0 the water flux at the root surface, b the buffer power, t the time and r the distance to the root surface.

The effective diffusion coefficient accounts for the retardation of ion movement in soil as compared with that in water and can be calculated as follows:

$$\mathrm{D}_\mathrm{e} = \frac{D_1 \theta f}{b} \tag{2}$$

where D_1 is the diffusion coefficient in water, θ the volumetric water content and f the impedance factor. The buffer power is a proportionality constant that describes the reaction between soil and soil solution. Several ways of defining the buffer power b are found in the literature (Barber, 1995; Syring and Claassen, 1996; Kirk, 1999; Darrah and Staunton, 2000):

$$\text{(i) } b = \frac{\mathrm{d}C_s}{\mathrm{d}C_1} \tag{3}$$

where b is the buffer power, C_s the concentration of ions adsorbed per unit volume of soil exchange complex and C_1 the concentration per unit volume of soil solution.

$$\text{(ii) } b = \frac{\mathrm{d}C_t}{\mathrm{d}C_1} \tag{4}$$

where C_t is the total concentration of diffusible ions (dissolved plus adsorbed) per unit volume of soil.

Cushman (1984) also presented the following definition:

$$\text{(iii) } b = \frac{\rho S}{C_1} + \theta \tag{5}$$

where ρ is the soil bulk density and S is the concentration of ions adsorbed per unit mass of dry soil. It is derived from the equation. $C_t = (\rho S/C_1 + \theta)C_1 = bC_1$

If sorption is nonlinear, b will not be constant but a function of C_1. One example is the nonlinear Freundlich isotherm

$$S = K_f C_1^m + d \tag{6}$$

where K_f and m are the regular Freundlich parameters. An additional parameter d can be introduced (Syring and Claassen, 1996). If $d \neq 0$, the isotherm does not pass through the origin. Note that this constant does not influence the model results for the solution concentration since the model used only the derivative dS/dC_l.

Ion uptake is accounted for by the boundary condition at the root–soil interface. Often, uptake is assumed to be primarily dependent on the solution concentration (Barber, 1995). De Willigen and Van Noorwijk (1994a) stated that this approach might overestimate uptake for conditions where not nutrient supply but plant demand regulates the uptake rate. These authors gave solutions for a zero-sink boundary condition to be applied when plant demand is not met by the supply, and for a constant-uptake boundary condition according to the plant demand, provided that the transport rate is sufficiently high. Zero-sink boundary conditions are used when solution concentrations are low and uptake is limited by the transport rate of the ion to the root surface (Geelhoed *et al.*, 1997)

$$t > 0 \quad r = r_0 \quad C_l = 0 \tag{7}$$

Influx J into the root is then calculated as

$$J = -D_l \theta f \frac{\partial C_l}{\partial r} \tag{8}$$

The constant flux boundary condition implies that sufficient ions are delivered to the root surface so that uptake is controlled only by plant demand (De Willigen and Van Noordwijk, 1994b):

$$J = -\frac{P}{2\pi L_{rv} l_r r_0} \tag{9}$$

where P is the plant demand for nutrients, L_{rv} the root length density per unit soil volume and l_r the root length. A way to describe the uptake rate in relation to the solution concentration is as intersecting lines (Tinker and Nye, 2000):

$$I = \begin{cases} 2\pi\alpha r_0 C_l, & C_{l0} < C_{lcrit} \quad (10) \\ I_{crit}, & C_{l0} > C_{lcrit} \quad (11) \end{cases}$$

where α is the root-absorbing power, αr_0 the root-demand coefficient, C_{l0} the solution concentration at the root surface, I the inflow into the root and I_{crit} and C_{lcrit} correspond to the value of I and C, respectively, where the two lines intersect. One of the frequently used equations for ion uptake is a modified version of the Michaelis–Menten kinetics originally used in enzyme kinetics. Using a minimum concentration at which uptake can take place, the boundary condition can be expressed as follows (Barber and Cushman, 1981):

$$t > 0, \quad r = r_0, \quad J = \frac{J_{\max}(C_1 - C_{\min})}{K_m + C_1 - C_{\min}} \tag{12}$$

where $J_{\max}$ is the maximum influx rate, K_m the Michaelis–Menten constant and $C_{\min}$ the minimum concentration where the influx is zero.

At the boundary away from the root surface (outer boundary), it can be assumed that the concentration remains constant. In this case, the soil is considered as a semi-infinite medium and no interroot competition occurs. This condition is mostly valid for immobile ions. It is represented by (Barber and Cushman, 1981)

$$t > 0, \quad r = r_1, \quad C_1 = \text{constant} \tag{13}$$

where r_1 is the radius of the cylinder, which can be exploited by the root. For mobile ions, inter-root competition is likely to occur. In this case, each root exploits a volume of surrounding soil, until it approaches a zero-transfer boundary situated at half the distance between two neighboring roots. Water is permitted to move across this cylinder. The boundary condition is (Barber and Cushman, 1981)

$$t > 0, \quad r = r_1, \quad J = 0 \tag{14}$$

To account for increasing root competition, Reginato *et al.* (1990, 2000) proposed to use a moving boundary approach:

$$t > 0, \quad r = R(t), \quad D_e b \frac{\partial C_1(R(t),t)}{\partial r} + q_0 C_1(\mathrm{R}(t),t) = 0 \tag{15}$$

$$t > 0, \quad R(t) = R_0 \sqrt{\frac{l_0}{l(t)}} \tag{16}$$

where $R(t)$ the variable half distance between root axes at time t, R_0 the initial half distance between root axes, $l(t)$ the root length as a function of time and l_0 the initial root length.

Table 1 shows a selection of assumptions most commonly found in rhizosphere models. None of them are strict; they can be modified on the expense of the number of input data and calculation time.

Five models of this type were applied to simulate the concentration gradients of Ni in the rhizosphere of *T. goesingense*. The mathematical equations of each model are described in sections 2.1–2.5. Models 1 and 2 have also been applied by Puschenreiter *et al.* (2005) where most of the necessary parameters were determined experimentally. Models 3–5 were developed for a theoretical study of the effect of their processes on simulation outputs. The necessary parameters were estimated from literature data. Simulations were performed with the software package FlexPDE (Schnepf *et al.*, 2002).

2.1. Model 1: Initial model

Model 1 is a classical single root model (Barber, 1995; Syring and Claassen, 1996) and it accounts for reactive transport in soil by diffusion, transpiration-induced convection and equilibrium sorption mechanisms between dissolved and adsorbed solutes. Model 1 was applied for the rhizobox experiments of Puschenreiter *et al.* (2005), referred to by them as "extended model." It was assumed that only Ni^{2+} is phytoavailable. This model is used as a starting point for further model developments. Owing to the geometry of the rhizobox, planar geometry instead of radial symmetry was used in the model equations, and in the outer boundary condition, a constant Ni concentration was assumed. Sorption was found to follow a composite Freundlich isotherm. It was also shown that due to complexation with dissolved organic matter (DOM), approximately 50% of the total Ni in the water extracts was present as Ni^{2+}, regardless of the distance from the root surface. This value was also found by other authors (e.g. Whiting *et al.*, 2003). Of the total Ni in solution, 50% was considered for potential uptake in this model. Nutrient influx into the root was corrected according to the coverage of the rhizobox membrane. The resulting set of equations is presented in Table 2.

2.2. Model 2: Including root hairs

Among the first to consider root hairs in their model were Bhat *et al.* (1976) and Itoh and Barber (1983), while attempting to explain experimentally obtained P uptake values exceeding those calculated by one of the previously available models. Three principal approaches to integrate root hairs into a rhizosphere model are found in the literature: (1) The boundary where exudation and uptake occurs is extended by the length of the root hairs (e.g. Kirk, 1999); (2) The continuity equation for root uptake is extended with a separate sink term (e.g. Geelhoed *et al.*, 1997); and (3) The transport equation is solved in a three-dimensional model with cylindrical coordinates (Geelhoed *et al.*, 1997).

Tinker and Nye (2000) and Whiting *et al.* (2003) used an analytical solution of an approximate steady-state root uptake model. According to Tinker and Nye (2000, p. 117), the flux in a cylindrical root is defined as

$$J = \alpha C_{10} \tag{17}$$

Table 1
Underlying assumptions of selected models

Assumptions concerning	Models								
	Syring and Claassen (1996)	Kirk (1999)	Bar-Yosef (1980)	Dunbabin *et al.* (2002a, b)	Geelhoed *et al.* (1997)	Nye and Marriot (1969)	Barber and Cushman (1981)	Darrah (1991)	Willigen and Noordwijk (1994b)
Soil medium									
Homogeneous and isotropic	x	x	x		x	x	x	x	x
Water flow									
Steady state	x	x			x	x	x	x	x
Transient			x	x					
Solute transport I									
Steady state									
Transient	x	x	x	x	x	x	x	x	x
Solute transport II									
Occurs by diffusion only		x			x			x	
By diffusion and convection	x		x	x		x	x		x
Geometry									
Planar		x							
Radial	x		x			x	x	x	x
Cylindrical 3 D						x			
Cartesian 3 D				x					
Presentation of roots									
Cylinders of uniform radius	x		x		x	x	x	x	x
Cylinders of various radii									
Root architecture described				x					
Surface in a planar system		x		x					
Root competition									
Not included		x				x		x	
Half-distance between neighboring roots	x		x	x	x		x		x

Root growth									
Not included		x	x		x	x		x	x
Linear/exponential with time	x						x		
Root system distribution model									
Dependent on resources and plant demand				x					
Root hairs									
Not included			x	x		x	x	x	x
Assumed to extend the root plane by the root hair length		x							
Sink term in model	x				x				
3-D model of root hairs					x				
Rhizo-deposition									
Not included	x			x	x	x	x		x
Soluble organic carbon (C)		x	x						
Soluble and insoluble organic C								x	
Soil microbes									
Not included	x			x	x	x	x		x
Via a decay rate constant		x	x						
As a pool in a carbon-flow model								x	
Ion uptake									
Not included								x	
Depends on the soil solution concentration at the root surface	x	x	x	x		x	x		
As a zero-sink boundary condition at the root surface					x				
Depends on plant demand									x

Table 2
Model 1: initial model

Continuity equation	$(\theta + \rho K_f m C_l^{m-1}) \frac{\partial C_l}{\partial t} = D_l f \theta \frac{\partial^2 C_l}{\partial x^2} + \frac{q}{b} \frac{\partial C_l}{\partial x}$
Initial condition $(t = 0, r > 0)$	$C_l = C_{li}$
Boundary condition at root surface $(t > 0, x = x_0)$	$J = \left(\frac{J_{max}(n_1 C_l)}{K_m + (n_1 C_l)} \right) n_2$
Outer boundary condition $(t > 0, x = x_1)$	$C_l = C_{li}$

Note: K_f and m are Freundlich isotherm parameters, n_1 is the fraction of free Ni ions in solution, and n_2 is the correction for the coverage of the rhizobox membrane.

at low values of C_{l0} ($C_{l0} << K_m$), α can be expressed by the Michaelis–Menten parameters J_{max} and K_m:

$$\alpha = \frac{J_{max}}{2\pi K_m r_0} \tag{18}$$

Expressing the same in term of fluxes,

$$\alpha = \frac{2\pi J_{max} r_0}{2\pi K_m r_0} = \frac{J_{max}}{K_m} \tag{19}$$

The radius of the soil cylinder that can be exploited by the root (r_1) is determined according to

$$r_1 = \frac{1}{\sqrt{\pi L_{rv}}} \tag{20}$$

where L_{rv} is the root length density. The steady-state approximation for the concentration at the root surface is, neglecting mass flow into roots (following Tinker and Nye, 2000, p. 298)

$$C_{l0} = \frac{C_{l1}}{\left(1 - \frac{\alpha r_0}{2D_l f\theta} + \frac{r_1^2 \alpha r_0}{2D_l f\theta} \frac{1}{r_1^2 - r_0^2}\right) \ln\left(\frac{r_1}{r_0}\right)} \tag{21}$$

where C_{l1} is the solution concentration in the bulk soil.

This steady-state solution of Tinker and Nye (2000) for the flux into a cylindrical root is used to approximate Ni-influx into root hairs, assuming the same uptake properties as for the root itself. The uptake by root hairs was introduced by the addition of a sink term into Model 1. The resulting set of equations is presented in Table 3.

The development of the following three models was based on the results of Puschenreiter *et al.* (2005), which suggest that root exudation and replenishment of soluble Ni from less available sources can be key processes governing metal uptake in plants.

2.3. Model 3: Including root exudation

The role of root exudates in the solubilization of nutrients has often been discussed (Bhat *et al.*, 1976; Claassen, 1990). Integrating root exudation into mechanistic rhizosphere models is quite complex owing to the difficulties in quantifying the nature and role of the exudates. The effect of soluble exudates such as phytosiderophores, phosphatases, mucilage and polycarboxylic acids on the availability of P, Fe and heavy metals has been recognized (Tinker and Nye, 2000). Mechanistic uptake models have considered the effect of polycarboxylic acids on P uptake (Hoffland, 1992; Kirk, 1999; Kirk *et al.*, 1999; Geelhoed *et al.*, 1999; Gerke *et al.*, 2000). Exudation rates are typically applied as a boundary condition at the root–soil interface. Two methods to treat the problem of simultaneous transport and interaction between root exudates and solutes have been employed: (1) coupled diffusion, and (2) separation of transport and reaction (equilibration).

Table 3
Model 2: model 1 including root hairs

Continuity equation	$(\theta + \rho K_f m C_1^{m-1}) \dfrac{\partial C_1}{\partial t} = D_1 f\theta \dfrac{\partial^2 C_1}{\partial x^2} + q\dfrac{\partial C_1}{\partial x} - aJ_{rh}$
	$J_{rh} = \alpha \dfrac{C_{11}}{(1-(\alpha r_{0,rh}/2D_1 f\theta)+((r_{1,rh}^2 \alpha r_0/2D_1 f\theta)(1/r_{1,rh}^2 - r_{0,rh}^2)))\ln(r_{1,rh}/r_{0,rh})}$
Initial condition ($t = 0, r > 0$)	$C_1 = C_{li}$
Boundary condition at root surface ($t > 0, x = x_0$)	$J = \left(\dfrac{J_{max}(n_1 C_1)}{K_m + (n_1 C_1)}\right) n_2$
Outer boundary condition ($t > 0, x = x_1$)	$C_1 = C_{li}$

Note: $r_{0,rh}$ is the root hair radius, $r_{1,rh}$ the half-mean distance between root hairs, J_{rh} the influx into root hairs, and a the root surface area per unit volume of soil.

The theory for the diffusion of two interacting solutes was proposed by Nye (1983). For this purpose, two partial differential equations are solved simultaneously. The interaction of the two solutes is expressed in terms of an interaction coefficient. The set of equations for a solute Y affecting the concentration of a solute X is as follows:

$$b_X \frac{\partial}{\partial t}(X_1 - \lambda Y_1) = D_{LX}\theta f \frac{\partial^2 X}{\partial x^2} \tag{22}$$

$$b_Y \frac{\partial Y_1}{\partial t} = D_{LY}\theta f \frac{\partial^2 Y_1}{\partial x^2} \tag{23}$$

where X and Y are the total concentrations of two different solutes in soil, X_1 and Y_1 are the concentrations in the soil solution, b_X and b_Y are the values of the buffer powers for both solutes, λ is the interaction coefficient $(\partial X_L/\partial Y_L)_X$ and D_{LX} and D_{LY} are the diffusion coefficients of X and Y in free solution. Nye's theory was applied by Kirk (1999, 2002) to simulate the effect of citrate (*Cit*) on the mobility and uptake of P by assuming uptake to correspond to a root absorbing power α, root exudation rate to be constant with time and *Cit* to be decomposed following first-order kinetics. Geelhoed *et al.* (1999) and Nietfeld's (2000) multicomponent model considered soil solution chemistry in more detail. Geelhoed *et al.* (1999) linked a transport model with the speciation model CD-Music and alternated between chemical equilibration and transport at each time step.

In Model 3, the approach of Kirk (1999) was adapted to the conditions of the rhizobox experiment. Planar geometry was used instead of radial, and transport was assumed to occur by diffusion and convection instead of diffusion only. Uptake conditions and sorption were treated as explained for Model 1. The resulting equations are presented in Table 4.

2.4. Model 4: Including slow-reacting Ni

There is experimental evidence that T. goesingense depletes more Ni pools that were less reactive and bioavailable than the labile (1 M NH4NO3-extractable) Ni (Puschenreiter et al., 2005). These might not be in equilibrium with the soil solution. In addition to assuming rapid and reversible sorption mechanisms (labile Ni), it must also be assumed that there is an additional kinetically controlled (slow and reversible) release from surfaces of soil constituents. A modeling approach to account for a slow sorption or dissolution mechanism was proposed by Darrah and Staunton (2000):

$$\frac{\partial C_1}{\partial t} = \frac{1}{r}\frac{\partial}{\partial r}\left(rD_e \frac{\partial C_1}{\partial r} + \frac{r_0 q_0 C_1}{\theta + b}\right) + \frac{\partial C_f}{\partial t} \tag{24}$$

$$\frac{\partial C_f}{\partial t} = k_+ \theta C_1 - k_- C_f \tag{25}$$

Table 4
Model 3: model 1 including root exudation

Continuity equations	$(\theta + \rho K_f m C_1^{m-1}) \dfrac{\partial (C_1 - \lambda C_{ex})}{\partial t} = D_1 f\theta \dfrac{\partial^2 C_1}{\partial x^2} + q\dfrac{\partial C_1}{\partial x}$
	$b_{ex}\dfrac{\partial C_{ex}}{\partial t} = D_{1,ex} f\theta \dfrac{\partial^2 C_{ex}}{\partial x^2} + q\dfrac{\partial C_{ex}}{\partial x} - \theta k C_{ex}$
Initial condition	$C_1 = C_{li}$
$(t = 0, r > 0)$	$C_{ex} = C_{exi}$
Boundary condition at root surface $(t > 0, x = x_0)$	$J = \left(\dfrac{J_{max}(n_1 C_1)}{K_m + (n_1 C_1)}\right) n_2$
	$J = -F_{ex}$
Outer boundary condition	$C_1 = C_{li}$
$(t > 0, x = x_1)$	$C_{ex} = 0$

Note: C_{ex} is the concentration of exudates in solution, F_{ex} the exudation rate, $D_{1,ex}$ the diffusion coefficient of exudates in water, and b_{ex} is the buffer power of exudates.

where C_f is the concentration in the slow-reacting (fixed) phase, and k_+ and k_- are the forward and backward reaction rate constants, respectively. Following this approach of Kirk and Staunton (1989), we included an additional slow sorption mechanism into Model 4. The resulting equations are presented in Table 5.

2.5. Model 5: Including two-stage sorption model

Selim *et al.* (1976) proposed a model where two types of exchange sites were present at the exchange complex, one governed by equilibrium (type 1) sorption, and the other by first-order kinetics (type 2). The total concentration of sorbed solutes S is assumed to be

$$S = \mathrm{Fr}\, S_1 + (1-\mathrm{Fr})S_2 \qquad (26)$$

where S_1 and S_2 are the concentrations of sorbed solute per unit sorbent mass of type 1 and type 2 sites, respectively, and Fr is the fraction of type 1 sites. This approach softens the assumption of instantaneous equilibrium between soil solution and the exchange complex. The resulting equations are presented in Table 6.

Table 5
Model 4: model 1 including slow reacting Ni

Continuity equation	$(\theta + \rho K_f m C_l^{m-1}) \dfrac{\partial C_l}{\partial t} + \dfrac{\partial C_f}{\partial t} = D_l f \theta \dfrac{\partial^2 C_l}{\partial x^2} + q \dfrac{\partial C_l}{\partial x}$
	$\dfrac{\partial C_f}{\partial t} = \theta k_+ C_l - \dfrac{k_+}{b_2} C_f$
Initial condition ($t = 0$, $r > 0$)	$C_l = C_{li}$ $C_f = C_{fi}$
Boundary condition at root surface ($t > 0$, $x = x_0$)	$J = \left(\dfrac{J_{max}(n_1 C_l)}{K_m + (n_1 C_l)} \right) n_2$
Outer boundary condition ($t > 0$, $x = x_1$)	$C_l = C_{li}$

Note: k_+ is the forward rate constant, C_f the concentration of slow reacting (fixed) Ni, and b_2 the buffer capacity of the fixed Ni.

Table 6
Model 5: model 1 including a two-stage sorption model

Continuity equation	$(\theta + \mathrm{Fr}\rho K_f^m C_l^{m-1}) \dfrac{\partial C_l}{\partial t} + \rho(1 - \mathrm{Fr}) \dfrac{\partial S_2}{\partial t} = D_l f \theta \dfrac{\partial^2 C_l}{\partial x^2} + q \dfrac{\partial C_l}{\partial x}$
	$\dfrac{\partial S_2}{\partial t} = \alpha_2 / 1 - \mathrm{Fr}\ ((K_f C_l^m + q) - S_2)$
Initial condition ($t = 0$, $r > 0$)	$C_l = C_{li}$ $S_2 = S_{2i}$
Boundary condition at root surface ($t > 0$, $x = x_0$)	$J = \left(\dfrac{J_{max}(n_1 C_l)}{K_m + (n_1 C_l)} \right) n_2$
Outer boundary condition ($t > 0$, $x = x_1$)	$C_l = C_{li}$

Note: α_2 is the rate constant of the two-stage sorption process.

2.6. Comparison of simulated and measured concentrations

Simulated and measured concentrations were compared by means of the normalized root mean squared error (NRMSE):

$$\mathrm{NRMSE} = \sqrt{\frac{\sum_1^n (x_{obs} - x_{sim})^2}{\sum_1^n (x_{obs} - \bar{x}_{obs})^2}} \tag{27}$$

where x_{obs} is the measured value, x_{sim} the simulated value, $\bar{x}_{obs}$ the mean of measured values and n the number of data points.

3. CASE STUDY WITH THE Ni HYPERACCUMULATOR *T. GOESINGENSE*

3.1. Rhizobox experiment and derived model input parameters

T. goesingense was grown in a rhizobox system developed and evaluated by Wenzel *et al.* (2001) (Puschenreiter *et al.*, 2005). A root monolayer developed after 20 days (Fig. 1), and was left to influence the adjacent soil for a further 12 days. Then, concentrations of 1 M NH_4NO_3-extractable (labile) and water-extractable Ni were measured in thin slices of soil parallel to the generated root layer. Simulation results of the models described in Section 2 were compared to this data set. One inherent feature of rhizoboxes is the fact that the root layer is made up of many roots of different ages, whereas single root models assume a boundary of "one root surface." Furthermore, the concentrations measured in the soil slices represent average concentrations in the whole membrane area, covered or uncovered by roots. In this experiment, it was visually observed that 60% of the rhizobox membrane was covered by roots. Hence, it was assumed in the models that all fluxes across the root surface were 60% of the values valid for a complete root surface.

Transport parameters, sorption parameters, including the composite Freundlich sorption isotherm, initial pool sizes, root uptake kinetics as well as root hair parameters were measured for the experimental conditions of the

(a) (b)

Fig. 1. Photo of the rhizobox: (a) side view, (b) front view with root layer.

rhizobox experiment (Puschenreiter *et al.*, 2005). Root morphological parameters of naturally grown *T. goesingense* plants were measured using the image-analysis approach by Himmelbauer *et al.* (2005). To examine the effect of including root exudation and chemical non-equilibrium in model results, literature data were used for the necessary model input parameters.

The concentration of the initially present slow-reacting Ni (C_f) was assumed to be the sum of extraction steps 3 and 5 (Zeien and Brümmer, 1989) measured in the bulk soil. They correspond to the specifically adsorbed ions as well as to the fraction bound by organic matter. For estimating the value of the buffer capacity b_2, it was assumed that the underlying kinetically controlled and reversible sorption process was at equilibrium in the bulk soil, and the quotient

$$b_2 = \frac{\partial C_f}{\theta \, \partial C_l} \tag{28}$$

was evaluated.

The values for the exudation rate (F_{ex}), interaction coefficient (λ), buffer power of exudate in soil (b_c) and the decomposition rate constant for the exudate (k_e) were adopted from Kirk (1999). The value of the forward rate constant k_+ was estimated from Scheckel and Sparks (2001), who evaluated kinetic adsorption data of Ni to different minerals where k_+ ranged from 2.5×10^{-6} to 9.78×10^{-6} s^{-1}. For the simulation, an average value of 5.00×10^{-6} was used. This value also coincides with the values that Kirk and Staunton (1989) suggested for the kinetic adsorption of C_d to soil, where the values ranged from 10^{-3} to 10^{-7} s^{-1}. This same value was assumed for the rate constant for the two-stage sorption model, α_2. The fraction of type 1 sites (F_r) was assumed to be 0.3. Table 7 summarizes all input parameter values.

3.2. Simulation results of Models 1–5 in *T. goesingense* case study

Resulting concentration gradients of labile and dissolved Ni produced by Models 1–5 are shown in Fig. 2. For model corroboration, 1 M NH_4NO_3-extractable Ni data shown in Fig. 2a are used. They are assumed to represent labile, readily reactive Ni. The curves that match the measured data best are those obtained by Model 1, Model 4 including release from a fixed Ni phase and Model 5, the two-stage sorption model. Similar results were observed for each of the three curves, indicating that the influence of the additionally included processes is relatively small under the assumptions of equilibrium between fixed phase and solution in the bulk soil. The models including both root hairs and exudation overestimate the depletion close to the root. Corresponding to the root hair length, additional depletion was simulated by Model 2 at < 1 mm distance from the root layer. Although *T. goesingense* is known to have root hairs abundantly, measured data do not show such a strong decrease of labile Ni. This is remarkable, since root hairs could penetrate the membrane of the rhizobox.

Table 7
Input parameter values

Parameter type	Symbol	Notation	Quantity
Transport parameters[a]	q	Water flux density	2.24×10^{-8} dm s^{-1}
	D_l	Diffusion coefficient in water	6.16×10^{-8} dm^2 s^{-1}
	θ	Volumetric water content	0.36
	f	Impedance factor	0.40
	ρ	Bulk density	1.15 kg dm^{-3}
Sorption parameters for labile pool (curve with three branches)[a]	m	Parameter of Freundlich isotherm	If C_l = 0–0.07 then 1, else if C_l = 0.07–0.13 then 0.12, else 0.60
	K_f	Parameter of Freundlich isotherm	If C_l = 0–0.07 then 270.53, else if C_l = 0.07–0.13 then 219.57, else 29.50 mg^m dm^{3m} kg^{-1}
	d	Parameter of Freundlich isotherm	If C_l = 0–0.07 then −21.89, else if C_l = 0.07–0.13 then −163.68, else 0 mg kg^{-1}
Sorption parameters for slow reacting Ni[b]	b_2	Buffer power of fixed fraction $[dC_f/(\theta dC_l)]$	882.30
	k_+	Forward rate constant	5.00×10^{-6} s^{-1}
Two-stage sorption[c]	α_2	Rate constant	5.00×10^{-6} s^{-1}
	Fr	Fraction of type 1 sites	0.3
Exudation[d]	F_{ex}	Exudation rate	5.00×10^{-6} mg dm^2 s^{-1}
	λ	Interaction coefficient	0.015
	b_{ex}	Buffer power of citrate	5.0
	$D_{l,ex}$	Diffusion coefficient of citrate in water	6.9×10^{-8} dm^2 s^{-1}
	k	Decomposition rate of citrate	2.00×10^{-5} s^{-1}
Root hairs[a]	$r_{0,rh}$	Root hair radius	5.00×10^{-5} dm
	l_{rh}	Loot hair length	0.008 dm
	n_{rh}	Number of root hairs per unit root length	1000 dm^{-1}
	a	Root hair surface area per unit volume of soil	100 m^2 m^{-3}

Table 7
(*Continued*)

Parameter type	Symbol	Notation	Quantity
Uptake kinetics[a]	K_m	Michaelis–Menten constant	2.12 mg dm^{-3}
	J_{max}	Maximum influx rate	1.01×10^{-5} mg dm^{-2} s^{-1}
Initial conditions[a]	C_{li}	In soil solution	0.25 mg dm^{-3}
	C_{ti}	Of total labile ions in soil	14.82 mg dm^{-3}
	C_{fi}	In fixed fraction	78.42 mg dm^{-3}
	C_{exi}	Of exudates	0.00 mg dm^{-3}

[a] Puschereiter *et al.* (2005). [b] Scheckel and Sparks, (2001). [c] Approximated. [d] Kirk (1999).

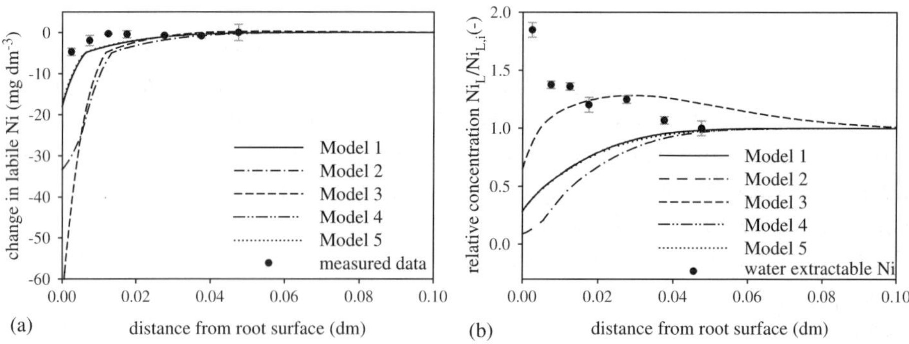

Fig. 2. Simulated curves of (a) labile and (b) dissolved Ni (Ni_L) vs. measurements. Model 1: initial model; Model 2: plus root hairs; Model 3: plus root exudation; Model 4: plus slow reacting Ni; Model 5: plus two-stage sorption; $Ni_{L,1}$ initial concentration.

Simulated concentration gradients of root exudates were highest at the root surface and declined with increasing distance to the root monolayer (data not shown). Regarding their effect on Ni, the additional mobilization caused higher uptake rates by roots. As a result, labile Ni was strongly depleted near the root, and the calculated curve is far from the measured data points. It is generally required that the model input parameters are measured independently for each experimental (soil/plant) setup. Use of cited parameter values gives rise to uncertainty. However, it is not uncommon that modeling work relies on literature data (Whiting *et al.*, 2003). Such an approach can be justified by the fact that it may help to test the effect of different hypotheses on model outputs at low cost.

From the experimental point of view, the effect of root exudation on Ni uptake by *T. goesingense* is still unclear and controversially discussed. Histidine and citrate are among the exudates found in the rhizosphere of *T. goesingense* (Salt *et al.*, 2000). For theoretical considerations, the model of Kirk (1999) was applied to this topic. Even if the absolute values do not match measurements, the calculated trends could help to assess the role of root exudates in Ni uptake by *T. goesingense*. Since no data for organic acid exudation by *T. goesingense* were available, values were taken from Kirk (1999) and Kirk *et al.* (1999). These values might not be valid under the studied soil and plant conditions, in particular regarding Ni sorption. For the experimental soil, a nonlinear Freundlich desorption isotherm was obtained that ensured that sorption behavior under the absence of exudates was well described. An exudate-influenced isotherm was not available. The obtained results were comparable to those of Kirk (1999) in shape of the curves and extend of depletion. However, the underlying mechanisms of Ni mobilization by exudates are different from those for P. Future experimental work should include determination of the quality and quantity of exudates as well as their influence on Ni sorption.

Contrary to the results for labile Ni, the gradients of relative concentrations of dissolved Ni obtained with the model including exudation (Model 3) matched water-extractable Ni data better (Fig. 2b). The curves produced by the other models were more discrepant. Water extracts may not represent real soil solutions and freezing of soil samples might have introduced additional artifacts. However, gradients of Ni concentration in soil solution were not available. Water-extractable Ni data in Fig. 2b are presented for orientation but cannot be used for proper model corroboration. All simulated curves were opposite to the trend implied by measured water-extractable Ni. Accumulation of solutes near the root is known for excluders or for situations where convective flow toward the root exceeds plant demand. This was not the case in our experimental system. In future experiments, the use of microsuction cups will enable comparison of simulated and measured dissolved Ni (Wenzel *et al.*, 2001).

Although none of the models could reproduce measured concentrations gradients, the results of Models 1, 4 (including slow-reacting Ni) and 5 (including two-stage sorption) matched experimental data best. This is confirmed by the error measure values (NRMSE) in Table 9. Consequently, sensitivity analysis was used to examine the effects of varying selected parameters on the resulting curves. Each value was changed between 0.5 and 2 times its original value, except for J_{max} and K_m. In Fig. 3, the effect of changing the value of the maximum influx rate (J_{max}) is presented. The higher the J_{max}, the higher was the uptake rate and hence the depletion of both labile and dissolved Ni. The impact of J_{max} changes was distinct at > 4 mm distance from the root surface. Reducing the original value of J_{max} up to a factor 5 reveals a curve of labile Ni closer to the measured data, compared with unchanged J_{max}.

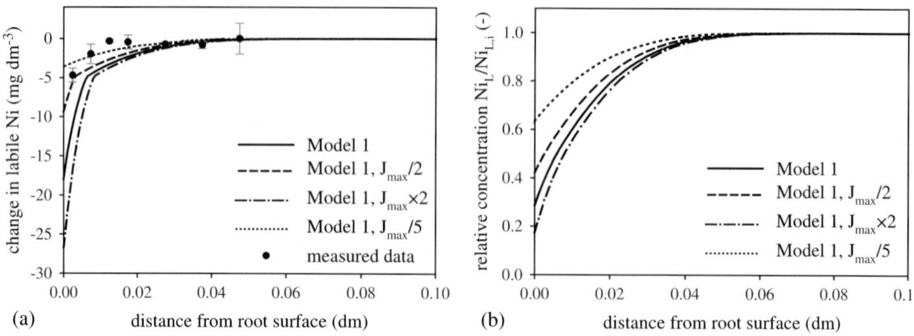

Fig. 3. Effect of varying the value of J_{max} on (a) labile Ni and (b) dissolved Ni (Ni_L) calculated by Model 1.

Such a deviation is clearly below the variation range usually considered with uptake rate data (Darrah and Staunton, 2000; Sadana and Claassen, 2000). Measured Ni uptake rates were obtained on a mass basis. For converting them to influx per unit root surface area, we considered three different ways that resulted in different values for J_{max}. For method 1, we simply assumed that the ratio between root fresh mass (RFM) and root surface area (RSA) in the rhizobox is the same as in the uptake experiment. An additional correction according to the coverage of the rhizobox membrane by roots was applied (see Section 2.1). In Method 2, data were converted based on geometrical considerations. We used the average root diameter, which we measured for *T. goesingense* (r_0 =0.088 mm, Himmelbauer *et al.*, 2005) and the widespread relation (e.g. Claassen *et al.*, 1986)

$$r_0 = \sqrt{\frac{\text{RFM}}{\pi l_r}} \tag{29}$$

where RFM is the root fresh mass and l_r the root length. For Method 3, root morphological characteristics of naturally grown *T. goesingense* were determined using an image analysis system approach (Himmelbauer *et al.*, 2005). The estimated RSA/RFM ratio was (a) 105.7 $cm^2\ g^{-1}$ for the whole root system and (b) 233.5 for fine roots (< 1 mm) only. The resulting fitting parameters J_{max} and K_m are presented in Table 8.

It was assumed that the best results are obtained with Method 3a due to the direct measurement of RSA/RFM. The range of J_{max} obtained with all these methods corresponds to the range used in the sensitivity analysis.

The same results were obtained when varying the value of K_m inversely proportional to the changes on J_{max} (Fig. 4). This is the case for low solution concentrations because of the following relations. If J_{max} is reduced or K_m increased

Table 8
Conversion of uptake rate data

	Method 1	Method 2	Method 3a	Method 3b
K_m [mg dm^{-3}]	2.12	2.12	2.12	2.12
J_{max} [mg dm^{-2} s^{-1}]	3.05E−7	4.70E−6	1.01E−5	4.58E−6

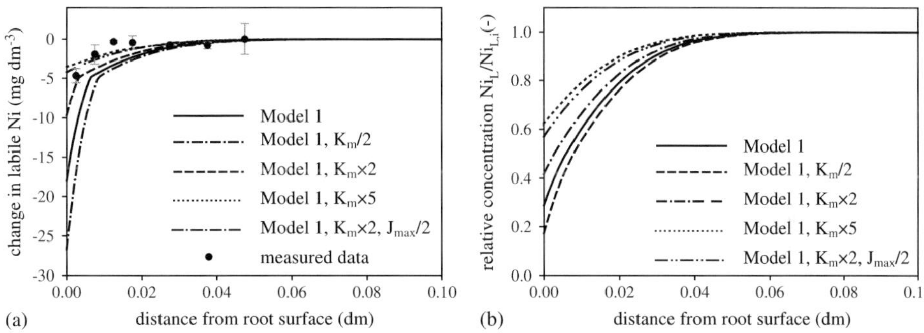

Fig. 4. Effect of varying the value of K_m on (a) labile Ni and (b) dissolved Ni (Ni_L) calculated by Model 1.

by a factor 2, respectively, the corresponding influxes are

$$J_1 = \frac{(J_{max}/2)C_1}{K_m + C_1} = \frac{1}{2}\frac{J_{max}C_1}{K_m + C_1} \quad \text{and} \quad J_2 = \frac{J_{max}C_1}{2K_m + C_1} = \frac{1}{2}\frac{J_{max}C_1}{K_m + (C_1/2)} \tag{30}$$

The higher the K_m, the lower is the uptake rate. Generally, increasing the original value of K_m up to a factor 5 results in a simulated curve of labile Ni close to the measured data. Contrary to J_{max}, however, the way of converting uptake per unit root dry mass to unit root did not affect the value of K_m. Therefore, this uncertainty cannot be taken into account here. Since J_{max} and K_m originate from the same experiment, i.e. the same solution and plant conditions, it can be expected that both values are afflicted with similar kinds of errors. Simultaneous variation of K_m and J_{max} by a factor 2 and 0.5, respectively, results in an overall reduction of the influx rate of a factor of approximately 4. The resulting curve was therefore close to the curve where K_m was increased by a factor 5.

Fig. 5 shows that the increased supply of Ni due to a higher D_e resulted in a more pronounced depletion of labile and dissolved Ni. At the root surface, the concentrations were higher when D_e was larger. The point of interception occurred approximately 0.5 mm from the root. The sharp bend at approximately 5 mg dm^{-3} change in labile Ni corresponds to the desorption isotherm obtained for this experimental soil (see Section 2.3). Reducing D_e resulted in a simulated concentration gradient close to the measured data. At the root surface, the concentration was still underestimated by the model.

Varying the values of the interaction coefficient (λ) and exudation rate (F_{ex}) resulted in similar concentration profiles (Fig. 6). The higher the values of F_{ex} or

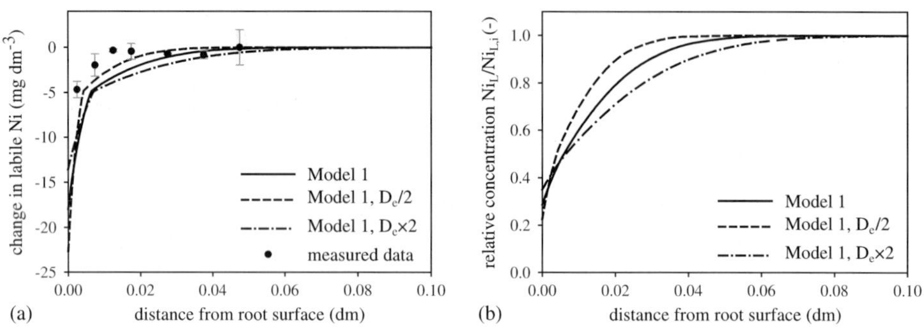

Fig. 5. Effect of varying the value of D_e on (a) labile Ni and (b) dissolved Ni (Ni_L) calculated by Model 1.

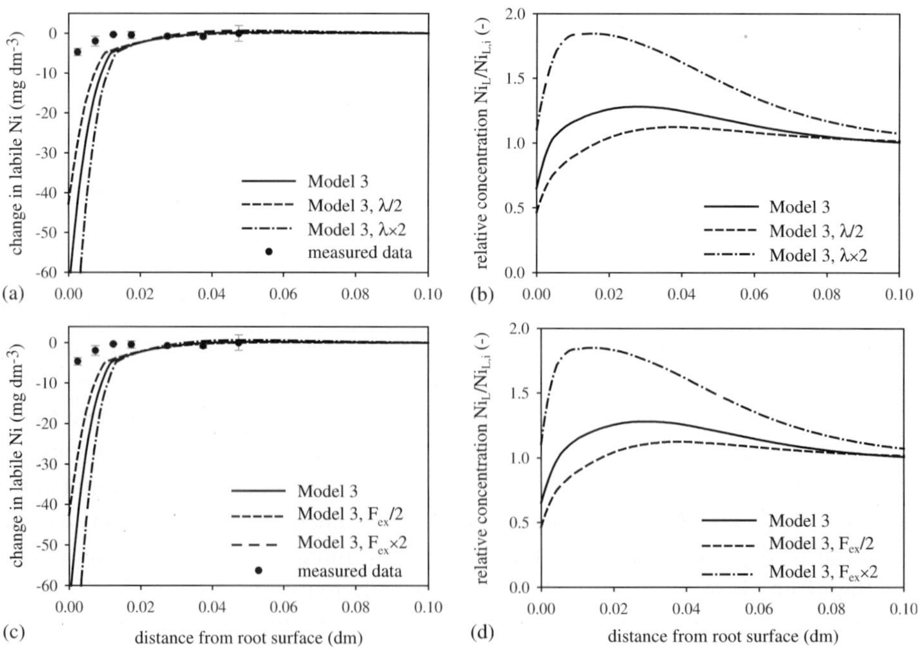

Fig. 6. Effect of varying the values of λ on (a) labile Ni and (b) dissolved Ni (Ni_L) and F_{ex} on (c) labile Ni and (d) dissolved Ni (Ni_L) calculated by Model 3: including root exudation.

λ, the higher were the concentrations of dissolved Ni. Therefore, the uptake rate was higher (data not shown). Due to this, a contrary effect was observed for labile Ni: the higher F_{ex} or λ, the lower was the concentration of labile Ni at the root surface. The effect on labile Ni extends only up to 1 mm from the root. Generally, increased root exudation rates caused larger discrepancies between simulated and experimentally obtained results. The model used here belongs to the class of models that use "lumped" parameters, meaning that several processes are described by one combined parameter. The parameter λ expresses the total effect of exudates on dissolved Ni. Increased Ni concentration could be due to several processes, but the contributions of each of them are not explicitly described. Furthermore, the solution species of Ni as well as other important ions are not explained.

Varying the values of the reaction rate constants for slow-reacting Ni (k_+) and the two-stage sorption model (α_2) also resulted in similar concentration profiles (Fig. 7). The differences among variations as well as the differences from Model 1 were not significant. Increasing k_+ as well as reducing α_2 by an order of magnitude resulted in an increase of labile Ni near the root

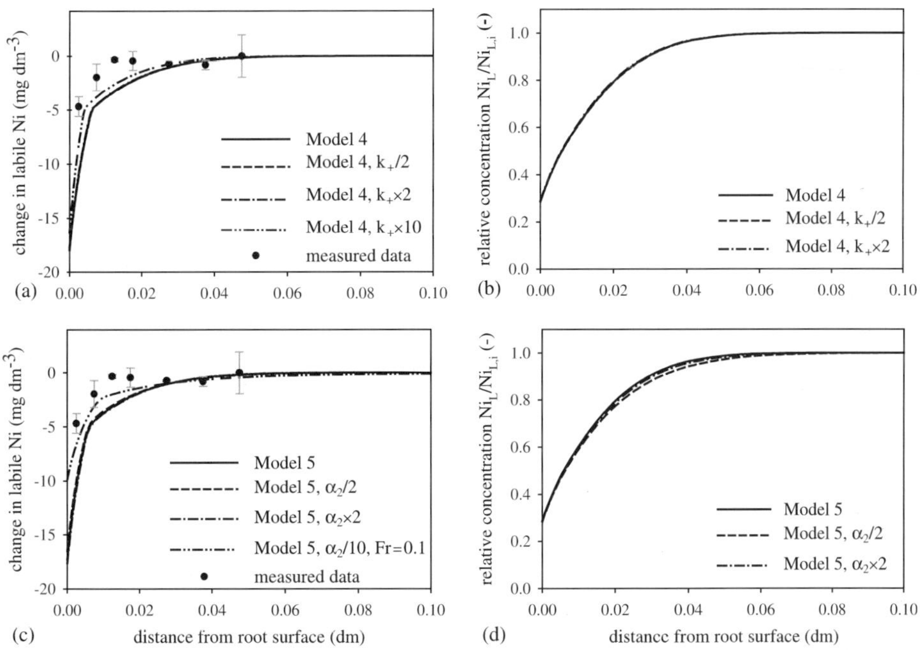

Fig. 7. Effect of varying the values of k_+ on (a) labile Ni and (b) dissolved Ni (Ni_L) calculated by Model 4: including slow reacting Ni and α on (c) labile Ni and (d) dissolved Ni (Ni_L) calculated by Model 5: including two-stage sorption.

surface and shifted the concentration gradients closer to the measured data. Such values are still within the range 10^{-3}–10^{-7} s^{-1}, as suggested by Kirk and Staunton (1989).

The deviation of simulated from measured concentrations (NRMSE) are given in Table 9.

Table 9
NRMSE of simulated and measured labile and water-extractable Ni

Model	NRMSE labile Ni	NRMSE water extractable Ni
Model 1: initial model	0.66	1.60
Model 2: incl. root hairs	0.88	1.47
Model 3: incl. root exudation	0.92	2.27
Model 4: incl. slow reacting Ni	0.65	1.60
Model 5: incl. two-stage sorption	0.63	1.60
Model 1, $J_{max}/2$	0.23	1.68
Model 1, $J_{max}\times 2$	0.78	1.55
Model 1, $J_{max}/5$	0.93	1.89
Model 1, $K_m/2$	0.78	1.55
Model 1, $K_m\times 2$	0.26	1.68
Model 1, $K_m\times 5$	0.83	1.87
Model 1, $K_m\times 2$; $J_{max}/2$	0.44	1.81
Model 1, $D_e/2$	0.60	1.62
Model 1, $D_e\times 2$	0.60	1.62
Model 3, $\lambda/2$	0.87	1.81
Model 3, $\lambda\times 2$	0.95	2.42
Model 3, $F_{ex}/2$	0.87	1.81
Model 3, $F_{ex}\times 2$	0.95	2.42
Model 4, $k_+/2$	0.66	1.60
Model 4, $k_+\times 2$	0.65	1.61
Model 5, $\alpha_2/2$	0.61	1.60
Model 5, $\alpha_2\times 2$	0.65	1.60

4. SUMMARY AND CONCLUSIONS

Model approaches to simulate solute uptake by a single root and their application to the simulation of Ni concentration gradients in the rhizosphere of *T. goesingense* were examined in detail, including adaptations to the rhizobox conditions in terms of geometry, coverage of membrane and uptake of free Ni only. Model 1 (Ni^{2+} uptake only) was the base of a modeling case study; further model approaches included root hair uptake, root exudation and chemical non-equilibrium. Model 1, Model 4 (including slow-reacting Ni) and Model 5 (including two-stage sorption) came closest to measured concentration gradients of labile Ni, but showed an opposite trend in the case of dissolved concentrations. These models might produce an increase concentrations of dissolved Ni when transport to the root surface exceeds the uptake rate. However, in this experimental setup, such an occurrence is not very likely. Therefore, we can conclude that we need to include another process that might explain this effect. Root exudates may contribute to the mobilization of Ni. When including root exudation in the calculations (i.e. Model 3), a much larger depletion of labile Ni was simulated than that measured. Simulated curves of dissolved Ni showed an increase near the root surface, but did not match with the measured peak. Using literature data can be critical even though they are in the typical range of cited values. With high uptake rate measured, it is not likely that any other values of the interaction coefficient (λ) or exudation rate (F_{ex}) would result in concentration gradients close to measured data. Depletion of labile Ni will always be overestimated. If model predictions do not agree with experimental observations, either the parameter values or the assumptions of the model have to be reconsidered. Single root model approaches could not fully explain observed concentration gradients in the rhizosphere of *T. goesingense*. Further work with respect to determination of parameter values is required. With regard to model assumptions, it could be beneficial to include more soil chemical processes. Such an approach would involve multi-ion and multispecies models.

ACKNOWLEDGMENTS

We gratefully acknowledge the financial support by the Austrian Science Foundation (FWF; P 15749, P 13454), the Ministry for Education, Science and Culture (GZ38.038/1-VIII/A/4/2000) and by the BOKU – University of Natural Resources and Applied Life Sciences, Vienna (BOKU Priority Research Area Project #16).

REFERENCES

Barber, S.A., 1995. Soil Nutrient Bioavailability: A Mechanistic Approach. Wiley, New York.

Barber, S.A., Cushman, J.H., 1981. Nitrogen uptake model for agronomic crops. In: Inskandar, I.K. (Ed.), Modeling Wastewater Renovation-Land Treatment. Wiley-Interscience, New York, pp. 382–409.

Bar-Yosef, B., Fishman, S., Talpaz, H., 1980. A model of zinc movement to single roots in soils. Soil Sci. Soc. Am. J. 44, 1272–1279.

Bhat, K.K.S., Nye, P.H., Baldwin, J.P., 1976. Diffusion of phosphate to plant roots in soil IV. The concentration distance profile in the rhizosphere of roots with root hairs in a low-P soil. Plant Soil 44, 63–72.

Brooks, R.R., 1998. Plants that Hyperaccumulate Heavy Metals. CAB International, New York.

Claassen, N., 1990. Fundamentals of soil-plant interactions as derived from nutrient diffusion in soil, uptake kinetics and morphology of roots. Proceedings of the Fourteenth International Congress of Soil Science, Kyoto, Japan, vol. II, pp. 118–123.

Claassen, N., Syring, K.M., Jungk, A., 1986. Verification of a mathematical model by simulating potassium uptake from soil. Plant Soil 95, 209–220.

Cushman, J.H., 1984. Nutrient transport inside and outside the root rhizosphere: generalized model. Soil Sci. 138, 164–171.

Darrah, P.R., 1991a. Models of the rhizosphere I. Microbial population dynamics around a root releasing soluble and insoluble carbon. Plant Soil 133, 187–199.

Darrah, P.R., 1991b. Models of the rhizosphere II. A quasi three-dimensional simulation of the microbial population dynamics around a growing root releasing soluble exudates. Plant Soil 138, 147–158.

Darrah, P.R., Roose, T., 2001. Modeling the Rhizosphere. In: Pinton, R., Varanini, Z., Nannipieri, P., (Eds.), The Rhizosphere. Biogeochemistry and Organic Substances at the Soil-Plant Interface. Marcel Dekker, New York, pp. 327–372.

Darrah, P.R., Staunton, S., 2000. A mathematical model of root uptake of cations incorporating root turnover, distribution within the plant, and recycling of absorbed species. Eur. J. Soil Sci. 51, 643–653.

Dunbabin, V.M., Diggle, A.J., Rengel, Z., 2002a. Simulation of field data by a basic three-dimensional model of interactive root growth. Plant Soil 239, 39–54.

Dunbabin, V.M., Diggle, A.J., Rengel, Z., van Hugten, R., 2002b. Modelling the interactions between water and nutrient uptake and root growth. Plant Soil 239, 19–38.

Geelhoed, J.S., Mous, S.L.J., Findenegg, G.R., 1997. Modeling zero sink nutrient uptake by roots with root hairs from soil: comparison of two models. Soil Sci. 162, 544–553.

Geelhoed, J.S., Van Riemsdijk, W.H., Findenegg, G.R., 1999. Simulation of the effect of citrate exudation from roots on the plant availability of phosphate adsorbed on goethite. Eur. J. Soil Sci. 50, 379–390.

Gerke, J., Beissner, L., Roemer, W., 2000. The quantitative effect of chemical phosphate mobilization by carboxylate anions on P uptake by a single root. I. The basic concept and determination of soil parameters. J. Plant Nutr. Soil Sci. 163, 207–212.

Gonnelli, C., Marsili-Libelli, S., Baker, A., Gabbrielli, R., 2000. Assessing plant phytoextraction potential trough mathematical modeling. Int J. Phytoremediation 2, 343–351.

Himmelbauer, M.L., Puschenreiter, M., Schnepf, A., Loiskandl, W., Wenzel, W.W., 2005. Root morphology of *Thlaspi goesingense* Hàlàcsy grown on a serpentine soil. J. Plant Nutr. Soil Sci. 168: 138–144.

Hoffland, E., 1992. Quantitative evaluation of the role of organic acid exudation in the mobilization of rock phosphate by rape. Plant Soil 140, 279–289.

Itoh, S., Barber, S.A., 1983. Phosphorus uptake by six plant species as related to root hairs. Agron. J. 75, 457–461.

Kirk, G.J.D., 1999. A model of phosphate solubilization by organic anion excretion from plant roots. Eur. J. Soil Sci. 50, 369–378.

Kirk, G.J.D., 2002. Modelling root-induced solubilization of nutrients. Plant Soil 255, 49–57.

Kirk, G.J.D., Santos, E.E., Findenegg, G.R., 1999. Phosphate solubilization by organic anion excretion from rice (*Oryza sativa* L.) growing in aerobic soil. Plant Soil 211, 11–18.

Kirk, G.J.D., Staunton, S., 1989. On predicting the fate of radioactive caesium in soil beneath grassland. J. Soil Sci. 40, 71–84.

Lide, D.R., 2000. Handbook of Chemistry and Physics, eighty-third ed. CRC Press LLC, Boca Raton, FL.

Marschner, H., 1995. Mineral Nutrition of Higher Plants, second ed. Academic Press, London.

McGrath, S.P., Zhao, F.J., Lombi, E., 2001. Plant and rhizosphere processes involved in phytoremediation of metal-contaminated soils. Plant Soil 232, 207–214.

Meeussen, J.C.L., 2003. ORCHESTRA: an object-oriented framework for implementing chemical equilibrium models. Environ. Sci. Technol. 37, 1175–1182.

Nietfeld, H.W.F., 2000. Modeling the dynamics of the rhizosphere aluminum chemistry in acid forest soils. In: Gobran, G.R., Wenzel, W.W., Lombi, E (Eds.), Trace Elements in the Rhizosphere. CRC Press, Boca Raton, FL, pp. 253–307.

Nye, P.H., 1983. The diffusion of two interacting solutes in soil. J. Soil Sci. 34, 677–691.

Nye, P.H., Marriott, F.H.C., 1969. A theoretical study of the distribution of substances around roots resulting from simultaneous diffusion and mass flow. Plant Soil 30, 459–472.

Puschenreiter, M., Schnepf, A., Molina Millàn, I., Fitz, W.J., Horak, O., Klepp, J., Schrefl, T., Lombi, E., Wenzel, W.W., 2005. Changes of Ni biogeochemistry in the rhizosphere of the hyperaccumulator *Thlaspi goesingense*. Plant Soil, (in press).

Reeves, R.D., Baker, R.R., 1984. Studies on metal uptake by plants from serprntine and non-serpentine populations of *Thlaspi goesingense* Hàlàcsy (Cruciferae). New Phytol. 98, 191–204.

Reginato, J.C., Palumbo, M.C., Moreno, I.S., Bernando, I.Ch., Tarzia, D.A., 2000. Modeling nutrient uptake using a moving boundary approach. Comparison with the Barber-Cushman model. Soil Sci. Soc. Am. J. 64, 1363–1367.

Reginato, J.C., Tarzia, D.A., Cantero, A., 1990. On the free boundary problem for the Michaelis-Menten absorption model for root growth. Soil Sci. 150, 722–729.

Regvar, M., Vogel, K., Irgel, N., Wraber, T., Hildebrandt, U., Wilde, P., Bothe, H., 2003. Colonization of pennycresses (*Thlaspi* spp.) of the Brassicaceae by arbuscular mycorrhizal fungi. J. Plant Physiol. 160(6), 615–626.

Rengel, Z., 1993. Mechanistic simulation models of nutrient uptake: a review. Plant Soil 152, 161–173.

Robinson, B., Fernandez, J.E., Madejon, P., Maranon, T., Murillo, J.M., Green, S., Clothier, B., 2003. Phytoextraction: an assessment of biogeochemical and economic viability. Plant Soil 249, 117–125.

Sadana, U.S., Claassen, N., 2000. Manganese dynamics in the rhizosphere and Mn uptake by different crops evaluated by a mechanistic model. Plant Soil 218, 233–238.

Salt, D.E., Kato, N., Krämer, U., Smith, R.D., Raskin, I., 2000. The Role of Root Exudates in Nickel Hyperaccumulation and Tolerance in Accumulator and Nonaccumulator Species of *Thlaspi*. In: Terry, N., Banuelos, G. (Eds.), Phytoremediation of Contaminated Soil and Water. CRC Press LLC, Boca Raton, FL, pp. 189–200.

Saltelli A., Chan, K., Scott, E.M., 2000. Sensitivity Analysis. Wiley, Chichester.

Selim, H.M., Davidson, J.H., Mansell, R.S., 1976. Evaluation of a two site adsorption desorption model for describing solute transport in soils. Proceedings of the Summer Computer Simulation Conference, Washington, DC, pp. 444–448.

Scheckel, K.G., Sparks, D.L., 2001. Dissolution kinetics of nickel surface precipitates on clay mineral and Oxide Surfaces. Soil Sci. Soc. Am. J. 65, 685–694.

Schnepf, A., Schrefl, T., Wenzel, W.W., 2002. The suitability of pde-solvers in rhizosphere modeling, exemplified by three mechanistic rhizosphere models. J. Plant Nutr. Soil Sci. 165, 713–718.

Syring K.N., Claassen, N., 1996. Model of nutrient uptake © NST 3.0. http://www.gwdg.de/~uaac/.

Tinker, P.B., Nye, P.H., 2000. Solute Movement in the Rhizosphere. Oxford University Press, Oxford.

Uren, N.C., 2001. Types, amounts, and possible functions of compounds released into the rhizosphere by soil-grown plants. In: Pinton, R., Varanini, Z., Nannipieri, P. (Eds.), The Rhizosphere. Marcel Dekker Inc., New York, p. 19–40.

Wenzel, W.W., Lombi, E., Adriano, D.C., 1999. Biogeochemical processes in the rhizosphere: role in phytoremediation of metal-polluted soils. In: Prasad, N.M.V., Hagemeyer, J. (Eds.), Heavy Metal Stress in Plants — from Molecules to Ecosystems. Springer, Heidelberg, pp. 273–303.

Wenzel, W.W., Wieshammer., G., Fitz, W.J., Puschenreiter, M., 2001. Novel rhizobox design to assess rhizosphere characteristics at high spatial resolution. Plant Soil 237, 37–45.

Whiting, S.N., Broadley, M.R., White, P.J., 2003. Applying a solute transfer model to phytoextraction: zinc acquisition by *Thlaspi caerulescens*. Plant Soil 249, 45–56.

Whiting, S.N., de Souza, M.P., Terry, N., 2001. Rhizosphere bacteria mobilize Zn for hyperaccumulation by *Thlaspi caerulescens*. Environ. Sci. Technol. 35, 3144–3150.

Willigen, P., de Van Noordwijk, M., 1994a. Mass flow and diffusion of nutrients to a root with constant or zero-sink uptake I. Zero-sink uptake. Soil Sci. 157, 171–175

Willigen, P., de Van Noordwijk, M., 1994b. Mass flow and diffusion of nutrients to a root with constant or zero-sink uptake II. Constant uptake. Soil Sci. 157, 162–170.

Zeien, H., Brümmer, G.W., 1989. Chemische Extraktionen zur Bestimmung von Schwermetallbindungsformen in Boden. Mitteilgn. Deutsch. Bodenkundl. Gesellsch. 59, 505–510.

Zhao, F.J., Hamon, R.E., McLaughlin, M.J., 2001. Root exudates of the hyperaccumulator *Thlaspi caerulescens* do not enhance metal mobilization. New Phytol. 151, 613–620.

Biogeochemistry of Trace Elements in the Rhizosphere
P.M. Huang and G.R. Gobran (Editors)

Chapter 14

Effect of arbuscular mycorrhizal (AM) fungi on heavy metal and radionuclide transfer to plants

C. Leyval

Laboratoire des Interactions Microorganismes-Minéraux-Matière Organique dans les sols, UMR 7137 CNRS-UHP, Université Henri Poincaré - Faculté des Sciences, BP239, 54506, Vandoeuvre-les-Nancy Cedex, France
E-mail: corinne.leyval@limos.uhp-nancy.fr

ABSTRACT

Heavy metal and radionuclide concentrations in soils increase due to man-made pollution. One of the first entry points of such elements into plant ecosystems is the rhizosphere, defined as the soil under the biological, physical and chemical influence of roots. Arbuscular mycorrhizal (AM) fungi, symbiotic microorganisms associated with the roots of many plant species, provide a direct link between soil and roots and affect metal transfer to plants. The present chapter includes recent laboratory work and some research aspects still to be adressed on the contribution of AM fungi to plant metal uptake. The necessity to develop new and adapted approaches, such as compartment devices and root-organ cultures, to separate AM to root contribution to metal uptake is emphasized. Available data may be difficult to compare because they were obtained under different experimental conditions. However, they suggest that the transfer of heavy metals from AM fungi to plants may be metal specific. Further research should focus on the mechanisms involved in reduced or improved uptake of metals by mycorrhizal plants, on AM tolerance to metals and radionuclides and on AM functional diversity in polluted soils. AM contribution to metal uptake should also be quantified to include data in models of plant uptake.

1. INTRODUCTION

Heavy metal and radionuclide concentrations in soils increase owing to man-made pollution related to industrial, agricultural or urban activities. Such concentrations can reach toxic levels and create major environmental and health problems. One of the first entry points of metals into plant ecosystems is the rhizosphere, defined as the soil under the biological, physical and chemical influence of roots. In the rhizosphere, the plant releases root exudates that soil microorganisms feed on, and take up water and nutrients. Among the rhizosphere microorganisms, arbuscular mycorrhizal (AM) fungi, ubiquitous root symbiotic fungi, can be present (Harley and Smith, 1983). Most terrestrial plants form arbuscular mycorrhizas under natural conditions, with the exception of a limited number of plant families, such as Chenopodiaceae, Caryophyllaceae, Brassicaceae, and Urticacae. However, with the combination of the classical trypan blue root staining and the newly developed molecular techniques, which allow the unequivocal detection and identification of AM fungi in roots, it has been shown that plant species of families such as Brassicaceae may be mycorrhizal.

AM fungi contribute to plant growth and plant nutrition as their extraradical hyphae increase the volume of the explored soil. These extraradical hyphae can extend up to more than 10 cm from the roots (Harley and Smith, 1983) and make up a hyphal density of 1–30 m g^{-1} soil (Smith and Read, 1997). Therefore, AM roots can more easily access soil nutrients localized outside the root depletion zone, especially the nonmobile elements such as P and Zn. Mycorrhizas also improve plant water uptake, stress tolerance, and affect the microbial community structure (Joner *et al.*, 2001). In polluted soils, mycorrhizal plants also benefit from the presence of the AM fungi.

Heavy metal tolerance of AM fungi and their potential use in phytoremediation treatments have been recently discussed (Leyval *et al.*, 1997, 2002). The contribution of AM fungi to the transfer of heavy metals from soil to plants and to their translocation from roots to shoots has been addressed in many laboratory studies. The bioavailability of heavy metals in the mycorrhizosphere was discussed in a review chapter by Leyval and Joner (2001). The present chapter includes recent laboratory work and some research aspects that still need to be addressed in order to fill in gaps in current knowledge of the contribution of AM fungi to plant metal uptake. The discussion focuses on metals which generally occur in trace amounts in environmental and ecological materials, such as Cd, Zn, Ni, as well as the so-called metalloids (Se) and radionuclides (Cs and U), all of them being potentially toxic in soil–plant systems. The main conclusions show that AM fungi affect the transfer of metals and radionuclides to plants and may protect them against heavy metal toxicity. Direct (transfer, binding of pollutants) and indirect effects (nutritional effects, selection of microbial communities) of AM fungi can affect the fate of pollutants in soil–plant systems.

2. DIFFERENT APPROACHES TO STUDY THE CONTRIBUTION OF AM FUNGI ON THE TRANSFER OF HEAVY METALS AND RADIONUCLIDES TO PLANTS

One difficulty related to studies on the contribution of AM fungi to the transfer of elements to plants is their obligate symbiotic growth. Since arbuscular mycorrhizas act as an extension of the root, the quantification of the soil-to-plant transport of the elements via the hyphae is difficult to estimate in pot experiments, where roots and hyphae have access to the elements. Another difficulty when studying metal uptake by mycorrhizal and nonmycorrhizal plants is the larger size of the mycorrhizal plants themselves, which affects their nutrient uptake rate. Differences in metal uptake by both types of plants could then be due to a dilution effect caused by different plant growth rates. Nonmycorrhizal plants with higher phosphorus concentrations have been used to obtain nonmycorrhizal and mycorrhizal plants of comparable size. However, although they may become comparable in size, they may be different in terms of root exudation, and this may affect metal speciation and availability. Unlike ectomycorrhizal fungi, which produce organic acids, have been shown to promote mineral weathering (Leyval *et al.*, 1990; Leyval and Berthelin, 1991) and impact the available pool of elements, such as P, in soils, AM plants seem to access the same available pool of nutrients as nonmycorrhizal plants. However, inoculation with AM fungi modifies the plant root exudation (Laheurte *et al.*, 1990) and could, therefore, affect the bioavailability of trace elements in the rhizosphere.

New techniques, such as compartment devices and transformed root systems, have been very useful in separating the effects of AM fungi from their root systems and in understanding the mechanisms involved. Compartment devices are composed of a main compartment containing the plant (RC, root compartment) and lateral compartments separated from the main compartment by a mesh (30 or 700 μm) allowing only hyphae (HC, hyphal compartment) or root and hyphae (RHC: root and hyphae compartment) to go through. Such compartment devices were initially designed to show the transfer of phosphorus to plants by AM hyphae (Jakobsen *et al.*, 1992). They have been adapted to study the transfer of trace elements, such as Cd (Joner and Leyval, 1997), using radiolabeled Cd added in the lateral compartments. It was then shown that AM fungal hyphae are able to transfer Cd to clover roots, but that the translocation from root to shoot was reduced, suggesting that the AM fungus was acting as a biological barrier to Cd toxicity (Joner and Leyval, 1997). Harvesting extraradical hyphae in compartment devices has also been used to study their surface-binding properties.

Compartment devices have also been used with root–organ cultures to study the contribution of AM hyphae to Cs and U uptake. In one instance, AM fungi and *Agrobacterium rhizogenes* (Ri-T-DNA)-transformed carrot roots (Declerck *et al.*, 1998) were cultivated in compartmentalized Petri plates, in liquid or gelified media, where hyphal and root uptake were compared. Such devices are very useful to study the potential uptake of elements by AM hyphae in the absence of any other microorganisms, and are very convenient to use. However, it has not been shown whether transformed roots have comparable absorption capacity when compared to normal roots and to whole plants.

Although compartment devices, including root–organ culture, are very useful for studying the relative contribution of AM and roots to element uptake, the results obtained in such conditions cannot be extrapolated to *in situ* conditions.

Often, the availability of the pollutants is not a limiting parameter under such conditions. Therefore, complementary approaches using pot and column experiments and field experiments, where the availability of the trace elements is an important parameter, should be performed to quantify the potential beneficial effect of AM presence or inoculation in polluted soils, and to provide useful data to include in plant uptake models.

3. METAL-SPECIFIC TRANSFER OF HEAVY METALS AND RADIONUCLIDES FROM SOIL TO PLANTS BY AM FUNGI

The studies on metal transfer to plants by AM fungi include different metallic trace elements, such as Cd, Zn and Ni, and radionuclides, such as Cs, Se or U. Previous studies (Leyval *et al.*, 1997) have shown that AM fungi affect the transfer of metals and radionuclides to plants and may protect them against heavy metal toxicity. However, the results cannot be generalized, and differ depending on plant species, heavy metals and their availability, and possibly other microbial components of the soils, although this last aspect has not been really investigated. AM fungi affect metal accumulation in several ways (see review by Leyval and Joner, 2001), by reducing the uptake of metals like Cd, Cu and Zn, or have no specific effects at all, depending on the types of fungi and plant species present under experimental conditions. It has often been discussed that without precise and standardized information on the availability of metal concentrations in the experiments, it is difficult to compare the data.

In compartment devices, where the heavy metals are available to the AM hyphae but not to the roots, it has been clearly shown that fungi can transport some heavy metals. Transport of Zn, Cu (Li *et al.*, 1991), and Cd (Joner and Leyval, 1997) has been demonstrated, whereas Ni (Guo *et al.*, 1996) was not transported by AM hyphae. Using root–organ cultures and labeling with ^{233}U, Rufyikiri *et al.* (2003) showed that U concentrations were 5–10 times higher in AM hyphae than in mycorrhizal and nonmycorrhizal roots, respectively. The extraradical hyphae had a relatively higher capacity for U uptake than the carrot roots, while the intraradical hyphae seemed to contribute to the immobilization of U in mycorrhizal roots. With Cs, no significant transfer through extraradical hyphae was observed with AM fungi in experiments in compartment devices performed in three different laboratories, while this took place with an ectomycorrhizal fungus (Joner *et al.*, 2004). Under root–organ culture conditions (Declerck *et al.*, 2003), *Glomus lamellosum* took up and translocated small amounts of radiocesium to the root compartment and within roots.

For elements such as Se, whose mobility, bioavailability, and toxicity in soil–plant systems are affected by oxydo–reduction processes controlling its speciation, there is very little data on the contribution of rhizosphere microorganisms to plant uptake. AM colonization of ryegrass roots by *Glomus mosseae* reduced root and shoot concentration of Se in a soil spiked with 10 mg kg^{-1} Se (unpublished data). For Cr, whose toxicity also depends on redox conditions (i.e. trivalent Cr (Cr(III)) is less toxic than Cr(VI) (as Cr_2O^{7-})), AM inoculation of sunflower with *G. intraradices* increased the ability of the plant to tolerate and hyperaccumulate Cr (Davies *et al.*, 2001). For As, there is, to our knowledge, even less data on the potential contribution of AM fungi to plant uptake. For the ericoid mycorrhizal fungus *Hymenoscyphus ericae,* it was shown that mycorrhizal *Calluna vulgaris* accumulated 100% more As than uninoculated plants (Sharples *et al.*, 2000). The authors suggested that the fungus acted as a filter preventing high As concentration in plant tissue. However, the results were obtained in hydroponic culture. Although all these data were obtained in different experimental conditions, with different host–fungus combinations, they may suggest that the transfer of heavy metals from AM fungi to plants may be metal specific.

4. MECHANISMS INVOLVED IN HEAVY METAL AND RADIONUCLIDE UPTAKE BY AM FUNGI AND MYCORRHIZAL PLANTS

Direct (transfer, binding of pollutants) and indirect effects (nutritional effects) of AM fungi can affect the fate of heavy metals and radionuclides in soil–plant systems. The precise mechanisms involved, which can be quite complex, are not yet clearly elucidated and need further investigation. Indeed, Joner and Leyval (2001) showed that sampling time, plant age, pot size (root density) and fungal species can affect the results. In the same experiment, with the same plant, soil (0.9 kg pots) and fungi, when plants (maize) were harvested after 6 weeks, mycorrhizal roots contained significantly less Cd than nonmycorrhizal ones, but after 9 weeks, the opposite trend was observed. When using different pot sizes, the results differed, and after 9 weeks in 2.5 kg instead of 0.9 kg pots, Cd concentrations were lower in mycorrhizal than nonmycorrhizal roots. Such results suggest that all parameters affecting root growth, root density, nutrient uptake, and root exudation can also affect AM contribution to metal uptake. For elements like Se, As and Cr, which are subject to oxydo–reduction processes possibly mediated in soil by microbial activity, changes in microbial community structure in the rhizosphere due to mycorrhizal colonization of roots (Marschner and Baumann, 2003) could be involved in AM effects on plant uptake.

Internal sequestration by extraradical hyphae (Berreck and Haselwandter, 2001), or adsorption onto negatively charged constituents of fungal mycelium (Joner *et al.*, 2000), could explain the reduced transfer of heavy metals or radionuclides to plants. Interactions including competition for uptake between analog elements, such as Cs–K or Zn–Cd, could also affect the results. For elements like U, high P concentrations and weakly acidic to neutral pH values in plants or fungal cells can promote the formation of U–phosphate complexes and precipitates, thereby lowering their accumulation in the mycorrhizal roots

(Rufyikiri *et al.*, 2003). Different uptake rates by roots and by AM hyphae could also explain reduced or increased uptake in mycorrhizal plants. For U, the translocation rate was shown to be higher in hyphae than in roots (Rufyikiri *et al.*, 2003). The ability of fungi to accumulate radiocesium at higher concentrations than green plants has been attributed to high influx and low efflux rates (Steiner *et al.*, 2002). However, such accumulation largely depends on the fungi since Cs transport by AM fungal hyphae was not always significant (Joner *et al.*, 2004). Adsorption experiments, potentiometric titration and surface functional group characterization showed that AM fungi have a higher binding capacity for metals like Cd, especially the isolates from contaminated soils, when compared with ectomycorrhizal or other fungi (Joner *et al.*, 2000). Cs could be translocated by AM hyphae via a motile and interconnected vacuole, or via cytoplasmic streaming (Declerck *et al.*, 2003). High cellular Zn levels could downregulate the expression of the activity of P transporter genes, since Zn deficiency upregulates P transporters, and disturb P uptake in sensitive plants, whereas mycorrhizal plants would be less affected (Adriaensen *et al.*, 2003). The same authors reported that Zn ions could affect membrane stability, the activity of membrane-bound ion pumps, proton channels, and therefore, nutrient acquisition. Molecular events involved in Cd tolerance in AM fungi and mycorrhizal plants, and in translocation via AM hyphae to plants, have rarely been considered. Using a targeted proteomic approach to investigate the modifications of cadmium-induced protein expression in pea roots colonized by *G. mosseae*, it was shown that a vacuolar H^+–ATPase synthase was induced by Cd and AM (Repetto *et al.*, 2003).

5. AM TOLERANCE TO METALS AND TRANSFER TO PLANTS

There is still little information about AM fungal ecotypes adapted to high heavy metal concentrations (Gildon and Tinker, 1981; Weissenhorn *et al.*, 1993; del Val *et al.*, 1999). Hildebrandt *et al.* (1999) isolated *Glomus* br1 from the rhizosphere and the root of a metal-tolerant plant, *Viola calaminaria*, and showed that the fungus allowed plants like maize, alfalfa, barley and others, to grow in metal-polluted soils. Similarly, Tonin *et al.* (2001) used a mixed AM fungi inoculum from the rhizosphere of the same plant to inoculate a nontolerant plant, clover, and showed that Cd and Zn increased in roots in comparison with uninoculated plants, but that their concentrations in shoots and plant biomass were not affected.

Since AM fungi cannot be cultivated without a host plant, it is difficult to screen a large number of fungal isolates from polluted and nonpolluted soils for their ability to tolerate elevated metal or radionuclide concentrations. The question whether these fungi keep their metal tolerance when they are subcultivated in the absence of the pollutant remains almost unanswered. Malcova *et al.* (2003) studied the Mn tolerance of *Glomus* sp. BEG140, isolated from a Mn-contaminated soil after a 2-year culture period either in metal-free substrate or in the original soil

(Malcova *et al.*, 2003). The fungus cultured in metal-free substrate was less tolerant to high Mn concentration (1000 μM) than the fungus kept on the Mn-contaminated soil. Plants inoculated with the fungus kept on the original soil had lower Mn concentrations in shoots than those inoculated with the fungus kept in metal-free substrate. However, nonmycorrhizal plants had lower Mn concentrations in shoots than mycorrhizal plants, and plant growth was not affected by the presence of Mn, suggesting that the tested concentration was not very toxic. The same research group (Rydlova and Vosatka, 2003) also investigated the Pb tolerance of *G. intraradices* subcultured in Pb-free substrate or in the original Pb-contaminated soil. Under such conditions, no difference in mycorrhizal colonization between the lineages was observed, but the plants inoculated with the lineage kept on the original polluted soil showed a larger shoot biomass and a lower shoot Pb concentration than plants inoculated with the lineage kept on Pb-free substrate, even though the Pb content of nonmycorrhizal plants was even lower. Such results suggest that metal tolerance for AM fungi estimated by root colonization ability may not reflect their potential to protect the plant against excessive metal accumulation. In fact, it is well known that root colonization by AM is not correlated with mycorrhizal activity. A *G. mosseae* isolated from a Cd-contaminated soil in Hungary was able to stimulate greater clover growth than an isolate of *G. mosseae* obtained from an AM fungi collection. In addition, *Brevibacillus* sp. isolated from the same polluted soil improved the beneficial effect of the Cd-tolerant *G. mosseae* on clover growth (Vivas *et al.*, 2003). Such results, supported by the demonstration that AM fungi change the microbial community structure in the rhizosphere (Joner *et al.*, 2001; Mansfeld-Giese *et al.*, 2002; Marschner and Baumann, 2003), suggest that other soil microbial components could interfere with AM fungi in polluted soils and could contribute to their effect on plant growth, metal mobility and plant uptake. The potential beneficial effect of tolerant AM fungi on the nutritional status of mycorrhizal plants has never been really discussed. On the contrary, with ectomycorrhizal fungi, a Zn-tolerant *Suillus bovinus* improved phosphorus and ammonium uptake rates of pines grown in the presence of high concentrations of Zn. A nontolerant one could not sustain the pines' acquisition of nutrients, but plant growth was not affected (Adriaensen *et al.*, 2003).

6. DIVERSITY OF AM FUNGI IN HEAVY METAL POLLUTED SOILS

Interspecific variation, which may be considerable with respect to P transport, is still poorly known in terms of AM transport of metals. Single AM fungal species are commonly used in most laboratory experiments, but this may not reflect the real situation *in situ*. The development of molecular techniques (Redecker *et al.*, 2003) during the last 10 years has opened new perspectives on the diversity of AM fungi in polluted sites, the maintenance of introduced AM fungi *in situ* and the genetic basis of heavy metal tolerance.

The presence and diversity of AM fungi associated with plant roots on heavy metal polluted sites has been intensively investigated during the recent years. Single or nested PCR (Sanders *et al.*, 1996; Van Tuinen *et al.*, 1998), combined PCR and restriction digest (Renker *et al.*, 2003) or specific primers (Helgason *et al.*, 1999), have been used to amplify AM fungi in roots. However, most of these techniques are not strictly specific, and the low quantity of AM fungi DNA in roots, as well as the presence of fungi other than AM, are still hampering the determination of AM fungi diversity in roots.

More precise observations of roots collected in polluted sites showed that members of Brassicaceae (considered for a long time to be nonmycorrhizal), such as *Biscutella laevigata* (Orlowska *et al.*, 2002) or *Cardaminopsis halleri* (Hildebrandt *et al.*, 1999), contained AM structures. These authors isolated and identified a *Glomus* sp. in the roots of the metallophyte *V. calaminaria*, using ITS4 and ITS5 primers (White *et al.*, 1990). As mentioned in Section 5, this same *Glomus* sp. increased the growth of non-metal-tolerant plants (Hildebrandt *et al.*, 1999). Tonin *et al.* (2001) used PCR amplification of 18S rDNA (Vandenkoornhuyse and Leyval, 1998) and showed that in the rhizosphere of *V. calaminaria*, a limited number of AM species (at least four *Glomus* spp.) were also found. When clover was inoculated with a mixture of spores of these fungi, Zn concentrations in roots increased in comparison with uninoculated ones, but plant growth was not affected. Although the results from two different groups working on the same plant were different, the findings suggest that AM fungi from polluted areas may be good potential candidates for phytoremediation applications. The presence of AM fungi in the Ni hyperaccumulating plant *Berkheya coddii*, and the ability of these fungi to improve *B. coddii* growth and Ni content (Turnau and Mesjasz-Przybylowicz, 2003), are also very promising for the development of mycorrhiza-assisted phytoremediation.

However, information on the diversity of AM fungi in heavy metal and radionuclide-polluted soils is still very scarce and needs to be further investigated. Furthermore, the *in situ* functional diversity of these fungi, including their respective role in transfer of heavy metals to host plants, is not known. In the roots of *Arrhenatherum elatius* growing on a Zn-contaminated site, the diversity of AM fungi was relatively low despite a large and unexpected fungal diversity (Vandenkoornhuyse *et al.*, 2002). The potential role of other root inhabitants has never been considered in heavy metal transfer to plants, either via direct or indirect possible competition with AM fungi.

7. CONCLUSIONS AND FUTURE RESEARCH

Arbuscular mycorrhizas are key components of the soil–root interface, where heavy metals and radionuclides are taken up by plants. It is possible to predict under what conditions a beneficial effect of AM fungi can be expected, but this requires a better understanding of the mechanisms involved, taking into account not only AM fungi, but the complexity of the microorganism–soil–plant systems. Further research should focus on the following aspects:

- The mechanisms involved in reduced or improved uptake of metals by mycorrhizal plants: this includes studies on the molecular mechanisms involved in metal tolerance and translocation in mycorrhizal plants (proteomic, genomic approaches), as well as surface physicochemical processes of AM fungi and AM roots in the presence of metallic pollutants via the characterization of their surface properties.
- Larger screening of plant–AM fungi associations to improve plant survival and reduce (or improve) metal uptake.
- The quantification of AM contribution to metal uptake to include data in models of plant uptake.
- Providing more information on the diversity of AM fungi in heavy metal and radionuclide-polluted soils and their tolerance to metals.
- Performance of demonstration experiments *in situ*.
- Contribution of AM fungi in polluted soils containing multiple pollutants including metallic and organic pollutants.

REFERENCES

Adriaensen, K., van der Lelie, D., Van Laere, A., Vangronsveld, J., Colpaert, J.V., 2003. A zinc-adapted fungus protect pines from zinc stress. New Phytol. 161, 549–555.

Berreck, M., Haselwandter, K., 2001. Effect of the arbuscular mycorrhizal symbiosis upon uptake of cesium and other cations by plants. Mycorrhiza 10, 275–280.

Davies, F.T.J., Puryear, J.D., Newton, R.J., Egilla, J.N., Saraiva Grossi, J.A., 2001. Mycorrhizal fungi enhance accumulation and tolerance of chromium in sunflower (*Hemianthus annuus*). J. Plant Physiol. 158, 777–786.

Declerck, S., Dupré de Boulois, H., Bivort, C., Delvaux, B., 2003. Extraradical mycelium of the arbuscular mycorrhizal fungus *Glomus lamellosum* can take up, accumulate and translocate radiocesium under root–organ culture conditions. Environ. Microbiol. 5, 510–516.

Declerck, S., Strullu, D.G., Plenchette, C., 1998. Monoxenic culture of the intraradical forms of *Glomus* sp. isolated from a tropical ecosystem: a proposed methodology for germplasm collection. Mycologia 90, 579–585.

del Val, C., Barea, J.M., Azcon Aguilar, C., 1999. Assessing the tolerance to heavy metals of arbuscular mycorrhizal fungi isolated from sewage sludge-contaminated soils. Appl. Soil Ecol. 11, 261–269.

Gildon, A., Tinker, P.B., 1981. A heavy metal-tolerant strain of a mycorrhizal fungus. Trans. Br. Mycol. Soc. 77, 648–649.

Guo, Y., George, E., Marschner, H., 1996. Contribution of an arbuscular mycorrhizal fungus to the uptake of cadmium and nickel in bean and maize plants. Plant Soil 184, 195–205.

Harley, J.L., Smith, S.E., 1983. Mycorrhizal Symbiosis, Academic Press, London, p. 483.

Helgason, T., Fitter, A., Young, J.P.W., 1999. Molecular diversity of arbuscular mycorrhizal fungi colonising *Hyacinthoides non-scripta* (bluebell) in a seminatural woodland. Mol. Ecol. 8, 659–666.

Hildebrandt, U., Kaldorf, M., Bothe, H., 1999. The zinc violet and its colonization by arbuscular mycorrhizal fungi. J. Plant Physiol. 154, 709–717.

Jakobsen, I., Abbott, L.K., Robson, A.D., 1992. External hyphae of vesicular–arbuscular mycorrhizal fungi associated with *Trifolium subterraneum* L. 1. Spread of hyphae and phosphorus inflow into roots. New Phytol. 120, 371–380.

Joner, E.J., Briones, R., Leyval, C., 2000. Metal binding capacity of arbuscular mycorrhizal mycelium. Plant Soil 226, 227–234.

Joner, E.J., Johansen, A., Loibner, A.P., De La Cruz, M., Szolar, O., Portal, J.M., Leyval, C., 2001. Rhizosphere effects on microbial community structure and dissipation and toxicity of polycyclic aromatic hydrocarbons (PAHs) in spiked soil. Environ. Sci. Technol. 35, 2773–2777.

Joner, E.J., Leyval, C., 1997. Uptake of ^{109}Cd by roots and hyphae of a *Glomus mosseae Trifolium subterraneum* mycorrhiza from soil amended with high and low concentrations of cadmium. New Phytol. 135, 353–360.

Joner, E.J., Leyval, C., 2001. Time-course of heavy metal uptake in maize and clover as affected by different mycorrhiza inoculation regimes. Biol. Fertil. Soils 33, 351–357.

Joner, E.J., Roos, P., Jansa, J., Frossard, E., Leyval, C., Jakobsen, I., 2004. No significant contribution of arbuscular mycorrhizal fungi to transfer of radiocesium from soil to plant. Appl. Environ. Microbiol. 70, 6512–6517.

Laheurte, F., Leyval, C., Berthelin, J., 1990. Root exudates of maize, pine, beech seedlings influenced by mycorrhizal and bacterial inoculation. Symbiosis 9, 111–116.

Leyval, C., Berthelin, J., 1991. Weathering of mica by roots and rhizospheric microorganisms of pine. Soil. Sci. Soc. Am. J. 55, 1009–1016.

Leyval, C., Joner, E.J., 2001. Bioavailability of heavy metals in the mycorrhizosphere. In: Gobran, G.R., Wenzel, W.W., Lombi, E., (Eds.), Trace Elements in the Rhizosphere, CRC Press, Boca Raton, FL, pp. 165–185.

Leyval, C., Joner, E.J., del Val, C., Haselwandter, K., 2002. Potential of arbuscular mycorrhizal fungi for bioremediation. In: Gianinazzi, S., Schüepp, H., Barea, J.M., Haselwandter, K., (Eds.), Mycorrhiza Technology in Agriculture: From Genes to Bioproducts, Birkhäuser Verlag, Basel, pp. 175–186.

Leyval, C., Laheurte, F., Belgy, G., Berthelin, J., 1990. Weathering of micas in the rhizospheres of maize, pine and beech seedlings influenced by mycorrhizal and bacterial inoculation. Symbiosis 9, 105–109.

Leyval, C., Turnau, K., Haselwandter, K., 1997. Effect of heavy metal pollution on mycorrhizal colonization and function: physiological, ecological and applied aspects. Mycorrhiza 7, 139–153.

Li, X.L., Marschner, H., George, E., 1991. Acquisition of phosphorus and copper by VA-mycorrhizal hyphae and root-to-shoot transport in white clover. Plant Soil 136, 49–57.

Malcova, R., Rydlova, J., Vosatka, M., 2003. Metal-free cultivation of *Glomus* sp. BEG 140 isolated from Mn-contaminated soil reduces tolerance to Mn. Mycorrhiza 13, 151–157.

Mansfeld-Giese, K., Larsen, J., Bodker, L., 2002. Bacterial populations associated with mycelium of the arbuscular mycorrhizal fungus *Glomus intraradices*. FEMS Microbiol. Ecol. 41, 133–140.

Marschner, P., Baumann, K., 2003. Changes in bacterial community structure induced by mycorrhizal colonisation in split-root maize. Plant Soil 251, 279–289.

Orlowska, E., Zubek, S., Jurkiewicz, A., Szarek-Jukaszewska, G., Turnau, K., 2002. Influence of restoration on arbuscular mycorrhiza of *Biscutella laevigata* L. (Brassicaceae) and *Plantago lanceolata* L. (Plantaginaceae) from calamine spoil mounds. Mycorrhiza 12, 153–159.

Redecker, D., Hijri, I., Wiemken, A., 2003. Molecular identification of arbuscular mycorrhizal fungi in roots: perspectives and problems. Folia Geobotanica 38, 113–124.

Renker, C., Heinrichs, J., Kaldorf, M., Buscot, F., 2003. Combining nested PCR and restriction digest of the internal transcribed spacer region to characterize arbuscular mycorrhizal fungi on roots from the field. Mycorrhiza 13, 191–198.

Repetto, O., Bestel-Corre, G., Dumas-Gaudot, E., Berta, G., Gianinazzi-Pearson, V., Gianinazzi, S., 2003. Targeted proteomics to identify cadmium-induced protein modifications in *Glomus mosseae*-inoculated pea roots. New Phytol. 157, 555–567.

Rufyikiri, G., Thiry, Y., Declerck, S., 2003. Contribution of hyphae and roots to uranium uptake and translocation by arbuscular mycorrhizal carrot roots under root–organ culture conditions. New Phytol. 158, 391–399.

Rydlova, J., Vosatka, M., 2003. Effect of *Glomus intraradices* isolated from Pb-contaminated soil on Pb uptake by *Agrostis capillaris* is changed by its cultivation in a metal-free substrate. Folia Geobotanica 38, 155–165.

Sanders, I., Clapp, J.P., Wiemken, A., 1996. The genetic diversity of arbuscular mycorrhizal fungi in natural ecosystems - a key to understanding the ecology and functioning of the mycorrhizal symbiosis. New Phytol. 133, 123–134.

Sharples, J.M., Meharg, A.A., Chambers, S.M., Cairney, J.W.G., 2000. Mechanism of arsenate resistance in the ericoid mycorrhizal fungus *Hymenoscyphus ericae*. Plant Physiol. 124, 1327–1334.

Smith, S., Read, D.J., 1997. Mycorrhizal Symbiosis, second ed. Academic Press, San Diego.

Steiner, M., Linkov, I., Yoshida, S., 2002. The role of fungi in the transfer and cycling of radionuclides in forest ecosystems. J. Environ. Radioact. 58, 217–241.

Tonin, C., Vandenkoornhuyse, P., Joner, E.J., Straczek, J., Leyval, C., 2001. Assessment of arbuscular mycorrhizal fungi diversity in the rhizosphere of *Viola calaminaria* and effect of these fungi on heavy metal uptake by clover. Mycorrhiza 10, 161–168.

Turnau, K., Mesjasz-Przybylowicz, J., 2003. Arbuscular mycorrhiza of *Berkheya coddii* and other Ni-hyperaccumulating members of Asteraceae from ultramafic soils in South Africa. Mycorrhiza 13, 185–190.

Van Tuinen, D., Jacquot, E., Zhao, B., Gollote, A., Gianinazzi-Pearson, V., 1998. Characterisation of root colonization profiles by a microcosm community of arbuscular mycorrhizal fungi using 25S-rDNA targeted nested PCR. Mol. Ecol. 7, 879–887.

Vandenkoornhuyse, P., Baldauf, S.L., Leyval, C., Straczek, J., Young, P.W., 2002. Extensive fungal diversity in plant roots. Science 285, 2051.

Vandenkoornhuyse, P., Leyval, C., 1998. SSU rDNA sequencing and PCR-fingerprinting reveal genetic variation within *Glomus mosseae*. Mycologia 90, 791–797.

Vivas, A., Vörös, A., Biro, B., Barea, J.M., Ruiz-Lozano, J., Azcon, R., 2003. Beneficial effects of indigenous Cd-tolerant and Cd-sensitive *Glomus mosseae* associated with a Cd-adapted strain of *Brevibacillus* sp. in improving plant tolerance to Cd contamination. Appl. Soil Ecol. 24, 177–186.

Weissenhorn, I., Leyval, C., Berthelin, J., 1993. Cd-tolerant arbuscular mycorrhizal (AM) fungi from heavy-metal polluted soils. Plant Soil 157, 247–256.

White, T.J., Bruns, S.L., Taylor, J., 1990. Amplification and direct sequencing of fungal ribosomal RNA genes for phylogenetics. In: Innes, M.A., Gelfrand, D.H., Sninsky, J.J., White, T.J., (Eds.), PCR-Protocols: A Guide to Methods and Applications, Academic Press, New York, pp. 315–322.

Biogeochemistry of Trace Elements in the Rhizosphere
P.M. Huang and G.R. Gobran (Editors)

Chapter 15

Uptake and translocation of uranium by arbuscular mycorrhizal fungi under monoxenic culture conditions

G. Rufyikiri[a], Y. Thiry[a], and S. Declerck[b]

[a]SCK•CEN (Studiecentrum voor Kernenergie – Centre d'étude de l'Energie Nucléaire, Foundation of Public Utility), Radiation Protection Research Department, Radioecology Section, Boeretang 200, 2400 Mol, Belgium
E-mail: grufyikiri@yahoo.com.fr

[b]Université catholique de Louvain, Mycothèque de l'Université catholique de Louvain (MUCL, Part of the Belgian Co-ordinated Collections of Micro-organisms (BCCM)), Unité de microbiologie, Place Croix du Sud 3, 1348 Louvain-la-Neuve, Belgium

ABSTRACT

Fundamental information on the transport of uranium by mycelium of arbuscular mycorrhizal (AM) fungi is reported and discussed in this chapter. The monoxenic cultivation system was shown to be convenient to investigate the processes of uptake and translocation of U by AM fungi, as it was already reported for some other elements. The evidence that extraradical AM hyphae can take up U and translocate it towards the host roots was clearly shown. Further research to understand the mechanisms involved and future prospects were also suggested.

1. INTRODUCTION

1.1. Uranium in the environment

The discovery of radioactivity was at the origin of a great development of nuclear technologies during the 20th century with numerous scientific and industrial applications. Today, public and scientific concerns for the protection of the environment have led to fundamental questions related to nuclear activities, nuclear waste disposal, U-mining relics and the risk of dissemination of natural and anthropogenic radionuclides into the biosphere. The pathways that are responsible for biota and human exposure depend on geochemical and biochemical processes. A better understanding of the biogeochemical cycle of radionuclides is necessary to identify the effective sinks and reservoirs in the environment and the source and extent of exposure. There are over 1000 naturally occurring and man-made radionuclides found in the environment in concentrations that range from 10^{-9} for ^{211}Po to 13 Bq L^{-1} for ^{40}K in ocean water up to activity levels of 440 Bq kg^{-1} for ^{40}K in the uppermost 100 m of the lithosphere (Santschi and Honeyman, 1989). Uranium is the most abundant of the naturally occurring actinides, the others being Ac, Th and Pa. There is no known biological function for U, but it is toxic to cells, even at low concentrations. The toxicity of U mainly derives from its chemical properties, since the specific radioactivity of U is generally too low to induce significant radiological effects (Meinrath *et al.*, 2003). Despite its high atomic number, U is by no means a rare element. Crustal abundance ranges from 1 to 4 μg g^{-1} in sedimentary rocks to tens or even hundreds of μg g^{-1} in phosphate-rich deposits (Langmuir, 1997; Krea and Khalaf, 2000; Khater *et al.*, 2001; Qureshi *et al.*, 2001). Its relative abundance compares with Ag and Au, and it is more common than Sn, Hg or Pb (Kabata-Pendias and Pendias, 2001; Meinrath *et al.*, 2003). Weathering of rocks and phosphate fertilizers are major sources of chronic enrichment in U of soil and natural waters. Besides, large amounts of U can be introduced in natural settings near fossil-fuel power plants and also from P fertilizer factories. The current residues from U mining, milling and refining most likely represent the greatest volume of radioactive waste containing U. The long-term containment of these wastes is of particular environmental concern. Various methods are being tested for removal of U or controlling its mobility in contaminated sites (Abdelouas *et al.*, 1999). Combined chemical and biological treatments, as well as phytoextraction and phytostabilization, have been suggested.

The soil–plant interactions are complex and involve physicochemical as well as biological factors. Despite numerous studies on U content in vegetation (see for Ref. e.g. Shahandeh *et al.*, 2001), little information is available on the mechanisms and rate of uptake by plants. In the soil, U, like other trace pollutants, undergoes a series of reactions that can increase or decrease its bioavailability to plants or even its toxicity. Microorganisms can further affect U mobility through immobilization or by facilitating the movement of adsorbed U (Suzuki and Banfield, 1999). Further investigations on soil–plant microbes interactions with U are needed to understand the U cycle in nature and to refine the remediation strategies.

1.2. Uranium in the mycorrhizosphere

The pH-dependent speciation of U in soil and aqueous systems has been extensively studied (Grenthe *et al.*, 1992; Fjeld *et al.*, 2002; Wang *et al.*, 2002), and several reports showed that the form of U greatly influences its uptake and distribution in plant tissues (Bondietti and Sweeton, 1977; Sheppard and Evenden, 1988; Mortvedt, 1994; Ebbs *et al.*, 1998) as well as its uptake and accumulation by microorganisms (Suzuki and Banfield, 1999). Uranium may also be adsorbed on soil particles or form organic and inorganic complexes in the soil solution. Soil carbonate and iron oxyhydroxide phases present strong affinity for U (Lee and Marsh, 1992); organic matter content and soil texture are also known to affect U uptake (Simon and Ibrahim, 1990; Fenton and Waite, 1996; Hossner *et al.*, 1998). Acid root exudation might thus increase the bioavailability of U due to solubilization and complexation of uranyl cation and, to a lesser extent, by changes in pH (Ebbs *et al.*, 1998), and thereby influence the success of phytoremediation.

Maximum interactions between soil, U and plant root take place in the rhizosphere, i.e. the volume of soil influenced by root activity. As stated by Shahandeh *et al.* (2001), there are no published studies on the mobility of U in the rhizosphere, although rhizosphere processes may play a significant role in U uptake, as was observed for other trace elements (Dufey *et al.*, 2001; Hinsinger, 2001a, b). The role of soil microorganisms in this interaction can also be significant because they can act directly on several processes, retarding or enhancing the element transport: solubilization, internal translocation, complexation, sequestration, transformation and precipitation. Among these organisms, the mycorrhizal fungi represent a key active compartment (Smith and Read, 1997). These root symbionts are present in undisturbed ecosystems, in man-made agricultural and forestry systems as well as in contaminated areas. Their unique location at the interface between soil and root makes them important actors in the soil–plant continuum. This leads to the use of the term "mycorrhizosphere," i.e. the volume of soil influenced by both roots and mycorrhizal fungi, while the term rhizosphere should be used for non-mycorrhizal roots. In addition to producing direct root effects, these symbionts are also believed to influence U mobility and further its biocycling. In this chapter, available data on the role of arbuscular mycorrhizal (AM) fungi on the transport of U are presented and discussed.

2. ARBUSCULAR MYCORRHIZAL FUNGI AT THE INTERFACE BETWEEN SOIL AND PLANT ROOT

2.1. The plant-to-soil mycorrhizal continuum

Arbuscular mycorrhizas are the most prevalent underground symbionts, and are formed in the roots of the vast majority of higher plants by obligatory symbiotic fungi (Smith and Read, 1997) belonging to the phylum *Glomeromycota* (Schüßler *et al.*, 2001). Their origin has been dated to approximately 400 Myr ago, supporting the hypothesis of a close co-evolution with ancient plants (Simon *et al.*, 1993). At present, AM symbiosis concerns a majority (80%) of plant species, including most plant crops in agriculture, horticulture and forestry.

The AM fungus develops two associated mycelial systems, one within the root, from which the other extends into the soil. This structural continuum at the interface between plant root and soil make these symbionts particularly relevant in plant-to-soil interactions. The extraradical mycelium develops links between roots and soil particles, hence participating in soil structural stability, and recruits minerals and water for the plant, greatly influencing plant nutrition. They further contribute to plant fitness and resistance against abiotic and biotic environmental stresses, and are thus thought to play a crucial role in sustainable plant–soil systems.

2.2. Role of arbuscular mycorrhizal fungi in mineral nutrition of plants

The importance and extent of hyphae growing into soil from AM roots has been appreciated for a long time, and their function in mineral nutrition has received more attention than any other aspect of the symbiosis (Smith and Read, 1997). It is estimated that AM fungal hyphae can extend several centimetres (up to 30 cm) under monoxenic[1] cultural conditions (Dupré de Boulois, Personal communication), beyond the zone of root exploration (Jakobsen *et al.*, 1992a), branching up to several orders to finely branched fans adapted to the exploration of micro-pores. The result is the formation of a dense absorptive hyphal network, estimated by some authors to be several metres (Jakobsen and Rosendhal, 1990; Sanders *et al.*, 1977) or tens of micrograms dry weight (Olsson *et al.*, 1999) per gram soil. There is clear evidence that this network absorbs poorly soluble nutrients such as phosphorus and translocates them to the plant. At the symbiotic interface within the root, the nutrients are transferred from fungal cell to root cell, resulting in the increased acquisition of nutrients by the plant and promotion of growth. Accumulation/sequestration of some heavy metals and toxic elements has also been reported (Leyval et al., 1997).

Most studies aimed to determine the contribution of AM fungi to the uptake of nutrients (Jakobsen *et al.*, 1992a, b; Clark and Zeto, 1996; Borie and Rubio, 1999; Kaya *et al.*, 2003), heavy metals (Weissenhorn *et al.*, 1995; Joner and Leyval, 1997) and radionuclides (Entry *et al.*, 1999; Berreck and Haselwandter, 2001) by plants were conducted in pot experiments with entire plants growing on soil. The inconveniences of such pot experiments are mainly the difficulty to maintain these experimental systems void of undesirable organisms other than the two symbionts, and the interferences with soil particles that may complicate the availability of mineral elements. These problems, coupled with the strict hypogenous nature of the mycobionts, have complicated the clear identification of the true role of AM fungi (in particular the extraradical mycelium) in element mobilization and transport and of the mechanisms involved. The recent finding of transporter genes for P from the extraradical mycelium of an AM fungus *Glomus intraradices* (Maldonado-Mendoza *et al.*, 2001) and of a plasma membrane Zn transporter from *Medicago truncatula* down-regulated by AM colonization (Burleigh *et al.*, 2003) has increased our understanding of the role of these symbionts in transport processes.

There is a need to study some fundamental relationships in AM fungal uptake and translocation of U without the interference of interactive soil effects.

[1] The monoxenic culture of AM fungi means the contaminant-free *in vitro* culture between a root organ and a Glomalean species.

During the last three decades, monoxenic culture systems have been developed (Strullu and Romand, 1986; Mugnier and Mosse, 1987; Bécard and Fortin, 1988; Diop *et al.*, 1992) for numerous applications of AM symbiosis (see review by Fortin *et al.*, 2002). Hormonal balance modification by genetic transformation renders roots particularly vigorous and allows greater growth potential on artificial media, making them more adaptable to different experimental conditions (Tepfer, 1989). *Daucus carota* L. (carrot) was among the earliest species to be genetically transformed using root-inducing transferred Ri T-DNA of *Agrobacterium rhizogenes* (Tepfer, 1984). These Ri T-DNA-transformed roots have since been used as host for AM fungi in a wide range of fundamental and applied studies (see review by Fortin *et al.*, 2002). This monoxenic culture system was progressively improved by the spatial separation of mycorrhizal roots and extraradical mycelia ramifying into a root-free compartment (St-Arnaud *et al.*, 1996). The system was adapted for different studies on the uptake and translocation of essential elements (Bago *et al.*, 1996; Joner *et al.*, 2000b; Nielsen *et al.*, 2002) as well as of radioactive isotopes of nonessential elements such as ^{137}Cs (Declerck *et al.*, 2003) and ^{233}U (Rufyikiri *et al.*, 2002, 2003, 2004).

Data on the role of AM fungi in the uptake and translocation of U were mainly obtained using this system and are presented below. To clarify some terms used, we referred to Cooper and Tinker (1978). These authors defined the term "transport of elements" as the total movement of elements from the source into the plant through the fungus. Thus, transport consists of three steps: (i) uptake, meaning the removal of elements from the source by absorption of the ions by the external hyphae; (ii) translocation, meaning the movement of the ions along the hyphae; and (iii) transfer, meaning the movement of the ions from the fungus to host tissues within the roots.

3. UPTAKE AND TRANSLOCATION OF URANIUM BY ARBUSCULAR MYCORRHIZAL FUNGI

3.1. Experimental design model: monoxenic culture in a two-compartment system

As outlined above, the monoxenic cultivation system in bicompartmental Petri plates offers an improved model to investigate transport processes. In the case of U, the two-compartment system consisted of a 50-mm Petri plate cover glued in a 90-mm Petri plate, thus separating a central compartment (CC) from a surrounding external compartment (EC) (Fig. 1). The Ri T-DNA-transformed carrot roots were grown in association with AM fungi in the CC on a gelled modified Strullu–Romand (MSR) medium (Declerck *et al.*, 1998; modified from Strullu and Romand, 1986). Depending on the objective of the experiments, three types of ECs were set up, allowing the growth of (i) hyphae (hyphal compartment), (ii) both roots and hyphae (root-hyphal compartment) and (iii) non-mycorrhizal roots (root compartment). Because of the negative geotropism of Ri T-DNA-transformed carrot roots, the Petri plates were commonly incubated horizontally in an inverted position

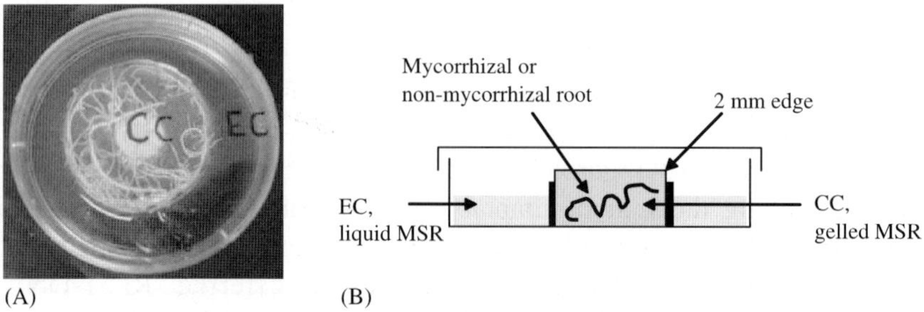

Fig. 1. (A) Illustration of a two-compartment monoxenic culture system allowing the spatial separation of a central compartment (CC) from a neighbouring external compartment (EC) (B). The MSR medium was solid in the CC and liquid in the EC. The CC was filled with the gel to reach a layer extending 2 mm above the edge of the 50-mm Petri plate, in order to facilitate hyphae and/or roots to cross the partition between the two compartments. Adapted with permission from Rufyikiri *et al.* (2002, 2003b).

at 27°C in the dark (Diop *et al.*, 1992; Plenchette *et al.*, 1996; St-Arnaud *et al.*, 1996). After 3 weeks of incubation, they were set upright and the EC filled with liquid MSR medium void of sucrose and vitamins, for an additional week. Numerous hyphae crossed the partition between the two compartments, owing to an improvement of the monoxenic culture system that was brought about by extending the gelled medium 2 mm above the physical separation between the two compartments.

Once in contact with the liquid MSR medium, a profusely branched mycelium developed. At this stage, the cultures were ready for radio-isotope labelling, and different experiments were carried out to quantify the uptake and translocation of U by AM fungi, and to study factors involved, such as the pH or AM fungal strain. At the end of the experiment, i.e. after 3 weeks of contact between the fungus and the liquid MSR medium, about 150 hyphae had crossed the partition between the two compartments (Rufyikiri *et al.*, 2003, 2004), and thousands of spores were produced, increasing six-fold the hyphal biomass reported in previous studies (Joner *et al.*, 2000b; Rufyikiri *et al.*, 2002).

3.2. Effect of pH on uranium speciation and uranium uptake

3.2.1. pH effects on uranium speciation and uranium uptake

The effects of pH on U speciation, and its uptake and translocation by extraradical hyphae of the AM fungus *G. intraradices* Schenck and Smith (MUCL 41833) was studied (Rufyikiri *et al.*, 2002). Liquid MSR medium was labelled with 0.1 μM ^{233}U, a concentration at which no precipitate was possible at any pH, according to speciation calculations (Fig. 2A) performed using the geochemical computer code, *The Geochemist's Workbench*® (Bethke, 2001), and the thermodynamic data of U (Grenthe *et al.*, 1992).

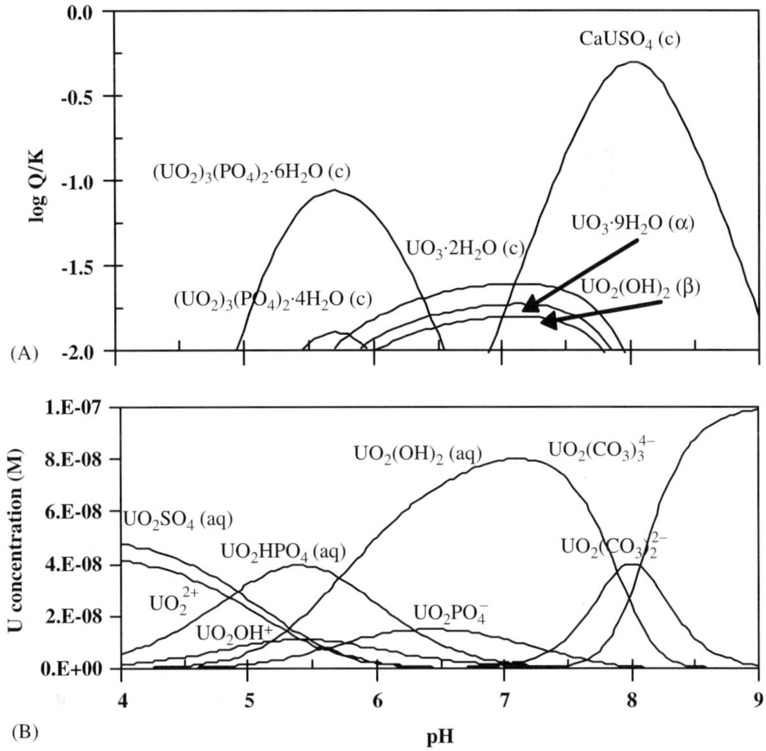

Fig. 2. The U solubility and speciation as a function of pH in the liquid MSR assuming an equilibrium with a partial pressure of CO_2 (g) of $10^{-3.5}$ atm: (A) U saturation indexes (log *Q*/*K*, where *Q* is the solubility product and *K* the solubility constant with precipitation when $Q > K$) and (B) calculated concentration of dominant U species. Total dissolved U = 0.1 μM. Reprinted with permission from Rufyikiri *et al.* (2002).

Results on U speciation (Fig. 2B) support other studies reporting that U speciation is highly pH-dependent (Langmuir, 1978; Mortvedt, 1994; Ebbs *et al.*, 1998). Numerous U species were formed, and among them, uranyl cation and uranyl-sulphate species were dominant in the solution below pH 4.8, phosphate species between pH 4.8 and 5.7 and hydroxyl species between pH 5.7 and 7.8. Above pH 7.8, solutions were dominated by anionic uranyl-carbonate species. The influence of these most representative U species was studied by adjusting the pH of the liquid MSR medium in the hyphal compartment, containing 0.1 μM ^{233}U, to 4.0, 5.5 and 8.0 (Rufyikiri *et al.*, 2002). Uranium content in the AM fungal hyphae and spores developed in the hyphal compartment was significantly higher at pH 5.5 than at pH 4 and 8, while U translocated via hyphae from the hyphal compartment to roots developing in the CC was significantly higher at pH 4 than at pH 5.5 and 8 (Fig. 3). Uranium translocated was positively correlated

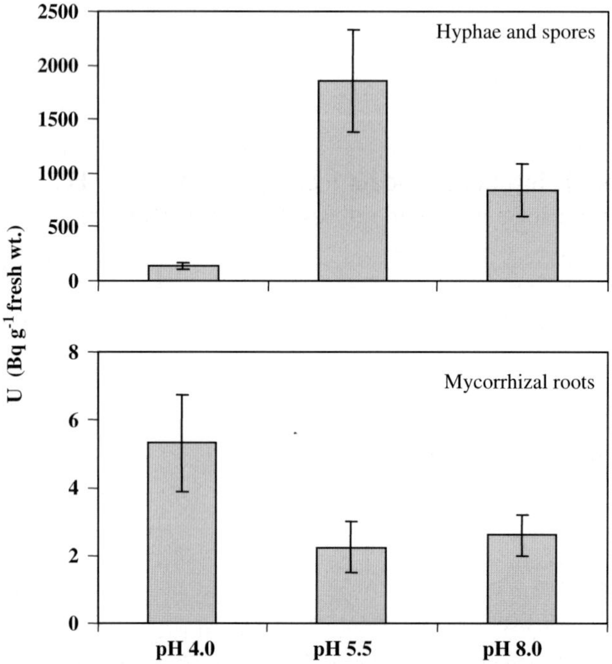

Fig. 3. Uranium activity concentrations in hyphae and spores developing in the hyphal compartment (top) and in the mycorrhizal roots (bottom) developing in the central root compartment for Ri T-DNA transformed carrot (*Daucus carota* L.) roots grown in association with *G. intraradices* in a two-compartment system with 0.1 μM ^{233}U added to the hyphal compartment set at pH 4, 5.5 and 8. Data used with permission from Rufyikiri *et al.* (2002).

with the number of hyphae crossing the partition between the two compartments for all pH treatments with linear regression coefficients r^2 of 0.86, 0.83 and 0.58 at pH 4, pH 5.5 and pH 8, respectively. It seems that soluble uranyl cations or uranyl-sulphate species that are stable under acidic conditions were translocated to a higher extent through fungal tissues, while phosphate and hydroxyl species dominating under acidic to near-neutral conditions, or carbonate species dominating under alkaline conditions, were rather immobilized by hyphal structures.

Similar results on the effects of U speciation on U bioavailability were reported in other studies involving other organisms. Ebbs *et al.* (1998) found that the U concentration was higher in shoots and lower in roots of peas (*Pisum sativum* cv. Sparkle) at pH 5.0 than at pH 6 and 8. The maximum uptake of U by several microbial species including bacteria, fungi, algae and lichens was most frequently observed in the pH range 4–5 (see review by Suzuki and Banfield, 1999). Under these pH conditions, free UO_2^{2+} and $(UO_2)_3OH_5^+$ were the dominant species in solutions containing 100 mg U L^{-1} dissolved in distilled water from $UO_2(NO_3)_2 \cdot 6H_2O$.

3.2.2. Fungal-induced pH changes

The development of extraradical mycelia of AM fungi can induce pH changes in the growth medium. Bago *et al.* (1996) used the pH indicator bromocresol purple and observed an increase in pH (up to 2 units) induced by the extraradical hyphae of *G. intraradices* (DAOM 197198) in the presence of NO_3^--N as source of N, but not in media lacking this N form. Feeding hyphae with another source of nitrogen should not result in such pH increase. The pH changes of growth media are common phenomena also induced by mycorrhizal and non-mycorrhizal plants (Rufyikiri *et al.*, 2000; Hinsinger, 2001b) and related to imbalances in the uptake of cations and anions (Marschner, 1995). Proton excretion resulting in a net acidification occurs when excess cations are absorbed over anions, while net alkalinization results from OH^- excretion due to an excess uptake of anions over cations (Rufyikiri *et al.*, 2001; Hinsinger *et al.*, 2003). Bago *et al.* (1996) suggested that the pH increase observed in his study was a consequence of the active uptake of NO_3^--N involving the NO_3^-/H^+ symport or NO_3^-/OH^- antiport mechanisms used by the fungus for NO_3^- uptake. These mechanisms would mask any other hyphal-promoted acidification, resulting in a net alkalinization. An increase of almost 1 pH unit was observed after 2 weeks of contact between hyphae and the liquid MSR medium containing 0.1 μM ^{233}U in the hyphal compartment when mycelia developed densely. However, roots induced a net acidification (decrease of 0.5 unit pH) under the same growth conditions (Rufyikiri *et al.*, 2003). In this liquid MSR medium, the source of N was NO_3^--N (more than 99%) with a negligible fraction as NH_4^+-N. This modification of pH in the presence of hyphae was an active process, because such an effect was not observed when their metabolic activity was inhibited by formaldehyde added to the solution (Rufyikiri *et al.*, 2003), or when the hyphal development was lower, as measured in a previous experiment (Rufyikiri *et al.*, 2002).

Although information on U speciation in the mycorrhizosphere is lacking in the literature, it is obvious that the development of extraradical hyphae can influence U speciation through hyphal-induced pH changes.

3.3. Efficiency of uranium transport by hyphae

Several studies showed that the influence of AM fungi in the acquisition of mineral elements by plants is dependent upon many factors, as increases, no effect and decreases have been observed as a function of the type of elements (Clark and Zeto, 2002) and of growth conditions such as element concentrations, plant densities, growth period and pot size (Leyval and Joner, 2001). Differences between element acquisition by mycorrhizal plants may also be due to differences in the hyphal efficiency for the transport of various elements, especially when it is a matter of essential elements versus nonessential ones.

Uranium has no known biological function (Suzuki and Banfield, 1999). Like other heavy metals, it is tolerated in small quantities but results in toxicity when accumulated in high concentration (Ebbs *et al.*, 1998). Given the fact that AM fungi often protect plants against the harmful effects of toxic metals by reducing their uptake by plants (Smith and Read, 1997), this raises the problem as to the understanding of the true role of AM fungi for U acquisition by plants. Although the results discussed above indicate that extraradical hyphae of *G. intraradices* can take up U and translocate it towards roots, we questioned the relative extent of both these processes. The magnitude of U uptake and translocation by fungal hyphae was therefore investigated, on the one hand, by comparing the contribution of hyphae to that of the host roots, and on the other hand, by comparing the hyphal efficiency for U to that of P used as reference (Rufyikiri *et al.*, 2004).

3.3.1. The efficiency of hyphae versus host roots

The uptake and translocation of U by extraradical hyphae were compared to those of carrot roots developing under the same conditions (Rufyikiri *et al.*, 2003). The Ri T-DNA-transformed carrot roots were grown in the two-compartment system described above in association with *G. intraradices* (MUCL 41833). For the EC, three scenarios were tested, as described in Section 3.1. After 2 weeks of contact between U and hyphae, mycorrhizal roots plus hyphae, and nonmycorrhizal roots, the biomass-specific U content was about 270 Bq g^{-1} fresh weight of AM fungal mycelia. This U concentration was 5.5 and 9.7 times larger than for mycorrhizal roots and nonmycorrhizal roots, respectively. The larger U concentration in fungal mycelia than in roots could partially be explained by differences in their respective cation exchange capacity (CEC), which was reported to be four times higher for AM mycelia (187 $cmol_c\ kg^{-1}$ dry weight) than for carrot roots (47 $cmol_c\ kg^{-1}$ dry weight) (Rufyikiri *et al.*, 2003). In fact, these authors observed that the amount of Cu-extractable U was 15 times higher in AM fungal mycelia than in carrot roots. Differences also exist in other mechanisms of U accumulation, since U remaining after successive extractions with 0.01 M $CuSO_4$, 0.01 M HCl and 0.1 M HCl was nine times higher in AM fungal mycelia than in carrot roots. These residual U represented 47 and 67% of the biomass-specific contents in carrot roots and AM fungal mycelia, respectively.

The uptake of U by the carrot roots was largely influenced by the presence or absence of AM fungus. The mycorrhizal roots grown in the root hyphal compartment accumulated 49 Bq g^{-1} fresh weight, and this was 1.8 times larger than that of the nonmycorrhizal roots grown in the root compartment. This could probably be explained by U uptake mechanisms being more active in the mycorrhizal roots than in the nonmycorrhizal ones, or by a marked contribution of the intraradical hyphae to the accumulation of U in the host roots. A higher concentration of U in intraradical fungal hyphae of an undefined AM fungal species than in the host root tissues was previously reported (Weiersbye *et al.*, 1999) and attributed to the particular chemical conditions prevailing in the intraradical fungal cells. Large P concentrations in the intraradical parts of AM fungi were recently observed

(Nielsen *et al.*, 2002; Pfeffer *et al.*, 2001), while intracellular pH values varying between 5.6 and 7.0 were reported for hyphae of *G. intraradices* (Jolicoeur *et al.*, 1998). Both high P concentrations and weakly acidic to neutral pH values are factors that can promote the formation of U-phosphate complexes and precipitates in the intraradical hyphae, thus favouring U accumulation in mycorrhizal roots.

Uranium was found in the gel and in the roots in both nonmycorrhizal and mycorrhizal cultures. This means that both roots and mycelia contain trans-located U. The total amount of U translocated from the ECs to the central root compartment significantly differed between the three EC systems. Hyphae and mycorrhizal roots together translocated the largest amount (19.2 Bq/Petri plate) and nonmycorrhizal roots the lowest (1.6 Bq/ Petri plate). Differences in U flux in hyphae and roots may explain such differences. Indeed, considering hyphae and roots as cylinders with an average diameter of 11 μm for a hyphae (Nielsen *et al.*, 2002) and of 1000 μm for a root (Rufyikiri *et al.*, 2003), the total cross-sectional area at the partition between the two compartments (A) was calculated as A = (diameter/2)^2 $\times$ π $\times$ number of hyphae/roots. For the hyphae and roots connecting the two compartments, the A value was about 0.014 mm^2 for the average 147 hyphae and 3.93 mm^2 for the average five roots, respectively. Although the total section area of roots was 281-fold larger than that observed for hyphae, U translocation by roots was lower than its translocation by hyphae. This indicates that U flux rate was larger in hyphae than in roots, most likely due to differences in exchange reactions between U and cell components as well as in other biochemical mechanisms involved in the transport.

3.3.2. Hyphal efficiency for uranium versus phosphorus

As for other experiments described above, this study used Ri T-DNA-transformed carrot roots grown in the two-compartment system in association with *G. intraradices* (MUCL 41833). The liquid MSR medium in the external hyphal compartment was labelled with 8.33 Bq ^{233}U mL^{-1} (= 0.1 μM) and 13.33 Bq ^{33}P mL^{-1} (Rufyikiri *et al.*, 2004). The concentration of stable ^{31}P in the liquid MSR medium was 50 μM. Table 1 shows that both ^{33}P and ^{233}U were taken up and translocated by the extraradical fungal hyphae. Yet, both the uptake and translocation were much higher for ^{33}P than for ^{233}U. The flux rates were estimated for the two elements on the basis of the cross-sectional area at the partition between the two compartments (0.013 mm^2, calculated as in the previous section, for the average 137 crossing hyphae). For the 14 days of contact between hyphae and the liquid medium, the average flux rate was 9.4×10^{-9} mol m^{-2} s^{-1} for U and 3.8×10^{-5} mol m^{-2} s^{-1} for P. By using similar experimental devices, Nielsen *et al.* (2002) found that the P flux rate was 1.5×10^{-3} mol m^{-2} s^{-1} during a day corresponding to the highest transfer rate of P from the hyphal compartment to the central root compartment. This value is about 40-fold the P flux rate reported above. However, these authors considered only active running hyphae in the calculation. Indeed, our calculation may underestimate the true hyphal capacity for translocation as the total

Table 1
Uranium and P activity contents (Bq/Petri plate) for Ri T-DNA-transformed carrot (*Daucus carota* L.) roots grown for 2 weeks in association with *G. intraradices* in a two-compartment system. Data used with permission from Rufyikiri *et al.* (2004)

	^{233}U		^{33}P	
	Average	% of input	Average	% of input
Hyphae/spores in the hyphal compartment	5.5 ± 1.3	4.4	32 ± 2	16
Gel with fungal biomass in the central root compartment	4.9 ± 0.3	3.9	14 ± 2	7
Mycorrhizal roots in the central root compartment	7.4 ± 1.2	5.9	143 ± 12	72

Note: The input solution contained 125 Bq/Petri plate for ^{233}U and 200 Bq/Petri plate for ^{33}P. Values are averages ± standard deviation of six replicates.

number of crossing hyphae was used despite the fact that some hyphae were probably not active for the whole period considered. Nevertheless, it was sufficient to demonstrate that the translocation rate was much higher for P than for U.

The high P uptake and translocation indicated that the system was functioning well and that the low uptake and translocation of U were not due to an experimental artefact. The relatively high efficiency of the extraradical hyphae to take up and to translocate P was also reported in numerous studies and some reviews give several references on the topic (Marschner, 1995; Jakobsen *et al.*, 2002). For instance, Cooper and Tinker (1978) compared the uptake and translocation of ^{32}P, ^{65}Zn and ^{35}S by the AM fungus *G. mosseae* (Nicol. and Gerd.) Gerd. and Trappe with *Trifolium repens* L. as host growing on sterile soil–agar split-plates. They found that the ratio of the molar amounts of P, S and Zn translocated was 35:5:1, and that the mean fluxes followed the ratio 50:8:1. Some reports have shown that mycorrhizal fungi alleviate metal toxicities considerably by reducing metal uptake in mycorrhizal plants exposed to toxic levels of these metals (Joner and Leyval, 1997; Rufyikiri *et al.*, 2000). The sequestration of metals in intraradical fungal hyphae (Joner and Leyval, 1997; Weiersbye *et al.*, 1999) was often evoked as a mechanism of AM fungal protection against metal toxicity for plants. The low translocation of U by the extraradical hyphae towards the host roots might be another important way to reduce heavy metal exposure to host root tissues. This suggests the existence in hyphal tissues of efficient mechanisms limiting the uptake and translocation of nonessential elements such as U. Little is known about these mechanisms.

3.4. Effect of mycorrhizal fungal species

Numerous studies have shown that the effectiveness of the extraradical hyphae in taking up elements varied greatly between fungal species. The distinct growth patterns of the mycelium and the specific variations in the efficiency of element uptake by AM fungi (Jakobsen *et al.*, 1992a, b) as well as in their metal-binding capacity (Joner *et al.*, 2000a; Gonzalez-Chavez *et al.*, 2002) were assumed to be the main causes of the differences observed between AM fungal species in the uptake and transport of elements. Metal sorption and accumulation by the extraradical mycelium were generally high, with large differences between AM fungal species, as it was reported for Cd (Joner *et al.*, 2000a) and Cu (Gonzalez-Chavez *et al.*, 2002).

The effects of four AM fungi on U uptake and translocation were investigated under monoxenic culture conditions in a two-compartment culture system. The four AM fungi were *G. intraradices* Schenk and Smith (MUCL 41833, # 1), *Glomus* sp.# 2, # 3 and # 4. The four AM fungi originated from different edapho-climatic conditions. Although numerous identical hyphae (range between 129 and 159 hyphae) crossed the partition between the central root compartment and the hyphal compartment and developed in the liquid medium labelled with 0.1 μM ^{233}U, the total biomass of mycelia produced was significantly different among the four AM fungi. This biomass production was about 18 mg fresh weight per Petri plate for *G. intraradices* # 1 and *Glomus* sp.# 2. It was 2.3 times larger for *Glomus* sp.# 3 and # 4. These differences were obviously due to differences in spore production. In fact, more than 3,500 spores were observed in the presence of *G. intraradices* # 1 and *Glomus* sp.# 2 in the hyphal compartment, while less than 30 spores were recorded in the presence of *Glomus* sp.# 3 and # 4, at the stage of harvest.

Results presented in Table 2 for ^{233}U uptake and translocation indicate that the absorption capacity was significantly different among the four AM fungi. The ^{233}U activity content in hyphae of *Glomus* sp.# 2 developing in the hyphal compartment was about two times larger than for *Glomus* sp.# 3, and three times larger than for *Glomus* sp.# 4 and *G. intraradices* # 1. Significant differences were also observed for the ^{233}U translocation. The total translocation of ^{233}U by hyphae amounted to 3.4–10% of the initial ^{233}U supply with the following order: *Glomus* sp.# 3 > *G. Intraradices* # 1 > *Glomus* sp.# 2 = *Glomus* sp.# 4. The rate of translocation was low for all AM fungi compared with the fraction of P translocated under the same growth conditions, as it was already reported for *G. intraradices* # 1 (see Table 1). It is also worth noticing that no significant differences were observed for P translocation among the four AM fungi. The rate of P translocation appeared to be related only to the number of runner hyphae, as this number was similar for the four AM fungi. For U, the rate of translocation was controlled by other factors, which might be fungal-dependent.

3.5. Mechanisms of uranium absorption and translocation

The accumulation of mineral elements by AM fungi may result from many mechanisms, including the metabolism and incorporation in tissues, as demonstrated for essential nutrients such as P and N (Pfeffer *et al.*, 2001), precipitation

Table 2
Uranium and P activity contents (Bq/Petri plate) in the hyphae and spores developed in the external hyphal compartment, and in the mycorrhizal roots and the gel with fungal biomass in the central root compartment for Ri T-DNA-transformed carrot (*Daucus carota* L.) roots grown in association with four AM fungi in a two-compartment system

AM fungi	Uptake by hyphae		Translocation			
	Means	% input	Roots		Gel	
			Means	% input	Means	% input
^{233}U activity						
G. intraradices # 1	4.24 b	3.4	6.49 b	5.2	4.22 a	3.4
Glomus sp.# 2	12.74 a	10.2	2.43 c	1.9	2.36 ab	1.9
Glomus sp.# 3	2.54 c	2.0	8.88 a	7.1	3.67 a	2.9
Glomus sp.# 4	1.78 c	1.4	2.34 c	1.9	1.90 b	1.5
^{33}P activity						
G. intraradices # 1	30.1 b	15.0	134.0 a	67.0	18.0 a	9.0
Glomus sp.# 2	51.0 a	25.5	121.1 a	60.6	21.0 a	10.5
Glomus sp.# 3	10.6 c	5.3	129.3 a	64.7	18.8 a	9.4
Glomus sp.# 4	17.0 c	8.5	147.0 a	73.5	25.7 a	12.9

Note: The initial supply was 125 ± 0.1 and 200 ± 0.2 Bq/ Petri plate for ^{233}U and ^{33}P, respectively. Values are averages of six replicates. Within columns, averages followed by the same letter are not significantly different ($P = 0.05$).

of nonessential metallic cations on or in the fungus assumed to occur with PO_4 (Turnau *et al*., 1993) and adsorption on negatively charged constituents of fungal tissues (Joner *et al*., 2000a). All these mechanisms possibly coexist for U.

The involvement of active mechanisms in absorption and translocation can be tested by the addition of metabolic inhibitors in the hyphal growth media to obtain negative control for metabolic activity of hyphae (Joner *et al*., 2000b). The addition of formaldehyde (2% v/v) to the solutions in the hyphal compartment in half Petri plates, 24 h before U was supplied, increased the U accumulation by AM hyphae by a factor of 3.7 at pH 4.0, in comparison with living hyphae. However, it decreased the U accumulation by AM hyphae by a factor of 2.9 at pH 5.5 (Rufyikiri *et al*., 2002, 2003). Differences between formaldehyde-killed hyphae and living hyphae were less marked at pH 8.0. Galun *et al.* (1983) reported that killing the fungal mycelia of *Penicillium digitatum* with boiling water, alcohols, dimethyl sulfoxide or KOH increased the uptake capability of U from aqueous solutions by a factor in the range 1.7–2.5. Killing the fungal mycelia with formaldehyde or with sodium azide did not enhance U uptake. These observations suggest that both metabolic-dependent and metabolic-independent mechanisms contribute to the uptake of U by fungal mycelia, while its translocation is a metabolic-dependent process, as U removed from the solution by formaldehyde-killed AM hyphae was not translocated to roots developing in the root compartment (Rufyikiri *et al.*, 2002, 2003, 2004). Cytoplasmic/protoplasmic streaming in active runner hyphae has been suggested as a mechanism of translocation of P in hyphae (Nielsen *et al.*, 2002), but whether this mechanism could be involved in the translocation of other elements, such as U, is not yet known.

Extraradical mycelia can take up U by adsorbing charged uranyl species onto functional groups of the cell wall. The potential binding sites of mycelia were estimated by the CEC, determined for some AM fungal species by different methods including (a) cations released after incubation in 0.01 M (equimolar) solution of $Ca(NO_3)_2$, $Mg(NO_3)_2$, KNO_3 and $CuCl_2$ (Gonzalez-Chavez *et al.*, 2002), (b) potentiometric titration (Joner *et al.*, 2000a) or (c) Cu desorption by 0.01 M HCl after incubation in 0.01 M $CuSO_4$ (Rufyikiri *et al.*, 2003). Whatever the methods used, results indicated that the CEC values of AM fungal mycelia are of the same order as those of ectomycorrhizal fungi, but much higher than those of plant roots (Table 3). As for ectomycorrhizal fungi and plant roots, significant differences could also be observed between AM fungal species.

Table 3
Cation-exchange capacity of some arbuscular mycorrhizal fungi in comparison with the CEC of ectomycorrhizal fungi and of plant species

Species	Value	Unit	References
AM fungi			
G. mosseae BEG132	230	$cmol_c\ kg^{-1}$ dry wt.	Gonzalez-Chavez *et al.* (2002)
G. caledonium BEG133	100	$cmol_c\ kg^{-1}$ dry wt.	Gonzalez-Chavez *et al.* (2002)
G. claroideum BEG134	250	$cmol_c\ kg^{-1}$ dry wt.	Gonzalez-Chavez *et al.* (2002)
G. mosseae BEG25	480	$cmol_c\ kg^{-1}$ dry wt.	Gonzalez-Chavez *et al.* (2002)
Glomus sp.	254	$cmol_c\ kg^{-1}$ fresh wt.	Joner *et al.*, 2000a
G. lamellosum	233	$cmol_c\ kg^{-1}$ fresh wt.	Joner *et al.*, 2000a
G. intraradices	187	$cmol_c\ kg^{-1}$ dry wt.	Rufyikiri *et al.*, 2003b
Ectomycorrhizal fungi			
Laccaria bicolor S238	200–300	$cmol_c\ kg^{-1}$ dry wt.	Marschner *et al.*, 1998
Paxillus involutus 533	80–120	$cmol_c\ kg^{-1}$ dry wt.	Marschner *et al.*, 1998
Plant species			
Monocotyledons	10–25	$cmol_c\ kg^{-1}$ dry roots	Crooke and Knight, 1971
Dicotyledons	15–60	$cmol_c\ kg^{-1}$ dry roots	Crooke and Knight, 1971

The high CEC of the fungal mycelia may be explained by their large surface area per unit weight (Marschner *et al.*, 1998). Wessels and Sietsma (1981) reported that fungal cell walls do not contain carboxylate groups of pectins, which contribute mainly to the CEC of higher plants. But in fungal cell walls, ions are bound predominantly to chitin and cellulose (Siegel *et al.*, 1990; Zhou, 1999), which lack acid groups (Marschner *et al.*, 1998). Chitin is a polysaccharide (poly[β-(1→4)-2-acetamido-2-deoxy-D-glucopyranose]) with alcohol and carbonyl functional groups. It is found in high proportions in the walls of AM fungal spores (Bonfante-Fasolo and Grippiolo, 1984; Bonfante-Fasolo *et al.*, 1990), but also in the walls of extraradical hyphae, intracellular coils and intercellular hyphae as laminated chitin fibrils (Smith and Read, 1997).

Chitin represents up to 60% of fungal cell walls (Muzzarelli and Tubertini, 1969; Muzzarelli, 1977). The binding capacity of chitin (and its derivative chitosan) for metal ions has been the subject of several studies because of its potential biogeochemical significance for localized metal accumulation, and its biotechnological significance for wastewater decontamination and metals recovery (Muzzarelli and Tubertini, 1969; Muzzarelli, 1977; Galun *et al.*, 1983; Chui *et al.*, 1996; Benguella and Benaissa, 2002). Some studies dealing with the role of cell wall chitin in the biosorption of U were also reported (Tsezos and Volesky, 1982; Tsezos, 1983; Tsezos and Mattar, 1986). The adsorption rate of U on chitin was compared that on cellulose phosphate, carboxymethyl cellulose and cellulose (Galun *et al.*, 1983). These wall-related biopolymers were packed in 5-mL syringe barrel microcolumns filled with 5 mL mg L^{-1} UO_2Cl_2 solution (61.7 mg L^{-1} U), and flushed with 50 ml water. Both these wall-related biopolymers appeared to be active in U retention. Uranium bound to chitin amounted to 311 μg g^{-1} dry weight and was of the same order as that bound to cellulose phosphate, but about threefold the amount of U retained by cellulose. This high binding capacity for chitin was also reported for several other metal ions (Muzzarelli and Tubertini, 1969).

Hyphae of AM fungi also produce an extracellular iron-containing glycoproteinaceous substance, glomalin (Wright *et al.*, 1996; Wright and Upadhyaya, 1996, 1998; Rillig *et al.*, 2001). This substance is produced by actively growing hyphae of all members of AM genera but not by other groups of soil fungi so far tested (Wright and Upadhyaya, 1996; Rillig *et al.*, 2001). Abundant information concerning this glycoprotein is available in relation to its role in soil aggregation (Wright and Anderson, 2000; Rillig *et al.*, 2002, 2003). Yet very little is known about other properties, such as its interaction with ions. However, this substance may play a role in the adsorption of ions and thus contribute to metal retention by AM fungal hyphae. A secreted glomalin from the hyphal compartment culture medium of monoxenic cultures of *G. intraradices* was partially purified and used to test its binding capacity for Sr in solution with various concentrations and pH values (Driver *et al.*, 2003). These authors observed that glomalin solution bound Sr. The ability to sequestrate potentially toxic elements such as Cu, Cd and Pb was reported from *in vivo* and monoxenic studies for glomalin extracted from hyphae, from soil or sand after removal of hyphae and from hyphae attached to roots (González-Chávez *et al.*, 2004). Glomalin seems to be a charged molecule capable of binding other cations, such as UO_2^{2+}. However, this assumption needs to be tested.

We determined the contribution of exchange sites of AM mycelia to the absorption of U on mycelia grown during 2 weeks in liquid MSR labelled with 0.1 μM ^{233}U in the hyphal compartment, as previously described, by sequential extraction using a method adapted from Dahlgren *et al.* (1991) and Dufey and Braun (1986). The elements desorbed from mycelia by Cu^{2+} (10^{-2} M $CuSO_4$ extract) were considered as "Cu-exchangeable elements" and those thereafter extracted by 10^{-2} and 10^{-1} M HCl as well as residual elements were considered as nonexchangeable elements. Data presented in Table 4 indicates that Cu-extractable U represented 15% of the total U contents in the mycelium. Additional U representing 12 and 6% of the total U contents in the mycelium were further extracted by 10^{-2} and 10^{-1} M HCl, respectively. These extractable U fractions were small since these procedures allowed to extract most of the Ca and Mg contents.

The pH of the bathing solution is a determinant factor on the process of U adsorption because of its simultaneous effects on surface charges and on U speciation. On the one hand, low pH promotes the dominance of free uranyl cations in liquid medium, but, on the other hand, most of the exchange sites of fungal hyphae are probably saturated by H^+ at low pH. This may result in low UO_2^{2+} adsorption on hyphae. Low rates of biosorption of metals at low pH due to a strong competition from hydrogen ions for binding sites were reported in other studies (Gadd, 1990; Zhou, 1999). Yang and Volesky (1999) observed that non-living biomass of the brown alga *Sargassum fluitans* sequestered uranyl ions

Table 4
Sequential extraction of U, Ca and Mg with 10^{-2} M $CuSO_4$, 10^{-2} M HCl and 10^{-1} M HCl for mycelium of *G. intraradices*. Data used with permission from Rufyikiri *et al.* (2003)

Variables	U (Bq g^{-1} f. wt)	Ca (mg g^{-1} f. wt)	Mg (mg g^{-1} f. wt)
$CuSO_4$ extract	41.0 (15.0)	1.015 (80.0)	0.276 (88.5)
10^{-2} M HCl extract	33.4 (12.2)	0.066 (5.2)	0.016 (5.1)
10^{-1} M HCl extract	17.2 (6.3)	0.088 (6.9)	0.006 (1.9)
Residual	182 (66.5)	0.100 (7.9)	0.014 (4.5)

Note: Values in parentheses indicate percentages of the total biomass-specific U contents.
f. wt: fresh weight

from aqueous solution, with the maximum U sorption capacity exceeding 560, 330 and 150 mg g^{-1} at pH 4.0, 3.2 and 2.6, respectively. Increasing the pH would increase negative charges by deprotonation of constituents of cell walls, with concurrent enhancement of the metallic cation adsorption capacity. However, for U, increasing the pH led to the formation of neutral and even negatively charged species at alkaline conditions, as already shown in Fig. 2. Thus, high pH would impair the bioadsorption of U.

The formation of stable complexes or precipitates is likely the main mechanism of U accumulation in fungal hyphae in contact with U in the ECs, and this was assumed to contribute to the low translocation of U (Rufyikiri *et al*., 2003, 2004). Various hyphal functional groups with high affinity for U such as hydroxyl, phosphate, and amino functions may be involved, resulting in complexation and precipitation. Uranium was shown to accumulate both extracellularly on the cell wall surface and intracellularly through the cytoplasm of the fungus *Saccharomyces cerevisiae* (for references see Suzuki and Banfield, 1999). Interaction of uranyl ions with amino ligands and polymers such as chitin by complexation and adsorption was reported by other authors (Guibal *et al*., 1996).

Chemical conditions prevailing in the fungal cells may favour the formation of precipitates. The first factor is P concentration which is very high in AM fungi. Total P contents in the range 19–32 mg g^{-1} dry weight in the intraradical hyphae and of 5–13 mg g^{-1} dry weight in the extraradical hyphae of *Gigaspora margarita* Becker and Hall MAFF 520054 were reported by Solaiman and Saito (2001). Phosphorus concentration as high as 338 mM in the translocated protoplasm volume of runner hyphae of *G. intraradices* was reported by Nielsen *et al.* (2002), while Pfeffer *et al.* (2001) estimated the concentration of mobile polyphosphate in AM hyphae of *G. etunicatum* to 10 mM. The second factor is pH. Jolicoeur *et al.* (1998) developed a method to perform real-time analysis of cytosolic pH of AM fungi. Cytosolic pH profile in hyphae measured under different culture conditions ranged from 6.5 to 7.2 for *Gi. margarita* and from 5.6 to 7.0 for *G. intraradices*. The pH profile along hyphae has been suggested to be maintained by the entry of proton ions via H^+ cotransport symports and by proton excretion by ATPase pumps. The external pH seemed to have no effects on this hyphal cytosolic pH, as it remained constant when the growth medium was adjusted at pH 5.5, 6.5 and 7.5. Both high P concentration and weakly acidic to neutral pH are factors that have the potential to promote the formation of U-phosphate complexes and precipitates in AM fungal tissues.

4. CONCLUSIONS AND FUTURE PROSPECTS

Fundamental information on the interactions between AM fungi and U is reported and discussed in this chapter. The monoxenic culture system was shown to be convenient to investigate the processes of uptake and translocation of U by AM fungi, as it was already reported for some other elements. The advantages of such a culture system include: (1) avoidance of interferences with undesirable organisms or with soil particles, (2) control of chemical conditions that are of a particular importance for the solubility and speciation of U; and (3) distinction and quantification of the respective contribution of AM fungi and of their host root on the two processes (uptake and translocation). The capacity of extraradical AM hyphae to take up U and translocate it towards the host roots was clearly shown, but more research is needed to understand the mechanisms involved. In the future, studies may focus on the identification of the main forms of U accumulation in the extraradical mycelia. These forms, assumed to be complexes and precipitates, appeared sufficiently stable to resist extractions by ion exchange (Cu^{2+} and H^{+}) or by solubilization with an acid solution (10^{-1} M HCl). This could partially help in understanding the low mobility of U within hyphae demonstrated by the low translocation rate of U in comparison with that of P. The other field of investigation is the assessment of the role of other parameters such as the levels of nutrients and the presence of other pollutants. In particular, since naturally U-contaminated sites often contain heavy metals above background levels (Weiersbye *et al.*, 1999; Donahue and Hendry, 2003; Schönbuchner *et al.*, 2002), it is interesting to test the behaviour of U in mycorrhizal fungi in a context of multipollution.

The study model presented here was suitable for the quantification of both uptake and translocation of U by AM fungi, but it could not determine if U was transferred into the root cells or if it was mainly immobilized in the intraradical fungal structures. A next step would be the determination of the role of intraradical AM structures by enhancing, for instance, the sink strength of the mycorrhizal host using entire plants grown monoxenically or *in vivo*, and to determine possible changes in U sequestration by roots or in U transfer to shoots linked to the presence of AM fungus in roots.

Finally, information obtained with the monoxenic cultivation approaches will have to be validated by experiments carried out under more realistic conditions, such as various interferences with soil components and other microorganisms. After this step, recommendations on the use of AM fungi in programmes of revegetation or of phytoremediation could be formulated.

ACKNOWLEDGMENTS

This work was supported by the Belgian Nuclear Research Centre (SCK•CEN) and the EU-MYRRH project contract FIGE-CT-2000-00014 "Use of mycorrhizal fungi for the phytostabilization of radio-contaminated environments". S. Declerck gratefully acknowledges the financial support from the Belgian Federal Office for Scientific, Technical and Cultural affairs (OSTC, contract BCCM C2/10/007) and thanks the director of MUCL for the facilities provided and for continual encouragement.

REFERENCES

Abdelouas, A., Lutze, W., Nuttall, H.E., 1999. Uranium contamination in the subsurface: characterization and remediation. In: Burns, P.C., Finch, R. (Eds.), Uranium: Mineralogy, Geochemistry and the Environment. Reviews in Mineralogy, Mineralogy Society of America, Washington, DC, pp. 433–473.

Bago, B., Vierheilig, H., Piché, Y., Azcon-Aguilar, C., 1996. Nitrate depletion and pH changes induced by the extraradical mycelium of the arbuscular mycorrhizal fungus *Glomus intraradices* grown in monoxenic culture. New Phytol. 133, 273–280.

Bécard, G., Fortin J.A., 1988. Early events of vascular-arbuscular mycorrhizal formation on Ri T-DNA transformed roots. New Phytol. 108, 211–218.

Benguella, B., Benaissa, H., 2002. Cadmium removal from aqueous solutions by chitin: kinetic and equilibrium studies. Water Res. 36, 2463–2474.

Berreck, M., Haselwandter, K., 2001. Effect of the arbuscular mycorrhizal symbiosis upon uptake of cesium and other cations by plants. Mycorrhiza 10, 275–280.

Bethke, C., 2001. The Geochemist's Workbench®, release 3.2, A User's Guide to Rxn, Act2, Tact, React, and Gtplot. Hydrogeology program, University of Illinois.

Bondietti, E.A., Sweeton, F.H., 1977. Transuranic speciation in the environment. In: White, M.G., Dunaway, P.B. (Eds.), Transuranics in Natural Environments, Springfield, VA, NTIS, pp. 449–476.

Bonfante-Fasolo, P., Faccio, A., Perotto, S., Schubert, A., 1990. Correlation between chitin distribution and cell wall morphology in the mycorrhizal fungus *Glomus versiforme*. Mycol. Res. 94, 157–165.

Bonfante-Fasolo, P., Grippiolo, R., 1984. Cytochemical and biochemical observations on the cell-wall of the spore of *Glomus epigaeum*. Protoplasma 123, 140–151.

Borie, F., Rubio, R., 1999. Effects of arbuscular mycorrhizae and liming on growth and mineral acquisition of aluminium-tolerant and aluminium-sensitive barley cultivars. J. Plant Nutr. 22, 121–137.

Burleigh, S.H., Kristensen, B.K., Bechmann, I.E., 2003. A plasma membrane zinc transporter from *Medicago truncatula* is up-regulated in roots by Zn fertilization, yet down-regulated by arbuscular mycorrhizal colonization. Plant Mol. Biol. 52, 1077–1088.

Chui, V.W.D., Mok, K.W., Ng, C.Y., Luong, B.P., Ma, K.K., 1996. Removal and recovery of copper(II), chromium(III), and nickel(II) from solutions using crude shrimp chitin packed in small columns. Environ. Int. 22, 463–468.

Clark, R.B., Zeto, S.K., 1996. Mineral acquisition by mycorrhizal maize grown on acid and alkaline soil. Soil Biol. Biochem. 28, 1495–1503.

Clark, R.B., Zeto, S.K., 2002. Arbuscular mycorrhizae: mineral nutrient and water acquisition. In: Sharma, A.K., Johri, B.N. (Eds.), Arbuscular Mycorrhizae: Interactions in Plants, Rhizosphere and Soils, Science Publishers, Enfield, NH, pp. 159–187.

Cooper, K.M., Tinker, P.B., 1978. Translocation and transfer of nutrients in vesicular-arbuscular mycorrhizas. II. Uptake and translocation of phosphorus, zinc and sulphur. New Phytol. 81, 43–52.

Crooke, W.M., Knight, A. H., 1971. Root cation-exchange capacity and organic acid content of tops as indices of varietal yield. J. Sci. Food Agric. 22, 389–392.

Dahlgren, R.A., Vogt, K.A., Ugolini, F.C., 1991. The influence of soil chemistry on fine root aluminium concentrations and root dynamics in a subalpine Spodosol, Washington State, USA. Plant Soil 133, 117–129.

Declerck, S., Dupré de Boulois, H., Bivort, C., Delvaux, B., 2003. Extraradical mycelium of the mycorrhizal fungus *Glomus lamellosum* can uptake, accumulate and translocate radiocesium under root-organ culture conditions. Environ. Microbiol. 5, 510–516.

Declerck, S., Strullu, D.G., Plenchette, C., 1998. Monoxenic culture of the intraradical forms of *Glomus* sp. isolated from a tropical ecosystem: a proposed methodology for germplasm collection. Mycologia 90, 579–585.

Diop, T.A., Bécard, G., Piché, Y., 1992. Long-term *in vitro* culture of an endomycorrhizal fungus, *Gigaspora margarita*, on Ri T-DNA transformed roots of carrot. Symbiosis 12, 249–259.

Donahue, R., Hendry, J., 2003. Geochemistry of arsenic in uranium mine mill tailings, Saskatchewan, Canada. Appl. Geochem. 18, 1733–1750.

Driver, J., Hamilton, M., Rillig, M., 2003. The secreted AM fungal protein glomalin binds strontium in solution. Book of Abstracts of the Fourth International Conference on Mycorrhizae ICOM4, Montreal, Canada, 10–15 August 2003, p. 225.

Dufey, J.E., Braun, R., 1986. Cation exchange capacity of roots: titration, sum of exchangeable cations, copper adsorption. J. Plant Nutr. 9, 1147–1155.

Dufey, J.E., Genon, J.E., Jaillard, B., Calba, H., Rufyikiri, G., Delvaux, B., 2001. Cation exchange on plant roots involving aluminium: experimental data and modelling. In: Gobran, G.R., Wenzel, W.W., Lombi, E. (Eds.), *Trace Elements in the Rhizosphere*, CRC Press LLC, New York , pp. 227–252.

Ebbs, S.D., Brady, D.J., Kochian, L.V., 1998. Role of uranium speciation in the uptake and translocation of uranium by plants. J. Exp. Bot. 49, 1183–1190.

Entry, J.A., Watrud, L.S., Reeves, M., 1999. Accumulation of ^{137}Cs and ^{90}Sr from contaminated soil by three grass species inoculated with mycorrhizal fungi. Environ. Pollut. 104, 449–457.

Fenton, B.R., Waite, T.D., 1996. A kinetic study of cation release from a mixed mineral assemblage: implications for determination of uranium uptake. Radiochim. Acta 74, 251–256.

Fjeld, R.A., Coates, J.T., Elzerman, A.W., Navratil, J.D., 2002. Mobilities of uranium species in sedimentary interbed from the Snake River Plain. In: Merkel, B.J., Planer-Friedrich, B., Wolkersdorfer, C. (Eds.), Uranium in the Aquatic Environment, Proceedings of the International Conference on Uranium Mining and Hydrology III and the International Mine Water Association Symposium, Freiberg, Germany, 15–21 September 2002. Springer, Berlin, Germany, pp. 293–302.

Fortin, J.A., Bécard, G., Declerck, S., Dalpé, Y., St-Arnaud, M., Coughlan, A.P., Piché, Y., 2002. Arbuscular mycorrhizae on root organ cultures: a review. Can. J. Bot. 80, 1–20.

Gadd, G., 1990. Heavy metal accumulation by bacteria and other microorganisms. Experientia 46, 834–840.

Galun, M., Keller, P., Malki, D., Feldstein, H., Galun, E., Siegel, S.M., Siegel, B.Z., 1983. Removal of uranium(VI) from solution by fungal biomass and fungal wall-related biopolymers. Science 219, 285–286.

González-Chávez, C., D'Haen, J., Vangronsveld, J., Dodd, J.C., 2002. Copper sorption and accumulation by the extraradical mycelium of different *Glomus* spp. (arbuscular mycorrhizal fungi) isolated from the same polluted soil. Plant Soil 240, 287–297.

González-Chávez, M.C., Carrillo-González, R., Wright, S.F., Nichols, K.A., 2004. The role of glomalin, a protein produced by arbuscular mycorrhizal fungi, in sequestering potentially toxic elements. Environ. Pollut. 130, 317–323.

Grenthe, I., Fuger, J., Konings, R.J.M., Lemire, R.J., Miller, A.B., Nguyen-Trung, C., Wanner, H., 1992. Chemical Thermodynamics of Uranium, OECD-NEA, Elsevier, Amsterdam, North-Holland.

Guibal, E., Roussy, J., LeCloirec, P., 1996. Photochemical reaction of uranium with glucosamine, acetylglucosamine and related polymers: chitin and chitosan. Water SA 22, 19–26.

Hinsinger, P., 2001a. Bioavailability of soil inorganic P in the rhizosphere as affected by root-induced chemical changes: a review. Plant Soil 237, 173–195.

Hinsinger, P., 2001b. Bioavailability of trace elements as related to root-induced chemical changes in the rhizosphere. In: Gobran, G.R., Wenzel, W.W., Lombi, E. (Eds.), Trace Elements in the Rhizosphere, CRC Press LLC, London, UK, pp. 25–41.

Hinsinger, P., Plassard, C., Tang, C., Jaillard, B., 2003. Origins of root-mediated pH changes in the rhizosphere and their responses to environmental constraints: a review. Plant Soil 248, 43–59.

Hossner, L.R., Loeppert, R.H., Newton, R.J., Szaniszlo, P.J., 1998. Litterature review: phytoaccumulation of chromium, uranium, and plutonium in plant systems, Amarillo National Resource Center for Plutonium, US Department of Commerce.

Jakobsen, I., Abbott, L.K., Robson, A.D., 1992a. External hyphae of vesicular-arbuscular mycorrhizal fungi associated with *Trifolium subterraneum* L. 1. Spread of hyphae and phosphorus inflow into roots. New Phytol. 120, 371–380.

Jakobsen, I., Abbott, L.K., Robson, A.D., 1992b. External hyphae of vesicular-arbuscular mycorrhizal fungi associated with *Trifolium subterraneum* L. 2. Hyphal transport of ^{32}P over defined distances. New Phytol. 120, 509–516.

Jakobsen, I., Rosendhal, S., 1990. Carbon flow into soil and external hyphae from roots of mycorrhizal cucumber plants. New Phytol. 115, 77–83.

Jakobsen, I., Smith, S.E., Smith, F.A., 2002. Function and diversity of arbuscular mycorrhizae in carbon and mineral nutrition. In: van der Heijden, M.G.A., Sanders, I. (Eds.), Mycorrhizal Ecology, Springer, Heidelberg, Berlin, Germany, pp. 75–92.

Jolicoeur, M., Germette, S., Gaudette, M., Perrier, M., Bécard, G., 1998. Intracellular pH in arbuscular mycorrhizal fungi. A symbiotic physiological marker. Plant Physiol. 116, 1279–1288.

Joner, E.J., Briones, R., Leyval, C., 2000a. Metal binding capacity of arbuscular mycorrhizal mycelium. Plant Soil 226, 227–234.

Joner, E.J., Leyval, C., 1997. Uptake of ^{109}Cd by roots and hyphae of a *Glomus mosseae/Trifolium subterraneum* mycorrhiza from soil amended with high and low concentrations of cadmium. New Phytol. 135, 353–360.

Joner, E.J., Ravnskov, S., Jakobsen, I., 2000b. Arbuscular mycorrhizal phosphate transport under monoxenic conditions using radio-labelled inorganic and organic phosphate. Biotechnol. Lett. 22, 1705–1708.

Kabata-Pendias, A., Pendias, H., 2001. Trace Metals in Soils and Plants, third ed. CRC Press, Boca Raton, FL.

Kaya, C., Higgs, D., Kirnak, H., Tas, I., 2003. Mycorrhizal colonisation improves fruit yield and water use efficiency in watermelon (*Citrullus lanatus* Thunb.) grown under well-watered and water-stressed conditions. Plant Soil 253, 287–292.

Khater, A.E.M., Higgy, R.H., Pimpl, M., 2001. Radiological impacts of natural radioactivity in Abu-Tartor phosphate deposits, Egypt. J. Environ. Radioact. 55, 255–267.

Krea, M., Khalaf, H., 2000. Liquid-liquid extraction of uranium and lanthanides from phosphoric acid using a synergistic DOPPA-TOPO mixture. Hydrometallurgy 58, 215–225.

Langmuir, D., 1978. Uranium solution-mineral equilibria at low temperatures with applications to sedimentary ore deposits. Geochim. Cosmochim. Acta 42, 547–569.

Langmuir, D., 1997. Aqueous Environmental Geochemistry, Prentice Hall, Upper Saddle River, New York.

Lee, L.Y., Marsh, J.D., Jr., 1992. Characterization of uranium contaminated soils from DOE Fernald Environmental Management Project site: results of phase I characterization. Oak Ridge National Laboratory Environmental Sciences, Division Publications N° 3786.

Leyval, C., Joner, E.J., 2001. Bioavailability of heavy metals in the mycorrhizosphere. In: Gobran, G.R., Wenzel, W.W., Lombi, E. (Eds.), Trace Elements in the Rhizosphere, CRC Press LLC, London, UK, pp. 165–185.

Leyval, C., Turnau, K., Haselwandter, K., 1997. Effect of heavy metal pollution on mycorrhizal colonization and function: physiological, ecological and applied aspects. Mycorrhiza 7, 139–153.

Maldonado-Mendoza, I.E., Dewbre, G.R., Harrison, M.J., 2001. A phosphate transporter gene from the extra-radical mycelium of an arbuscular mycorrhizal fungus *Glomus intraradices* is regulated in response to phosphate in the environment. Mol. Plant-Microbe Interact. 14, 1140–1148.

Marschner, H., 1995. *Mineral Nutrition of Higher Plants*, Second ed. Academic Press, London, UK.

Marschner, P., Jentschke, G., Godbold, D.L., 1998. Cation exchange capacity and lead sorption in ectomycorrhizal fungi. Plant Soil 205, 93–98.

Meinrath, A., Schneider, P., Meinrath, G., 2003. Uranium ores and depleted uranium in the environment, with a reference to uranium in the biosphere from the Erzgebirge/Sachsen, Germany. J. Environ. Radioact. 64, 175–193.

Mortvedt, J.J., 1994. Plant and soil relationships of uranium and thorium decay series radionuclides – a review. J. Environ. Qual. 23, 643–650.

Mugnier, J., Mosse, B., 1987. Vesicular-arbuscular mycorrhizal infections in transformed Ri T-DNA roots grown axenically. Phytopathology 77, 1045–1050.

Muzzarelli, R.A.A., 1977. Chitin, Pergamon, New York, USA.

Muzzarelli, R.A.A., Tubertini. O., 1969. Chitin and chitosan as chromatographic supports and adsorbents for collection of metal ions from organic and aquatic solutions and sea-water. Talanta 16, 1571–1577.

Nielsen, J.S., Joner, E.J., Declerck, S., Olsson, S., Jakobsen, I., 2002. Phospho-imaging as a tool for visualization and noninvasive measurement of P transport dynamics in arbuscular mycorrhizas. New Phytol. 154, 809–820.

Olsson, P.A., Thingstrup, I., Jakobsen, I., Baath, F., 1999. Estimation of the biomass of arbuscular mycorrhizal fungi in a linseed field. Soil Biol. Biochem. 31, 1879–1887.

Pfeffer, P.E., Bago, B., Shachar-Hill, Y., 2001. Exploring mycorrhizal function with NMR spectroscopy. New Phytol. 150, 543–553.

Plenchette, C., Declerck, S., Diop, T.A., Strullu, D.G., 1996. Infectivity of monoaxenic subcultures of the arbuscular mycorrhizal fungus *Glomus versiforme* associated with Ri-T-DNA-transformed carrot root. Appl. Microbiol. Biotechnol. 46, 545–548.

Qureshi, A.A., Khattak, N.U., Sardar, M., Tufail, M., Akram, M., Iqbal, T., Khan, H.A., 2001. Determination of uranium contents in rock samples from Kakul phosphate deposit, Abbotabad (Pakistan), using fission-track technique. Radiat. Meas. 34, 355–359.

Rillig, M.C., Ramsey, P.W., Morris, S., Paul, E.A., 2003. Glomalin, an arbuscular-mycorrhizal fungal soil protein, responds to land-use change. Plant Soil 253, 293–299.

Rillig, M.C., Wright, S.F., Eviner, V., 2002. The role of arbuscular mycorrhizal fungi and glomalin in soil aggregation: comparing effects of five plant species. Plant Soil 238, 325–333.

Rillig, M.C., Wright S.F., Nichols, K.A., Schmidt, W.F., Torn, M.S., 2001. Large contribution of arbuscular mycorrhizal fungi to soil carbon pools in tropical forest soils. Plant Soil 233, 167–177.

Rufyikiri, G., Declerck, S., Dufey, J.E., Delvaux, B., 2000. Arbuscular mycorrhizal fungi might alleviate aluminium toxicity in banana plants. New Phytol. 148, 343–352.

Rufyikiri, G., Declerck, S., Thiry, Y., 200(4). Comparison of ^{233}U and ^{33}P uptake and translocation by the arbuscular mycorrhizal fungus *Glomus intraradices* in root organ culture conditions. Mycorrhiza 13, 202–209.

Rufyikiri, G., Dufey, J.E., Nootens, D., Delvaux, B., 2001. Effect of aluminium on bananas (*Musa* spp.) cultivated in acid solutions. II. Water and nutrient uptake. Fruits 56, 3–14.

Rufyikiri, G., Thiry, Y., Declerck, S., 2003. Contribution of hyphae and roots to uranium uptake and translocation by arbuscular mycorrhizal carrot roots under root-organ culture conditions. New Phytol. 158, 391–399.

Rufyikiri, G., Thiry, Y., Wang, L., Delvaux, D., Declerck, S., 2002. Uranium uptake and translocation by the arbuscular mycorrhizal fungus, *Glomus intraradices,* under root-organ culture conditions. New Phytol. 156, 275–281.

Sanders, F.E., Tinker, P.B., Black, R.L.B., Palmerley, S.M., 1977. The development of endomycorrhizal root systems. I. Spread of infection and growth promoting effects with four species of vesicular-arbuscular mycorrhizas. New Phytol. 78, 257–268.

Santschi, P.H., Honeyman, B.D., 1989. Radionuclides in aquatic environments. Radiat. Phys. Chem. 34, 213–240.

Schönbuchner, H., Leiterer, M., Machelett, B., Bergmann, H., 2002. Mobility and plant availability of heavy metals in soils of uranium mining dumps. In: Merkel, B.J., Planer-Friedrich, B., Wolkersdorfer, C. (Eds.), Uranium in the Aquatic Environment, Proceedings of the International Conference on Uranium Mining and Hydrology III and the International Mine Water Association Symposium, Freiberg, Germany, 15–21 September 2002. Springer, Berlin, Germany, pp. 529–536.

Schüβler, A., Schwarzott, D., Walker, C., 2001. A new fungal phylum, the *Glomeromycota*: phylogeny and evolution. Mycol. Res. 105, 1413–1421.

Shahandeh, H., Lee, J.H., Hossner, L.R., Loeppert, R.H., 2001. Bioavailability of uranium and plutonium to plants in soil-water systems and the potential of phytoremediation. In: Gobran, G.R., Wenzel, W.W., Lombi, E. (Eds.), Trace Elements in the Rhizosphere, CRC Press LLC, London, UK, pp. 93–124.

Sheppard, S.C., Evenden, W.G., 1988. Critical compilation and review of plant/soil concentration ratios for uranium, thorium and lead. Critical compilation and review of plant/soil concentration ratios for uranium, thorium and lead. J. Environ. Radioact. 8, 255–285.

Siegel, S.M., Galun, M., Siegel, B.Z., 1990. Filamentous fungi as metal biosorbents: a review. Water, Air, Soil Pollut. 53, 335–344.

Simon, L., Bousquet, J., Levesque, R.C., Lalonde, M., 1993. Origin and diversification of endomycorrhizal fungi and coincidence with vascular landplants. Nature 363, 67–69.

Simon, S.L., Ibrahim, S.A., 1990. Biological uptake of radium by terrestrial plants. In: IAEA (Ed.), Environmental Behaviour of Radium, IAEA, Vienna, pp. 545–599.

Smith, S.E., Read, D., 1997. *Mycorrhizal Symbiosis*, Second ed. Academic Press. San Diego, CA.

Solaiman, M.Z., Saito, M., 2001. Phosphate efflux from intraradical hyphae of *Gigaspora margarita in vitro* and its implication for phosphorus translocation. New Phytol. 151, 525–533.

St-Arnaud, M., Hamel, C., Vimard, B., Caron, M., Fortin, J.A., 1996. Enhanced hyphal growth and spore production of the arbuscular mycorrhizal fungus *Glomus intraradices* in an *in vitro* system in the absence of host roots. Mycol. Res. 100, 328–332.

Strullu, D.G., Romand, C., 1986. Méthode d'obtention d'endomycorhizes à vésicules et arbuscules en conditions axéniques. Comptes Rendus l'Académie Sci. 303, 245–250.

Suzuki, Y., Banfield, J.F., 1999. Geomicrobiology of uranium. In: Burns, P.C., Finch, R. (Eds.), Uranium: Mineralogy, Geochemistry and the Environment, Reviews in Mineralogy, Vol. 33. Mineralogy Society of America, Washington, DC, pp. 393–432.

Tepfer, D., 1984. Transformation of several species of higher plants by *Agrobacterium rhizogenes*: sexual transmission of the transformed genotype and phenotype. Cell 37, 959–967.

Tepfer, D., 1989. Ri T-DNA from *Agrobacterium rhizogenes*: a source of genes having applications in rhizosphere biology and plant development, ecology and evolution. In: Kosuge, T., Nester, E.W. (Eds.), Plant-Microbe Interactions, Vol. 3. McGraw-Hill Publishing, New York, pp. 294–342.

Tsezos, M., 1983. The role of chitin in uranium adsorption by *R. arrhizus*. Biotechnol. Bioeng. 25, 2025–2040.

Tsezos, M., Mattar, S., 1986. A further insight into the mechanism of biosorption of metals by examining chitin EPR spectra. Talanta 33, 225–232.

Tsezos, M., Volesky, B., 1982. The mechanism of uranium biosorption by *Rhizopus arrhizus*. Biotechnol. Bioeng. 24, 385–401.

Turnau, K., Kottke, I., Oberwinkler, F., 1993. Element localisation in mycorrhizal roots of *Pteridium aquilinum* (L.) Kuhn collected from experimental plots treated with cadmium dust. New Phytol. 123, 313–324.

Wang, X., Ma, T., Xu, L., 2002. A 3-D numerical modeling of reactive transport of U(VI) in shallow groundwaters: a case study at a uranium mill tailing site in southern China. In: Merkel, B.J., Planer-Friedrich, B., Wolkersdorfer, C. (Eds.), Uranium in the Aquatic Environment, Proceedings of the International Conference on Uranium Mining and Hydrology III and the International Mine Water Association Symposium, Freiberg, Germany, 15–21 September 2002. Springer, Berlin, Germany, pp. 303–310.

Weiersbye, I.M., Straker, C.J., Przybylowicz, W.J., 1999. Micro-PIXE mapping of elemental distribution in arbuscular mycorrhizal roots of the grass, *Cynodon dactylon*, from gold and uranium mine tailings. Nucl. Instrum. Methods Phys. Res., Sect. B. 158, 335–343.

Weissenhorn, I., Leyval, C., Belgy, G., Berthelin, J., 1995. Arbuscular mycorrhizal contribution to heavy metal uptake by maize (*Zea mays* L.) in pot culture with contaminated soil. Mycorrhiza 5, 245–251.

Wessels, J.G.H., Sietsma, J.H., 1981. Fungal walls: a survey. In: Tanner, W., Loewus, F.A. (Eds.), Enzyclopedia of Plant Physiology, Vol. 13B. Springer, Berlin, pp. 352–384.

Wright, S.F., Anderson, R.L., 2000. Aggregate stability and glomalin in alternative crop rotations for the central Great Plains. Biol. Fert. Soils 31, 249–253.

Wright, S.F., Franke-Snyder, M., Morton, J.B., Upadhyaya, A., 1996. Time-course study and partial characterization of a protein on hyphae of arbuscular mycorrhizal fungi during active colonization of roots. Plant Soil 181, 193–203.

Wright, S.F., Upadhyaya, A., 1996. Extraction of an abundant and unusual protein from soil and comparison with hyphal protein of arbuscular mycorrhizal fungi. Soil Sci. 161, 575–586.

Wright, S.F., Upadhyaya, A., 1998. A survey of soils for aggregate stability and glomalin, a glycoprotein produced by hyphae of arbuscular mycorrhizal fungi. Plant Soil 198, 97–107.

Yang, J., Volesky, B., 1999. Biosorption of uranium on *Sargassum* biomass. Water Res. 33, 3357–3363.

Zhou, J.L., 1999. Zn biosorption by *Rhizopus arrhizus* and other fungi. Appl. Microbiol. Biotechnol. 51, 686–693.

Subject Index